D0849771

Exactly Solved Models
in Statistical Mechanics

Exactly Solved Models
in Statistical Mechanics

Rodney J. Baxter F.R.S.
Department of Theoretical Physics,
Research School of Physical Sciences,
The Australian National University,
Canberra, A.C.T., Australia

1982

ACADEMIC PRESS
A Subsidiary of Harcourt Brace Jovanovich, Publishers

London   New York
Paris   San Diego   San Francisco   São Paulo
Sydney   Tokyo   Toronto

ACADEMIC PRESS INC. (LONDON) LTD
24–28 Oval Road
London NW1

US edition published by
ACADEMIC PRESS INC.
111 Fifth Avenue
New York, New York 10003

British Library Cataloguing in Publication Data

Baxter, R.J.
    Exactly solved models in statistical mechanics.
    1. Statistical mechanics
    I. Title
    530.1'3   QC174.8

    ISBN 0-12-083180-5
    LCCN 81-68965

Printed in Great Britain by
Page Bros (Norwich) Ltd

PREFACE

This book was conceived as a slim monograph, but grew to its present size as I attempted to set down an account of two-dimensional lattice models in statistical mechanics, and how they have been solved. While doing so I have been pulled in opposite directions. On the one hand I remembered the voice of the graduate student at the conference who said 'But you've left out all the working—how *do* you get from equation (81) to (82)?' On the other hand I knew from experience how many sheets of paper go into the waste-paper basket after even a modest calculation: there was no way they could all appear in print.

I hope I have reached a reasonable compromise by signposting the route to be followed, without necessarily giving each step. I have tried to be selective in doing so: for instance in Section 8.13 I discuss the functions $k(\alpha)$ and $g(\alpha)$ in some detail, since they provide a particularly clear example of how elliptic functions come into the working. Conversely, in (8.10.9) I merely quote the result for the spontaneous staggered polarization P_0 of the F-model, and refer the interested reader to the original paper: its calculation is long and technical, and will probably one day be superseded when the eight-vertex model conjecture (10.10.24) is verified by methods similar to those used for the magnetization result (13.7.21).

There are 'down-to-earth' physicists and chemists who reject lattice models as being unrealistic. In its most extreme form, their argument is that if a model can be solved exactly, then it must be pathological. I think this is defeatist nonsense: the three-dimensional Ising model is a very realistic model, at least of a two component alloy such as brass. If the predictions of universality are corrected, then they should have exactly the same critical exponents. Admittedly the Ising model has been solved only in one and two dimensions, but two-dimensional systems do exist (see Section 1.6), and can be quite like three-dimensional ones. It is true that the two-dimensional Ising model has been solved only for zero magnetic field, and that this case is quite unlike that of non-zero field; but physically this means Onsager solved the most interesting and tricky case. His solution vastly helps us understand the full picture of the Ising model in a field.

In a similar way, the eight-vertex model helps us understand more complicated systems and the variety of behaviour that can occur. The hard hexagon model is rather special, but needs no justification: It is a perfectly good lattice gas and can be compared with a helium monolayer adsorbed onto a graphite surface (Riedel, 1981).

There is probably also a feeling that the models are 'too hard' mathematically. This does not bear close examination: Ruelle (1969) rightly says in the preface to his book that if a problem is worth looking at at all, then no mathematical technique is to be judged too sophisticated.

Basically, I suppose the justification for studying these lattice models is very simple: they are relevant and they can be solved, so why not do so and see what they tell us?

In the title the phrase 'exactly solved' has been chosen with care. It is not necessarily the same as 'rigorously solved'. For instance, the derivation of (13.7.21) depends on multiplying and diagonalizing the infinite-dimensional corner transfer matrices. It ought to be shown, for instance, that the matrix products are convergent. I have not done this, but believe that they are (at least in a sense that enables the calculation to proceed), and that as a result (13.7.21) is exactly correct.

There is of course still much to be done. Barry McCoy and Jacques Perk rightly pointed out to me that whereas much is now known about the correlations of the Ising model, almost nothing is known about those of the eight-vertex and hard hexagon models.

There are many people to whom I am indebted for the opportunity to write this book. In particular, my interest in mathematics and theoretical physics was nurtured by my father, Thomas James Baxter, and by Sydney Adams, J. C. Polkinghorne and K. J. Le Couteur. Elliott Lieb initiated me into the complexities of the ice-type models. Louise Nicholson and Susan Turpie worked wonders in transforming the manuscript into immaculate typescript. Paul Pearce has carefully read the proofs of the entire volume. Most of all, my wife Elizabeth has encouraged me throughout, particularly through the last turbulent year of writing.

R. J. Baxter

Canberra, Australia
February 1982

CONTENTS

4   Ising Model on the Bethe Lattice

5   The Spherical Model

6   Duality and Star – Triangle Transformations of Planar Ising Models

7   Square-Lattice Ising Model

8   Ice-Type Models

9   Alternative Way of Solving the Ice-Type Models

10 Square Lattice Eight-Vertex Model

11 Kagomé Lattice Eight-Vertex Model

12 Potts and Ashkin – Teller Models

13   Corner Transfer Matrices

14   Hard Hexagon and Related Models

15   Elliptic Functions

1

BASIC STATISTICAL MECHANICS

1.1 Phase Transitions and Critical Points

As its name implies, statistical mechanics is concerned with the average properties of a mechanical system. Obvious examples are the atmosphere inside a room, the water in a kettle and the atoms in a bar magnet. Such systems are made up of a huge number of individual components (usually molecules). The observer has little, if any, control over the components: all he can do is specify, or measure, a few average properties of the system, such as its temperature, density or magnetization. The aim of statistical mechanics is to predict the relations between the observable macroscopic properties of the system, given only a knowledge of the microscopic forces between the components.

For instance, suppose we knew the forces between water molecules. Then we should be able to predict the density of a kettleful of water at room temperature and pressure. More interestingly, we should be able to predict that this density will suddenly and dramatically change as the temperature is increased from 99°C to 101°C: it decreases by a factor of 1600 as the water changes from liquid to steam. This is known as a *phase transition*.

Yet more strange effects can occur. Consider an iron bar in a strong magnetic field, H, parallel to its axis. The bar will be almost completely magnetized: in appropriate units we can say that its magnetization, M, is +1. Now decrease H to zero: M will decrease, but not to zero. Rather, at zero field it will have a *spontaneous magnetization* M_0.

On the other hand, we expect molecular forces to be invariant with respect to time reversal. This implies that reversing the field will reverse the magnetization, so M must be an odd function of H. It follows that

1

$M(H)$ must have a graph of the type shown in Fig. 1.1(a), with a discontinuity at $H = 0$.

This discontinuity in the magnetization is very like the discontinuity in density at a liquid – gas phase transition. In fact, in the last section of this chapter it will be shown that there is a precise equivalence between them.

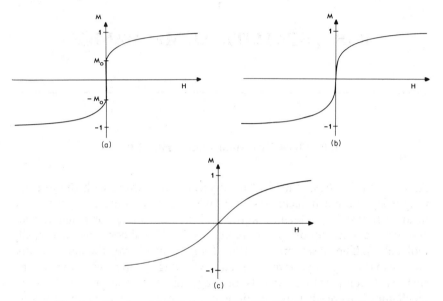

Fig. 1.1. Graphs of $M(H)$ for (a) $T < T_c$, (b) $T = T_c$, (c) $T > T_c$.

The iron bar can be regarded as undergoing a phase transition at $H = 0$, changing suddenly from negative to positive magnetization. In an actual experiment this discontinuity is smeared out and the phenomenon of hysteresis occurs: this is due to the bar not being in true thermodynamic equilibrium. However, if the iron is soft and subject to mechanical disturbances, a graph very close to that of Fig. 1.1(a) is obtained (Starling and Woodall, 1953, pp. 280–281; Bozorth, 1951, p. 512).

The above remarks apply to an iron bar at room temperature. Now suppose the temperature T is increased slightly. It is found that $M(H)$ has a similar graph, but M_0 *is decreased*. Finally, if T is increased to a *critical* value T_c (the Curie point), M_0 vanishes and $M(H)$ becomes a continuous function with infinite slope (susceptibility) at $H = 0$, as in Fig. 1.1(b).

If T is further increased, $M(H)$ remains a continuous function, and becomes analytic at $H = 0$, as in Fig. 1.1(c).

These observations can be conveniently summarized by considering a (T, H) plane, as in Fig. 1.2. There is a cut along the T axis from 0 to T_c. The magnetization M is an analytic function of both T and H at all points in the right-half plane, *except those on the cut*. It is discontinuous across the cut.

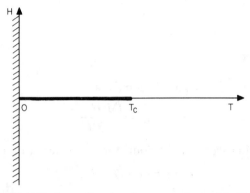

Fig. 1.2. The (T, H) half-plane, showing the cut across which M is discontinuous. Elsewhere M is an analytic function of T and H.

The cut is a line of phase transitions. Its endpoint $(T_c, 0)$ is known as a *critical point*. Clearly the function $M(H, T)$ must be singular at this point, and one of the most fascinating aspects of statistical mechanics is the study of this singular behaviour near the critical point.

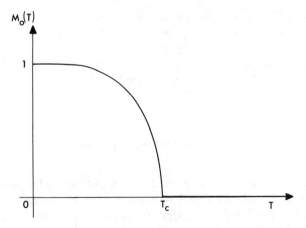

Fig. 1.3. The spontaneous magnetization M_0 as a function of temperature.

The spontaneous magnetization is a function of T and can be defined as

$$M_0(T) = \lim_{H \to 0^+} M(H, T), \tag{1.1.1}$$

the limit being taken through positive values of H. It has a graph of the type shown in Fig. 1.3, being positive for $T < T_c$ and identically zero for $T > T_c$.

Critical Exponents

The *susceptibility* of a magnet is defined as

$$\chi(H, T) = \frac{\partial M(H, T)}{\partial H}. \tag{1.1.2}$$

When considering critical behaviour it is convenient to replace T by

$$t = (T - T_c)/T_c. \tag{1.1.3}$$

Then the thermodynamic functions must have singularities at $H = t = 0$. It is expected that these singularities will normally be simple non-integer powers; in particular, it is expected that

$$M_0(T) \quad \sim (-t)^\beta \qquad \text{as } t \to 0^-, \tag{1.1.4}$$

$$M(H, T_c) \sim H^{1/\delta} \qquad \text{as } H \to 0, \tag{1.1.5}$$

$$\chi(0, T) \quad \sim t^{-\gamma} \qquad \text{as } t \to 0^+, \tag{1.1.6}$$

$$\chi(0, T) \quad \sim (-t)^{-\gamma'} \qquad \text{as } t \to 0^-. \tag{1.1.7}$$

Here the notation $X \sim Y$ means that X/Y tends to a non-zero limit. The power-law exponents β, δ, γ, γ' are numbers, independent of H and T: they are known as *critical exponents*.

For brevity, the phrase 'near T_c' will be frequently used in this book to mean 'near the critical point', it being implied that H is small, if not zero.

1.2   The Scaling Hypothesis

It is natural to look for some simplified form of the thermodynamic functions that will describe the observed behaviour near T_c. Widom (1965) and Domb and Hunter (1965) suggested that certain thermodynamic functions might be homogeneous. In particular, Griffiths (1967) suggested that H

might be a homogeneous function of $M^{1/\beta}$ and t. Since H is an odd function of M, this means that near T_c

$$H/kT_c = M|M|^{\delta-1}h_s(t|M|^{-1/\beta}), \qquad (1.2.1)$$

where β and δ are numbers (as yet undefined), k is Boltzmann's constant, and $h_s(x)$ is a dimensionless *scaling function*. A typical graph of $h_s(x)$ is shown in Fig. 1.4: it is positive and monotonic increasing in the interval $-x_0 < x < \infty$, and vanishes at $-x_0$.

Note that (1.2.1) implies that H is an odd function of M, as it should be.

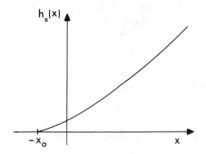

Fig. 1.4. The scaling function $h_s(x)$ for the square-lattice Ising model (Gaunt and Domb, 1970).

The scaling hypothesis predicts certain relations between the critical exponents. To see this, first consider the behaviour on the cut in Fig. 1.2. Here $H = 0$, $t < 0$ and $M = \pm M_0$. From (1.2.1) the function $h_s(x)$ must be zero, so $x = -x_0$, i.e.

$$t = -x_0|M|^{1/\beta}. \qquad (1.2.2)$$

The relation (1.1.4) follows, so β in (1.2.1) is the critical exponent defined in (1.1.4).

Now set $t = 0$ (1.2.1). Since $h_s(0)$ is non-zero, this implies that near T_c

$$H \sim M^\delta, \qquad (1.2.3)$$

in agreement with (1.1.5). Hence the δ in (1.2.1) is the same as that in (1.1.5).

Differentiate (1.2.1) with respect to M, keeping t fixed. From (1.1.2) this gives

$$(kT_c\chi)^{-1} = |M|^{\delta-1}[\delta h_s(x) - \beta^{-1}xh_s'(x)] \qquad (1.2.4)$$

where

$$x = t|M|^{-1/\beta}. \tag{1.2.5}$$

Again consider the behaviour on the cut in Fig. 1.2. Here x has the fixed value $-x_0$, so

$$\chi^{-1} \sim |M|^{\delta-1} \sim (-t)^{\beta(\delta-1)}. \tag{1.2.6}$$

This agrees with (1.1.7), and predicts that the critical exponent γ' is given by

$$\gamma' = \beta(\delta - 1). \tag{1.2.7}$$

To obtain (1.1.6) from the scaling hypothesis, we need the large x behaviour of the scaling function $h_s(x)$. This can be obtained by noting that for fixed positive t, we must have

$$H \sim M \qquad \text{as } M \to 0. \tag{1.2.8}$$

Comparing this with (1.2.1), we see that

$$h_s(x) \sim x^{\beta(\delta-1)} \qquad \text{as } x \to \infty. \tag{1.2.9}$$

From (1.2.1) and (1.2.9), it follows that for arbitrary small positive t,

$$H \sim t^{\beta(\delta-1)}M \qquad \text{as } M \to 0, \tag{1.2.10}$$

so from (1.1.1),

$$\chi(0, T) \sim t^{-\beta(\delta-1)} \qquad \text{as } t \to 0^+. \tag{1.2.11}$$

Comparing this with (1.1.6), and using (1.2.7), we see that the scaling hypothesis predicts the exponent relations

$$\gamma = \gamma' = \beta(\delta - 1). \tag{1.2.12}$$

Other exponents α, ν, ν', η, μ will be defined in Section 1.7, but for completeness the various scaling predictions are listed here:

$$\alpha + 2\beta + \gamma' = 2, \tag{1.2.13}$$

$$\nu = \nu', (2 - \eta)\nu = \gamma, \tag{1.2.14}$$

$$\mu + \nu = 2 - \alpha, \tag{1.2.15}$$

$$d\nu = 2 - \alpha, \tag{1.2.16}$$

where d is the dimensionality of the system.

A partial derivation of (1.2.14) will be given in Section 1.7, but it is beyond the scope of this book to attempt to justify all these relations: the interested reader is referred to the articles by Widom (1965), Fisher (1967), Kadanoff *et al.* (1967), Hankey and Stanley (1972), Stanley (1971) and

Vicentini-Missoni (1972). Their relevance here is that exactly solved models can be used to test the relations, and indeed we shall find that scaling passes every possible test for the models to be discussed.

The scaling relations (1.2.12)–(1.2.15) are in good agreement with available experimental and theoretical results, and the scaling function $h_s(x)$ has been obtained approximately for a number of systems (see for example Gaunt and Domb, 1970).

The last relation (1.2.16) involves the dimensionality d. It is derived by making further assumptions, known as 'strong scaling' or 'hyperscaling'. It is expected to be valid for $d \leq 4$, but there is some question whether it is consistent with available numerical results for three- and four-dimensional models (Baker, 1977). The total set of equations (1.2.12)–(1.2.16) is sometimes known as 'two exponent' scaling, since if two independent exponents (such as δ and β) are given, then all other exponents can be obtained from the equations.

1.3 Universality

Consider a system with conservative forces. Let s denote a state (or configuration) of the system. Then this state will have an *energy* $E(s)$, where the function $E(s)$ is the *Hamiltonian* of the system.

The thermodynamic properties, such as $M(H, T)$ and T_c, are of course expected to depend on the forces in the system, i.e. on $E(s)$. However, it is believed (Fisher, 1966; Griffiths, 1970) that the critical exponents are 'universal', i.e. independent of the details of the Hamiltonian $E(s)$.

They will, of course, depend on the dimensionality of the system, and on any symmetries in the Hamiltonian. To see the effect of these, suppose $E(s)$ can be written as

$$E(s) = E_0(s) + \lambda E_1(s) , \tag{1.3.1}$$

where $E_0(s)$ has some symmetry (such as invariance under spatial reflection) and $E_1(s)$ has not. The critical exponents are then supposed to depend on λ only in so far as they have one value for $\lambda = 0$ (symmetric Hamiltonian), and another fixed value for $\lambda \neq 0$ (non-symmetric). For example, there would be two numbers β_0, β_1 such that

$$\beta = \beta_0 \qquad \text{if } \lambda = 0$$
$$= \beta_1 \qquad \text{if } \lambda \neq 0 , \tag{1.3.2}$$

β being discontinuous at $\lambda = 0$.

On the other hand, if $E_0(s)$ is some simple Hamiltonian and $E_1(s)$ is very complicated, but they have the same dimensionality and symmetry, then β should be completely constant, even at $\lambda = 0$. The implications of this are far reaching. One could take a realistic and complicated Hamiltonian $E(s)$, 'strip' it to a highly idealized Hamiltonian $E_0(s)$, and still obtain exactly the same critical exponents. For instance, on these grounds it is believed that carbon dioxide, xenon and the three-dimensional Ising model should all have the same critical exponents. To within experimental error, this appears to be the case (Hocken and Moldover, 1976).

There are some difficulties: there is usually more than one way of describing a system, in particular of labelling its states. In one of these there may be an obvious symmetry which occurs for some special values of the parameters. In another formulation this symmetry may not be obvious at all. Thus if the second formulation were used, and these special values of the parameters were accidentally chosen, then the critical exponents could be unexpectedly different from those appropriate to other values.

Also, in this book the solution of the two-dimensional 'eight-vertex' model will be presented. This has exponents that vary continuously with the parameters in the Hamiltonian. This violates the universality hypothesis, but it is now generally believed that such violations only occur for very special classes of Hamiltonians.

It should be noted that scaling and universality, while commonly grouped together, are independent assumptions. One may be satisfied and the other not, as in the case of the eight-vertex model, where universality fails but scaling appears to hold.

1.4   The Partition Function

How do we calculate thermodynamic functions such as $M(H, T)$ from the microscopic forces between the components of the system? The answer was given by John Willard Gibbs in 1902. Consider a system with states s and Hamiltonian $E(s)$. Form the partition function

$$Z = \sum_s \exp[-E(s)/kT], \tag{1.4.1}$$

where k is Boltzmann's constant and the summation is over all allowed states s of the system. Then the free energy F is given by

$$F = -kT \ln Z. \tag{1.4.2}$$

Also, the probability of the system being in a state s is

$$Z^{-1} \exp[-E(s)/kT], \qquad (1.4.3)$$

so if X is some observable property of the system, such as its total energy or magnetization, with value $X(s)$ for state s, then its observed average thermodynamic value is

$$\langle X \rangle = Z^{-1} \sum_s X(s) \exp[-E(s)/kT]. \qquad (1.4.4)$$

In particular, the internal energy is

$$U = \langle E \rangle$$
$$= Z^{-1} \sum_s E(s) \exp[-E(s)/kT], \qquad (1.4.5)$$

and by using the above definitions (1.4.1) and (1.4.2) we can verify that

$$U = kT^2 \frac{\partial}{\partial T} \ln Z$$
$$= -T^2 \frac{\partial}{\partial T}(F/T), \qquad (1.4.6)$$

in agreement with standard thermodynamics.

The basic problem of equilibrium statistical mechanics is therefore to calculate the sum-over-states in (1.4.1) (for continuum systems this sum becomes an integral, for quantum mechanical ones a trace). This will give Z and F as functions of T and of any variables that occur in $E(s)$, such as a magnetic field. The thermodynamic properties can then be obtained by differentiation.

Unfortunately, for any realistic interacting system of macroscopic size, including the examples mentioned above, the evaluation of Z is hopelessly difficult. One is therefore forced to do one or both of the following:

A. Replace the real system by some simple idealization of it: this idealization is known as a *model*. Mathematically, it consists of specifying the states s and the energy Hamiltonian function $E(s)$.
B. Make some approximation to evaluate the sum-over-states (1.4.1).

1.5 Approximation Methods

Let us consider the step (B) above. Some of the better-known approximation schemes are:

(i) Cell or cluster approximations. In these the behaviour of the whole system is extrapolated from that of a very few components inside some 'cell', approximations being made for the interaction of the cell with the rest of the system. Examples are the mean-field (Bragg and Williams, 1934; Bethe, 1935), quasi-chemical (Guggenheim, 1935) and Kikuchi (1951) approximations. They have the advantage of being fairly simple to solve; they predict the correct qualitative behaviour shown in Figs. 1.1 to 1.3, and are reasonably accurate except near the critical point (Domb, 1960, pp. 282–293; Burley, 1972).

(ii) Approximate integral equations for the correlation functions, notably the Kirkwood (1935), hyper-netted chain (van Leeuwen *et al.*, 1959) and Percus-Yevick (Percus and Yevick, 1958; Percus, 1962) equations. These give fairly good numerical values for the thermodynamic properties of simple fluids.

(iii) Computer calculations on systems large on a microscopic scale (e.g. containing a few hundred atoms), but still not of macroscopic size. These calculations evaluate Z by statistically sampling the terms on the RHS of (1.4.1), so are subject to statistical errors, usually of a few per cent. For this reason they are really 'approximations' rather than 'exact calculations'.

(iv) Series expansions in powers of some appropriate variable, such as the inverse temperature or the density. For very realistic models these can only be obtained to a few terms, but for the three-dimensional Ising model expansions have been obtained to as many as 40 terms (Sykes *et al.*, 1965, 1973a).

The approximation schemes (i) to (iii) can give quite accurate values for the thermodynamic properties, *except near the critical point.* There is a reason for this: they all involve neglecting in some way the correlations between several components, or two components far apart. However, near T_c the correlations become infinitely-long ranged, all components are correlated with one another, and almost any approximation breaks down. This means that approximations like (i), (ii) and (iii) are of little, if any, use for determining the interesting cooperative behaviour of the system near T_c.

Method (iv) is much better: if sufficient terms can be obtained then it is possible, with considerable ingenuity, to obtain plausible guesses as to the nature of the singularities of the thermodynamic functions near the critical point. In particular, the best estimates to date of critical exponents in three dimensions have been obtained by the series expansion method. However, an enormous amount of work is required to obtain the series, and the resulting accuracy of the exponents is still not as good as one would like.

(v) There is another approach, due to Kadanoff (1966) and Wilson (1971) (see also Wilson and Kogut, 1974; Fisher, 1974): this is the so-called *renormalization group*. In this method the sum over states (1.4.1) is evaluated in successive stages, a 'renormalized' Hamiltonian function $E(s)$ being defined at each stage. This defines a mapping in Hamiltonian space. If one makes some fairly mild assumptions about this mapping, notably that it is analytic, then it follows that the thermodynamic functions do have branch-point singularities such as (1.1.4) at T_c, that the scaling hypothesis (1.2.1) and the relations (1.2.12)–(1.2.16) are satisfied, and that the exponents of the singularities should normally be universal (Fisher, 1974, p. 602).

In principle, the renormalization group approach could be carried through exactly. However, this is more difficult than calculating the partition function directly, so to obtain actual numerical results some approximation method is needed for all but the very simplest models. The fascinating result is that quite crude cell-type approximations give fairly accurate values of the critical exponents (Kadanoff *et al.*, 1976). The reason for this is not yet fully understood.

To summarize: approximate methods (step B) either fail completely near T_c, or require considerable acts of faith in the assumptions made.

1.6 Exactly Solved Models

Another approach is to use step A to the fullest, and try to find models for which $E(s)$ is sufficiently simple that the partition function (1.4.1) can be calculated exactly. This may not give useful information about the values of the thermodynamic functions of real systems, but it will tell us qualitatively how systems can behave, in particular near T_c. In fact if we could solve a model with the same dimensionality and symmetry as a real system, universality asserts that we should obtain the *exact* critical exponents of the real system.

There is a further condition for universality, which was not mentioned in Section 1.3. In most physical systems the intermolecular forces are effectively short ranged: in inert gases they decay as r^{-7}, r being the distance between molecules; in crystals it may be sufficient to regard each atom as interacting only with its nearest neighbour. The infinite-range correlations that occur at a critical point are caused by the cooperative behaviour of the system, not by infinite-range interactions.

If, on the other hand, sufficiently long-range interactions are included in $E(s)$, they clearly can affect the way the correlations become infinite near T_c, and it comes as no surprise that critical exponents can be altered in this way. Thus universality only applies to systems with the same range of interactions. To obtain the correct critical behaviour, a model of a real system should not introduce non-physical long-range interactions.

Unfortunately no short-range genuinely three-dimensional model has been solved. The simplest such model is the three-dimensional Ising model (which will be defined shortly): this has been extensively investigated using the series expansion method (Gaunt and Sykes, 1973), but no exact solution obtained.

The models of interacting systems for which the partition function (1.4.1) has been calculated exactly (at least in the limit of a large system) can generally be grouped into the following four classes.

One-Dimensional Models

One-dimensional models can be solved if they have finite-range, decaying exponential, or Coulomb interactions. As guides to critical phenomena, such models with short-range two-particle forces (including exponentially decaying forces) have a serious disadvantage: they do not have a phase transition at a non-zero temperature (van Hove, 1950; Lieb and Mattis, 1966). The Coulomb systems also do not have a phase transition, (Lenard, 1961; Baxter, 1963, 1964 and 1965), though the one-dimensional electron gas has long-range order at all temperatures (Kunz, 1974).

Of the one-dimensional models, only the nearest-neighbour Ising model (Ising, 1925; Kramers and Wannier, 1941) will be considered in this book. It provides a simple introduction to the transfer matrix technique that will be used for the more difficult two-dimensional models. Although it does not have a phase transition for non-zero temperature, the correlation length does become infinite at $H = T = 0$, so in a sense this is a 'critical point' and the scaling hypothesis can be tested near it.

A one-dimensional system can have a phase transition if the interactions involve infinitely many particles, as in the cluster interaction model (Fisher and Felderhof, 1970; Fisher, 1972). It can also have a phase transition if the interactions become infinitely long-ranged, but then the system really belongs to the following class of 'infinite-dimensional' models.

'Infinite Dimensional' Models

To see what is meant by an 'infinite dimensional' system, one needs a working definition of the effective dimensionality of a Hamiltonian. For a system with finite or short-range interactions in all available directions there is usually no problem: the dimensionality is that of the space considered.

For other systems, a useful clue is to note that the dimensionality of a lattice can be defined by starting from a typical site and counting the number of sites that can be visited in a walk of n steps. For a d-dimensional regular lattice and for n large, this is proportional to the volume of a box of side n, i.e. to n^d. The larger the dimensionality, the more close neighbours there are to each site.

If the number of neighbours becomes infinite, then the system is effectively infinite-dimensional. Such a system is the mean-field model discussed in Chapter 3. In Chapter 4 the Ising model on the Bethe lattice is considered. This 'lattice' has the property that the number of neighbours visited in n steps grows exponentially with n. This is a faster rate of growth than n^d, no matter how large d is, so again this model is infinite-dimensional.

The results for these two models are the same as those obtained from the mean-field and Bethe approximations, respectively, for regular lattices (Section 1.5). Thus these two approximations are equivalent to replacing the original Hamiltonian by an infinite-dimensional model Hamiltonian.

Kac *et al.* (1963/4) considered a solvable one-dimensional particle model with interactions with a length scale R. For such a model it is appropriate to define 'close neighbours' as those particles within a distance R of a given particle. They then let $R \rightarrow \infty$ and found that in this limit (and *only* in this limit) there is a phase transition. From the present point of view this is not surprising: by letting $R \rightarrow \infty$ the number of close neighbours becomes infinite and the system effectively changes from one-dimensional to infinite-dimensional. A remarkable feature of this system is that the equation of state is precisely that proposed phenomenologically by van der Waals in 1873 (eq. 1.10.1). All these three 'infinite-dimensional' models satisfy the scaling hypothesis (1.2.1), and have *classical exponents* (see Section 1.10).

The Spherical Model

As originally formulated (Montroll, 1949; Berlin and Kac, 1952), this model introduces a constraint coupling all components equally, no matter how far apart they are. Thus it is 'unphysical' in that it involves infinite range interactions. However, Stanley (1968) has shown that it can be regarded as a limiting case of a system with only nearest neighbour interactions. The model is discussed in Chapter 5. It is interesting in that its exponents are *not* classical in three dimensions.

Two-Dimensional Lattice Models

There are a very few two-dimensional models that have been solved (i.e. their free energy calculated), notably the Ising, ferroelectric, eight-vertex and three-spin models. These are all 'physical' in that they involve only finite-range interactions; they exhibit critical behaviour. The main attention of this book will be focussed on these models.

It is of course unfortunate that they are only two-dimensional, but they still provide a qualitative guide to real systems. Indeed, there are real crystals which have strong horizontal and weak vertical interactions, and so are effectively two-dimensional. Examples are K_2NiF_4 and Rb_2MnF_4 (Birgenau *et al.*, 1973; Als-Nielsen *et al.*, 1975). The models may provide a very good guide to such crystals.

What is probably more unfortunate is that most of the two-dimensional models have only been solved in zero field ($H = 0$), so only very limited information on the critical behaviour has been obtained and the scaling functions $h(x)$ have not been calculated. The one exception is the ferroelectric model in the presence of an electric field, but this turns out to have an unusual and atypical behaviour (Section 7.10).

1.7   The General Ising Model

Most of the models to be discussed in this book can be regarded as special cases of a general Ising model, which can be thought of as a model of a magnet. Regard the magnet as made up of molecules which are constrained to lie on the sites of a regular lattice. Suppose there are N such sites and molecules, labelled $i = 1, \ldots, N$.

Now regard each molecule as a microscopic magnet, which either points along some preferred axis, or points in exactly the opposite direction. Thus each molecule i has two possible configurations, which can be labelled by a 'spin' variable σ_i with values $+1$ (parallel to axis) or -1 (anti-parallel). The spin is said to be 'up' when σ_i has value $+1$, 'down' when it has value -1. Often these values are written more briefly as $+$ and $-$. Let

$$\sigma = \{\sigma_1, \ldots, \sigma_N\}$$

denote the set of all N spins. Then there are 2^N values of σ, and each such value specifies a state of the system. For instance, Fig. 1.5 shows a system of 9 spins in the state

$$\sigma = \{+, +, +, -, +, -, +, -, -\}. \tag{1.7.1}$$

The Hamiltonian is now a function $E(\sigma_1, \ldots, \sigma_N)$ of the N spins $\sigma_1, \ldots, \sigma_N$, or more briefly a function $E(\sigma)$ of σ. It is made up of two parts:

$$E(\sigma) = E_0(\sigma) + E_1(\sigma), \tag{1.7.2}$$

where E_0 is the contribution from the intermolecular forces inside the magnet, and $E_1(\sigma)$ is the contribution from the interactions between the

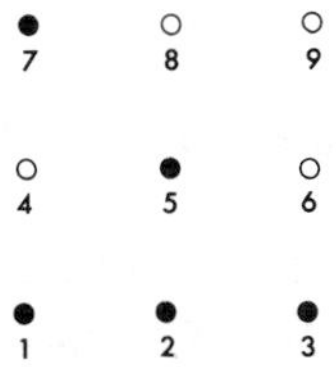

Fig. 1.5. An arrangement of spins on a square lattice with labelled sites. Full circles denote up (positive) spins, open circles denote down (negative) spins.

spins and an external magnetic field. Since σ_i is effectively the magnetic moment of molecule i, $E_1(\sigma)$ can be written as

$$E_1(\sigma) = -H \sum_i \sigma_i, \tag{1.7.3}$$

where H is proportional to the component of the field in the direction of the preferred axis. From now on we shall refer to H simply as 'the magnetic field'. The sum in (1.7.3) is over all sites of the lattice, i.e. over $i = 1, \ldots, N$.

In a physical system we expect the interactions to be invariant under time reversal, which means that E is unchanged by reversing all fields and magnetizations, i.e. by negating H and $\sigma_1, \ldots, \sigma_N$. It follows that E_0 must be an even function of σ, i.e.

$$E_0(\sigma_1, \ldots, \sigma_N) = E_0(-\sigma_1, \ldots, -\sigma_N). \tag{1.7.4}$$

These relations define a quite general Ising model, special cases of which have been solved. From a physicist's point of view it is highly simplified, the obvious objection being that the magnetic moment of a molecule is a vector pointing in any direction, not just up or down. One can build this property in, thereby obtaining the classical Heisenberg model (Stanley, 1974), but this model has not been solved in even two dimensions.

However, there are crystals with highly anisotropic interactions such that the molecular magnets effectively point only up or down, notably $FeCl_2$ (Kanamori, 1958) and $FeCO_3$ (Wrege et al., 1972). The three-dimensional

Ising model should give a good description of these, in fact universality implies that it should give exactly correct critical exponents.

The gaps in Sections 1.1, 1.2 and 1.4, notably a statistical-mechanical definition of $M(H, T)$ and the critical exponents α, ν, η, μ, can now be filled in. From (1.4.1), (1.7.2) and (1.7.3), the partition function is a function of N, H and T, so can be written

$$Z_N(H, T) = \sum_{\sigma} \exp\{-[E_0(\sigma) - H \sum_i \sigma_i]/kT\}. \qquad (1.7.5)$$

Free Energy and Specific Heat

Physically, we expect the free energy of a large system to be proportional to the size of the system, i.e. we expect the *thermodynamic limit*

$$f(H, T) = -kT \lim_{N \to \infty} N^{-1} \ln Z_N(H, T) \qquad (1.7.6)$$

to exist, f being the free energy per site.

We also expect this limit to be independent of the way it is taken. For example, it should not matter whether the length, breadth and height of the crystal go to infinity together, or one after the other: so long as they do all ultimately become infinite.

From (1.4.6), the internal energy per site is

$$u(H, T) = -T^2 \frac{\partial}{\partial T}[f(H, T)/T]. \qquad (1.7.7)$$

The specific heat per site is defined to be

$$C(H, T) = \frac{\partial}{\partial T} u(H, T). \qquad (1.7.8)$$

It has been usual to define two critical exponents α and α' by asserting that near T_c the zero-field specific heat diverges as a power-law, i.e.

$$\begin{aligned} C(0, T) &\sim t^{-\alpha} & \text{as } t \to 0^+, \\ &\sim (-t)^{-\alpha'} & \text{as } t \to 0^-, \end{aligned} \qquad (1.7.9)$$

where t is defined by (1.1.3).

The difficulty with this definition is that $C(0, T)$ may remain finite as t goes to zero through positive (or negative) values, even though it is not an analytic function at $t = 0$. For instance $C(0, T)$ may have a simple jump discontinuity at $t = 0$, as in the mean-field model of Chapter 3.

To obtain an exponent which characterizes such behaviour it is better to proceed as follows.

Let $f_+(0, T)$ and $f_-(0, T)$ be the zero-field free energy functions for $T > T_c$ and $T < T_c$, respectively. Analytically continue these functions into the complex T plane and define the 'singular part' of the free energy to be

$$f_s(0, T) = f_+(0, T) - f_-(0, T). \qquad (1.7.10a)$$

Near $T = T_c$ this usually vanishes as a power law, and α can be defined by

$$f_s(0, T) \sim t^{2-\alpha} \qquad \text{as } t \to 0. \qquad (1.7.10b)$$

This definition is equivalent to (1.7.9) (with $\alpha' = \alpha$) for those cases where $u(0, T)$ is continuous and $C(0, T)$ diverges both above and below T_c.

It used to be thought that the only possible singularity in $f(0, T)$ was a jump-discontinuity in some derivative of f. If the first $r - 1$ derivatives were continuous, but the rth derivative discontinuous, then it was said that the system had a 'transition of order r'. In particular, a discontinuity in u (i.e. latent heat) is called a first-order transition.

While it is now known that this classification is not exhaustive, such behaviour is included in (1.7.10): a transition of order r corresponds to $2 - \alpha = r$. In particular, $\alpha = 1$ for a first-order transition.

From (1.7.8), the definition (1.7.10) implies that $u(0, T)$ contains a term proportional to $t^{1-\alpha}$. Since $u(0, T)$ is usually bounded, it follows that

$$\alpha \leq 1. \qquad (1.7.11)$$

The exponent α may be negative.

Magnetization

The magnetization is the average of the magnetic moment per site, i.e., using (1.4.4),

$$M(H, T) = N^{-1}\langle \sigma_1 + \ldots + \sigma_N \rangle, \qquad (1.7.12)$$

$$= N^{-1} Z_N^{-1} \sum_{\sigma} (\sigma_1 + \ldots + \sigma_N)$$

$$\times \exp\left\{ -\left[E_0(\sigma) - H \sum_i \sigma_i \right] / kT \right\}. \qquad (1.7.13)$$

Differentiating (1.7.5) with respect to H, and using (1.7.6), one obtains that in the thermodynamic limit $(N \to \infty)$

$$M(H, T) = -\frac{\partial}{\partial H} f(H, T). \qquad (1.7.14)$$

Since the summand in (1.7.5) is unchanged by negating H and σ, Z_N and f are even functions of H, so M is an odd function, i.e.

$$M(-H, T) = -M(H, T).\tag{1.7.15}$$

From (1.7.12) it lies in the interval

$$-1 \leqslant M(H, T) \leqslant 1.\tag{1.7.16}$$

Differentiating (1.7.13) with respect to H and using (1.1.1) and (1.4.4), the susceptibility is

$$\chi = \frac{\partial M}{\partial H}$$

$$= (NkT)^{-1}\{\langle \mathcal{M}^2 \rangle - \langle \mathcal{M} \rangle^2\},\tag{1.7.17}$$

where

$$\mathcal{M} = \sum_i \sigma_i.\tag{1.7.18}$$

Using only the fact that the average of a constant is the same constant, (1.7.17) can be written

$$\chi = (NkT)^{-1}\langle[\mathcal{M} - \langle \mathcal{M} \rangle]^2\rangle.\tag{1.7.19}$$

Thus χ is the average of a non-negative quantity, so

$$\chi = \frac{\partial M}{\partial H} \geqslant 0.\tag{1.7.20}$$

The magnetization M is therefore an odd monotonic increasing function of H, lying in the interval (1.7.16), as indicated in Fig. 1.1.

Note that for finite N, Z is a sum of analytic positive functions of H, so f and M are also analytic. The discontinuity in Fig. 1.1(a), and the singularity in Fig. 1.1(b), can only occur when the thermodynamic limit is taken.

The critical exponents β, δ, γ, γ' associated with the magnetization have been defined in Section 1.1. The scaling relations (1.2.13) can be obtained by integrating (1.7.14), using the scaling hypothesis (1.2.1).

Correlations

The correlation between spins i and j is

$$g_{ij} = \langle \sigma_i \sigma_j \rangle - \langle \sigma_i \rangle \langle \sigma_j \rangle.\tag{1.7.21}$$

If $E_0(\sigma)$ is translation invariant, as is usually the case, $\langle \sigma_i \rangle$ is the same for all sites i, so from (1.7.12),

$$\langle \sigma_i \rangle = \langle \sigma_j \rangle = M(H, T).\tag{1.7.22}$$

Also, g_{ij} will depend only on the vector distance $\mathbf{r}_{ij}$ between sites i and j, i.e.

$$g_{ij} = g(\mathbf{r}_{ij}) , \qquad (1.7.23)$$

where $g(\mathbf{r})$ is the *correlation function*.

Away from T_c the function $g(\mathbf{r})$ is expected to decay exponentially to zero as $\mathbf{r}$ becomes large. More precisely, if $\mathbf{k}$ is some fixed unit vector, we expect that

$$g(x\mathbf{k}) \sim x^{-\tau} e^{-x/\xi} \qquad \text{as } x \to \infty , \qquad (1.7.24)$$

where τ is some number and ξ is the *correlation length* in the direction $\mathbf{k}$.

The correlation length is a function of H and T, and is expected to become infinite at T_c. In fact, this property of an infinite correlation length can be regarded as the hallmark of a critical point. In particular, it is expected that

$$\xi(0 , T) \sim t^{-\nu} \qquad \text{as } t \to 0^+ \qquad (1.7.25)$$
$$\sim (-t)^{-\nu'} \qquad \text{as } t \to 0^- ,$$

where ν and ν' are the *correlation length critical exponents*.

It is a little unfortunate that ξ also depends on the direction $\mathbf{k}$. However, near T_c this dependence is expected to disappear and the large-distance correlations to become isotropic (see for example McCoy and Wu, 1973, p. 306). Thus the exponents ν and ν' should not depend on the direction in which ξ is defined.

At the critical point itself, the correlation function $g(\mathbf{r})$ still exists, but instead of decaying exponentially decays as the power law

$$g(\mathbf{r}) \sim r^{-d+2-\eta} , \qquad (1.7.26)$$

where η is a critical exponent.

In scaling theory, these properties are simple corollaries of the correlation scaling hypothesis, which is that near T_c, for $r \sim \xi$,

$$g(\mathbf{r}) \sim r^{-d+2-\eta} D(r/\xi , t|H|^{-1/\beta\delta}) . \qquad (1.7.27)$$

The susceptibility χ can be expressed in terms of $g(\mathbf{r})$. To do this, simply sum (1.7.21) over all sites i and j. From (1.7.17) it immediately follows that

$$\chi = (NkT)^{-1} \sum_i \sum_j g_{ij} . \qquad (1.7.28)$$

For a translation-invariant system,

$$\sum_j g_{ij} = \sum_j g(\mathbf{r}_{ij}) = \text{independent of } i , \qquad (1.7.29)$$

so (1.7.28) becomes

$$\chi = (kT)^{-1} \sum_j g(\mathbf{r}_{0j}) , \tag{1.7.30}$$

where 0 is some fixed site in the lattice.

Near T_c the function $g(\mathbf{r})$ is an isotropic bounded slowly varying function of $\mathbf{r}$, so the summation can be replaced by an integration, giving

$$\chi \sim \int_0^\infty g(r) \, r^{d-1} \, dr . \tag{1.7.31}$$

Making the substitution $r = x\xi$ and using (1.7.27), it follows that near T_c

$$\chi \sim \xi^{2-\eta} . \tag{1.7.32}$$

The scaling relations (1.2.14) now follow from the definitions of γ, γ', ν, ν' and the equality of γ and γ'.

Interfacial Tension

This quantity is defined only on the cut in Fig. 1.2, i.e. for $H = 0$ and $T < T_c$. If the cut is approached from above, i.e. H goes to zero through positive values, the equilibrium state is one in which most spins are up. If the cut is approached from below, most spins are down.

At $H = 0$ these two equilibrium states can coexist: the crystal may consist of two large domains, one in one state, the other in the other. The total free energy is then

$$F = Nf + Ls , \tag{1.7.33}$$

where Nf is the normal bulk free energy and Ls is the total surface free energy due to the interface between the domains. If L is the area of this interface, then s is the interfacial tension per unit area.

It will be shown in Section 1.9 that there is a correspondence between the magnetic model used here and a model of a liquid – gas transition. In the latter teminology, s is the surface tension of a liquid in equilibrium with its vapour, e.g. water and steam at 100°C.

The interfacial tension is not usually emphasized in the theory of critical phenomena, but it is one of the thermodynamic quantities that can be calculated for the exactly soluble two-dimensional models, so is of interest here. It is a function of the temperature T.

As T approaches T_c from below, the two equilibrium states become the same, so s goes to zero. It is expected that near T_c

$$s(T) \sim (-t)^\mu , \tag{1.7.34}$$

where μ is yet another critical exponent, the last to be defined in this book. Widom (1965) used scaling arguments to suggest that near T_c

$$s(T) \propto \xi(0, T) M^2(0, T)/\chi(0, T), \qquad (1.7.35)$$

from which the scaling relation (1.2.15) follows. He also obtained the hyper-scaling relation (1.2.16).

1.8 Nearest-Neighbour Ising Model

The discussion of Section 1.7 applies for any even Hamiltonian $E_0(\sigma)$, subject only to some implicit assumptions such as the existence of the thermodynamic limit (1.7.6) and a ferromagnetic critical point.

The simplest such Hamiltonian is one in which only nearest neighbours interact, i.e.

$$E_0(\sigma) = -J \sum_{(i,j)} \sigma_i \sigma_j \qquad (1.8.1)$$

where the sum is over all nearest-neighbour pairs of sites in the lattice. This is the normal Ising model mentioned in Section 1.6. If J is positive the lowest energy state occurs when all spins point the same way, so the model is a ferromagnet.

A great deal is known about this model, even for those cases where it has not been exactly solved, such as in three dimensions, or in two dimensions in the presence of a field. For instance, one can develop expansions valid at high or low temperatures.

From (1.7.5), the partition function is

$$Z_N = \sum_{\sigma} \exp\left[K \sum_{(i,j)} \sigma_i \sigma_j + h \sum_i \sigma_i \right], \qquad (1.8.2)$$

where

$$K = J/kT, \quad h = H/kT, \qquad (1.8.3)$$

so Z_N can be thought of as a function of h and K. From (1.7.6) and (1.7.14) the magnetization per site is

$$M = \frac{\partial}{\partial h} \lim_{N \to \infty} N^{-1} \ln Z_N(h, K). \qquad (1.8.4)$$

It is easy to produce a plausible, though not rigorous, argument that M should have the behaviour shown in Fig. 1.1, and that there should be a critical point at $H = 0$ for some positive value T_c of T. This will now be done.

For definiteness, consider a square lattice (but the argument applies to any multi-dimensional lattice). The RHS of (1.8.2) can be expanded in powers of K, giving

$$Z_N = (2 \cosh h)^N \{1 + 2NKt^2$$
$$+ NK^2 [(2N - 7) t^4 + 6t^2 + 1] + \mathcal{O}(K^3)\}, \quad (1.8.5)$$

where

$$t = \tanh h. \quad (1.8.6)$$

Substituting this expansion into (1.8.4) gives

$$M = \tanh h \{1 + 4 \operatorname{sech}^2 h [K + (3 - 7t^2) K^2 + \mathcal{O}(K^3)]\}. \quad (1.8.7)$$

All terms in this expansion are odd analytic bounded functions of h. Assuming that the expansion converges for sufficiently small K, i.e. for sufficiently high temperatures, it follows that for such temperatures $M(H, T)$ has the graph shown in Fig. 1.1(c). In particular, it is continuous at $H = 0$ and

$$M_0(T) = M(0, T) = 0, \ T \text{ sufficiently large}. \quad (1.8.8)$$

Alternatively, at low temperatures K is large and the RHS of (1.8.2) can be expanded in powers of

$$u = \exp(-4K). \quad (1.8.9)$$

The leading term in this expansion is the contribution to Z from the state with all spins up (or all down). The next term comes from the N states with one spin down and the rest up (or vice versa); the next from the $2N$ states with two adjacent spins down (or up), the next term comes from either states with two non-adjacent spins, or a spin and two of its neighbours, or four spins round a square, reversed; and so on. This gives

$$Z_N = e^{2NK + Nh}\{1 + Nu^2 e^{-2h}$$
$$+ 2Nu^3 e^{-4h} + \tfrac{1}{2}N(N - 5) u^4 e^{-4h}$$
$$+ 6N u^4 e^{-6h} + Nu^4 e^{-8h} + \mathcal{O}(u^5)\}$$
$$+ e^{2NK - Nh}\{1 + Nu^2 e^{2h}$$
$$+ 2Nu^3 e^{4h} + \tfrac{1}{2}N(N - 5) u^4 e^{4h}$$
$$+ 6Nu^4 e^{6h} + Nu^4 e^{8h} + \mathcal{O}(u^5)\}. \quad (1.8.10)$$

The first series in curly brackets is the contribution from states with almost all spins up, the second from states with almost all spins down.

Equation (1.8.10) can be written

$$Z_N = e^{N\psi(h,K)} + e^{N\psi(-h,K)} , \qquad (1.8.11)$$

where

$$\psi(h,K) = 2K + h + u^2 e^{-2h}$$
$$+ 2u^3 e^{-4h} + u^4(-2\tfrac{1}{2} e^{-4h} + 6 e^{-6h} + e^{-8h})$$
$$+ \mathcal{O}(u^5) . \qquad (1.8.12)$$

To any order in the u-expansion, $\psi(h,K)$ is independent of N, provided N is sufficiently large.

If h is positive, the first term on the RHS of (1.8.11) will be larger than the second. In the limit of N large it will be the dominant contribution to Z_N, so from (1.8.4)

$$M = \frac{\partial}{\partial h} \psi(h,K)$$
$$= 1 - 2u^2 e^{-2h} - 8u^3 e^{-4h}$$
$$- u^4(-10 e^{-4h} + 36 e^{-6h} + 8 e^{-8h})$$
$$- \mathcal{O}(u^5) \qquad \text{if } h > 0 , \qquad (1.8.13)$$

and the spontaneous magnetization is

$$M_0(T) = \lim_{h \to o^+} M$$
$$= 1 - 2u^2 - 8u^3 - 34u^4 - \mathcal{O}(u^5) . \qquad (1.8.14)$$

If these expansions converge for sufficiently small u (i.e. sufficiently low temperatures), then M_0 is positive for small enough u. Remembering that $M(H,T)$ is an odd function of H, it follows that at low temperatures $M(H,T)$ has the graph shown in Fig. 1.1(a), with a discontinuity at $H = 0$.

The function $M_0(T)$ is therefore identically zero for sufficiently large T, but strictly positive for sufficiently small T. At some intermediate temperature T_c it must change from zero to non-zero, as indicated in Fig. 1.3, and at this point must be a non-analytic function of T. Thus there must be a 'critical point' at $H = 0$, $T = T_c$, where the thermodynamic functions become non-analytic, as indicated in Fig. 1.2.

This argument does not preclude further singularities in the interior of the (H, T) half-plane, but Figs. 1.1 to 1.3 are the simplest picture that is consistent with it.

Parts of the argument, or variants of them, can be made quite rigorous.

For instance, as long ago as 1936 Peierls proved that $M_0(T)$ is positive for sufficiently low temperatures (see also Griffiths, 1972, p. 59).

The argument fails for the one-dimensional Ising model. This is because the next-to-leading term in the low temperature u expansion comes from states such as that shown in Fig. 1.6, where a line of adjacent spins are all

Fig. 1.6. An arrangement of spins in a one-dimensional Ising model that contributes to next-to-leading order in a low-temperature expansion. Full circles denote up spins, open circles down spins.

reversed, rather than just a single spin. There are $\frac{1}{2}N(N-1)$ such states, instead of N, so even to this order Z_N is *not* of the form (1.8.11). This of course is consistent with the fact that the one-dimensional model does not have a phase transition at non-zero temperatures.

1.9   The Lattice Gas

As well as being a model of a magnet, the Ising model is also a model of a fluid.

To see this rather startling fact, consider a fluid composed of molecules interacting via some pair potential $\phi(r)$. Typically this potential will have a hard-core (or at least very strong short-range repulsion), an attractive well and a fairly rapidly decaying tail. The usual example is the Lennard – Jones potential

$$\phi(r) = 4\varepsilon[(r_0/r)^{12} - (r_0/r)^6] \qquad (1.9.1)$$

shown in Fig. 1.7(a).

Instead of allowing the molecules to occupy any position in space, restrict them so that their centres lie only on the sites of some grid, or lattice. If the grid is fairly fine this is a perfectly reasonable step: indeed it is a necessary one in almost any numerical calculation.

Since $\phi(r)$ is infinitely repulsive at $r = 0$, no two molecules can be centred on the same site. With each site i associate a variable s_i which is zero if the site is empty, one if it is occupied. If there are N sites, then any spatial arrangement of the molecules can be specified by $s = \{s_1, \ldots, s_N\}$. The number of molecules in such an arrangement is

$$n = s_1 + \ldots + s_N, \qquad (1.9.2)$$

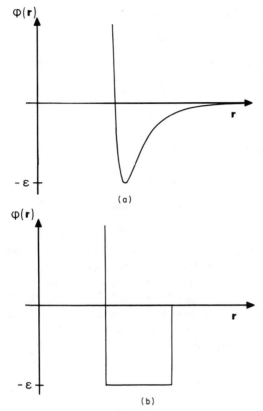

Fig. 1.7. Interaction potentials for a model fluid: (a) Lennard–Jones, (b) square-well.

and the total potential energy is

$$E = \sum_{(i,j)} \phi_{ij} s_i s_j \, , \tag{1.9.3}$$

where the sum is over all pairs of sites on the lattice (not necessarily nearest neighbours) and $\phi_{ij} = \phi(r_{ij})$ is the interaction energy between molecules centred on sites i and j.

The grand-canonical partition function is then

$$Z = \sum_{s} \exp[(n\mu - E)/kT] \, , \tag{1.9.4}$$

where μ is the effective chemical potential (for classical systems the contribution of the integrations in momentum space can be incorporated into μ).

In appropriate units, the pressure is

$$P = N^{-1} kT \ln Z ,$$ (1.9.5)

the density is the average number of molecules per site, i.e.

$$\rho = \langle n \rangle / N$$

$$= \frac{\partial P}{\partial \mu} ,$$ (1.9.6)

and the compressibility is

$$k_T = \frac{1}{\rho} \frac{\partial \rho}{\partial P}$$

$$= \frac{1}{\rho^2} \frac{\partial \rho}{\partial \mu} ,$$ (1.9.7)

the differentiations being performed at constant temperature.

The Lennard – Jones potential (1.9.1) is a fairly realistic one, but the qualitative features of the liquid – gas transition are not expected to depend on the details of the potential: it should be sufficient that it have short-range repulsion and an attractive well. Thus ϕ_{ij} should be large and positive when sites i and j are close together: negative when they are a moderate distance apart; and zero when they are far apart. The simplest such choice is

$$\phi_{ij} = +\infty \quad \text{if } i = j ,$$

$$= -\varepsilon \quad \text{if } i \text{ and } j \text{ are nearest neighbours} ,$$

$$= 0 \quad \text{otherwise} .$$ (1.9.8)

This corresponds to the 'square well' potential shown in Fig. 1.7(b), which is often used in model calculations.

Letting $\phi_{ii} = +\infty$ is equivalent to taking the potential to be infinitely repulsive if two molecules come together, i.e. to prohibiting two molecules from occupying the same site. This feature has already been built into the formulation, so if ϕ_{ij} is given by (1.9.8), then from (1.9.3) the energy is

$$E = -\varepsilon \sum_{(i,j)} s_i s_j ,$$ (1.9.9)

the sum now being only over nearest-neighbour pairs of sites on the lattice.

It is now trivial to show that (1.9.4) is the partition function of a nearest-neighbour Ising model in a field. Replace each s_i by a 'spin' σ_i, where

$$\sigma_i = 2s_i - 1 .$$ (1.9.10)

Thus $\sigma_i = -1$ if the site is empty, $+1$ if it is full. If each site has q neighbours, there are $\frac{1}{2}Nq$ nearest-neighbour pairs, and eliminating n, E, $s_1, \ldots, s_N$ between equations (1.9.2), (1.9.4), (1.9.9) and (1.9.10) gives

$$Z = \sum_\sigma \exp\left\{\left[\varepsilon \sum_{(i,j)} \sigma_i \sigma_j + (2\mu + \varepsilon q) \sum_i \sigma_i \right.\right.$$
$$\left.\left. + N(\tfrac{1}{2}\varepsilon q + 2\mu)\right]\Big/4kT\right\}. \tag{1.9.11}$$

Comparing this with (1.8.2) and (1.8.3), it is obvious that, apart from a trivial factor, Z is the partition function of an Ising model with

$$J = \varepsilon/4, \quad H = (2\mu + \varepsilon q)/4. \tag{1.9.12}$$

Using also (1.9.5)–(1.9.7), (1.7.6), (1.7.14) and (1.7.18), one can establish the following expressions for the lattice gas variables in terms of those of the Ising model:

$$\varepsilon = 4J, \tag{1.9.13}$$

$$\mu = 2H - 2qJ, \tag{1.9.14}$$

$$P = -\tfrac{1}{2}qJ + H - f, \tag{1.9.15}$$

$$\rho = \tfrac{1}{2}(1 + M), \tag{1.9.16}$$

$$\rho^2 k_T = \tfrac{1}{4}\chi. \tag{1.9.17}$$

The known general behaviour of the Ising model can now be used to obtain the form of the equation of state of the lattice gas. To do this, consider a fixed value of T. Then (1.9.15) and (1.9.16) define P and ρ as functions of H. Using also (1.7.14) and (1.7.20), it is easily seen that

$$\frac{\partial P}{\partial H} = 1 + M > 0, \quad \frac{\partial \rho}{\partial H} = \tfrac{1}{2}\chi \geq 0, \tag{1.9.18}$$

so both P and ρ are monotonic increasing functions of H. When H is large (positive or negative) the dominant term in the Ising model partition function is one in which all spins are alike, so

$$f \to -\tfrac{1}{2}qJ - |H| \quad \text{as } H \to \pm\infty. \tag{1.9.19}$$

From (1.7.14), (1.9.15) and (1.9.16) it follows that

$$P \to 0 \quad \text{and } \rho \to 0 \qquad \text{as } H \to -\infty, \tag{1.9.20}$$

$$P \sim 2H \text{ and } \rho \to 1 \qquad \text{as } H \to +\infty. \tag{1.9.21}$$

Since P and ρ are monotonic increasing functions of H, from (1.9.20) they must be positive.

For $T > T_c$, f and M, and hence P and ρ, are continuous functions of H. Thus P is a monotonic increasing function of ρ, and a monotonic decreasing function of the volume per molecule

$$v = \rho^{-1}.$$ (1.9.22)

As v increases from 1 to ∞, P decreases from infinity to zero.

For $T < T_c$, M is a discontinuous function of H as shown in Fig. 1.1(a). Thus ρ and v have a discontinuity (but P does not).

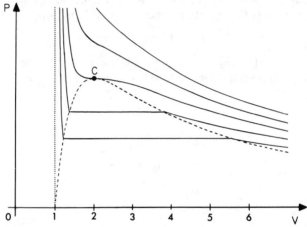

Fig. 1.8. Typical (P, v) isotherms for a simple fluid whose intermolecular interactions have a hard core. The upper two isotherms are for temperatures greater than T_c, the middle one is the critical isotherm ($T = T_c$), and the lower two are for temperatures less than T_c.

Noting also that the expansion coefficient

$$v^{-1}\left(\frac{\partial v}{\partial T}\right)_P$$

of a fluid is usually positive (an exception is water between 0°C and 4°C), it follows that the (P, v) isotherms of the lattice gas (in any dimension greater than one) have the general structure indicated in Fig. 1.8. These are typical isotherms of a fluid in which the intermolecular potential has a hard core.

The point C in this figure is the critical point, and corresponds to the critical point $H = 0$, $T = T_c$ in Fig. 1.2.

Since $M = 0$ at this point, we see from (1.9.14) and (1.9.16) that the critical values of μ, ρ and v for the lattice gas are

$$\mu_c = -2qJ, \quad \rho_c = \tfrac{1}{2}, \quad v_c = 2.$$ (1.9.23)

At $T = T_c$, from (1.1.5) and (1.9.16) we expect that

$$v_c - v \sim H^{1/\delta} \qquad \text{as } H \to 0. \tag{1.9.24}$$

Since $P - P_c$ is proportional to H for small H, it follows that near C the equation of the critical isotherm is

$$P - P_c \sim (v_c - v)^\delta. \tag{1.9.25}$$

For $T < T_c$ an isotherm breaks up into three parts: that part to the left of the broken curve in Fig. 1.8, corresponding to fairly high densities and to a liquid state; the low-density part to the right, corresponding to a gas; and the horizontal line in between, corresponding to the two-phase region where the liquid can co-exist with its vapour. The broken curve is known as the *co-existence curve*. It corresponds to the cut in Fig. 1.2, where $H = 0$ and $M = \pm M_0(T)$. From (1.9.16) and (1.9.23), we see that on this curve

$$|\rho - \rho_c| = M_0(T). \tag{1.9.26}$$

From (1.1.3), (1.1.4) and (1.9.22), it follows that near T_c, the equation of the co-existence curve in the (v, T) plane is

$$|v - v_c| \sim (T_c - T)^\beta. \tag{1.9.27}$$

Near the critical point $P - P_c$ is proportional to t, so from (1.9.27) the equation of the co-existence curve in the (v, P) plane is

$$P_c - P \sim |v - v_c|^{1/\beta}. \tag{1.9.28}$$

Equations (1.9.25) and (1.9.28) relate the exponents δ and β to the liquid – gas critical point. To do the same for α, γ and γ', first note that $M = 0$ on the line segment $H = 0$, $T > T_c$ in Fig. 1.2. From (1.9.16) this line segment therefore corresponds to the critical isochore $v = v_c$. From (1.7.7)–(1.7.9) and (1.9.15), and (1.1.6) and (1.9.17), it follows that

$$\partial^2 P/\partial T^2 \sim t^{-\alpha}, \quad k_T \sim t^{-\gamma} \tag{1.9.29a}$$

as C is approached from above along the critical isochore $v = v_c$.

The line segment $H = 0$, $T < T_c$ in Fig. 1.2 corresponds to the co-existence curve in Fig. 1.8, so

$$\partial^2 P/\partial T^2 \sim (-t)^{-\alpha'}, \quad k_T \sim (-t)^{-\gamma'} \tag{1.9.29b}$$

as C is approached along the co-existence curve, the differentiation being performed on this curve.

These definitions (1.9.29) of α and α' are the analogue of (1.7.9), and suffer from the same difficulties. If $\partial P/\partial T$ is not continuous, or if $\partial^2 P/\partial T^2$

does not diverge at C, it is better to use the analogue of (1.7.10) and define a single exponent α as follows.

Let $P_+(T)$ be the pressure when $v = v_c$ and $T > T_c$; $P_-(T)$ be the pressure when v lies on the co-existence curve and $T < T_c$. Analytically continue these functions into the complex T-plane and define $P_s(T)$ and α by

$$P_s(T) = P_+(T) - P_-(T) \sim t^{2-\alpha}. \tag{1.9.30}$$

To summarize this section: the Ising model of a magnet is also a model of a lattice gas; it merely depends whether one uses 'magnetic language' (spins up or down) or 'particle language' (sites occupied or empty). In the second language the critical exponents δ, β, γ, γ', α are defined by (1.9.25), and (1.9.28)–(1.9.30).

The magnetic language is more convenient in theoretical calculations: it clearly exhibits the symmetries of the Hamiltonian and the thermodynamic functions, notably the relation $M(-H) = -M(H)$.

1.10   The van der Waals Fluid and Classical Exponents

There are phenomenological equations of state, notably that proposed for continuum fluids by van der Waals (1873):

$$P = kT/(v - b) - a/v^2 \tag{1.10.1}$$

where a and b are constants. This equation is valid only outside the co-existence curve, which curve is defined by the Maxwell equal area construction (Pathria, 1972, p. 376) which ensures that P and μ are continuous along any isotherm. As we remarked in Section 1.6, it is the exact equation of state of a model solved by Kac et al. (1963/4).

The critical exponent definitions (1.9.25), (1.9.28–30) apply to any liquid – gas critical point, not just that of the simple lattice gas of Section 1.9. Equations such as van der Waals predict that near T_c the critical isotherm is a cubic curve, and the coexistence curve a parabola. From (1.9.25) and (1.9.28) this implies

$$\delta = 3 , \quad \beta = \tfrac{1}{2} . \tag{1.10.2}$$

Also, the van der Waals equation (1.10.1) has a critical point at

$$T_c = 8a/27bk , \quad v_c = 3b . \tag{1.10.3}$$

Near this point it is readily verified that $k_T \sim t^{-1}$, so

$$\gamma = \gamma' = 1 . \tag{1.10.4}$$

On the critical isochore it is easily seen from (1.10.1) that

$$P - P_c = 4at/27b^2 , \qquad (1.10.5)$$

while on the coexistence curve a more complicated calculation gives

$$P - P_c = (4a/27b^2) [t + 6t^2/5 + \mathbb{O}(t^3)] \qquad (1.10.6)$$

Thus $\partial^2 P/\partial T^2$ is finite at C but has a jump discontinuity on going from the critical isochore to the co-existence curve. The definitions (1.9.29) of α and α' fail, but (1.9.30) gives

$$\alpha = 0 . \qquad (1.10.7)$$

The values (1.10.2), (1.10.4), (1.10.7) of the critical exponents are known as the *classical* values. They satisfy the scaling relations (1.2.12) and (1.2.13), and are the values given by the simple 'infinite dimensional' mean field and Bethe lattice models (Chapters 3 and 4). They are *not* correct for the nearest-neighbour Ising model in two or three dimensions, but it is now generally believed (Fisher, 1974, p.607) that they are correct in four or more dimensions.

2

THE ONE-DIMENSIONAL ISING MODEL

2.1 Free Energy and Magnetization

Ising proposed his model in 1925 and solved it for a one-dimensional system. The solution is presented in this chapter, partly because it provides an introduction to the transfer matrix technique that will be used in later chapters, as well as for the intrinsic interest of a simple exactly soluble model. The one-dimensional model does not have a phase transition at any non-zero temperature, but it will be shown that it has a critical point at $H = T = 0$, that critical exponents can be sensibly defined, and that the scaling hypothesis and relevant scaling relations are satisfied.

Fig. 2.1. The one-dimensional lattice of N sites.

Consider an Ising model on a line of N sites, labelled successively $j = 1, \ldots, N$, as shown in Fig. 2.1. Then the energy of the model is given by (1.7.2), (1.7.3) and (1.8.1), i.e.

$$E(\sigma) = -J \sum_{j=1}^{N} \sigma_j \sigma_{j+1} - H \sum_{j=1}^{N} \sigma_j. \tag{2.1.1}$$

Here site N is regarded as being followed by site 1, so that σ_{N+1} in (2.1.1) is to be interpreted as σ_1. This is equivalent to joining the two ends of the line so as to form a circle, or to imposing periodic boundary conditions on the system. This is often a useful device, partly because it ensures that all sites are equivalent and that the system is translationally invariant. In particular,

$$\langle \sigma_1 \rangle = \langle \sigma_2 \rangle = \ldots = \langle \sigma_N \rangle \,, \tag{2.1.2}$$

so from (1.7.12) the magnetization per site is

$$M(H, T) = \langle \sigma_1 \rangle \,, \tag{2.1.3}$$

where 1 is any particular site of the lattice. This result is true for any translationally invariant system.

From (1.8.2), the partition function is

$$Z_N = \sum_\sigma \exp\left\{ K \sum_{j=1}^N \sigma_j \sigma_{j+1} + h \sum_{j=1}^N \sigma_j \right\} \tag{2.1.4}$$

where

$$K = J/kT \,, \quad h = H/kT \,. \tag{2.1.5}$$

Now we make a vital observation: the exponential in (2.1.4) can be factored into terms each involving only two neighbouring spins, giving

$$Z_N = \sum_\sigma V(\sigma_1, \sigma_2) V(\sigma_2, \sigma_3) V(\sigma_3, \sigma_4) \ldots$$

$$\ldots V(\sigma_{N-1}, \sigma_N) V(\sigma_N, \sigma_1) \,, \tag{2.1.6}$$

where

$$V(\sigma, \sigma') = \exp[K\sigma\sigma' + \tfrac{1}{2}h(\sigma + \sigma')] \,. \tag{2.1.7}$$

This is not the only possible choice of V: it could be multiplied by $\exp[a(\sigma - \sigma')]$ (for any a) without affecting (2.1.6). However, this choice (in which each $h\sigma_j$ is shared equally between two V's) ensures that

$$V(\sigma, \sigma') = V(\sigma', \sigma) \,, \tag{2.1.8}$$

which we shall see is a useful symmetry property.

Now look at the RHS of (2.1.6): regard the $V(\sigma, \sigma')$ as elements of a two-by-two matrix

$$\mathbf{V} = \begin{pmatrix} V(+,+) & V(+,-) \\ V(-,+) & V(-,-) \end{pmatrix} = \begin{pmatrix} e^{K+h} & e^{-K} \\ e^{-K} & e^{K-h} \end{pmatrix} \,. \tag{2.1.9}$$

Then the summations over $\sigma_2, \sigma_3, \ldots, \sigma_N$ in (2.1.6) can be regarded as successive matrix multiplications, and the summation over σ_1 as the taking of a trace, so that

$$Z_N = \text{Trace } \mathbf{V}^N \,. \tag{2.1.10}$$

At each stage in the procedure, matrix multiplication by $\mathbf{V}$ corresponds to summing over the configurations of one more site of the lattice. The matrix V is known as the *transfer matrix*. In later chapters we shall see that transfer matrices can be defined for two- and higher dimensional models. Equation (2.1.10) is then still satisfied, but unfortunately V becomes an extremely large matrix.

Let x_1, x_2 be the two eigenvectors of V, and λ_1, λ_2 the corresponding eigenvalues. Then

$$V x_j = \lambda_j x_j \, , j = 1, 2 \, . \tag{2.1.11}$$

Let P be the two-by-two matrix with column vectors x_1, x_2, i.e.

$$P = (x_1 \, , x_2) \, . \tag{2.1.12}$$

Then from (2.1.11)

$$V P = P \begin{pmatrix} \lambda_1 & 0 \\ 0 & \lambda_2 \end{pmatrix} \, . \tag{2.1.13}$$

Since V is a symmetric matrix, it must be possible to choose x_1 and x_2 orthogonal and linearly independent. Doing so, it follows that the matrix P is non-singular, i.e. it has an inverse P^{-1}. Multiplying (2.1.13) on the right by P^{-1} gives

$$V = P \begin{pmatrix} \lambda_1 & 0 \\ 0 & \lambda_2 \end{pmatrix} P^{-1} \, . \tag{2.1.14}$$

Substituting this expression for V into (2.1.10), the matrix P cancels out, leaving

$$Z_N = \text{Trace} \begin{pmatrix} \lambda_1 & 0 \\ 0 & \lambda_2 \end{pmatrix}^N = \lambda_1^N + \lambda_2^N \, . \tag{2.1.15}$$

Let λ_1 be the larger of the two eigenvalues and write (2.1.15) as

$$N^{-1} \ln Z_N = \ln \lambda_1 + N^{-1} \ln [1 + (\lambda_2/\lambda_1)^N] \, . \tag{2.1.16}$$

Since $|\lambda_2/\lambda_1| < 1$, the second term on the RHS tends to zero as $N \to \infty$. Thus from (1.7.6) the free energy per site does tend to a limit as $N \to \infty$, namely

$$f(H, T) = -kT \lim_{N \to \infty} N^{-1} \ln Z_N$$

$$= -kT \ln \lambda_1$$

$$= -kT \ln [e^K \cosh h + (e^{2K} \sinh^2 h + e^{-2K})^{\frac{1}{2}}] \, . \tag{2.1.17}$$

Differentiating this result with respect to h, using (1.7.14) and (2.1.5), gives

$$M(H, T) = \frac{e^K \sinh h}{[e^{2K} \sinh^2 h + e^{-2K}]^{\frac{1}{2}}} \, . \tag{2.1.18}$$

The free energy is an analytic function of H and T for all real H and positive T. The magnetization $M(H, T)$ is an analytic function of H, with a graph of the type shown in Fig. 1.1(c). Thus the system does not have a phase transition for any positive temperature.

2.2   Correlations

From (1.4.3), (2.1.1), (2.1.7), the probability of the system being in the state $\sigma \equiv \{\sigma_1, \ldots, \sigma_N\}$ is

$$Z_N^{-1} V(\sigma_1, \sigma_2) V(\sigma_2, \sigma_3) V(\sigma_3, \sigma_4) \ldots V(\sigma_N, \sigma_1). \tag{2.2.1}$$

Thus the average value of (say) $\sigma_1 \sigma_3$ is

$$\langle \sigma_1 \sigma_3 \rangle = Z_N^{-1} \sum_\sigma \sigma_1 V(\sigma_1, \sigma_2) V(\sigma_2, \sigma_3) \sigma_3$$

$$V(\sigma_3, \sigma_4) \ldots V(\sigma_N, \sigma_1). \tag{2.2.2}$$

This can also be written in terms of matrices: let $\mathbf{S}$ be the diagonal matrix

$$\mathbf{S} = \begin{pmatrix} 1 & 0 \\ 0 & -1 \end{pmatrix}, \tag{2.2.3}$$

i.e. $\mathbf{S}$ has elements

$$S(\sigma, \sigma') = \sigma \, \delta(\sigma, \sigma'). \tag{2.2.4}$$

Then the RHS of (2.2.2) can be written as

$$Z_N^{-1} \operatorname{Trace} \mathbf{SVVSV} \ldots \mathbf{V}, \tag{2.2.5}$$

so

$$\langle \sigma_1 \sigma_3 \rangle = Z_N^{-1} \operatorname{Trace} \mathbf{SV}^2 \mathbf{SV}^{N-2}. \tag{2.2.6}$$

Similarly, if $0 \leqslant j - i \leqslant N$,

$$\langle \sigma_i \sigma_j \rangle = Z_N^{-1} \operatorname{Trace} \mathbf{SV}^{j-i} \mathbf{SV}^{N+i-j}, \tag{2.2.7}$$

$$\langle \sigma_i \rangle = Z_N^{-1} \operatorname{Trace} \mathbf{SV}^N. \tag{2.2.8}$$

Note that the translation invariance of the system is explicitly shown in these equations: $\langle \sigma_i \rangle$ is independent of i and $\langle \sigma_i \sigma_j \rangle$ depends on i and j only via their difference $j - i$.

Define a number ϕ by the equation

$$\cot 2\phi = e^{2K} \sinh h , \quad 0 < \phi < \frac{\pi}{2}. \tag{2.2.9}$$

Then a direct calculation of the eigenvectors of V, using (2.1.9), (2.1.11) and (2.1.12), reveals that the matrix P can be chosen to be orthogonal, being given by

$$P = \begin{pmatrix} \cos\phi & -\sin\phi \\ \sin\phi & \cos\phi \end{pmatrix}. \tag{2.2.10}$$

The expressions (2.2.7), (2.2.8) are unchanged by applying the similarity transformation (2.1.14) to both V and S, i.e. replacing V, S by

$$P^{-1}VP = \begin{pmatrix} \lambda_1 & 0 \\ 0 & \lambda_2 \end{pmatrix},$$

$$P^{-1}SP = \begin{pmatrix} \cos 2\phi & -\sin 2\phi \\ -\sin 2\phi & -\cos 2\phi \end{pmatrix}, \tag{2.2.11}$$

respectively.

Substituting these expressions into (2.2.7) and (2.2.8), and taking the limit $N \rightarrow \infty$ (keeping $j - i$ fixed), we obtain

$$\langle \sigma_i \sigma_j \rangle = \cos^2 2\phi + \sin^2 2\phi \left(\frac{\lambda_2}{\lambda_1}\right)^{j-i}, \tag{2.2.12}$$

$$\langle \sigma_i \rangle = \cos 2\phi . \tag{2.2.13}$$

Together with (2.1.3), this second equation gives us an alternative derivation of the magnetization $M(H, T)$. The result is of course the same as (2.1.18) above.

From (1.7.21), (2.2.12) and (2.2.13), the correlation function g_{ij} can now be evaluated. It is

$$g_{ij} = \langle \sigma_i \sigma_j \rangle - \langle \sigma_i \rangle \langle \sigma_j \rangle$$

$$= \sin^2 2\phi \, (\lambda_2/\lambda_1)^{j-i} \tag{2.2.14}$$

for $j \geq i$.

Since $|\lambda_2/\lambda_1| < 1$, we see immediately that g_{ij} does tend exponentially to zero as $j - i$ becomes large, and from (1.7.24) the correlation length ξ is given (in units of the lattice spacing) by

$$\xi = [\ln(\lambda_1/\lambda_2)]^{-1}. \tag{2.2.15}$$

2.3 Critical Behaviour near $T = 0$

It is true that $|\lambda_2/\lambda_1| < 1$ for all positive temperatures T and all real fields H. However, if $H = 0$, then

$$\lim_{T \to 0^+} (\lambda_2/\lambda_1) = 1 .$$

The correlation length ξ therefore becomes infinite at $H = T = 0$. We remarked in Section 1.7 that a critical point can be defined as a point at which $\xi = \infty$, so in this sense $H = T = 0$ is a critical point of the one-dimensional Ising model.

This is interesting because it enables us to make some tests of the scaling hypotheses discussed in Sections 1.2 and 1.7. We shall find that the tests are satisfied.

The scaling hypothesis (1.2.1) is formulated in terms of M, H and $t = (T - T_c)/T_c$. However, if $T_c = 0$ it is more sensible to replace these by the variables $M, h = H/kT$, and

$$t = \exp(-2K) = \exp(-2J/kT). \tag{2.3.1}$$

Then h and t measure the deviation of the field and temperature, respectively, from their critical values.

The scaling hypothesis (1.2.1) is equivalent to stating that the relation between M, h and t is unchanged by replacing them by

$$\lambda^\beta M , \lambda^{\beta\delta} h , \lambda t$$

for any positive number λ. Thus another way of writing (1.2.1) is (for h, t small)

$$M = h|h|^{\delta^{-1}-1} \phi(t|h|^{-1/\beta\delta}) , \tag{2.3.2}$$

where $\phi(x)$ is another scaling function, related to $h(x)$.

For the one-dimensional Ising model, we see from (2.1.18) and (2.3.1) that if $|h| \ll 1$, then

$$M = h/(t^2 + h^2)^{\frac{1}{2}}. \tag{2.3.3}$$

Clearly M is a function only of t/h, so the scaling hypothesis (2.3.2) is indeed satisfied, with

$$\beta\delta = 1 , \delta = \infty , \tag{2.3.4}$$

and

$$\phi(x) = (x^2 + 1)^{-\frac{1}{2}}. \tag{2.3.5}$$

The exponent relations (1.2.12) and (1.2.13) are consequences of the scaling hypothesis, so must be satisfied. From these and (2.3.4) it follows that

$$\alpha = 1 , \beta = 0 , \gamma = 1 . \tag{2.3.6}$$

Also, if $h = 0$ we see from (2.1.9) that the eigenvalues of $\mathbf{V}$ are

$$\lambda_1 = 2 \cosh K , \lambda_2 = 2 \sinh K , \tag{2.3.7}$$

so from (2.3.1)

$$\lambda_1/\lambda_2 = (1 + t)/(1 - t) . \tag{2.3.8}$$

When $t \ll 1$, equation (2.2.15) therefore becomes

$$\xi \sim (2t)^{-1} , \tag{2.3.9}$$

which is of the scaling form (1.7.25), with

$$\nu = 1 . \tag{2.3.10}$$

At the critical point $\lambda_1 = \lambda_2$, so from (2.2.14) the correlation function g_{ij} is a constant. This is of the scaling form (1.7.26), with

$$\eta = 1 . \tag{2.3.11}$$

We can now use these values of the exponents to test the scaling relation (1.2.16) and the second of the relations (1.2.14). They are indeed satisfied.

The other relations $\nu = \nu'$, $\mu + \nu = 2 - \alpha$ cannot be tested, since they involve functions defined in the ordered state $0 < T < T_c$ and $h = 0$. This state does not exist for this model.

The definition (2.3.1) of t is somewhat arbitrary: the RHS could be replaced by any positive power of $\exp(-2K)$. The effect of this would be to multiply each of $2 - \alpha$, γ and ν by the same factor. In view of this, we can only say of the critical exponents of the one-dimensional Ising model that they satisfy

$$2 - \alpha = \gamma = \nu , \tag{2.3.12}$$
$$\beta = 0 , \delta = \infty , \eta = 1 .$$

Despite the fact that $T_c = 0$, these exponents are still of interest: they can be compared with the Ising model exponents for 2, 3 and higher dimensions.

3

THE MEAN FIELD MODEL

3.1 Thermodynamic Properties

In any statistical mechanical system each component interacts with the external field and with the neighbouring components. In the mean-field model the second effect is replaced by an average over all components.

Consider a nearest-neighbour Ising model of N spins, with Hamiltonian given by (1.7.2), (1.7.3) and (1.8.1). If each spin σ_i has q neighbours, then the total field acting on it is

$$H + J \sum \sigma_j, \qquad (3.1.1)$$

where the sum is over the q neighbouring sites j. In the mean-field model this is replaced by

$$H + (N-1)^{-1} q J \sum_{j \neq i} \sigma_j, \qquad (3.1.2)$$

the sum now being over all $N-1$ sites j other than i. This is equivalent to replacing the Hamiltonian by

$$E(\sigma) = -\frac{qJ}{N-1} \sum_{(i,j)} \sigma_i \sigma_j - H \sum_{i=1}^{N} \sigma_i, \qquad (3.1.3)$$

where the first sum is over *all* the $\frac{1}{2}N(N-1)$ distinct pairs (i,j).

This 'mean-field' Hamiltonian (3.1.3) is the one that will be considered in this chapter. As was remarked in Section 1.6, it is in a sense 'infinite-dimensional', since each spin interacts equally with every other. It also has the unphysical property that the interaction strength depends on the number of particles. Nevertheless, it does give moderately sensible thermodynamic properties.

For a given configuration of spins, the total magnetization is

$$\mathcal{M} = \sum_{i=1}^{N} \sigma_i, \tag{3.1.4}$$

and (3.1.3) can be written (using $\sigma_i^2 = 1$) as

$$E(\sigma) = -\tfrac{1}{2}qJ(\mathcal{M}^2 - N)/(N - 1) - H\mathcal{M}. \tag{3.1.5}$$

Thus in this model $E(\sigma)$ depends on $\sigma_1, \ldots, \sigma_N$ only via $\mathcal{M}$. This is a great simplification: the sum over spin-values in the partition function can be replaced by a sum over the allowed values of $\mathcal{M}$, weighted by the number of spin configurations for each value.

From (3.1.4), if r of the spins are down (value -1) and $N - r$ are up (value $+1$), then

$$\mathcal{M} = N - 2r. \tag{3.1.6}$$

There are $\binom{N}{r}$ such arrangements of spins, so from (1.7.5) the partition function is

$$Z = \sum_{r=0}^{N} c_r, \tag{3.1.7}$$

where

$$c_r = \frac{N!}{r!(N-r)!} \exp\{\tfrac{1}{2}\beta qJ[(N - 2r)^2 - N]/(N - 1)$$
$$+ \beta H(N - 2r)\}, \tag{3.1.8}$$

and

$$\beta = 1/kT. \tag{3.1.9}$$

Also, from (1.4.4), the average magnetization per site is

$$M = N^{-1}\langle \mathcal{M} \rangle = \langle 1 - 2r/N \rangle = Z^{-1} \sum_{r=0}^{N} (1 - 2r/N) \, c_r. \tag{3.1.10}$$

The properties of $c_0, \ldots, c_N$ are most readily obtained by considering $d_r = c_{r+1}/c_r$. From (3.1.8)

$$d_r = \frac{c_{r+1}}{c_r} = \frac{N - r}{r + 1} \exp\{-2\beta qJ(N - 2r - 1)/(N - 1) - 2\beta H\}. \tag{3.1.11}$$

We are interested in the case when N is large. As r increases from 0 to $N - 1$, the RHS of (3.1.11) increases from large values (of order N) to small values (of order N^{-1}). Provided βqJ is not too large, this decrease must be monotonic. Then there must be a single integer r_0 such that

$$d_r > 1 \qquad \text{for } r = 0, \ldots, r_0 - 1$$

$$d_{r_0} \leq 1 \tag{3.1.12}$$

$$d_r < 1 \qquad \text{for } r = r_0 + 1, \ldots, N - 1.$$

Since $c_{r+1} = d_r c_r$, it follows that c_r increases as r goes from 0 to r_0, decreases as r goes from $r_0 + 1$ to N, and that c_{r_0} is the largest c_r.
When N and r are both large, (3.1.11) can be written

$$d_r = c_{r+1}/c_r = \phi(1 - 2r/N), \tag{3.1.13}$$

where, for $-1 < x < 1$,

$$\phi(x) = \frac{1 + x}{1 - x} \exp[-2\beta qJx - 2\beta H]. \tag{3.1.14}$$

Let x_0 be the solution of the equation

$$\phi(x_0) = 1. \tag{3.1.15}$$

Then, when N is large, r_0 is given by

$$1 - 2r_0/N = x_0. \tag{3.1.16}$$

Regarded as a function of r, c_r has a peak at $r = r_0$, the width of the peak being proportional to $N^{\frac{1}{2}}$. Although this width is large compared to one, it is small compared to N. Thus across this peak $1 - 2r/N$ in (3.1.10) can be replaced by $1 - 2r_0/N$. Since values of r outside the peak give a negligible contribution to the sums in (3.1.7) and (3.1.10), it follows that the magnetization per site is

$$M = 1 - 2r_0/N = x_0. \tag{3.1.17}$$

From (3.1.14) and (3.1.15), M is given by $\phi(M) = 1$, i.e.

$$M = \tanh[(qJM + H)/kT]. \tag{3.1.18}$$

This equation defines M as a function of H and T. It was first obtained by Bragg and Williams (1934). The free energy can now be obtained by integration, using (1.7.14), or more directly by arguing that when N is large the sum in (3.1.7) is dominated by values of r close to r_0, so

$$-\beta f = \lim_{N \to \infty} N^{-1} \ln Z$$

$$= \lim_{N \to \infty} N^{-1} \ln c_{r_0}. \tag{3.1.19}$$

Using (3.1.8), Stirling's approximation

$$n! \sim (2\pi)^{\frac{1}{2}} e^{-n} n^{n+\frac{1}{2}}, \tag{3.1.20}$$

and (3.1.17) and (3.1.18), it follows that

$$-f/kT = \tfrac{1}{2} \ln[4/(1 - M^2)] - \tfrac{1}{2} q J M^2/kT. \tag{3.1.21}$$

This gives f as a function of M and T.

3.2   Phase Transition

From (3.1.18),

$$H = -qJM + kT \operatorname{artanh}(M). \tag{3.2.1}$$

This equation can be used to plot H as a function of M, for $-1 < M < 1$. The graph can then of course be reversed to give M as a function of H. If $qJ < kT$, then the resulting graph is similar to Fig. 1.1(c), i.e. a typical high-temperature graph, with no spontaneous magnetization.

However, if $qJ > kT$, the graph looks like that in Fig. 3.1(a). This graph is not sensible, since for sufficiently small H it allows 3 possible values of M, whereas M is defined by (1.7.12) or (1.7.14) to be a single-valued function of H.

The source of this contradiction is in the statements preceding equation (3.1.12). If $qJ > kT$, then the RHS of (3.1.11) is *not* a monotonic decreasing function of r: instead it behaves as indicated in Fig. 3.2.

If H is sufficiently small, then there are three solutions of the equation $d_r = 1$, as indicated in Fig. 3.2. This means that c_r has two maxima, as

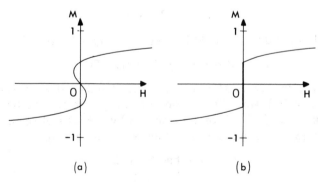

(a)   (b)

Fig. 3.1. M as a function of H for $T = 0.94 \, T_c$; (a) shows all solutions of (3.1.18), (b) is the correct graph obtained by rejecting spurious solutions.

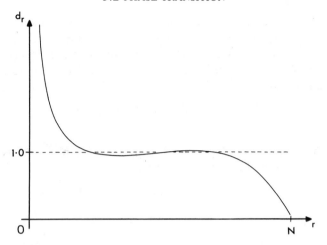

Fig. 3.2. d_r as a function of r for $T = 0.94\,T_c$, $\beta H = 0.006$ and N large.

shown in Fig. 3.3. Together with the intervening minimum, these correspond to the three solutions for M of equation (3.1.18). If H is positive (negative), then the left-hand (right-hand) peak is the greater.

It is still true that the sum in (3.1.7) is dominated by values of r close to r_0, where r_0 is the value of r that maximizes (absolutely) c_r. Thus if (3.1.18) has three solutions and H is positive, we must choose the solution

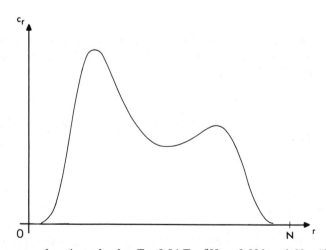

Fig. 3.3. c_r as a function of r for $T = 0.94\,T_c$, $\beta H = 0.006$ and $N = 100$. As N increases, the maximum becomes larger and more sharply peaked. The other two turning values correspond to the spurious solutions of (3.1.18).

with the smallest value of r_0, i.e. the largest value of M. Conversely if H is negative. Doing this, the multi-valued graph of Fig. 3.1(a) becomes the single-valued graph of Fig. 3.1(b). This is similar to the typical low-temperature graph of $M(H)$ shown in Fig. 1.1. In particular, there is a spontaneous magnetization M_0 given by

$$M_0 = \tanh(qJM_0/kT) , \quad M_0 > 0 , \qquad (3.2.2)$$

provided that $qJ > kT$.

Thus the mean-field model has a ferromagnetic phase transition for temperatures below the Curie temperature

$$T_c = qJ/k . \qquad (3.2.3)$$

3.3   Zero-Field Properties and Critical Exponents

Spontaneous Magnetization and β

Set

$$t = (T - T_c)/T_c ; \qquad (3.3.1)$$

then, using (3.2.3), the equation (3.2.2) can be written as

$$M_0 = (1 + t) \operatorname{artanh} M_0 . \qquad (3.3.2)$$

For T just less than T_c the spontaneous magnetization M_0 is small but non-zero, so $\operatorname{artanh} M_0$ can be approximated by $M_0 + M_0^3/3$. Solving the resulting equation for M_0 gives

$$M_0 = (-3t)^{\frac{1}{2}} \{1 + \mathbb{O}(t)\} . \qquad (3.3.3)$$

Thus M_0 is effectively proportional to $(-t)^{\frac{1}{2}}$. From (1.1.4) the critical exponent β exists and is given by

$$\beta = \tfrac{1}{2} . \qquad (3.3.4)$$

Free Energy and α

Let $H \to 0$ for $T > T_c$. Then $M \to 0$ and from (3.1.21) the free energy is given very simply by

$$-f/kT = \ln 2 . \qquad (3.3.5)$$

On the other hand, if $T < T_c$ then $M \to M_0$. For M_0 small it follows from (3.1.21) that

$$-f/kT = \ln 2 + \tfrac{1}{2}M_0^2(1 - qJ/kT)$$

$$+ M_0^4/4 + \mathcal{O}(M_0^6). \qquad (3.3.6)$$

Using (3.2.3), (3.3.1) and (3.3.3), when t is small and negative the free energy is therefore given by

$$-f/kT = \ln 2 + 3t^2/4 + \mathcal{O}(t^3). \qquad (3.3.7)$$

From (1.7.7), (1.7.8), (3.3.5) and (3.3.7), we see that the free energy and internal energy are continuous at $T = T_c$, but the specific heat has a jump discontinuity. The definition (1.7.9) of the exponents α and α' is meaningless, but the alternative definition (1.7.10) gives

$$\alpha = 0. \qquad (3.3.8)$$

Susceptibility and γ, γ'

Hold T fixed and differentiate (3.2.1) with respect to H. Using (1.7.17), (3.2.3) and (3.3.1), it follows that the susceptibility χ is given exactly by

$$\chi = (1 - M^2)/[qJ(t + M^2)]. \qquad (3.3.9)$$

Now let $H \to 0$. If $T > T_c$ then $M \to 0$, giving

$$\chi = (qJt)^{-1}. \qquad (3.3.10a)$$

If $T < T_c$ then $M \to M_0$. Using the approximate relation (3.3.3) we then obtain that near T_c

$$\chi \simeq (-2qJt)^{-1}. \qquad (3.3.10b)$$

Thus at T_c the zero-field susceptibility becomes infinite, diverging as t^{-1}. From (1.1.6) and (1.1.7) the exponents γ and γ' are given by

$$\gamma = \gamma' = 1. \qquad (3.3.11)$$

3.4   Critical Equation of State

Using (3.2.3) and (3.3.1) the exact equation of state can be written as

$$H/kT_c = -M + (1 + t)\,\text{artanh}\,M. \qquad (3.4.1)$$

Near the critical point M is small. Taylor expanding the function artanh M, (3.4.1) gives

$$H/kT_c \simeq M^2 (\tfrac{1}{3} + tM^{-2}) , \qquad (3.4.2)$$

neglecting terms of order tM^3 or M^5.

Comparing this result with (1.2.1), we see that the scaling hypothesis is indeed satisfied for this model, with

$$h_s(x) = \tfrac{1}{3} + x , \qquad (3.4.3)$$

$$\beta = \tfrac{1}{2}, \quad \delta = 3 . \qquad (3.4.4)$$

This agrees with (3.3.4) and it is easy to verify that the scaling relations (1.2.12) and (1.2.13) are satisfied. Indeed they should be, since they are consequences of the scaling hypothesis.

The values (3.3.4), (3.3.8), (3.3.11), (3.4.4) of the exponents are the same as those of the van der Waals fluid discussed in Section 1.10, i.e. they are the *classical* values.

Since each spin interacts equally with every other, correlations are not distance dependent, nor can the model have two physically separated coexisting phases. Thus the exponents ν, η and μ are not defined for this model.

3.5 Mean Field Lattice Gas

Regarding a 'down' spin as an empty site and an 'up' spin as a site containing a particle, the above model is also one of a lattice gas. Making the substitutions (1.9.13)–(1.9.16) in (3.2.1) and (3.1.21), we find that the chemical potential μ and pressure P are given by

$$\mu = -q\varepsilon\rho + kT \ln[\rho/(1 - \rho)] , \qquad (3.5.1)$$

$$P = -kT \ln(1 - \rho) - \tfrac{1}{2}q\varepsilon\rho^2 . \qquad (3.5.2)$$

Here ρ is the density, i.e. the mean number of particles per site. It must lie in the range $0 < \rho < 1$.

Equation (3.5.2) is the equation of state of the mean-field lattice gas. Comparing it with (1.9.31), and noting that $v = \rho^{-1}$, we see that it is very similar to the van der Waals equation. Both equations are of the form

$$P = kT \, \phi(\rho) - a\rho^2 , \qquad (3.5.3)$$

where a is a constant and the function $\phi(\rho)$ is independent of the temperature T. Indeed, there are solvable models which have exactly the van der Waals equation of state (Kac *et al.*, 1963/4).

4

ISING MODEL ON THE BETHE LATTICE

4.1 The Bethe Lattice

Another simple model that can be exactly solved is the Ising model (or indeed any model with only nearest-neighbour interactions) on the Bethe lattice. Like the mean-field model, this is equivalent to an approximate treatment of a model on, say, a square or cubic lattice (Bethe, 1935). However, it can be defined as an exactly solvable model, and this is what we shall do here.

Consider the graph constructed as follows: start from a central point 0 and add q points all connected to 0. Call the set of these q points the 'first shell'. Now create further shells by taking a point in shell r and connecting $q - 1$ new points to it. Do this for all points in shell r and call the set of all the new points 'shell $r + 1$'.

Proceeding interatively in this way, construct shells $2, 3, \ldots, n$. This gives a graph like that shown in Fig. 4.1. There are $q(q - 1)^{r-1}$ points in shell r and the total number of points in the graph is

$$q[(q - 1)^n - 1]/(q - 2) \tag{4.1.1}$$

We call the points in shell n 'boundary points'. They are exceptional in that each has only one neighbour, while all other points (interior points) each have q neighbours.

Such a graph contains no circuits and is known as a Cayley tree. From our point of view it can be thought of as a regular 'lattice' of coordination number q (i.e. q neighbours per site), *provided the boundary sites can be ignored.*

There is a problem here: normally the ratio of the number of boundary sites to the number of interior sites of a lattice becomes small in the

47

thermodynamic limit of a large system. Here it does not, since both numbers grow exponentially like $(q-1)^n$. To overcome this problem we here consider only local properties of sites *deep within the graph* (i.e. infinitely far from the boundary in the limit $n \to \infty$). Such sites should all be equivalent, each having coordination number q, and can be regarded as forming the *Bethe lattice*. (This distinction between the Cayley tree and the Bethe lattice is not always made, but does seem to be useful terminology. I am grateful to Professor J. Nagle for suggesting it to me and drawing my attention to a relevant article [Chen *et al.*, 1974].)

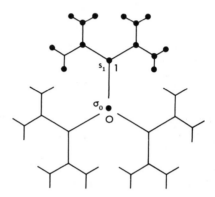

Fig. 4.1. A Cayley tree (with $q = 3$ and $n = 4$), divided at the central site 0 into three sub-trees. They are identical, but here the upper sub-tree is distinguished by indicating its sites with solid circles. Each sub-tree is rooted at 0. The site 1 adjacent to 0 in the upper sub-tree is shown. The spin at 0 is σ_0, that at 1 is s_1.

Put another way, if we construct an Ising model on the complete Cayley tree, then the partition function Z contains contributions from both sites deep within the graph, and sites close to or on the boundary. The contribution from the latter is not negligible, even in the thermodynamic limit.

If one considers the total partition function, then one is considering the 'Ising model on the Cayley tree'. This problem has been solved (Runnels, 1967; Eggarter, 1974; Müller-Hartmann and Zittartz, 1974) and has some quite unusual properties. We shall not, however, consider this problem here. Instead we shall effectively consider only the contribution to Z from sites deep within the graph, i.e. from the Bethe lattice.

Some motivation for this choice is given by series expansions. If one makes a low temperature expansion as in Section 1.8 for any regular lattice, then to second order the only properties of the lattice that one needs to know are the number of sites and the coordination number. To third order one needs the number of triangles in the lattice, to fourth order the number of tetrahedra (i.e. clusters of 4 sites all connected to one another) and other highly connected 4-point sub-graphs, and so on. An interesting simple case is when there are no circuits at all, and hence no triangles, tetrahedra, etc. Then one obtains the Ising model on the Bethe lattice as defined here.

4.2 Dimensionality

Consider any regular lattice. Let $m_1(=q)$ be the number of neighbours per site, m_2 the number of next-nearest neighbours, m_3 the number of next-next-nearest neighbours, etc. Then $c_n = 1 + m_1 + m_2 + \ldots + m_n$ is the number of sites within n steps of a given site. For the hyper-cubic lattices it is easy to see that

$$\lim_{n \to \infty} (\ln c_n)/\ln n = d , \qquad (4.2.1)$$

where d is the dimensionality of the lattice.

The relation (4.2.1) is also true for all the regular two and three-dimensional lattices, and can be regarded as a definition of the dimensionality d.

Now return to considering the Bethe lattice. In this case c_n is given by (4.1.1). Substituting this expression into (4.2.1) gives $d = \infty$, so in this sense the Bethe lattice is 'infinite-dimensional'.

4.3 Recurrence Relations for the Central Magnetization

Consider an Ising model on the complete Cayley tree (but we shall later ignore boundary terms, thereby reducing it to the Bethe lattice). The partition function is given by (1.8.2), i.e. by

$$Z = \sum_{\sigma} P(\sigma) , \qquad (4.3.1)$$

where

$$P(\sigma) = \exp\left[K \sum_{(i,j)} \sigma_i \sigma_j + h \sum_i \sigma_i \right] . \qquad (4.3.2)$$

The first summation in (4.3.2) is over all edges of the graph, the second over all sites. The $P(\sigma)$ can be thought of as an unnormalized probability distribution: in particular, if σ_0 is the spin at the central site 0, then the local magnetization there is

$$M = \langle \sigma_0 \rangle = \sum_\sigma \sigma_0 P(\sigma)/Z .$$ (4.3.3)

From Fig. 4.1 it is apparent that if the graph is cut at 0, then it splits up into q identical disconnected pieces. Each of these is a rooted tree (with root 0). This implies that the expression (4.3.2) factors:

$$P(\sigma) = \exp(h\sigma_0) \prod_{j=1}^{q} Q_n(\sigma_0|s^{(j)}) ,$$ (4.3.4)

where $s^{(j)}$ denotes all the spins (other than σ_0) on the jth sub-tree, and

$$Q_n(\sigma_0|s) = \exp\left[K \sum_{(i,j)} s_i s_j + K s_1 \sigma_0 + h \sum_i s_i \right],$$ (4.3.5)

s_i being the spin on site i of the sub-tree (other than the root, which has spin σ_0). Site 1 is the site adjacent to 0, as in the upper sub-tree of Fig. 4.1. The first summation in (4.3.5) is over all edges of the sub-tree other than $(0,1)$; the second is over all sites other than 0. The suffix n denotes the fact that the sub-tree has n shells, i.e. n steps from the root to the boundary sites.

Further if the upper sub-tree in Fig. 4.1 is cut at the site 1 adjacent to 0, then it too decomposes into q pieces: one being the 'trunk' $(0, 1)$, the rest being identical branches. Each of these branches is a sub-tree like the original, but with only $n - 1$ shells. Thus

$$Q_n(\sigma_0|s) = \exp(K\sigma_0 s_1 + h s_1) \prod_{j=1}^{q-1} Q_{n-1}(s_1|t^{(j)})$$ (4.3.6)

where $t^{(j)}$ denotes all the spins (other than s_1) on the jth branch of the sub-tree.

These factorization relations (4.3.4) and (4.3.6) make it easy to calculate M. Let

$$g_n(\sigma_0) = \sum_s Q_n(\sigma_0|s) .$$ (4.3.7)

Then from (4.3.1) and (4.3.4),

$$Z = \sum_{\sigma_0} \exp(h\sigma_0) [g_n(\sigma_0)]^q .$$ (4.3.8)

Similarly, from (4.3.3) and (4.3.4),

$$M = Z^{-1} \sum_{\sigma_0} \sigma_0 \exp(h\sigma_0) [g_n(\sigma_0)]^q .$$ (4.3.9)

Let

$$x_n = g_n(-)/g_n(+) \, . \tag{4.3.10}$$

Then from (4.3.8) and (4.3.9),

$$M = \frac{e^h - e^{-h} x_n^q}{e^h + e^{-h} x_n^q} \, . \tag{4.3.11}$$

Thus M is known if x_n is. To obtain x_n we sum (4.3.6) over all the spins s, i.e. over s_1 and the $t^{(j)}$, to give, using only (4.3.7):

$$g_n(\sigma_0) = \sum_{s_1} \exp(K\sigma_0 s_1 + h s_1) \, [g_{n-1}(s_1)]^{q-1} \tag{4.3.12}$$

Remembering that σ_0 and s_1 are single spins, with values $+1$ and -1, performing the summation in (4.3.12) for $\sigma_0 = +1$ or -1, taking ratios and using (4.3.10), we obtain

$$x_n = y(x_{n-1}) \, , \tag{4.3.13}$$

where the function $y(x)$ is given by

$$y(x) = [e^{-K+h} + e^{K-h} x^{q-1}]/[e^{K+h} + e^{-K-h} x^{q-1}] \, . \tag{4.3.14}$$

Equation (4.3.13) is a recurrence relation between x_n and x_{n-1}. It is easy to see that

$$x_0 = g_0(\sigma_0) = 1 \, , \tag{4.3.15}$$

so (4.3.13) defines x_n, and (4.3.11) defines M.

4.4 The Limit $n \to \infty$

Hereafter we consider the ferromagnetic case, $K > 0$. Then $y(x)$ increases monotonically from $\exp(-2K)$ to $\exp(2K)$ as x goes from 0 to ∞.

The recurrence relation (4.3.13) can be thought of graphically by simultaneously plotting $y = y(x)$ and $y = x$.

Let P_{n-1} be the point $(x_{n-1}, y(x_{n-1}))$ in the (x, y) plane. To construct P_n draw a horizontal line through P_{n-1} to intercept the line $y = x$ at a point Q_n. Now draw a vertical line through Q_n. Its intercept with $y = y(x)$ is the point P_n.

There are two cases to consider: either the line $y = x$ crosses the curve $y = y(x)$ once, or it crosses it three times, as shown in Fig. 4.2. In the former case the point P_n will always monotonically approach the cross-over point A as $n \to \infty$, as indicated in Fig. 4.2(a). Thus x_n and M tend to a limit as n becomes large, as we expect. This M is therefore the local magnetization of a site deep within the Cayley tree, i.e. the magnetization per site of the Bethe lattice.

If there are three cross-over points, then the outer two (A and C in Fig. 4.2(b)) are stable limit points of (4.3.13), while the centre one (B) is unstable. If P_0 lies to the left (right) of B, then P_n tends to A (C). Thus again P_n tends to a limit, giving the magnetization M for the Bethe lattice.

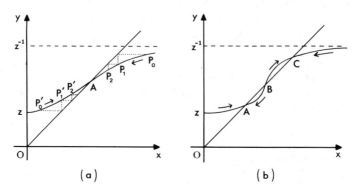

(a)   (b)

Fig. 4.2. Typical sketches of the function $y(x)$ given by (4.3.14), with $z = \exp(-2K)$. In (a) the curve intercepts the straight line $y = x$ only once, at A. Two typical sequences of points $P_n = (x_n, y(x_n))$ are shown, one starting to the right of A, the other $\{P_0', P_1', P_2', \ldots\}$ to the left. All such sequences converge to the limit point A. In (b) there are three intersections A, B, C. A sequence $\{P_n\}$ grows in the direction of the arrows, never crossing A, B or C. Thus A and C are stable limit points, B is an unstable fixed point.

We need some more convenient rule to determine which stable fixed point, A or C, is the one approached. The borderline case is when P_0 is the point B, i.e. when $x = 1$ is a solution of the equation $x = y(x)$. From (4.3.14) this occurs when, and only when, $h = 0$. If $h > 0$, then P_0 lies to the left of B so P_n tends to A. Conversely, if $h > 0$, then P_n tends to C.

Summarizing, when $n \to \infty$ the magnetization is given, using (4.3.11), by

$$M = \frac{e^{2h} - x^q}{e^{2h} + x^q},\qquad (4.4.1)$$

where x is a solution of

$$x = y(x).\qquad (4.4.2)$$

If there are three solutions, the smallest must be chosen for $h > 0$, the largest for $h < 0$.

These equations can be written in a more conventional form by defining

$$z = e^{-2K}, \quad \mu = e^{-2h}, \quad \mu_1 = \mu x^{q-1}. \quad (4.4.3)$$

Then, using (4.3.14), (4.4.2) gives

$$x = (z + \mu_1)/(1 + \mu_1 z). \quad (4.4.4)$$

From (4.4.3), (4.4.4) and (4.4.1) it follows that

$$\mu_1/\mu = [(z + \mu_1)/(1 + \mu_1 z)]^{q-1}, \quad (4.4.5a)$$

$$M = (1 - \mu_1^2)/(1 + \mu_1^2 + 2\mu_1 z). \quad (4.4.5b)$$

The first of the equations (4.4.5) defines μ_1; the second gives the magnetization M. These are the same as the results of the Bethe approximation for a lattice of coordination number q (Domb, 1960, pp. 251–254).

4.5 Magnetization as a Function of H

Now suppose T, and hence K, is fixed and consider the variation of x and M with $h = H/kT$. Using (4.3.14) the equation (4.4.2) can be written

$$e^{2h} = x^{q-1}(e^{2K} - x)/(e^{2K}x - 1). \quad (4.5.1)$$

All the x_n are positive, and so is the limit point x. For the RHS of (4.5.1) to be positive it follows that x must lie in the interval

$$e^{-2K} < x < e^{2K}. \quad (4.5.2)$$

Clearly (4.5.1) defines h as a function of x, for fixed K. (This function is of course not the same as the scaling function $h_s(x)$ of Section 1.2.) Differentiating (4.5.1) logarithmically gives

$$2x \frac{dh}{dx} = q - 1 - \frac{2 \sinh 2K}{2 \cosh 2K - x - x^{-1}}. \quad (4.5.3)$$

For x in the interval (4.5.2), the RHS of (4.5.3) has its maximum at $x = 1$. If this maximum is negative, i.e. if $K < K_c$, where

$$K_c = \tfrac{1}{2}\ln[q/(q - 2)], \quad (4.5.4)$$

then h decreases monotonically from ∞ to 0 as x increases from $\exp(-2K)$ to $\exp(2K)$. Hence for given real h, (4.5.1) has one and only one real positive solution for x, and x is an analytic function of h for $-\infty < h < \infty$.

If, on the other hand, $K > K_c$, then dh/dx is positive for x sufficiently close to one. From (4.5.1), $h = 0$ when $x = 1$, so the function $h(x)$ has a graph of the type shown in Fig. 4.3.

For sufficiently small h, (4.5.1) therefore has three solutions for x. From the discussions of Section 4.4, if $h > 0$ the limit point of the sequence given by (4.3.1) corresponds to the smallest solution for x. If $h < 0$ it corresponds to the largest solution.

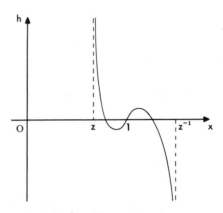

Fig. 4.3. A typical sketch of h as a function of x for $T < T_c$.

Considering the behaviour as h decreases from $+\infty$ through zero to $-\infty$, it is therefore apparent from Fig. 4.3 that x is an analytic function of h, except at $h = 0$, where it jumps discontinuously from the smallest to the largest solution.

In all cases x is a decreasing function of h, satisfying

$$x(-h) = 1/x(h).\tag{4.5.5}$$

From (4.4.1) it follows that M is an odd function of h. It increases monotonically from -1 to 1 as h increases from $-\infty$ to ∞ and is analytic if $K < K_c$. If $K > K_c$, then it is analytic apart from a jump discontinuity at $h = 0$.

This is precisely the typical behaviour of a ferromagnet that was outlined in Section 1.1. Thus the Ising model on the Bethe lattice exhibits ferromagnetism, with a critical point at $H = 0$, $T = T_c$, where

$$J/kT_c = \tfrac{1}{2}\ln\left[q/(q - 2)\right].\tag{4.5.6}$$

4.6 Free Energy

The total free energy of the Cayley tree is

$$F = -kT \ln Z , \qquad (4.6.1)$$

where Z is given by (4.3.1) and (4.3.2). Differentiating these equations with respect to $H = hkT$ gives

$$-\frac{\partial F}{\partial H} = \sum_i M_i , \qquad (4.6.2)$$

where the summation is over all sites i and

$$M_i = \langle \sigma_i \rangle$$

is the local magnetization at site i. Each M_i is a function of H, and hence h, for given temperature T. To show this we shall sometimes write it as $M_i(h)$.

If H is large and positive the summation in (4.3.1) is dominated by the state with all spins up, so in this limit

$$F/kT = -KN_e - hN , \qquad (4.6.3)$$

N_e being the number of edges and N the number of sites. Also, in this limit $\langle \sigma_i \rangle = 1$ for $i = 1, \dots , N$.

We can now integrate (4.6.2) with respect to H, using (4.6.3) to obtain the integration constant. This gives

$$F/kT = -KN_e - hN + \sum_i \int_h^\infty [M_i(h') - 1] \, dh' . \qquad (4.6.4)$$

Alternatively, if q_i is the number of sites adjacent to site i, then $\sum_i q_i = 2N_e$, and (4.6.4) can be written

$$F = \sum_i f_i ,$$

where

$$f_i/kT = -\tfrac{1}{2}Kq_i - h + \int_h^\infty [M_i(h') - 1] \, dh' . \qquad (4.6.5)$$

Each f_i can be thought of as the free energy of site i. For an homogeneous lattice the f_i are all equal to the usual free energy f, and on differentiating (4.6.5) one regains the usual relation (1.7.14).

As we remarked above, the difficulty with the Cayley tree is that it is not homogeneous, there being a significant number of boundary or near-boundary sites that have properties different from the interior. However, all sites deep inside the graph have the same local magnetization M, and hence the same local free energy f, given by (4.6.5). This free energy is therefore the free energy of the Ising model on the Bethe lattice. It is given by setting $q_i = q$, $M_i = M$ in (4.6.5), and using the equations (4.5.1), (4.4.1) for x and M as functions of h.

Noting that x is a monotonic differentiable function of h for $h > 0$, one can change the integration variable in (4.6.5) from h' to $x' = x(h')$. This gives [dropping the suffixes i and using $z = \exp(-2K)$]

$$f/kT = -\tfrac{1}{2}Kq - h - \int_z^x [M(x') - 1] \left[\frac{dh}{dx}\right]_{x=x'} dx' \qquad (4.6.6)$$

provided $h > 0$ (or $K < K_c$).

Substituting the expression (4.5.1) for $\exp(2h)$ into (4.4.1), and using (4.5.3), the integrand in (4.6.6) can be written, after a little re-arrangement, as

$$\frac{z}{1 - zx'} - (q - 2)\frac{x' - z}{1 + x'^2 - 2x'z}. \qquad (4.6.7)$$

This can be easily integrated to give, eliminating h by using (4.5.1),

$$f/kT = -\tfrac{1}{2}Kq - \tfrac{1}{2}q \ln(1 - z^2)$$
$$+ \tfrac{1}{2}\ln[z^2 + 1 - z(x + x^{-1})] + \tfrac{1}{2}(q - 2)\ln(x + x^{-1} - 2z). \qquad (4.6.8)$$

Negating h has the effect of inverting x, which leaves (4.6.8) unchanged. Since f must be an even function of h, it follows that (4.6.8) is true for all real h. Together with the equation (4.5.1) for x, it gives the free energy per site of the Ising model on the Bethe lattice.

4.7 Low-Temperature Zero-Field Results

A problem arises with *any* ferromagnetic Ising model if $H = 0$ and $T < T_c$. In this case the spins do not know whether to be mostly up, or mostly down. If just the boundary spins are fixed to be up, every spin will have a greater probability of being up than down. In a sense the 'thermodynamic limit' does not exist, since the bulk properties depend on the boundary conditions.

This is particularly evident in the present model: if $H = 0$ then it is obvious from (4.3.13)–(4.3.15) that $x_n = 1$, for all n. If $T < T_c$ this means that all the points $P_n = (x_n, y_n)$ are the point B in Fig. 4.2(b). However, this is an unstable fixed point of (4.3.13): if x_0 is not one, but just less than one, then the sequence $\{P_n\}$ will converge not to B, but to the stable limit point A.

There are at least two ways round this difficulty: one is to take $H = 0$ and fix all boundary spins up; the other to take $H > 0$, let $n \to \infty$, and *then* let $H \to 0_+$. In either case the sequence $\{P_n\}$ will converge to A and the limiting value of x is, from (4.4.2) and (4.3.14), the smallest positive solution of the equation

$$z = e^{-2J/kT} = x \frac{1 - x^{q-2}}{1 + x^{q-2}}. \tag{4.7.1}$$

If $T < T_c$, this value of x is less than one. From (4.4.1) and (4.6.8) the spontaneous magnetization M and free energy f are then given by

$$M = \frac{1 - x^q}{1 + x^q}, \tag{4.7.2}$$

$$e^{-f/kT} = (1 + x^q) \left[\frac{(1 - x^q)(1 - x^{2q-2})^2}{x(1 - x^{q-2})(1 - x^{2q})^2} \right]^{q/4}. \tag{4.7.3}$$

It is interesting to compare these results with those of the two-dimensional Ising model. This will be done in Section 11.8.

4.8 Critical Behaviour

Set $x = \exp(-2s)$, then (4.5.1) becomes

$$h = -(q - 1)s + \tfrac{1}{2} \ln[\sinh(K + s)/\sinh(K - s)], \tag{4.8.1}$$

which makes it clear that h is an odd function of s. Taylor expanding, we obtain

$$h = [\coth K - q + 1]s + \tfrac{1}{3} \coth K \operatorname{cosech}^2 K \, s^3 + \ldots. \tag{4.8.2}$$

The critical value of K is given by (4.5.4), i.e. by $\coth K_c = q - 1$. Setting as usual

$$t = (T - T_c)/T_c, \tag{4.8.3}$$

and using $K = J/kT$, it follows that for t small

$$\coth K - q + 1 = q(q - 2)K_c t + \mathbb{O}(t^2). \tag{4.8.4}$$

Using this result in (4.8.2), together with $h = H/kT$, gives (for t and s small):

$$H/kT_c = q(q - 2)\{K_c ts + \tfrac{1}{3}(q - 1)s^3$$
$$+ \mathcal{O}(t^2 s, ts^3, s^5)\}. \qquad (4.8.5)$$

From (4.4.1), the magnetization M is given by

$$M = \tanh(h + qs). \qquad (4.8.6)$$

From (4.8.5), h is much less than s, which is itself small, so $M \simeq qs$, or conversely

$$s = q^{-1}M + \mathcal{O}(h, M^3). \qquad (4.8.7)$$

Substituting this result into (4.8.5) and neglecting terms of order $t^2 M$, tM^3 or M^5, we obtain

$$H/kT_c = M^3 h_s(t/M^2), \qquad (4.8.8)$$

where

$$h_s(x) = \tfrac{1}{2}(q - 2)x \ln[q/(q - 2)] + (q - 1)(q - 2)/(3q^2). \qquad (4.8.9)$$

Comparing (1.2.1) and (4.8.8), we see that the scaling hypothesis is satisfied for this model, $h_s(x)$ being the scaling function. It is linear, and critical exponents β and δ have the values

$$\beta = \tfrac{1}{2}, \quad \delta = 3. \qquad (4.8.10)$$

Thus all the exponents β, δ, α, α', γ, γ' must have the same values as those of the mean-field model (Section 3.3), i.e. the 'classical' values.

All the above results are very similar to those of the mean-field model of Chapter 3. (In fact they are the same in the limit $q \to \infty$, qK finite.) However, the Bethe-lattice model is really much more respectable than the mean-field one: its interactions are independent of the size of the system, and each spin interacts only with its nearest neighbours.

4.9 Anisotropic Model

The key equations (4.3.14), (4.4.2), (4.4.1), (4.6.8) of the above working can be summarized (using the first two to eliminate z from the last) as

$$z = \exp(-2K) = (x - \mu x^{q-1})/(1 - \mu x^q), \qquad (4.9.1)$$

$$M = (1 - \mu x^q)/(1 + \mu x^q), \qquad (4.9.2)$$

$$-f/kT = h + \tfrac{1}{2}qK + \ln(1 + \mu x^q)$$
$$+ \tfrac{1}{2}q \ln[(1 - \mu^2 x^{2q-2})/(1 - \mu^2 x^{2q})] . \qquad (4.9.3)$$

The edges of the Bethe lattice can be grouped into classes $1, \ldots, q$, so that each site lies on just one edge of each class. Then the interaction coefficient K can be given a different value for different classes of edges. If K_r is its value for class r (where $r = 1, \ldots, q$), then this anisotropic model can also be solved by the above methods.

The equations (4.9.1)–(4.9.3) generalize to

$$z_r = \exp(-2K_r) = (x_r - tx_r^{-1})/(1 - t), \quad r = 1, \ldots, q, \qquad (4.9.4a)$$

$$\mu = \exp(-2h) = t/(x_1 \ldots x_q), \qquad (4.9.4b)$$

$$M = (1 - t)/(1 + t) \qquad (4.9.5)$$

$$-f/kT = h + \tfrac{1}{2}(K_1 + \ldots + K_q) + \ln(1 + t) + \tfrac{1}{2}\sum_{r=1}^{q} \ln \frac{1 - t^2 x_r^{-2}}{1 - t^2}. \qquad (4.9.6)$$

These define M, f as functions of $K_1, \ldots, K_q, h$; the parameters $x_1, \ldots, x_q, t$ being defined by (4.9.4). The critical point occurs when $h = 0$ and $x_1, \ldots, x_q, t$ are infinitesimally different from one. From (4.9.4) this implies that

$$\exp(-2K_1) + \ldots + \exp(-2K_q) = q - 2. \qquad (4.9.7)$$

[This result is derived in (11.8.37)–(11.8.42).]

5

THE SPHERICAL MODEL

5.1 Formulation of the Model

In 1952, Berlin and Kac solved another model of ferromagnetism, the spherical model. This is similar to the Ising model of Section 1.8. One considers a lattice $\mathcal{L}$ in space (e.g. the simple cubic lattice), containing N sites. To each site j of $\mathcal{L}$ one assigns a spin σ_j which interacts with its neighbours and with an external field. However, instead of taking only the values $+1$ or -1, each σ_j can now take all real values, subject only to the constraint that

$$\sum_{j=1}^{N} \sigma_j^2 = N. \tag{5.1.1}$$

For an homogeneous system this constraint ensures that the average of the square of any spin is one, as in the usual Ising model.

The partition function is again given by (1.8.2), except that the σ-summation is replaced by an integration subject to the constraint (5.1.1), so

$$Z_N = \int_{-\infty}^{\infty} \ldots \int_{-\infty}^{\infty} d\sigma_1 \ldots d\sigma_N \exp\left[K \sum_{(j,l)} \sigma_j \sigma_l \right.$$
$$\left. + h \sum_j \sigma_j \right] \delta\left[N - \sum_j \sigma_j^2 \right]. \tag{5.1.2}$$

The first summation in (5.1.2) is over all edges (j, l) of $\mathcal{L}$; the other two are over all sites j. As usual, $K = J/kT$ and $h = H/kT$.

Berlin and Kac regarded this as an approximation to the usual Ising model. They argued that in the Ising model the σ-summation can be viewed as a sum over all corners of an N-dimensional hyper-cube in σ-space. In the spherical model this is replaced by an integration over the surface of a hyper-sphere passing through all such corners.

While this is mathematically plausible, it is still true that the constraint (5.1.1) is unphysical in that it implies an equal coupling, or interaction, between all spins, no matter how far apart on $\mathscr{L}$ they may be.

Fortunately Stanley (1968) has shown that the spherical model is a special limiting case of another model (the n-vector model) which has only nearest-neighbour interactions. This equivalence has since been proved rigorously by Kac and Thompson (1971) and Pearce and Thompson (1977). It effectively removes the above objection and establishes the spherical model as a physically acceptable model of critical behaviour.

Many papers have been written on the spherical model (see Joyce (1972) and references therein) covering many aspects of it. This is because it is one of the few (if not the only) model of ferromagnetism that can be solved exactly in a field, and exhibits non-classical critical behaviour.

In this chapter we shall not attempt to consider all facets of the model, but shall outline the derivation of the equation of state and discuss the critical behaviour of the thermodynamic functions.

5.2 Free Energy

To evaluate (5.1.2), first note that it is unchanged if an extra factor

$$\exp\left[aN - a \sum_j \sigma_j^2 \right] \tag{5.2.1}$$

is introduced into the integrand, since the delta function ensures that this is unity. Now use the identity

$$\delta(x) = (2\pi)^{-1} \int_{-\infty}^{\infty} \exp(isx)\, ds \tag{5.2.2}$$

to obtain

$$Z_N = (2\pi)^{-1} \int_{-\infty}^{\infty} \ldots \int_{-\infty}^{\infty} d\sigma_1 \ldots d\sigma_N \int_{-\infty}^{\infty} ds \exp\left[K \sum_{(j,l)} \sigma_j \sigma_l \right.$$
$$\left. + h \sum_j \sigma_j + (a + is)N - (a + is) \sum_j \sigma_j^2 \right]. \tag{5.2.3}$$

The argument of the exponential in (5.2.3) is the sum of quadratic and linear forms in $\sigma_1, \ldots, \sigma_N$. It is useful to introduce a matrix notation to handle these.

Let $\boldsymbol{\sigma}$ be the N-dimensional vector with elements $\sigma_1, \ldots, \sigma_N$. Let $\mathbf{V}$ be the N by N symmetric matrix such that

$$\boldsymbol{\sigma}^T \mathbf{V} \boldsymbol{\sigma} = \sum_j \sum_l \sigma_j V_{jl} \sigma_l$$

$$= (a + is) \sum_j \sigma_j^2 - K \sum_{(j,l)} \sigma_j \sigma_l . \qquad (5.2.4)$$

Finally, let $\mathbf{h}$ be the N-dimensional vector with every element equal to h. Then (5.2.3) can be written more neatly as

$$Z_N = (2\pi)^{-1} \int_{-\infty}^{\infty} \ldots \int_{-\infty}^{\infty} d\boldsymbol{\sigma} \int_{-\infty}^{\infty} ds \exp[-\boldsymbol{\sigma}^T \mathbf{V} \boldsymbol{\sigma} + \mathbf{h}^T \boldsymbol{\sigma} + (a + is)N] . \qquad (5.2.5)$$

Choose the arbitrary constant $\mathbf{a}$ sufficiently large to ensure that all the eigenvalues of $\mathbf{V}$ have positive real part. Then (and only then) the order of the $\boldsymbol{\sigma}$ and s integrations can be interchanged. The $\boldsymbol{\sigma}$ integration can be performed by first changing variables from $\boldsymbol{\sigma}$ to

$$\mathbf{t} = \boldsymbol{\sigma} - \tfrac{1}{2} \mathbf{V}^{-1} \mathbf{h} \qquad (5.2.6)$$

and then rotating axes in $(t_1, \ldots, t_N)$ space to make $\mathbf{V}$ diagonal. This gives

$$Z_N = \tfrac{1}{2} \pi^{\frac{1}{2}N - 1} \int_{-\infty}^{\infty} ds \, [\det \mathbf{V}]^{-\frac{1}{2}} \exp[(a + is)N$$

$$+ \mathbf{h}^T \mathbf{V}^{-1} \mathbf{h}/4] . \qquad (5.2.7)$$

The matrix $\mathbf{V}$ depends on s and the structure of the lattice on which the spins are placed. Let us take $\mathcal{L}$ to be the d-dimensional hyper-cubic lattice, contained in a box with each side of length L lattice spacings. Then

$$N = L^d . \qquad (5.2.8)$$

Impose periodic boundary conditions. Then $\mathbf{V}$ is cyclic and from (5.2.4) its eigenvalues can be found to be

$$\lambda(\omega_1, \ldots, \omega_d) = a + is - K(\cos \omega_1 + \ldots + \cos \omega_d) , \qquad (5.2.9)$$

where each ω_j can take the values $0, 2\pi/L, 4\pi/L, \ldots, 2\pi(L - 1)/L$. The determinant of $\mathbf{V}$ is the product of its eigenvalues, so

$$\ln \det \mathbf{V} = \sum_{\omega_1} \ldots \sum_{\omega_d} \ln \lambda(\omega_1, \ldots, \omega_d) . \qquad (5.2.10)$$

In the thermodynamic limit L is large and the summations in (5.2.10) become integrations. Using (5.2.8) and (5.2.9), it follows that

$$\ln \det \mathbf{V} = N[\ln K + g(z)] , \qquad (5.2.11)$$

where

$$z = (a + is - Kd)/K, \qquad (5.2.12)$$

$$g(z) = (2\pi)^{-d} \int_0^{2\pi} \cdots \int_0^{2\pi} d\omega_1 \ldots d\omega_d \ln[z$$
$$+ d - \cos \omega_1 - \ldots - \cos \omega_d]. \qquad (5.2.13)$$

Also, since V is cyclic, $\mathbf{h}$ is the eigenvector of V corresponding to its minimum eigenvalue $a + is - Kd = Kz$. Thus

$$\mathbf{h}^T V^{-1} \mathbf{h} = (Kz)^{-1} \mathbf{h}^T \mathbf{h} = Nh^2/Kz. \qquad (5.2.14)$$

Using (5.2.11) and (5.2.14), and changing the integration variables from s to z, the equation (5.2.7) can now be written as

$$Z_N = (K/2\pi i)\,(\pi/K)^{\frac{1}{2}N} \int_{c-i\infty}^{c+i\infty} dz \, \exp[N\phi(z)], \qquad (5.2.15)$$

where

$$\phi(z) = Kz + Kd - \tfrac{1}{2}g(z) + h^2/4Kz, \qquad (5.2.16)$$

and $c = (a - Kd)/K$. From (5.2.9) it is apparent that all the eigenvalues of V have positive real part only if $a > Kd$, so c must be positive. The function $\phi(z)$ is analytic for $\mathrm{Re}(z) > 0$, so the RHS of (5.2.15) is the same for all positive values of c.

In the limit of N large, the integral in (5.2.15) can be evaluated by the method of steepest descent (Courant and Hilbert, 1953). First consider the function $\phi(z)$ for z real and positive. Provided $K > 0$ and $h \neq 0$, the function tends to plus infinity as z tends to either zero or infinity. Thus in between $\phi(z)$ must have a minimum at some positive value z_0 of z. Further, it is easy to see from (5.2.16) and (5.2.13) that $\phi''(z) > 0$, so there is only one such minimum.

Take the arbitrary constant c to be z_0. Then along the path of integration in (5.2.15), $\phi(z)$ has a *maximum* at $z = z_0$. In the limit of N large this maximum will give the dominant contribution to the integral, so

$$-f/kT = \lim_{N \to \infty} N^{-1} \ln Z_N$$
$$= \tfrac{1}{2} \ln(\pi/K) + \phi(z_0). \qquad (5.2.17)$$

Here f is as usual the free energy per site. The parameter z_0 is of course defined by the condition that $\phi'(z_0)$ be zero, i.e., using (5.2.16),

$$K - h^2/4Kz_0^2 = \tfrac{1}{2}g'(z_0). \qquad (5.2.18)$$

There is one and only one positive solution for z_0, so (5.2.16)–(5.2.18) define f as a function of K and h, provided $K > 0$ and $h \neq 0$.

5.3   Equation of State and Internal Energy

The parameter z_0 can be simply related to the magnetization. To do this, hold K fixed and differentiate (5.2.17) with respect to h, using (5.2.16). Remembering that z_0 itself depends on h, this gives

$$-\frac{d}{dh}\left(\frac{f}{kT}\right) = \frac{h}{2Kz_0} + \phi'(z_0)\frac{dz_0}{dh}. \qquad (5.3.1)$$

However, z_0 is defined so that $\phi'(z_0)$ is zero. Using (1.7.14) and (1.8.3), the equation (5.3.1) therefore simplifies to

$$M = h/2Kz_0 = H/2Jz_0, \qquad (5.3.2)$$

M being the magnetization per site.

We can now eliminate z_0 between (5.2.18) and (5.3.2). Using the definitions (1.8.3) of K and h, we obtain

$$2J(1 - M^2) = kTg'(H/2JM). \qquad (5.3.3)$$

This is the exact equation of state (i.e. the relation between M, H and T) of the spherical model.

From (1.7.7), the internal energy per site is

$$u = -T^2\frac{\partial}{\partial T}(f/T), \qquad (5.3.4)$$

where the differentiation is performed holding J and H fixed. Using (1.8.3), (5.2.16) and (5.2.17), it follows that

$$U = \tfrac{1}{2}kT - J(z_0 + d) - H^2/4Jz_0 + kT^2\phi'(z_0)(\partial z_0/\partial T). \qquad (5.3.5)$$

Again we note that $\phi'(z_0)$ is zero. Using (5.3.2) to eliminate z_0, we obtain

$$u = \tfrac{1}{2}kT - Jd - \tfrac{1}{2}H(M + M^{-1}). \qquad (5.3.6)$$

This is an exact relation between the internal energy and magnetization.

5.4 The Function $g'(z)$

The equation of state (5.3.3) involves the function $g'(z)$. This can be obtained by differentiating (5.2.13), but the result is a rather unwieldy multi-dimensional integral. It is useful to simplify it as follows.

Differentiate (5.2.13) and use the formula

$$\lambda^{-1} = \int_0^\infty \exp(-\lambda t)\, dt \tag{5.4.1}$$

to write the result as

$$g'(z) = (2\pi)^{-d} \int_0^{2\pi} \cdots \int_0^{2\pi} d\omega_1 \ldots d\omega_d \int_0^\infty dt$$

$$\times \exp\{-t[z + d - \cos\omega_1 - \ldots - \cos\omega_d] . \tag{5.4.2}$$

Provided $\mathrm{Re}(z) > 0$, the integrals converge and may be re-ordered. The $\omega_1, \ldots, \omega_d$ integrations can then be performed by using the formula

$$J_0(it) = (2\pi)^{-1} \int_0^{2\pi} \exp(t \cos\omega)\, d\omega , \tag{5.4.3}$$

$J_0(x)$ being the usual Bessel function (Courant and Hilbert, 1953, p. 474). This gives

$$g'(z) = \int_0^\infty \exp[-t(z+d)]\,[J_0(it)]^d\, dt . \tag{5.4.4}$$

This expression for $g'(z)$ is convenient when considering the dependence of the thermodynamic functions on the dimensionality d of the lattice $\mathscr{L}$. In fact, d need no longer be restricted to integer values, but can be allowed to take any positive value.

This concept of continuously variable dimensionality is quite common in modern statistical mechanics (e.g. Wilson and Fisher, 1972; Fisher, 1974). It can be quite useful in discussing the dependence of critical exponents on d, as we shall see in Section 5.6.

To discuss the behaviour of $g'(z)$ we need to consider the convergence of the infinite integral in (5.4.4). To do this we can use the large t relation

$$J_0(it) = (2\pi t)^{-\frac{1}{2}} e^t [1 + \mathcal{O}(t^{-1})] . \tag{5.4.5}$$

From this we see that the integral (5.4.4) converges if $\mathrm{Re}(z) > 0$, so $g'(z)$ is analytic in the right half-plane. In particular, for real positive z it is analytic and decreases monotonically to zero as $z \to \infty$.

We shall find that the critical properties depend on the behaviour of

$g'(z)$ for small positive z. If z is zero we see from (5.4.5) that the integral in (5.4.4) diverges if $d > 2$, but converges if $d \leqslant 2$. Thus

$$g'(0) = \infty \quad \text{if} \quad 0 < d \leqslant 2,$$
$$< \infty \quad \text{if} \quad d > 2. \tag{5.4.6}$$

For $d > 2$ we shall need the dominant small z behaviour of $g'(0) - g'(z)$. To obtain this, first differentiate $g'(z)$ and then apply the same reasoning as above. This gives

$$g''(0) = \infty \quad \text{if} \quad 0 < d \leqslant 4,$$
$$< \infty \quad \text{if} \quad d > 4. \tag{5.4.7}$$

If $d < 4$, the dominant small z behaviour of $g''(z)$ is obtained by simply neglecting the terms of relative order t^{-1} in (5.4.5) and substituting into (5.4.4) to give

$$g''(z) \simeq - (2\pi)^{-\frac{1}{2}d} \int_0^\infty t^{1 - \frac{1}{2}d} \, e^{-tz} \, dt$$

$$\simeq - (2\pi)^{-\frac{1}{2}d} \, \Gamma(2 - \tfrac{1}{2}d) \, z^{\frac{1}{2}d - 2} . \tag{5.4.8a}$$

For $d = 4$ a slightly more subtle calculation gives

$$g''(z) \simeq - (2\pi)^{-2} \ln z . \tag{5.4.8b}$$

Define a (positive) coefficient A_d by

$$A_d = (2\pi)^{-\frac{1}{2}d} (\tfrac{1}{2}d - 1)^{-1} \Gamma(2 - \tfrac{1}{2}d), \quad 2 < d < 4,$$
$$= (2\pi)^{-2}, \quad d = 4,$$
$$= - g''(0), \quad d > 4. \tag{5.4.9}$$

Then from (5.4.7) and (5.4.8) it follows for z small that

$$g'(0) - g'(z) \simeq A_d z^{\frac{1}{2}d - 1}, \quad 2 < d < 4,$$
$$\simeq A_4 z \ln(1/z), \quad d = 4,$$
$$\simeq A_d z, \quad d > 4. \tag{5.4.10}$$

5.5   Existence of a Critical Point for $d > 2$

Suppose T, and hence K, is fixed. From (5.3.2) and (1.8.3) the function $M(H)$ can be obtained from z_0 as a function of h. The behaviour of these functions can be understood by plotting both sides of (5.2.18) as functions of z_0 (or rather z) for non-negative z. Typical graphs are sketched in Fig. 5.1.

Let P be the intersection of the two curves in the graph. Then its z-coordinate is the solution z_0 of (5.2.18). Provided $h \neq 0$, z_0 is non-zero and varies smoothly with h: in fact z_0 is an even analytic function of h. Hence M is an odd analytic function of H, provided $H \neq 0$.

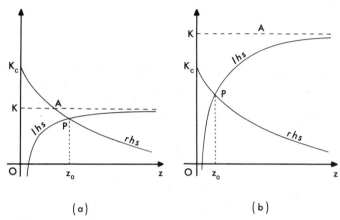

(a)   (b)

Figure 5.1. The z_0 in eq. (5.2.18) is here replaced by z, and typical sketches given of the LHS and RHS as functions of z. The intersection P corresponds to the solution $z = z_0$ of the equation.

Now suppose that h^2 is decreased to zero. The graph of the LHS of (5.2.18) becomes the step-function OKA in Fig. 5.1. Thus P moves to the left, its limiting position being the intersection of OKA with the graph of $\frac{1}{2}g'(z)$. There are two cases to consider, depending on whether the limit of P lies on the horizontal line KA (as in Fig. 5.1(a)), or the vertical line OK (Fig. 5.1(b)).

Define K_c, T_c by

$$K_c = J/kT_c = \tfrac{1}{2}g'(0) \,. \tag{5.5.1}$$

Then the first case arises if $T > T_c$; the second if $T < T_c$.

$T > T_c$

Suppose that $T > T_c$, i.e. $K < K_c = \frac{1}{2}g'(0)$, as in Fig. 5.1(a). As $h^2 \to 0$, $P \to A$, so from (5.2.18) z_0 tends to a non-zero value w given by

$$\tfrac{1}{2}g'(w) = K \,. \tag{5.5.2}$$

For sufficiently small h the term $h^2/4Kz_0^2$ in (5.2.18) can be treated as a perturbation and the equation solved iteratively to give z_0 as a non-zero even analytic function of h. From (5.3.2) and (1.8.3), M is therefore an odd analytic function of H at $H = 0$, and its graph must be similar to that of Fig. 1.1(c). There is no spontaneous magnetization and no phase transition across $h = 0$.

If $d \leqslant 2$, $g'(0)$ and K_c are infinite, so K is always less than K_c. Thus *the spherical model has no transition for $d \leqslant 2$.*

$T < T_c$

Now suppose that $d > 2$, so that K_c is finite, and that $K > K_c$, i.e. $T < T_c$. Then the graphs of (5.2.18) take the form sketched in Fig. 5.1(b). As h^2 tends to zero, P tends to the point $(0, K_c)$ and z_0 tends to zero.

More strongly, the RHS of (5.2.18) tends to K_c, so

$$\lim_{h \to 0} |h|/z_0 = [4K(K - K_c)]^{\frac{1}{2}}. \tag{5.5.3}$$

From (5.3.2) and (1.8.3) it follows that

$$\lim_{h \to 0} M = \mathrm{sgn}\,(H)\, M_0, \tag{5.5.4}$$

where M_0 is given by

$$M_0 = (1 - T/T_c)^{\frac{1}{2}}. \tag{5.5.5}$$

Thus in this case $M(H)$ has a jump-discontinuity across $H = 0$, as in Fig. 1.1(a). There is a non-zero spontaneous magnetization M_0, given by the remarkably simple exact formula (5.5.5).

Thus for $d > 2$ the spherical model exhibits the typical ferromagnetic behaviour outlined in Section 1.1. There is a Curie point (i.e. a critical point) at $H = 0$, $T = T_c$, where T_c is given by (5.5.1).

5.6   Zero-Field Properties: Exponents α, β, γ, γ'

Internal Energy and α

Let $H \to 0$ in (5.3.6). If $T < T_c$ then M tends to the non-zero value M_0, so

$$u = u_- = \tfrac{1}{2}kT - Jd \qquad \text{if } T < T_c. \tag{5.6.1a}$$

If $T > T_c$, then M tends to zero and from (5.3.2) H/M tends to $2Jw$, where w is given by (5.5.2). Thus

$$u = u_+ = \tfrac{1}{2}kT - Jd - Jw. \tag{5.6.1b}$$

Clearly the low-temperature function $u(T)$ defined by (5.6.1a) is analytic at T_c, so the definition (1.7.9) fails and we must use (1.7.10) to define α. Let u_s, the singular part of the internal energy be

$$u_s(T) = u_+(T) - u_-(T). \tag{5.6.2}$$

As in (1.1.3), set

$$t = (T - T_c)/T_c. \tag{5.6.3}$$

Then for t small (1.7.10) implies that

$$u_s(T) \sim t^{1-\alpha}. \tag{5.6.4}$$

This defines α.

Now use the results (5.6.1) in (5.6.2) (taking $T > T_c$). This gives

$$u_s(T) = -Jw. \tag{5.6.5}$$

From (5.4.10), (5.5.1) and (5.5.2), w vanishes as $t \to 0$, its asymptotic behaviour being given by

$$
\begin{aligned}
w &\sim t^{2/(d-2)} & &\text{if } 2 < d < 4, \\
&\sim t/\ln(t^{-1}) & &\text{if } d = 4, \\
&\sim t & &\text{if } d > 4.
\end{aligned}
\tag{5.6.6}
$$

Thus for $d \neq 4$ the relation (5.6.4) is satisfied, with

$$
\begin{aligned}
\alpha &= -(4 - d)/(d - 2) & &\text{if } 2 < d < 4, \\
&= 0 & &\text{if } d > 4.
\end{aligned}
\tag{5.6.7}
$$

Spontaneous Magnetization and β

The spontaneous magnetization has been calculated in (5.5.5). Comparing this exact result with (1.1.4) it is obvious that

$$\beta = \tfrac{1}{2}. \tag{5.6.8}$$

Susceptibility and γ, γ'

The susceptibility χ is defined by (1.1.2). Differentiating (5.3.3) with respect to H and using (5.3.2), it follows that

$$\chi^{-1} = 2Jz_0 - 8JKM^2/g''(z_0) .\qquad(5.6.9)$$

Let H tend to zero. If $T > T_c$, then M tends to zero and z_0 to w, giving

$$\chi^{-1} = 2Jw .\qquad(5.6.10)$$

From (5.6.6) it follows that χ becomes infinite as $T \to T_c$ from above. Provided $d \neq 4$, its asymptotic behaviour has the power-law form (1.1.6), with

$$\begin{aligned}\gamma &= 2/(d-2) \qquad &&\text{if } 2 < d < 4 , \\ &= 1 \qquad &&\text{if } d > 4 .\end{aligned}\qquad(5.6.11)$$

On the other hand, if $H \to 0$ for some (fixed) $T < T_c$, then $z_0 \to 0$. From (5.4.7), $g''(0)$ is infinite if $d \leq 4$, so from (5.6.9)

$$\chi \to \infty \qquad \text{as } H \to 0 .\qquad(5.6.12)$$

This result is qualitatively different from that of the classical mean-field and Bethe lattice models. When $d \leq 4$ the zero-field susceptibility is infinite for all temperatures less than T_c. The usual definition (1.1.7) of the exponent γ' has no meaning.

If $d > 4$, then $g''(0)$ is finite and (5.6.9) gives

$$\chi^{-1} = -8JKM_0^2/g''(0) .\qquad(5.6.13)$$

For t small χ is therefore effectively proportional to M_0^{-2}, i.e. using (5.5.5), to $(-t)^{-1}$. Thus it does have the power-law behaviour (1.1.7), with

$$\gamma' = 1, \quad d > 4 .\qquad(5.6.14)$$

5.7   Critical Equation of State

Using (5.5.1) and (5.6.3), the exact equation of state can be written as

$$g'(0) - g'(H/2JM) = 2J(M^2 + t)/kT .\qquad(5.7.1)$$

When H and t are small, so are both sides of (5.7.1). The T on the RHS can be replaced by T_c and the LHS approximated by (5.4.10). Solving the resulting equation for H gives

$$H \simeq 2JM[2K_c(M^2 + t)/A_d]^{2/(d-2)}, \qquad 2 < d < 4,$$

$$\simeq (4JK_c/A_4)\, M(M^2 + t)/\ln[(M^2 + t)^{-1}], \qquad d = 4, \qquad (5.7.2)$$

$$\simeq (4JK_c/A_d)\, M(M^2 + t), \qquad\qquad d > 4.$$

The quantities J, K_c, A_d are constants, so this critical equation of state is of the form (1.2.1), with $\beta = \frac{1}{2}$ (in agreement with (5.6.8)) and

$$\delta = (d + 2)/(d - 2), \quad 2 < d < 4, \qquad\qquad (5.7.3)$$
$$= 3, \quad d > 4.$$

Using (5.6.11) and neglecting multiplicative constants, the scaling function $h_s(x)$ is given by

$$h_s(x) = (1 + x)^\gamma, \qquad\qquad (5.7.4)$$

provided $d \neq 4$.

The scaling hypothesis is therefore satisfied, so the scaling relations (1.2.12) and (1.2.13) should be also. Indeed they are, as is evident from (5.7.3) and the results of the previous section, subject to the proviso that γ' does not exist for $d \leq 4$.

If $d < 4$ most of the exponents vary with d, but for $d > 4$ they all take the classical constant values. This is perhaps the most interesting result of the spherical model, for it is generally believed that the same is true of the usual nearest-neighbour Ising model, but with different values of the exponents for $d < 4$ (Fisher, 1974, first two lines of p. 607).

6

DUALITY AND STAR – TRIANGLE TRANSFORMATIONS OF PLANAR ISING MODELS

6.1 General Comments on Two-Dimensional Models

In this and the remaining chapters of this book I shall consider the few Ising-type models that have been solved exactly in two dimensions. As I remarked in Section 1.6, it is unfortunate that they are only two dimensional, and even more so that they have only been solved in the absence of external fields. Even so, they do contain the essential prerequisite for a 'physical' model of a magnet or a fluid, namely short-range non-zero interactions, and they do have critical points. They can therefore be used to obtain insight into the behaviour (particularly the critical behaviour) of real systems.

In particular, the two-dimensional exactly solvable models provide extremely valuable tests of general theories and assumptions, such as the scaling and universality hypotheses. For instance, the first evidence of universality was provided by the solution of the square lattice Ising model by Onsager in 1944. Onsager allowed the interactions to have different strengths J and J' in the horizontal and vertical directions, but his solution showed that for T near T_c the specific heat diverges as $\ln|T - T_c|$, *independently of the ratio J'/J*. The evidence for universality accumulated in the next twenty-five years. It took another exact solution, that of the eight-vertex model (Baxter, 1972b), to show that there are exceptions to universality.

6.2 Duality Relation for the Square Lattice Ising Model

Three years before Onsager solved the square lattice model, Kramers and Wannier (1941) located its critical temperature. Their argument can be simplified to the following.

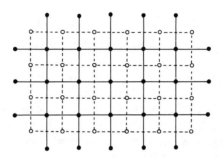

Fig. 6.1. The square lattice $\mathcal{L}$ (solid circles and lines), and its dual lattice $\mathcal{L}_D$ (open circles and broken lines).

Consider an Ising model on the square lattice $\mathcal{L}$ shown in Fig. 6.1. At each site there is a spin σ_i, with two possible values: $+1$ or -1. Two nearest-neighbour spins σ_i and σ_j contribute a term $-J\sigma_i\sigma_j$ to the Hamiltonian if they are horizontal neighbours, $-J'\sigma_i\sigma_j$ if they are vertical neighbours, where J and J' are some fixed energies. If there is no external magnetic field, then the Hamiltonian is simply the sum of such terms, one for each nearest-neighbour pair of sites (i.e. an edge) of the lattice $\mathcal{L}$. From (1.4.1), for a lattice of N sites the analogue of equation (1.8.2) for the partition function is

$$Z_N = \sum_{\sigma} \exp\left[K \sum_{(i,j)} \sigma_i\sigma_j + L \sum_{(i,k)} \sigma_i\sigma_k \right], \qquad (6.2.1)$$

where the first sum inside the brackets is over all horizontal edges (i, j), the second is over all vertical edges (i, k), the outer sum is over all values of all the spins, and

$$K = J/k_BT, \quad L = J'/k_BT, \qquad (6.2.2)$$

k_B being Boltzmann's constant and T the temperature.

To locate the critical temperature, one notes that Z_N can be represented graphically in two different ways, but with a similar form:

'Low-Temperature' Graphical Representation

For a given set of values of the spins (a spin 'configuration'), let r be the number of unlike nearest-neighbour vertical pairs, and s the number of horizontal ones. Let M be the total number of vertical edges of $\mathscr{L}$, and suppose $\mathscr{L}$ has as many columns as rows, so that M is also the number of horizontal edges. Then there are $M - r$ like vertical pairs, and $M - s$ like horizontal pairs, and the summand in (6.2.1) has the value

$$\exp[K(M - 2s) + L(M - 2r)]. \tag{6.2.3}$$

In particular, it depends only on the numbers of unlike nearest-neighbour pairs.

A useful concept in two-dimensional lattice models is that of the *dual lattice*: from any planar lattice $\mathscr{L}$ one can form another lattice by placing points at the centres of the faces of $\mathscr{L}$ and connecting points in faces that are 'adjacent' (i.e. have an edge in common). These points and their connections are the sites and edges of the dual lattice $\mathscr{L}_D$.

The dual $\mathscr{L}_D$ of the square lattice $\mathscr{L}$ in Fig. 6.1 is also shown therein and is also a square lattice. It differs from $\mathscr{L}$ in being shifted a half-lattice spacing in both directions.

Instead of regarding the spins as being on the sites of $\mathscr{L}$, we can just as well regard them as being on the faces of $\mathscr{L}_D$. Given a spin configuration, we can then represent unlike nearest-neighbours pairs by lines on $\mathscr{L}_D$, as follows: If two adjacent spins are different, draw a line on the edge of $\mathscr{L}_D$ between them; if they are the same, do nothing. Do this for all nearest-neighbour spin pairs.

This generates a set of r horizontal lines and s vertical lines on $\mathscr{L}_D$. There must be an even number of lines into each site, since there must be an even number of successive spin changes between the four surrounding faces. The lines can therefore be joined up to form polygons, as in Fig. 6.2.

Conversely, these polygons divide the plane into up-spin domains and down-spin domains, as is evident in Fig. 6.2. For any such set of polygons there are just two corresponding spin configurations, one being obtained from the other by negating all spins.

Using (6.2.3), it follows that the expression (6.2.1) for Z_N can equivalently be written as

$$Z_N = 2 \exp[M(K + L)] \sum_P \exp(-2Lr - 2Ks), \tag{6.2.4}$$

where the summation is over all polygon configurations on $\mathscr{L}_D$, i.e. over all sets of lines with an even number of lines into each site. The r and s are the numbers of horizontal and vertical lines, respectively.

The expression (6.2.4) is useful when developing low-temperature series expansions, since then K and L are large and the dominant term comes from the case $r = s = 0$, i.e. no lines at all. For this reason it is convenient to call it a 'low-temperature' representation, but note that it is exact for all temperatures.

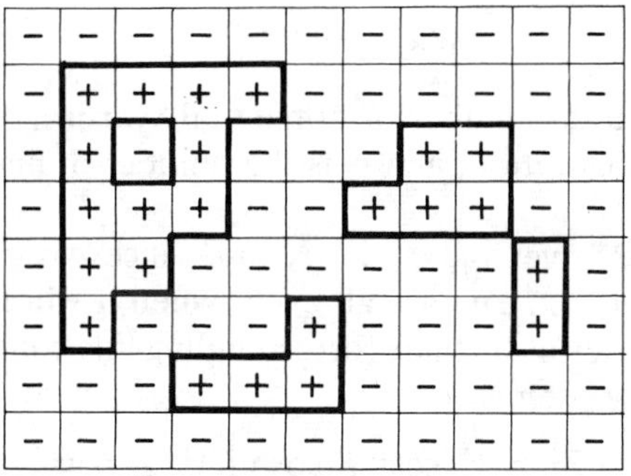

Fig. 6.2. A configuration of spins on the faces of a square lattice, showing the polygons that separate + and − spins.

'High-Temperature' Graphical Representation

Another form for Z_N can be obtained by noting that, since $\sigma_i \sigma_j$ can only take the values of $+1$ or -1:

$$\exp[K\sigma_i\sigma_j] = \cosh K + \sinh K \, \sigma_i\sigma_j. \tag{6.2.5}$$

Using this identity, and its analogue with K replaced by L, the definition (6.2.1) can be written

$$Z_N = (\cosh K \cosh L)^M \sum_\sigma \prod_{(i,j)} (1 + v\sigma_i\sigma_j) \prod_{(i,k)} (1 + w\sigma_i\sigma_k), \tag{6.2.6}$$

where

$$v = \tanh K, \quad w = \tanh L, \tag{6.2.7}$$

the first product is over all the M horizontal edges of $\mathscr{L}$; the second is over all the M vertical edges.

Now expand the combined product in the summand of (6.2.6). Since there are $2M$ factors (one for each edge), each with two terms, there are 2^{2M} terms in this expansion. Each such term can be represented graphically as follows:

Draw a line on the edge (i, j) if from the corresponding factor one selects the term $v\sigma_i\sigma_j$, or $w\sigma_i\sigma_j$. Draw no line if one takes the term 1. Do this for all edges of $\mathscr{L}$.

This gives a one-to-one correspondence between terms in the expansion and line configurations on the edges of $\mathscr{L}$. Each term in the expansion is of the form

$$v^r w^s \sigma_1^{n_1} \sigma_2^{n_2} \sigma_3^{n_3} \ldots , \tag{6.2.8}$$

where $r(s)$ is the total number of horizontal (vertical) lines in the corresponding line configuration, and n_i is the number of lines with site i as an end-point.

Now sum (6.2.8) over $\sigma_1, \sigma_2, \ldots, \sigma_N$. Since $\sigma_i = \pm 1$, the result will vanish unless $n_1, n_2, \ldots, n_N$ are all even, when it will be $v^r w^s 2^N$.

Classifying such terms by their corresponding line configurations, (6.2.6) can therefore be written as

$$Z_N = 2^N (\cosh K \cosh L)^M \sum_P v^r w^s , \tag{6.2.9}$$

where the sum is over all line configurations on $\mathscr{L}$ having an even number of lines into each site, i.e. over polygon configurations on $\mathscr{L}$.

Duality

Let $k_B T \psi$ be the free energy per site, i.e. from (1.7.6),

$$-\psi = \lim_{N \to \infty} N^{-1} \ln Z_N . \tag{6.2.10}$$

From (6.2.1), ψ is a function of K and L, so we can write it as $\psi(K, L)$.

The summations in (6.2.4) and (6.2.9) are similar, but not quite identical as the first is a sum over polygon configurations on $\mathscr{L}_D$, while the second is over polygon configurations on $\mathscr{L}$. For finite square lattices $\mathscr{L}_D$ and $\mathscr{L}$ differ at their boundaries.

However, in the thermodynamic limit this should have no effect on the free energy. Also, in this limit $M/N = 1$, so (6.2.4), (6.2.9) and (6.2.10) give

$$-\psi(K, L) = K + L + \Phi(e^{-2L}, e^{-2K}), \tag{6.2.11}$$

$$= \ln[2 \cosh K \cosh L] + \Phi(v, w) , \tag{6.2.12}$$

where

$$\Phi(v, w) = \lim_{N \to \infty} N^{-1} \ln \left\{ \sum_P v^r w^s \right\} . \tag{6.2.13}$$

Replacing K, L in (6.2.12) by K^*, L^*, where

$$\tanh K^* = e^{-2L}, \quad \tanh L^* = e^{-2K}, \quad (6.2.14a)$$

and comparing with (6.2.11), it becomes obvious that the function Φ can be eliminated, leaving

$$\psi(K^*, L^*) = K + L + \psi(K, L) - \ln[2 \cosh K^* \cosh L^*]. \quad (6.2.14b)$$

If K, L are large, then K^*, L^* are small. Thus (6.2.14) relates the free energy at a low temperature to that at a high temperature, and is known as a duality relation. It can be written in the more symmetric form

$$\sinh 2K^* \sinh 2L = 1, \quad \sinh 2L^* \sinh 2K = 1,$$

$$\psi(K^*, L^*) = \psi(K, L) + \tfrac{1}{2} \ln(\sinh 2K \sinh 2L). \quad (6.2.15)$$

which makes it clear that it is a reciprocal relation.

To locate the critical point, consider first the isotropic case when $J = J'$, so $K = L$ and $K^* = L^*$. At a critical point the free energy is non-analytic, so ψ will be a non-analytic function of T, and hence of K. Suppose this happens at some value K_c of K, then from (6.2.15) it will also be true that ψ is non-analytic when $K^* = K_c$. Normally this will correspond to a different value of K, so there will be two critical points. If we *assume* that there is only one critical point, then it must occur when $K^* = K$, i.e. K_c is given by

$$\sinh 2K_c = 1, \quad K_c = 0.44068679 \ldots . \quad (6.2.16)$$

The argument is similar for the anisotropic case: the mapping $(K, L) \to (K^*, L^*)$ takes the region I in Fig. 6.3 into the region II, and vice-versa. It leaves all points on the curve AB unchanged. Thus if there is a line of critical points inside I, there must be another such line inside II.

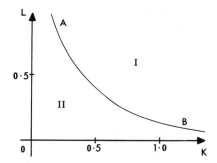

Fig. 6.3. Square lattice duality: the mapping (6.2.15) interchanges regions I and II, and leaves unaltered all points on the graph AB of $\sinh 2K \sinh 2L = 1$.

If there is only one line of critical points, it must be the boundary line AB, the equation of which is

$$\sinh 2K \sinh 2L = 1 . \qquad (6.2.17)$$

In the next chapter it will be shown that this is indeed the criticality condition for the square lattice Ising model.

6.3 Honeycomb-Triangular Duality

One can construct Ising models on any lattice, in particular on the honeycomb and triangular lattices shown in Fig. 6.4.

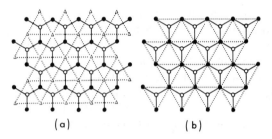

(a)   (b)

Fig. 6.4. The honeycomb lattice (solid lines) and associated triangular lattices (dotted lines) formed by: (a) duality; (b) the star-triangle transformation.

Consider first the honeycomb lattice, with N sites. The edges can be grouped into three classes: those parallel to the edges marked L_1 in Fig. 6.5; those parallel to edges marked L_2; and those parallel to edges marked L_3. Let the energy of two adjacent spins σ, σ' be $-k_B T L_r \sigma \sigma'$, if the edge between them is of the L_r class. Then in zero magnetic field the analogue of equation (1.8.2) for the partition function is

$$Z_N^H\{L\} = \sum_\sigma \exp\left[L_1 \sum \sigma_i \sigma_l + L_2 \sum \sigma_j \sigma_l + L_3 \sum \sigma_k \sigma_l \right]. \qquad (6.3.1)$$

Here L denotes the set of three 'interaction coefficients' L_1, L_2, L_3, and the three summations within the exponential are over all edges of the classes L_1, L_2, L_3, respectively. For instance, the last summation is over all vertical edges (k, l) of the lattice.

Similarly, for the triangular lattice with N sites the partition function is

$$Z_N^T\{K\} = \sum_\sigma \exp\left[K_1 \sum \sigma_j \sigma_k + K_2 \sum \sigma_k \sigma_i + K_3 \sum \sigma_i \sigma_j \right], \qquad (6.3.2)$$

where the first summation inside the exponential is over all edges (j, k) parallel to that marked K_1 in Fig. 6.5, the second is over edges (k, i) parallel to that marked K_2, and the third is over edges (i, j) parallel to that marked K_3.

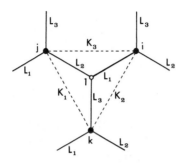

Fig. 6.5. A star $ijkl$ on the honeycomb lattice and its associated (dotted) triangle. The interaction coefficients for the various edges are shown.

One can readily use the technique of Section 6.2 to obtain a duality relation between these two partition functions. First apply the 'low-temperature' procedure of Section 6.2 to the honeycomb Ising model. As is shown in Fig. 6.4(a), the dual of a honeycomb lattice of $2N$ sites is a triangular lattice of N sites. It follows that the analogue of (6.2.4) is

$$Z_{2N}^H\{L\} = \exp[N(L_1 + L_2 + L_3)] \sum_P \exp[-2L_1 r_1 - 2L_2 r_2 - 2L_3 r_3], \quad (6.3.3)$$

where the P-summation is over all polygon configurations on the triangular lattice, r_j being the number of lines on edges of type j.

(A factor 2 corresponding to the leading term in (6.2.4) has been ignored, and the number of edges of each class has been taken to be N, which ignores boundary effects. These approximations have no effect on the free energy in the thermodynamic limit.)

Also, apply the 'high-temperature' procedure of Section 6.2 to the triangular lattice. This gives

$$Z_N^T\{K\} = (2 \cosh K_1 \cosh K_2 \cosh K_3)^N \sum_P v_1^{r_1} v_2^{r_2} v_3^{r_3}, \quad (6.3.4)$$

where

$$v_j = \tanh k_j, \quad j = 1, 2, 3, \quad (6.3.5)$$

and in the thermodynamic limit the P-summation has the same meaning as in (6.3.3).

Comparing (6.3.3) and (6.3.4), it follows that if

$$\tanh K_j = \exp(-2L_j) \, , \quad j = 1, 2, 3 \, , \tag{6.3.6}$$

then

$$Z_{2N}^H\{L\} = (2s_1 s_2 s_3)^{N/2} \, Z_N^T\{K\} \, , \tag{6.3.7}$$

where

$$s_j = \tfrac{1}{2} \exp(2L) \, \mathrm{sech}^2 K_j$$

$$= \sinh 2L_j = 1/\sinh 2K_j \, . \tag{6.3.8}$$

If K_1, K_2, K_3 are large and positive, then L_1, L_2, L_3 are small; and vice-versa. Thus the duality relation (6.3.7) maps a low-temperature (high-temperature) model on the triangular lattice to a high-temperature (low-temperature) one on the honeycomb lattice.

This is not sufficient to locate the critical temperature: to do this we need some more information so as to be able to map a low-temperature model to a high-temperature one *on the same lattice*. This information will be supplied in the following two sections.

6.4   Star – Triangle Relation

In addition to the duality relation (6.3.7), there exists another relation, known as the 'star – triangle' relation, between the partition functions of the triangular and honeycomb Ising models. Onsager (1944) referred in passing to it in the introduction to his paper on the solution of the square lattice Ising model. Wannier (1945) wrote it down, and it has subsequently been represented by many authors (e.g. Houtappel, 1950).

To derive it, first note that the honeycomb lattice is 'bi-partite', i.e. its sites can be divided into two classes A and B such that all neighbours of an A site are B sites, and vice-versa. This is indicated in Fig. 6.4, where the A sites are indicated by solid circles and the B sites by open ones.

The summand in (6.3.2) can therefore be written as

$$\prod_l W(\sigma_l | \sigma_i \, , \sigma_j \, , \sigma_k) \, , \tag{6.4.1}$$

where

$$W(\sigma_l | \sigma_i \, , \sigma_j \, , \sigma_k) = \exp[\sigma_l(L_1 \sigma_i + L_2 \sigma_j + L_3 \sigma_k)] \, , \tag{6.4.2}$$

the product in (6.4.1) is over all B sites l, and i, j, k are the A-site neighbours of l, arranged as in Fig. 6.5.

The important point about (6.4.1) is that each B-spin σ_l occurs in one and only one factor. Using the form (6.4.1) of the summand, it follows that the summations in (6.3.2) over the B-spins can be performed at once, giving

$$Z_N^H\{L\} = \sum_{\sigma_A} \prod_{(i,j,k)} w(\sigma_i , \sigma_j , \sigma_k) , \qquad (6.4.3)$$

where

$$w(\sigma_i , \sigma_j , \sigma_k) = \sum_{\sigma_l} W(\sigma_l|\sigma_i , \sigma_j , \sigma_k)$$

$$= 2 \cosh(L_1\sigma_i + L_2\sigma_j + L_3\sigma_k) , \qquad (6.4.4)$$

and the summation in (6.4.3) is over the remaining A-spins.

Since $w(\sigma_i , \sigma_j , \sigma_k)$ is unaltered by negating all of σ_i, σ_j, σ_k, and since σ_i, σ_j, σ_k only take the values $+1$ or -1, there must exist parameters R, K_1, K_2, K_3 such that

$$w(\sigma_i , \sigma_j , \sigma_k) = R \exp(K_1\sigma_j\sigma_k + K_2\sigma_k\sigma_i + K_3\sigma_i\sigma_j) , \qquad (6.4.5)$$

for all values of σ_i, σ_j, σ_k. Substituting this expression for w into (6.4.3) then gives

$$Z_N^H\{L\} = R^{N/2} \sum_{\sigma_A} \prod_{(i,j,k)} \exp(K_1\sigma_j\sigma_k + K_2\sigma_k\sigma_i + K_3\sigma_i\sigma_j) . \qquad (6.4.6)$$

The summation is over $\frac{1}{2}N$ spins on the A-sites of the honeycomb lattice. As is indicated in Fig. 6.4(b) these form a triangular lattice. The product in (6.4.6) is over all down-pointing triangles (i , j , k) of this triangular lattice.

The sum in (6.4.6) is therefore precisely the partition function of the Ising model on a triangular lattice of $N/2$ spins. Replacing N by $2N$, and comparing with (6.3.1), it is obvious that

$$Z_{2N}^H\{L\} = R^N Z_N^T\{K\} . \qquad (6.4.7)$$

This relation between the honeycomb and triangular lattice partition functions is known as the star – triangle relation, since it is obtained by summing over the centre spin of a star (Fig. 6.5) to obtain a triangle.

Relations Between Interaction Coefficients

Given L_1, L_2, L_3, the parameters R, K_1, K_2, K_3 are defined by the four equations obtained by equating (6.4.4) and (6.4.5) for all values of σ_i, σ_j, σ_k. These are

$$2 \cosh(L_1 + L_2 + L_3) \quad = R \exp(K_1 + K_2 + K_3) \qquad (6.4.8a)$$

$$2 \cosh(-L_1 + L_2 + L_3) = R \exp(K_1 - K_2 - K_3) \qquad (6.4.8b)$$

$$2 \cosh(L_1 - L_2 + L_3) \quad = R \exp(-K_1 + K_2 - K_3) \qquad (6.4.8c)$$

$$2 \cosh(L_1 + L_2 - L_3) \quad = R \exp(-K_1 - K_2 + K_3) . \qquad (6.4.8d)$$

Multiplying the first two of these equations, and dividing by the second two, gives

$$cc_1/c_2c_3 = \exp(4K_1) , \qquad (6.4.9)$$

where, for all permutations i, j, k of 1, 2, 3,

$$c = \cosh(L_1 + L_2 + L_3) , \quad c_i = \cosh(-L_i + L_j + L_k) . \qquad (6.4.10)$$

Using standard hyperbolic function identities, it follows from (6.4.9) that

$$\exp(4K_1) - 1 = \sinh 2L_2 \, \sinh 2L_3 / c_2c_3 , \qquad (6.4.11)$$

and hence that

$$\sinh 2K_1 \sinh 2L_1 = \frac{\sinh 2L_1 \, \sinh 2L_2 \, \sinh 2L_3}{2(cc_1c_2c_3)^{\frac{1}{2}}} . \qquad (6.4.12)$$

Clearly the original set of star – triangle relations (6.4.8) is invariant under permutation of the suffixes 1, 2, 3, so two other equations can be obtained from (6.4.12) by such permutations. However, the RHS is a symmetric function of L_1, L_2, L_3, so remains unchanged. Defining k^{-1} to be its value, it follows that

$$\sinh 2K_j \sinh 2L_j = k^{-1}, \quad j = 1, 2, 3 . \qquad (6.4.13)$$

This is a remarkable and very important property of the star – triangle relations: the products $\sinh 2K_j \sinh 2L_j$, for $j = 1, 2, 3$, all have the same value.

Multiplying the four equations (6.4.8), using (6.4.12) to eliminate $cc_1c_2c_3$, and then using (6.4.13), one obtains

$$R^2 = 2k \sinh 2L_1 \, \sinh 2L_2 \, \sinh 2L_3$$

$$= 2/(k^2 \sinh 2K_1 \, \sinh 2K_2 \, \sinh 2K_3) . \qquad (6.4.14)$$

These last three equations define K_1, K_2, K_3, k and R as functions of L_1, L_2, L_3. Alternatively, one can obtain equations for L_1, k, etc. as functions of K_1, K_2, K_3. For instance, eliminating R, L_2 and L_3 between the equations (6.4.8) gives, after a little algebra,

$$\sinh 2K_1 \cosh 2K_2 \cosh 2K_3 + \cosh 2K_1 \sinh 2K_2 \sinh 2K_3$$

$$= \sinh 2K_1 \cosh 2L_1 . \qquad (6.4.15)$$

Eliminating L_1 between this identity and (6.4.13) gives

$$k = \frac{(1 - v_1^2)(1 - v_2^2)(1 - v_3^2)}{4[(1 + v_1 v_2 v_3)(v_1 + v_2 v_3)(v_2 + v_3 v_1)(v_3 + v_1 v_2)]^{\frac{1}{2}}}, \qquad (6.4.16)$$

where v_1, v_2, v_3 are given by (6.3.5).

Operator Form

In two-dimensional lattice problems it is often useful to consider a row of spins $\sigma_1, \ldots, \sigma_N$, and operators that build up the lattice by adding sites and/or edges. These operators are 2^N by 2^N matrices, with rows labelled by $(\sigma_1, \ldots, \sigma_N)$, and columns by $(\sigma_1', \ldots, \sigma_N')$. Two important simple sets of operators are $s_1, \ldots, s_N$, and $c_1, \ldots, c_N$, where

$$(s_i)_{\sigma\sigma'} = \sigma_i \delta(\sigma_1, \sigma_1') \, \delta(\sigma_2, \sigma_2') \ldots \delta(\sigma_N, \sigma_N')$$

$$(c_i)_{\sigma\sigma'} = \delta(\sigma_1, \sigma_1') \ldots \delta(\sigma_{i-1}, \sigma_{i-1}') \, \delta(\sigma_i, -\sigma_i') \, \delta(\sigma_{i+1}, \sigma_{i+1}')$$

$$\ldots \delta(\sigma_N, \sigma_N'), \qquad (6.4.17)$$

and σ denotes $(\sigma_1, \ldots, \sigma_N)$, σ' denotes $(\sigma_1', \ldots, \sigma_N')$.

Thus s_i is a diagonal matrix with entries σ_i, c_i is the operator that reverses the spin in position i. Writing (x, y) for the commutator $xy-yx$ of two operators x, y, and I for the identity operator, it is readily seen that

$$s_i^2 = c_i^2 = I, \quad s_i c_i + c_i s_i = 0, \qquad (6.4.18)$$

$$(s_i, s_j) = (s_i, c_j) = (c_i, c_j) = 0 \quad \text{for } i \neq j.$$

In the Ising model, the two basic sets of operators are

$$P_1(K), \ldots, P_{N-1}(K), Q_1(L), \ldots, Q_N(L),$$

where

$$[P_i(K)]_{\sigma\sigma'} = \exp(K\sigma_i\sigma_{i+1}) \, \delta(\sigma_1, \sigma_1') \ldots \delta(\sigma_N, \sigma_N')$$

$$[Q_i(L)]_{\sigma\sigma'} = \delta(\sigma_1, \sigma_1') \ldots \delta(\sigma_{i-1}, \sigma_{i-1}') \exp(L\sigma_i\sigma_i')$$

$$\times \, \delta(\sigma_{i+1}, \sigma_{i+1}') \ldots \delta(\sigma_N, \sigma_N'). \qquad (6.4.19)$$

The effect of the operator $P_i(K)$ is to introduce an edge, with interaction coefficient K, between sites i and $i + 1$. The effect of $Q_i(L)$ is to introduce a new site in position i, linked to the old site by an edge with interaction coefficient L. If we regard $\sigma_1, \ldots \sigma_N$ as being a horizontal row of spins; then $P_i(K)$ adds a horizontal edge, $Q_i(L)$ a vertical one.

Using (6.4.17), the definitions (6.4.19) can be written more compactly as

$$P_i(K) = \exp(Ks_i s_{i+1})$$

$$Q_i(L) = \exp(L)\,I + \exp(-L)\,c_i. \tag{6.4.20}$$

Since $c_i^2 = I$, it follows that

$$\exp(Lc_i) = (\cosh L)I + (\sinh L)c_i \tag{6.4.21}$$

for all complex numbers L. Thus $Q_i(L)$ can be written as

$$Q_i(L) = (2 \sinh 2L)^{\frac{1}{2}} \exp(L^* c_i)\,, \tag{6.4.22}$$

where L^* is related to L by

$$\tanh L^* = \exp(-2L)\,. \tag{6.4.23}$$

(This is the same as the relation (6.2.14a) which occurs in the duality transformation.)

It is useful to interlace the operators P_i, Q_i in the order Q_1, P_1, Q_2, $P_2, \ldots, Q_N$ and to define a corresponding set of operators $U_1, U_2, \ldots, U_{2N-1}$, dependent on two interaction coefficients K and L, by

$$U_i(K, L) = P_j(K) = \exp(Ks_j s_{j+1}) \quad \text{if } i = 2j\,,$$

$$= (2 \sinh 2L)^{-\frac{1}{2}} Q_j(L) = \exp(L^* c_j) \quad \text{if } i = 2j - 1. \tag{6.4.24}$$

These are all the operators that are needed to construct a square-lattice Ising model with horizontal interaction coefficient K and vertical coefficient L.

Let K_1, K_2, K_3, L_1, L_2, L_3 be related by the star-triangle relations (6.4.8)–(6.4.15). Then, using (6.4.19) to directly expand the matrix products, and using (6.4.13) and (6.4.14), the star – triangle relation (6.4.4–5) is found to imply that

$$U_{i+1}(K_1, L_1)\, U_i(L_2, K_2)\, U_{i+1}(K_3, L_3)$$

$$= U_i(K_3, L_3)\, U_{i+1}(L_2, K_2)\, U_i(K_1, L_1) \tag{6.4.25}$$

for $i = 1, \ldots, 2N - 2$. It is also obvious that

$$U_i(K, L)\, U_j(K', L') = U_j(K', L')\, U_i(K, L) \tag{6.4.26}$$

for all complex numbers K, L, K', L', provided $|i - j| \geq 2$.

Like the P_i and Q_i, the $U_i(K, L)$ are operators that add edges to the lattice. If i is even, the LHS of (6.4.25) has the effect of adding an (L_1, L_2, L_3) star, the RHS of adding a (K_1, K_2, K_3) triangle; and vice versa if i is odd. Thus (6.4.25) is a simple operator form of the star – triangle relation.

If we write $U_i(K_1, L_1)$, $U_i(L_2, K_2)$, $U_i(K_3, L_3)$ simply as U_i, U_i', U_i'', respectively, then (6.4.25) is simply

$$U_{i+1} U_i' U_{i+1}'' = U_i'' U_{i+1}' U_i, \qquad (6.4.27)$$

which makes its structure rather more obvious. Also, (6.4.26) implies that

$$U_i U_j' = U_j' U_i \quad \text{if } |i - j| \geqslant 2. \qquad (6.4.28)$$

Significance of the Star – Triangle Relation

The star – triangle relation turns out to have very significant consequences. Consider two square-lattice Ising models as in Section 6.2, with different values of K and L, but the same value of sinh $2K$ sinh $2L$. Onsager (1971) noted that the star – triangle relation implies that their diagonal-to-diagonal transfer matrices commute, providing cyclic boundary conditions are imposed.

A proper derivation of this is given in Section 7.3, but a partial demonstration follows readily from (6.4.25). Consider the operator

$$V(K, L) = U_1(K, L) U_2(K, L) \ldots U_n(K, L), \qquad (6.4.29)$$

where $n = 2N - 1$. This corresponds to adding a vertical edge in column N, then a horizontal one between columns $N - 1$ and N, then a vertical one in column $N - 1$, and so on, going downwards as this proceeds. Altogether this adds a 'staircase' to the usual square lattice. Apart from boundary conditions (and a trivial normalization factor), $V(K, L)$ is therefore the diagonal-to-diagonal transfer matrix of the square lattice.

Let us again take $K_1, K_2, K_3, L_1, L_2, L_3$ to satisfy the star – triangle relations (6.4.8)–(6.4.15). Write $V(K_1, L_1)$, $V(L_2, K_2)$ simply as V, V'. Then

$$V = U_1 U_2 \ldots U_n, \qquad (6.4.30a)$$

$$V' = U_1' U_2' \ldots U_n'. \qquad (6.4.30b)$$

By repeated use only of (6.4.27) and (6.4.28), it is easily verified that

$$VV' (U_n'^{-1} U_n'' U_n) = (U_1 U_1'' U_1'^{-1}) V'V. \qquad (6.4.31)$$

The bracketted terms are 'boundary terms' involving only operators acting on the end spins. It is therefore perhaps not surprising that these terms disappear when the cyclic boundary conditions are treated properly (as is done in Section 7.3), leaving

$$VV' = V'V. \tag{6.4.32}$$

Thus $V(K_1, L_1)$ and $V(L_2, K_2)$ commute providing K_3, L_3 can be chosen to satisfy (6.4.8)–(6.4.15). This is so if $\sinh 2K_1 \sinh 2L_1 = \sinh 2L_2 \sinh 2K_2$.

More generally, if we have any lattice model whose transfer matrix V can be written in the form (6.4.30a), and if we can also construct operators $U'_1, \ldots, U'_n = U''_1, \ldots, U''_n$ satisfying (6.4.27) and (6.4.28), then the pseudo-commutation rule (6.4.31) is satisfied by V and V'. In Chapters 9 and 10 it is shown that this can be done for the six- and eight-vertex models. The corresponding commutation relation is a vital first step in the solution of the eight-vertex model.

To obtain exact commutation relations it is necessary to use an explicit representation of the operators so as to handle the cyclic boundary conditions. Also, to cast the transfer matrix into a form like (6.4.30a), it would be necessary to introduce an irritating cyclic shift in the spin-labelling from row to row. For these reasons the commutation relations of the Ising six- and eight-vertex models will be obtained directly, instead of by invoking (6.4.30)–(6.4.31). Even so, in each case the commutation relations are a direct consequence of the appropriate 'star – triangle' relation, and this will be emphasized.

Further, in Section 11.7 it is shown that for the Ising model the transfer matrix formalism can be dispensed with altogether: the free energy is obtained solely from the star-triangle relation and its corollaries!

6.5   Triangular – Triangular Duality

If L_1, L_2, L_3 in (6.4.8) are small, then so are K_1, K_2, K_3. The star – triangle relation (6.4.7) therefore maps a high-temperature model on the triangular lattice to a high-temperature one on the honeycomb lattice.

Now apply the duality transformation (6.3.6)–(6.3.8). This maps the high-temperature honeycomb model to a low-temperature triangular one.

Taken together in this way, the star – triangle and duality transformations therefore give the following self-duality relation for the triangular Ising model:

$$Z_N^T(K) = k^{-N/2} Z_N^T(K^*), \tag{6.5.1}$$

where

$$\sinh 2K_j^* = k \sinh 2K_j, \quad j = 1, 2, 3, \tag{6.5.2}$$

and k is given in terms of K_1, K_2, K_3 by (6.4.16) and (6.3.5). Alternatively, in terms of K_1^*, K_2^*, K_3^*, it is given by

$$k^{-1} = \frac{(1 - v_1^{*2}) (1 - v_2^{*2}) (1 - v_3^{*2})}{4[(1 + v_1^* v_2^* v_3^*) (v_1^* + v_2^* v_3^*) (v_2^* + v_3^* v_1^*) (v_3^* + v_1^* v_2^*)]^{\frac{1}{2}}}, \tag{6.5.3}$$

where

$$v_j^* = \tanh K_j^*, \quad j = 1, 2, 3. \tag{6.5.4}$$

Clearly this mapping is reciprocal: i.e. it maps a point $K = (K_1, K_2, K_3)$ to the point $K^* = (K_1^*, K_2^*, K_3^*)$, and the point K^* back again to K. From (6.5.2) there is a surface of self-dual points in (K_1, K_2, K_3) space, corresponding to $k = 1$. We can therefore argue as in Section 6.2: if there is only one critical surface in (K_1, K_2, K_3) space, then it must be the self-dual surface, in which case the condition for criticality must be

$$k = 1. \tag{6.5.5}$$

This is in fact true, as is shown in Chapter 11. For the isotropic triangular model, with $K_1 = K_2 = K_3 = K$, it implies that the critical point $K = K_c$ is given by

$$\sinh 2K_c = 3^{-\frac{1}{2}}, \quad K_c = 0.27465307 \ldots \tag{6.5.6}$$

From (6.3.6), the isotropic honeycomb model therefore has its critical point at $L_1 = L_2 = L_3 = L_c$, where

$$\sinh 2L_c = 3^{\frac{1}{2}}, \quad L_c = 0.6584789 \ldots \tag{6.5.7}$$

7
SQUARE-LATTICE ISING MODEL

7.1 Historical Introduction

The free energy of the two-dimensional Ising model in zero field was first obtained by Onsager in 1944. He diagonalized the transfer matrix by looking for irreducible representations of a related matrix algebra. His student, Bruria Kaufman, simplified this derivation in 1949 by showing that the transfer matrix belongs to the group of spinor operators.

Since then many alternative derivations have been given. The transfer matrix method has been used by Schultz et al., Lieb (1964), Thompson (1965), Baxter (1972b) and Stephen and Mittag (1972).

A completely different technique was discovered by Kac and Ward (1952), who used combinatorial arguments to write the partition function as a determinant which could be easily evaluated. This method was refined by Potts and Ward (1955).

Hurst and Green (1960), and Kasteleyn (1963) also used combinatorial arguments, but this time to write the partition function as a Pfaffian. Another combinatorial solution was obtained by Vdovichenko (1965), and is given by Landau and Lifshitz (1968).

Quite recently, Hilhorst et al (1978), and Baxter and Enting (1978), have shown that the planar Ising models can be solved quite directly by using the star – triangle relation of Section 6.4 as a recurrence relation.

It is quite beyond the scope of this book to discuss all these approaches in detail. The one given in this chapter may be called the 'commuting transfer matrices' method. It has the advantage that it can be generalized to solve the eight-vertex model, as is shown in Chapter 9.

The basic idea is to regard the diagonal-to-diagonal transfer matrix as a function of the two interaction coefficients K and L. It is easily established that two such matrices commute if they have the same value of $k = (\sinh 2K$ $\sinh 2L)^{-1}$, and for any such matrix another one can be found which is effectively its inverse. These properties are basically sufficient to obtain the eigenvalues of the transfer matrix. From these, the free energy, interfacial tension and correlation length are derived.

The result for the spontaneous magnetization M_0 is also given (in Section 7.10), but is not derived in this chapter as its calculation is rather technical: five years elapsed between Onsager's derivation of the free energy f and his announcement at a conference in Florence of the result for M_0 (Onsager, 1949 and 1971). The first published derivation was given by Yang in 1952, while Montroll *et al.* obtained it by the simpler Pfaffian method in 1963. A derivation based on corner transfer matrices is given in Section 13.7 for the more general eight-vertex model.

In Sections 7.7–7.12 the cases $k < 1$, $k = 1$ and $k > 1$ are distinguished. As is shown in Section 7.12, these correspond to the low-temperature $(T < T_c)$, critical temperature $(T = T_c)$ and high-temperature $(T > T_c)$ cases, respectively.

7.2 The Transfer Matrices V, W

Consider the square lattice zero-field Ising model, as defined in Section 6.2, but draw the lattice diagonally as in Fig. 7.1. The partition function is still given by (6.2.1), but now the first summation inside the brackets is over all edges parallel to those marked K in Fig. 7.1, and the second summation is over all edges parallel to those marked L.

Group the sites into horizontal rows: for instance, the sites denoted by solid circles in Fig. 7.1 form a row. As is indicated in Fig. 7.1, these rows can be classified into two types A and B (open circles and solid circles). A row of type A is above one of type B, and vice-versa.

Let m be the number of such rows in the lattice. Label them so that row r is below row $r + 1$, and impose cyclic boundary conditions so that row m is below row 1. This means that m must be even.

Let n be the number of sites in each row and label them from left to right. Again impose cyclic boundary conditions, this time to ensure that site n is to the left of site 1, as indicated in Fig. 7.1. (Taken together, these cyclic boundary conditions are equivalent to drawing the lattice on a torus: they are known as 'toroidal'.)

Let ϕ_r denote all the spins in row r, so ϕ_r has 2^n possible values. The summand in (6.2.1) can be thought of as a function of $\phi_1, \ldots, \phi_m$. Since each spin interacts only with spins in adjacent rows, this function factorizes and (6.2.1) can be written as

$$Z_N = \sum_{\phi_1} \sum_{\phi_2} \ldots \sum_{\phi_m} V_{\phi_1,\phi_2} \, W_{\phi_2,\phi_3} \, V_{\phi_3,\phi_4} \, W_{\phi_4,\phi_5} \ldots W_{\phi_m,\phi_1} \,. \qquad (7.2.1)$$

Here $V_{\phi_j,\phi_{j+1}}$ contains all the Boltzmann weight factors in the summand that involve only spins in adjacent rows j and $j + 1$, the lower row j being of type A. The same is true for $W(\phi_j, \phi_{j+1})$, except that the lower row is of type B.

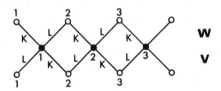

Fig. 7.1. Three successive rows of the square-lattice (drawn diagonally).

Consider two typical successive rows. Let $\phi = \{\sigma_1, \ldots, \sigma_n\}$ be the spins in the lower row, and $\phi' = \{\sigma_1', \ldots, \sigma_n'\}$ the spins in the upper row. Then from (6.2.1) and Fig. 7.1, it is clear that

$$V_{\phi,\phi'} = \exp\left[\sum_{j=1}^{n} (K\sigma_{j+1}\sigma_j' + L\sigma_j\sigma_j') \right],$$

$$W_{\phi,\phi'} = \exp\left[\sum_{j=1}^{n} (K\sigma_j\sigma_j' + L\sigma_j\sigma_{j+1}') \right], \qquad (7.2.2)$$

where $\sigma_{n+1} \equiv \sigma_1$ and $\sigma_{n+1}' \equiv \sigma_1'$.

These observations parallel those made in Chapter 2 for the one-dimensional Ising model. Again $V(\phi, \phi')$ can be regarded as the element ϕ, ϕ' of a matrix V, and similarly for W. Then (7.2.1) can be written as

$$Z_N = \text{Trace } VWVW \ldots W$$

$$= \text{Trace } (VW)^{m/2}. \qquad (7.2.3)$$

The main difference from the one-dimensional case is that ϕ and ϕ' have 2^n values, so V and W are 2^n by 2^n matrices, rather than 2 by 2. It is also no longer true that VW is symmetric: even so, the working of (2.1.11) to (2.1.15) can be generalized to show that (7.2.3) implies

$$Z_N = \Lambda_1^m + \Lambda_2^m + \ldots + \Lambda_{2^n}^m \,, \qquad (7.2.4)$$

where $\Lambda_1^2, \Lambda_2^2, \ldots$ are the eigenvalues of VW.

We are interested primarily in the thermodynamic limit, when m and n are large. To obtain this it is permissible to first let $m \to \infty$, keeping n fixed. From (7.2.4) it then immediately follows that

$$Z_N \sim (\Lambda_{max})^m, \qquad (7.2.5)$$

where Λ_{max}^2 is the numerically largest eigenvalue of VW.

The matrices V and W are known as *transfer matrices*. The problem now is to calculate the maximum eigenvalue of VW.

7.3 Two Significant Properties of V and W

Commutation

From (7.2.2) it is obvious that the matrices V and W are functions of the interaction coefficients K and L. It is convenient here to exhibit this explicitly and to write V and W as $V(K, L)$ and $W(K, L)$.

In this section, I shall establish two properties of these matrix functions, and in subsequent sections will show that these properties enable Λ_{max} to be evaluated. Although indirect, this presentation has the advantage of showing that it is basically only local properties of the lattice model that are being used.

In (7.2.3) one is interested in the matrix product $V(K, L)\, W(K, L)$. Let us generalize this and consider the product.

$$V(K, L)\, W(K', L'), \qquad (7.3.1)$$

where for the moment K, L, K', L' can be any complex numbers. This product matrix has a lattice model meaning: it is the transfer matrix for going from the lower row of open circles in Fig. 7.1 to the upper one, provided the interaction coefficients K, L of edges above the solid circles are replaced by K', L', respectively.

Each element of the product matrix is therefore the product of the Boltzmann weights of the complete edges shown in Fig. 7.1, summed over all the spins $\sigma_1'', \ldots, \sigma_n''$ on the intermediate solid circles. Let $\phi = \{\sigma_1, \ldots, \sigma_n\}$ be the spins on the lower row of open circles, and $\phi' = \{\sigma_1', \ldots, \sigma_n'\}$ the spins on the upper row. Then the element ϕ, ϕ' of the matrix product (7.3.1) is

$$\sum_{\sigma_1''} \cdots \sum_{\sigma_n''} \prod_{j=1}^{n} \exp\left[\sigma_j'' (K\sigma_{j+1} + L\sigma_j + K'\sigma_j' + L'\sigma_{j+1}')\right]. \qquad (7.3.2)$$

The exponential in (7.3.2) is simply the Boltzmann weight of the four edges in Fig. 7.1 that have solid circle j as an end-point. Since σ_j'' enters only this term, the summation over σ_j'' is readily performed. Doing this for $\sigma_1'', \ldots, \sigma_n''$, (7.3.2) becomes

$$\prod_{j=1}^{n} X(\sigma_j, \sigma_{j+1}; \sigma_j', \sigma_{j+1}') \,, \tag{7.3.3}$$

where, for $a, b, c, d = \pm 1$,

$$X(a, b; c, d) = \sum_{f=\pm 1} \exp[f(La + Kb + K'c + L'd)] \,. \tag{7.3.4}$$

Now ask the following question: suppose we interchange the interaction coefficients K and K', and the coefficients L and L': does this change the product matrix (7.3.1)? Explicitly: is the equation

$$V(K, L) W(K', L') = V(K', L') W(K, L) \tag{7.3.5}$$

true?

This is a generalized commutation relation. If it is true, then it is shown in the next section that the transfer matrices all commute, and this property will be used to obtain the free energy. For the moment, however, let us just ask whether (7.3.5) can be satisfied.

Clearly (7.3.3) is unchanged by replacing $X(a, b; c, d)$ by

$$e^{Mac} X(a, b; c, d) e^{-Mbd} \,, \tag{7.3.6}$$

since the exponential factors from adjacent terms cancel. Thus (7.3.5) will certainly be true if there exists a number M such that

$$e^{Mac} X(a, b; c, d) = X'(a, b; c, d) e^{Mbd} \,, \tag{7.3.7}$$

where X' is obtained from X by interchanging K with K', and L with L'.

This equation can be interpreted graphically as in Fig. 7.2(a). It is equivalent to requiring that the Boltzmann weights of both figures therein, summed over the centre spin, be the same for all values (± 1) of the exterior spins a, b, c, d.

The condition (7.3.7) can be examined directly, but to link with the remarks of Section 6.4, it is better to proceed as follows.

In both figures there is a (L, K', M) triangle. Define K_1, K_2, K_3 by

$$K_1 = L, \quad K_2 = K', \quad K_3 = M \,, \tag{7.3.8a}$$

and convert these triangles to stars by using the star – triangle relation (6.4.4), (6.4.5). Then L_1, L_2, L_3 are defined by (6.4.8)–(6.4.15), and, to within a common factor R, we see that the Boltzmann weights are those of the figures in Fig. 7.2(b), each summed over the two internal spins.

Clearly these weights are the same if

$$L_1 = K, \quad L_2 = L' . \tag{7.3.8b}$$

From (6.4.14), it follows that K, L, K', L' must satisfy

$$\sinh 2K \sinh 2L = \sinh 2K' \sinh 2L' . \tag{7.3.9}$$

This condition can be obtained more directly by making the substitutions (7.3.8) into (6.4.8) and eliminating R, L_3 and M. Provided (7.3.9) is true, we can choose R, L_3 and M so that (6.4.8) is satisfied. The relations (7.3.7) and (7.3.5) follow immediately.

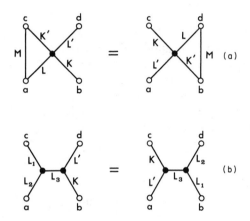

Fig. 7.2. (a) The lattice segments whose weights (summed over the spin on the solid circle) are the left- and right-hand sides of (7.3.7); (b) the same segments after applying a triangle-to-star transformation.

We have therefore used the star – triangle relation to establish that (7.3.9) is a sufficient condition for the exact commutation relation (7.3.5) to be satisfied. It is also necessary.

Inversion

The other property that will be needed can be thought of as a relation for the inverse of V, or W. It can be approached by asking the question: given K, L; can K', L' be chosen to ensure that the product (7.3.1) is a diagonal, or 'near-diagonal', matrix?

Since the elements of (7.3.1) are of the form (7.3.3), this property would be satisfied if $X(a, b; c, d)$ vanished when $a \neq c$ (or $b \neq d$). This requirement is too strong: it cannot in general be satisfied.

What can be satisfied is a weaker condition: namely that $X(a, b; c, d)$ vanish if $a \neq c$ *and* $b = d$. From (7.3.4) this is equivalent to the two equations

$$\cosh (L + K - K' + L') = 0 \qquad (7.3.10)$$
$$\cosh (L - K - K' - L') = 0.$$

These equations have no real solutions, but they do have the complex solution

$$K' = L + \tfrac{1}{2}i\pi, \quad L' = -K. \qquad (7.3.11)$$

What is the effect of the requirement that $X(a, b; c, d)$ vanish if $a \neq c$ and $b = d$? From (7.3.3) it implies that for non-zero elements of VW, if σ_j and σ_j' are unlike, then σ_{j+1} and σ_{j+1}' must also be unlike. Since $j = 1, \ldots, n$, with cyclic boundary conditions, this implies that either all pairs (σ_j, σ_j') are like, or they are all unlike.

If they are like, then we are interested only in $X(a, b; c, d)$ for $a = c$ and $b = d$. From (7.3.4) and (7.3.11), all such values of X are

$$X_{\text{like}} = 2i \sinh 2L. \qquad (7.3.12\text{a})$$

If they are unlike, then $a \neq c$ and $b \neq d$, and

$$X_{\text{unlike}} = -2i\, ab \sinh 2K. \qquad (7.3.12\text{b})$$

Substituting these expressions into (7.3.3), it follows that (7.3.3) is now the same as the expression

$$(2i \sinh 2L)^n \, \delta(\sigma_1, \sigma_1') \, \delta(\sigma_2, \sigma_2') \ldots \delta(\sigma_n, \sigma_n') \qquad (7.3.13)$$
$$+ (-2i \sinh 2K)^n \, \delta(\sigma_1, -\sigma_1') \, \delta(\sigma_2, -\sigma_2') \ldots \delta(\sigma_n, -\sigma_n')$$

Thus $V(K, L) \, W(K', L')$ is the matrix with elements (7.3.13). Let I be the identity matrix of dimension 2^n, and R the matrix with elements

$$R_{\phi, \phi'} = \delta(\sigma_1, -\sigma_1') \ldots \delta(\sigma_n, -\sigma_n'). \qquad (7.3.14)$$

Then we see that we have established the matrix identity

$$V(K, L) \, W(L + \tfrac{1}{2}i\pi, -K)$$
$$= (2i \sinh 2L)^n I + (-2i \sinh 2K)^n R. \qquad (7.3.15)$$

Since

$$R^2 = I, \qquad (7.3.16)$$

the RHS of (7.3.15) is easily inverted, so (7.3.15) could be used to obtain the inverse of the matrix $V(K, L)$. For this reason I shall sometimes call it the 'inversion identity'.

7.4 Symmetry Relations

In addition to the commutation and inversion properties established above, we shall need some simple symmetry properties of the transfer matrices.

Interchanging K with L, and each σ_j with each σ_j', in (7.2.2) is equivalent to interchanging V and W. This means that

$$W(K, L) = V^T(L, K), \tag{7.4.1}$$

and

$$V(K, L) W(K, L) = [V(L, K) W(L, K)]^T. \tag{7.4.2}$$

Also, negating K and L in (7.2.2) is equivalent to negating $\sigma_1, \ldots, \sigma_n$, or $\sigma_1', \ldots, \sigma_n'$. This implies

$$V(-K, -L) = R V(K, L) = V(K, L) R, \tag{7.4.3}$$

and similarly for W.

Finally, let r be the number of unlike pairs of spins $(\sigma_{j+1}, \sigma_j')$, and s the number of unlike pairs (σ_j, σ_j'). Then $r + s$ is the number of changes of sign in the sequence $\sigma_1, \sigma_1', \sigma_2, \sigma_2', \ldots, \sigma_n'$. This means that $r + s$ must be even and, from (7.2.2),

$$V_{\phi,\phi'} = \exp[(n - 2r) K + (n - 2s) L] \tag{7.4.4}$$

We are interested in the thermodynamic limit, when n is large. It should not matter how n becomes large, so we can restrict n to be even. This slightly simplifies the following discussion, so from now on in this chapter let us set

$$n = 2p , \tag{7.4.5}$$

where p is an integer.

The equation (7.4.4) can now be written as

$$V_{\phi,\phi'} = \exp[\pm 2r' K \pm 2s' L] , \tag{7.4.6}$$

where r' and s' are non-negative integers in the range $(0, p)$. They are either both even or both odd, so that RHS is unchanged by negating both of $\exp(2K)$ and $\exp(2L)$. This means that the matrix $V(K, L)$ satisfies

$$V(K \pm \tfrac{1}{2} \pi i, L \pm \tfrac{1}{2} \pi i) = V(K, L) , \tag{7.4.7}$$

and similarly for W.

7.5   Commutation Relations for Transfer Matrices

The relation (7.3.5) is true if the condition (7.3.9) is satisfied. It will now be shown that this implies that $V(K, L)$, $W(K, L)$, $V(K', L')$, $W(K', L')$ all commute.

Define C to be the 2^n by 2^n matrix with elements

$$\delta(\sigma_1, \sigma_2') \, \delta(\sigma_2, \sigma_3') \ldots \delta(\sigma_n, \sigma_1') . \tag{7.5.1}$$

This operator C shifts the columns of the lattice to the left or the right: for instance, applying the transformation $A \to C^{-1}AC$ to any matrix A has the effect of replacing the spin labels $1, \ldots, n$ by $2, \ldots, n, 1$. From (7.2.2) it is obvious that this leaves $V(K, L)$ and $W(K, L)$ unchanged, so

$$V(K, L) = C^{-1} V(K, L) \, C$$
$$W(K, L) = C^{-1} W(K, L) \, C . \tag{7.5.2}$$

Also, from (7.2.2),

$$W(K, L) = V(K, L) \, C . \tag{7.5.3}$$

Thus $C, V(K, L), W(K, L)$ all commute with one another.
Now substitute (7.5.3) into (7.3.5). We obtain at once

$$V(K, L) \, V(K', L') = V(K', L') \, V(K, L) \tag{7.5.4a}$$

provided (7.3.9) is satisfied, i.e.

$$\sinh 2K \, \sinh 2L = \sinh 2K' \, \sinh 2L' . \tag{7.5.4b}$$

Thus $V(K, L)$, $V(K', L')$, and hence $W(K, L)$, $W(K', L')$, do all commute, as was asserted.

Using (7.5.3), the matrix W can be eliminated from the identity (7.3.15) to give

$$V(K, L) \, V(L + \tfrac{1}{2} i\pi, -K) \, C$$
$$= (2i \sinh 2L)^n \, I + (-2i \sinh 2K)^n \, R . \tag{7.5.5}$$

Finally, from (7.3.14) the transformation $A \to R^{-1}AR$ is equivalent to negating all spins $\sigma_1, \ldots, \sigma_n, \sigma_1', \ldots, \sigma_n'$. This leaves (7.2.2) unchanged, so

$$V(K, L) = R^{-1} V(K, L) \, R . \tag{7.5.6}$$

Thus $V(K, L)$ also commutes with R. So does $W(K, L)$.

7.6 Functional Relation for the Eigenvalues

Let

$$k = (\sinh\ 2K\ \sinh 2L)^{-1}. \qquad (7.6.1)$$

Suppose k is a given fixed real number, regard K and L as complex variables subject to the constraint (7.6.1). Then an infinite set of transfer matrices $V(K, L)$ can be generated by so varying K and L.

From (7.5.4), all such matrices commute. From (7.5.2) and (7.5.6) they also commute with C and R, and hence with $W(K, L)$. It follows that all these matrices, for all values of K and L satisfying (7.6.1), have a common set of eigenvectors.

Let x be one such eigenvector. It cannot depend on K or L, so long as (7.6.1) is satisfied. It can (and does) depend on k, so can be written as $x(k)$.

Let $v(K, L)$, c, r be the corresponding eigenvalues of $V(K, L)$, C, R. Then, for all K, L satisfying (7.6.1),

$$V(K, L)\,x(k) = v(K, L)\,x(k)$$
$$C\,x(k) = c\,x(k) \qquad (7.6.2)$$
$$R\,x(k) = r\,x(k).$$

Since $C^n = R^2 = I$, the eigenvalues c, r are unimodular constants, satisfying

$$c^n = r^2 = 1. \qquad (7.6.3)$$

Note that if K, L satisfy (7.6.1) so do the K', L' defined by (7.3.11). Now pre-multiply $x(k)$ by both sides of (7.5.5). It follows at once that

$$v(K, L)\,v(L + \tfrac{1}{2}i\pi, -K)\,c$$
$$= (2i\sinh 2L)^n + (-2i\sinh 2K)^n\,r. \qquad (7.6.4)$$

The squares of the Λ_js in Section 7.2 are the eigenvalues of $V(K, L)\,W(K, L)$. From (7.5.3) and (7.6.2), $x(k)$ is also an eigenvector of this matrix. Let $\Lambda(K, L)$ be the corresponding Λ_j. Then

$$\Lambda^2(K, L) = v^2(K, L)\,c. \qquad (7.6.5)$$

Thus $\Lambda(K, L)$ can be defined to be

$$\Lambda(K, L) = v(K, L)\,c^{\frac{1}{2}}. \qquad (7.6.6)$$

Since c, and hence $c^{\frac{1}{2}}$, are constants, the relation (7.5.4) can therefore be written in the c-independent form

$$\Lambda(K, L) \, \Lambda(L + \tfrac{1}{2} i\pi, -K)$$
$$= (2i \sinh 2L)^n + (-2i \sinh 2K)^n r . \qquad (7.6.7)$$

7.7 Eigenvalues Λ for $T = T_c$

The equation (7.6.7) is a functional relation for the function $\Lambda(K, L)$. This relation is very useful: together with some simple analytic properties of $\Lambda(K, L)$, it determines $\Lambda(K, L)$ completely. There are of course many solutions, corresponding to the different eigenvalues.

To see this, it is helpful to first consider the case $k = 1$. As was remarked in the previous chapter, and will later be shown in this chapter, this is the case when the temperature T has its critical value T_c.

Parametrization of K, L

When $k = 1$, rather than working with K and L, it is convenient to use a variable u defined by

$$\sinh 2K = \tan u , \qquad (7.7.1)$$
$$\sinh 2L = \cot u .$$

The condition (7.6.1) is then automatically satisfied. If K and L are real and positive, then u lies in the interval $(0, \tfrac{1}{2}\pi)$.

Clearly $\Lambda(K, L)$ can be thought of now as a function of u, so let us write it as $\Lambda(u)$. Then (7.6.7) becomes

$$\Lambda(u) \, \Lambda(u + \tfrac{1}{2}\pi) = (2i \cot u)^n + (-2i \tan u)^n r . \qquad (7.7.2)$$

The usefulness of working with the variable u lies in the fact that not only is (7.6.1) satisfied, but also $\exp(\pm 2K)$ and $\exp(\pm 2L)$ are 'simple' functions of u. In fact

$$\exp(2K) = (1 + \sin u)/\cos u ,$$
$$\exp(-2K) = (1 - \sin u)/\cos u , \qquad (7.7.3)$$
$$\exp(2L) = (1 + \cos u)/\sin u ,$$
$$\exp(-2L) = (1 - \cos u)/\sin u .$$

To be more precise, these functions (regarding u as a complex variable) have the following properties:

(a) They are single-valued.
(b) They are meromorphic, i.e. their only singularities are poles (in fact simple poles).
(c) They are periodic, of period 2π.

The Form of the Function $\Lambda(u)$

Substituting the forms (7.7.3) of $\exp(\pm 2K)$ and $\exp(\pm 2L)$ into (7.4.6), it is obvious that every matrix element $V_{\phi,\phi'}$ is of the form

$$V_{\phi,\phi'} = t(u)/(\sin u \, \cos u)^p , \qquad (7.7.4)$$

where $t(u)$ is a polynomial in $\sin u$ and $\cos u$, of combined degree $2p$. Thus for any particular element, $t(u)$ can be written as

$$t(u) = e^{-2ipu}(c_0 + c_1 e^{iu} + \ldots + c_{2n} e^{4ipu}) . \qquad (7.7.5)$$

Now consider the first vector equation in (7.6.2). This is really 2^n scalar equations, any one of which can be regarded as expressing the eigenvalue $v(K, L)$ as a linear combination of the elements of the matrix $V(K, L)$. The coefficients are ratios of the elements of $x(k)$. The crucial point to remember is that (for the commutativity reasons discussed in Section 7.6) *these ratios depend only on k. They are independent* of u.

Thus $v(K, L)$ is a linear combination of functions of the form (7.7.4), with constant coefficients. Clearly it also must be of this form. From (7.6.6), so must $\Lambda(K, L)$, now called $\Lambda(u)$, be of this form.

This form can be simplified by using the symmetry relations. Suppose u is replaced by $u + \pi$. From (7.7.3) this is equivalent to replacing K and L by $-K \pm \frac{1}{2}\pi i$ and $-L \pm \frac{1}{2}\pi i$. From (7.4.3) and (7.4.6), this is equivalent to multiplying V by R. Writing $v(K, L)$ as $v(u)$, the first of the equations (7.6.2) therefore becomes

$$V(K, L) R x(k) = v(u + \pi) x(k) , \qquad (7.7.6)$$

again using the u-independence of $x(k)$.

Using the first and the last of the equations (7.6.2), it follows at once that

$$v(u + \pi) = r v(u) , \qquad (7.7.7)$$

and hence, from (7.6.6),

$$\Lambda(u + \pi) = r \Lambda(u) . \qquad (7.7.8)$$

Thus $\Lambda(u)$ is of the form given by (7.7.4), and satisfies the periodicity relation (7.7.8). The polynomial in (7.7.5) therefore only has non-zero even coefficients if $r = +1$, odd ones if $r = -1$. Factoring this polynomial, the resulting expression for $\Lambda(u)$ is

$$\Lambda(u) = \rho \, (\sin u \cos u)^{-p} \prod_{j=1}^{l} \sin (u - u_j) \,, \qquad (7.7.9a)$$

where $\rho, u_1, \ldots, u_l$ are constants (as yet unknown) and

$$
\begin{aligned}
l &= 2p \qquad \text{if } r = +1 \\
&= 2p - 1 \quad \text{if } r = -1 \,.
\end{aligned}
\qquad (7.7.9b)
$$

Zeros of $\Lambda(u)$

Now substitute this expression for $\Lambda(u)$ into the relation (7.7.2). This gives, using (7.4.5)

$$\rho^2 \prod_{j=1}^{l} \sin(u - u_j) \cos(u - u_j) = 2^{2p} \left[\cos^{4p} u + r \sin^{4p} u\right] . \qquad (7.7.10)$$

This must be an identity, true for all values of u. It is most easily understood by writing it in terms of the variables

$$z = \exp(2iu) \,, \quad z_j = \exp(2iu) \,. \qquad (7.7.11)$$

Then (7.7.10) becomes

$$\rho^2 (i/4)^l \prod_{j=1}^{l} [(z^2 - z_j^2)/z_j] = 2^{-2p} z^{l-2p} \left[(z + 1)^{4p} + r(z - 1)^{4p}\right] . \qquad (7.7.12)$$

From (7.7.9b), both sides are polynomials of degree l in z^2, so the constants $\rho, z_1, \ldots, z_l$ can indeed be chosen to ensure that (7.7.12) is satisfied identically. Clearly $z_1^2, \ldots, z_l^2$ are the l distinct zeros of the RHS, which are readily found to be

$$z_j^2 = -\tan^2(\theta_j/2) \,, \qquad (7.7.13)$$

where, for $j = 1, \ldots, l$,

$$
\begin{aligned}
\theta_j &= \pi(j - \tfrac{1}{2})/2p \quad \text{if } r = 1 \\
&= \pi j/2p \qquad\quad \text{if } r = -1 \,.
\end{aligned}
\qquad (7.7.14)
$$

These θ_j all lie within the interval $(0 \,, \pi)$. Define $\phi_1, \ldots, \phi_l$ by

$$\phi_j = \tfrac{1}{2} \ln \tan(\theta_j/2) \,, \quad j = 1, \ldots, l. \qquad (7.7.15)$$

Then, from (7.7.11) and (7.7.13),

$$u_j = \mp \tfrac{1}{4}\pi - i\phi_j, \quad j = 1, \ldots, l. \tag{7.7.16}$$

There are other solutions, but they correspond to incrementing u_j by an integer multiple of π. From (7.7.9a), this leaves $\Lambda(u)$ unchanged (to within an irrelevant sign), so the only truly distinct solutions are given by (7.7.16).

Since the sign in (7.7.16) can be chosen independently for each value of j, we appear to have 2^l possible solutions. However, not quite all of these are allowed.

Suppose $u \to \pm i \infty$. Then from (7.7.3), $\exp(2K)$ and $\exp(-2L) \to \pm i$. However, from (7.4.7) the elements of the transfer matrices are unchanged by negating $\exp(2K)$ and $\exp(2L)$. Thus

$$\Lambda(i\infty) = \Lambda(-i\infty). \tag{7.7.17}$$

From (7.7.9), this condition is automatically satisfied if $r = -1$. If $r = +1$ it implies that

$$(u_1 + \ldots + u_{2p})/\pi = \text{integer} + \tfrac{1}{2}p, \tag{7.7.18}$$

so only $2p - 1$ of the signs in (7.7.16) can be chosen independently. For both $r = +1$ and $r = -1$ there are therefore 2^{2p-1} eigenvalues Λ, as expected.

Substituting the values (7.7.16) of the u_j, (7.7.9a) becomes

$$\Lambda(u) = \rho \, (\sin u \cos u)^{-p} \prod_{j=1}^{l} \sin(u + i\phi_j + \tfrac{1}{4}\gamma_j\pi), \tag{7.7.19}$$

where $\gamma_1, \ldots, \gamma_l$ have values ± 1, and if $r = +1$,

$$\gamma_1 + \ldots + \gamma_{2p} = 2p - 4 \times \text{integer}. \tag{7.7.20}$$

Clearly the constant ρ can now be evaluated (to within an irrelevant sign) by substituting the expression (7.7.19) for $\Lambda(u)$ into the identity (7.7.2). I shall not proceed further with this calculation, since it is a limiting case of that of the next section. The main point has been made: when $k = 1$ the eigenvalues of VW are determined by the commutation relations and the inversion identity (7.6.7), and can be calculated by ordinary algebra.

7.8   Eigenvalues Λ for $T < T_c$

I have presented the solution for the case $k = 1$ in some detail because the derivation can then be carried out solely in terms of elementary functions.

There were three main steps:

  (i) For the given value of k, find a parametrization of (7.6.1) so that $\exp(\pm 2K)$ and $\exp(\pm 2L)$ are single-valued meromorphic functions of a variable u.

 (ii) Note that (7.4.6) implies that every element of V is also a single-valued meromorphic function of u. From (7.6.2), so therefore is any eigenvalue $v(u)$, and hence $\Lambda(u)$.

(iii) The zeros of $\Lambda(u)$ must be contained in the zeros of the known function on the RHS of (7.6.7). They can therefore be evaluated. There will be many choices of the zeros, corresponding to different eigenvalues. The normalization of $\Lambda(u)$ can then be determined (to within a sign) from (7.6.7).

Parametrization of K, L

Can this programme be used when $T \neq T_c$, i.e. $k \neq 1$? From (7.6.1), an obvious first step is to introduce an intermediate variable x such that

$$\sinh 2K = x \tag{7.8.1}$$
$$\sinh 2L = (kx)^{-1}.$$

Solving these equations for $\exp(2K)$ and $\exp(2L)$ gives

$$\exp(2K) = x + (1 + x^2)^{\frac{1}{2}}$$
$$\exp(2L) = (kx)^{-1}[1 + (1 + k^2x^2)^{\frac{1}{2}}]. \tag{7.8.2}$$

This is a parametrization of $\exp(2K)$ and $\exp(2L)$ satisfying (7.6.1), but it is not single-valued and meromorphic, due to the presence of the square roots of $1 + x^2$ and $1 + k^2x^2$.

When $k = 1$ these can be eliminated by setting $x = \tan u$, as in Section 7.7. Then $1 + x^2$ is a perfect square, and $\exp(2K)$, $\exp(2L)$ become meromorphic functions of u.

For general values of k there is no parametrization using elementary functions that simultaneously makes $1 + x^2$ and $1 + k^2x^2$ perfect squares. However, such a parametrization can be made by using elliptic functions. In Chapter 15 the meromorphic functions sn u, cn u, dn u are defined and shown to satisfy the relations (15.4.4) and (15.4.5), i.e.

$$\mathrm{cn}^2\, u = 1 - \mathrm{sn}^2\, u$$
$$\mathrm{dn}^2\, u = 1 - k^2 \,\mathrm{sn}^2\, u. \tag{7.8.3}$$

Comparing (7.8.2) and (7.8.3), it is obvious that if we set

$$x = -i \operatorname{sn}(iu), \qquad (7.8.4)$$

then

$$\exp(\pm 2K) = \operatorname{cn} iu \mp i \operatorname{sn} iu,$$

$$\exp(\pm 2L) = ik^{-1}(\operatorname{dn} iu \pm 1)/\operatorname{sn} iu. \qquad (7.8.5)$$

In (15.1.6) the functions sn, cn, dn are expressed in terms of the theta functions H, H_1, Θ, Θ_1. From (7.8.5) and (7.8.3) it follows that

$$\exp(\pm 2K) = [k'^{\frac{1}{2}}H_1(iu) \mp iH(iu)]/[k^{\frac{1}{2}}\Theta(iu)],$$

$$\exp(\pm 2L) = i[k'^{\frac{1}{2}}\Theta_1(iu) \pm \Theta(iu)]/[k^{\frac{1}{2}}H(iu)]. \qquad (7.8.6)$$

The theta functions are entire (i.e. analytic everywhere), so (7.8.6) explicitly gives $\exp(\pm 2K)$ and $\exp(\pm 2L)$ as ratios of entire functions of u, i.e. as meromorphic functions.

These elliptic functions occur also in solving the six-vertex and eight-vertex models in the following chapters. Provided one has some knowledge of elementary complex variable theory they are not at all difficult to use: in fact they are delightfully easy. At this stage I suggest the reader looks through Chapter 15, paying particular attention to the three theorems in Sections 15.3. Once these are understood, all the various identities that follow are easily obtained.

From (7.8.1) and (7.8.4), the relation between the interaction coefficient K and the parameter u can be written

$$\operatorname{sn} iu = \sin 2iK. \qquad (7.8.7)$$

From (15.5.7) and (15.5.8), setting $\alpha = i\beta$, it follows that

$$u = \int_0^{2K} \frac{d\beta}{(1 + k^2 \sinh^2\beta)^{\frac{1}{2}}}, \qquad (7.8.8)$$

so if K and L are real and positive, then u is real, and $0 < u < I'$.

If $k = 1$, the integral (7.8.8) can be evaluated, giving the first of the equations (7.7.3). In fact, (7.8.6) reduces to (7.7.3) when $k = 1$, and most of the equations of this section then become precisely those of Section 7.7. In making such comparisons, note that if $k = 1$, then $I = \infty$, $I' = \frac{1}{2}\pi$, $\operatorname{sn} iu = i \tan u$, $H(iu) \propto i \sin u$ and $\Theta(iu) \propto \cos u$.

In Chapter 15 the elliptic functions are defined only for

$$0 < k < 1, \qquad (7.8.9)$$

so for definiteness it will be supposed in this section that this is so, i.e. that $T < T_c$. In the next section this restriction will be removed.

The Form of the Function $\Lambda(u)$

It is now quite straightforward to generalize the programme of Section 7.7, as outlined at the beginning of this section. Step (i) has been performed in equation (7.8.6). From this and (7.4.6) it is evident that every element of V is of the form

$$V_{\phi,\phi'} = \frac{\cdots}{[h(iu)]^p} , \tag{7.8.10}$$

where

$$h(u) = H(u)\,\Theta(u) \tag{7.8.11}$$

and the . . . in (7.8.10) denotes an entire function of u.

From (7.6.2) and (7.6.6), each eigenvalue Λ is a linear combination of elements of V, with coefficients that depend on k but *not* on u. Writing Λ as $\Lambda(u)$, it follows from (7.8.10) that

$$\Lambda(u) = \frac{\cdots}{[h(iu)]^p} , \tag{7.8.12}$$

where again . . . stands for an entire function.

Now consider the effect of incrementing u by $2I'$, and $-2iI$, where I, I' are the half-period magnitudes of the elliptic functions. (This unconventional notation, instead of K, K', is used to avoid confusion with the interaction coefficients.) From (15.2.5), incrementing u by $2I'$ in (7.8.5) is equivalent to replacing K, L by $-K \pm \frac{1}{2}\pi i$, $-L \pm \frac{1}{2}\pi i$. As was shown in Section 7.7, this replaces Λ by $r\Lambda$ where r ($= \pm 1$) is the corresponding eigenvalue of the spin-reversal matrix R.

Thus

$$\Lambda(u + 2I') = r\Lambda(u) . \tag{7.8.13}$$

Also from (15.2.5), incrementing u by $-2iI$ in (7.8.5) is equivalent to replacing K, L by $K \pm \frac{1}{2}\pi i$, $L \pm \frac{1}{2}\pi i$. Since $r' + s'$ in (7.4.6) is even, this leaves the matrix elements of V unchanged, so

$$\Lambda(u - 2iI) = \Lambda(u) . \tag{7.8.14}$$

Now we can use the vital theorem 15c of Section 15.3. From (7.8.13) and (7.8.14), $\Lambda(u)$ is doubly periodic, while from (7.8.12) it has $2p$ poles per period rectangle. It follows that

$$\Lambda(u) = \rho\,e^{\lambda u}\,[h(iu)]^{-p} \prod_{j=1}^{2p} H(iu - iu_j) , \tag{7.8.15}$$

where $u_1, \ldots , u_{2p}$ are the zeros of $\Lambda(u)$ within a period rectangle, ρ and λ are constants, and λ must be chosen to ensure (7.8.13) and (7.8.14).

This expression (7.8.15) is the required generalization of (7.7.9). Step (ii) is completed.

Zeros of $\Lambda(u)$

The next step is to determine the zeros $u_1, \ldots, u_{2p}$ of $\Lambda(u)$ from the identity (7.6.7).

First replace u by $u + I'$ in (7.8.5). Using (15.2.6) this is found to be equivalent to replacing K, L by $L + \frac{1}{2}\pi i, -K$. Using also (7.8.1) and (7.8.4), the identity (7.6.7) therefore becomes

$$\Lambda(u) \Lambda(u + I') = \left(\frac{-2}{k \, \text{sn} \, iu}\right)^n + (-2 \, \text{sn} \, iu)^n r. \tag{7.8.16}$$

This is the generalization of (7.7.2). Using (7.4.5), (7.8.11), (7.8.15) and (15.1.6), it becomes

$$\rho^2 \exp[\lambda(2u + I')] \prod_{j=1}^{2p} H(iu - iu_j) \, H(iu - iu_j + iI')$$

$$= (4/k)^p[\Theta^{4p}(iu) + r \, H^{4p}(iu)], \tag{7.8.17}$$

which is the generalization of (7.7.10).

The zeros of the RHS of (7.8.17) occur when

$$(k \, \text{sn}^2 \, iu)^{2p} + r = 0. \tag{7.8.18}$$

The expression on the LHS of (7.8.18) is a doubly periodic function of iu, with periods $2I$, $2iI'$. It has one pole, of order $4p$, per period rectangle, so from theorem 15b it has $4p$ zeros per period rectangle.

To locate these zeros, set

$$u = -\tfrac{1}{2}I' - i\phi. \tag{7.8.19}$$

Then, using (15.4.12), (7.8.18) becomes

$$\exp[4ip\text{Am}(\phi)] + r = 0. \tag{7.8.20}$$

Define θ_j as in (7.7.14). Then (7.8.20) will certainly be satisfied if $\phi = \phi_j$, where

$$\text{Am}(\phi_j) = \theta_j - \tfrac{1}{2}\pi, \quad j = 1, \ldots, 2p. \tag{7.8.21}$$

As is shown in Section 15.4, the function $\text{Am}(\phi)$ is real and increases monotonically from $-\tfrac{1}{2}\pi$ to $\tfrac{1}{2}\pi$ as ϕ increases from $-I$ to I. Since

$0 < \theta_j \leqslant \pi$, (7.8.21) therefore has a unique real solution, with $-I <$ $\phi_j \leqslant I$. Solutions with different values of j are distinct.

From (15.2.6), if u is a solution of (7.8.18), then so is $u + I'$. Thus (7.8.18) has $4p$ solutions

$$u = \mp \tfrac{1}{2} I' - i\,\phi_j\,, \quad j = 1, \ldots, 2p\,. \tag{7.8.22}$$

To moduli $2I'$, $2iI$ these are distinct, so we have found the $4p$ zeros of the RHS of (7.8.17). Their locations in the iu-plane are shown in Fig. 7.3.

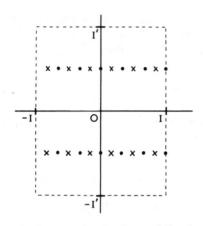

Fig. 7.3. The locations in the complex iu-plane of the $4p$ zeros of the RHS of (7.8.17), namely $iu = \phi_j \mp \tfrac{1}{2}iI'$. (Here p is 3.) The crosses are the zeros for $r = +1$; the circles are the zeros for $r = -1$. The broken line is the perimeter of the period rectangle.

The LHS of (7.8.17) has $4p$ zeros, in pairs u_j and $u_j - I'$. Thus the general solution of (7.8.17) is

$$u_j = -\tfrac{1}{2}\,\gamma_j I' - i\phi_j\,, \tag{7.8.23}$$

where

$$\gamma_j = \pm 1\,, \quad j = 1, \ldots, 2p\,. \tag{7.8.24}$$

This is the generalization of (7.7.16). (The value $j = 2p$ when $r = -1$ is excluded in Section 7.7, because when $k \to 1$, I and ϕ_{2p} then tend to infinity.)

As in Section 7.7, not all solutions of (7.8.23) are allowed. The reason for this is actually easier to see now than it was then: theorem 15c imposes the restriction (15.3.7) on the locations of the zeros of a doubly periodic function. Applied to equations (7.8.12)–(7.8.15), this restriction becomes

$$u_1 + \ldots + u_{2p} = (p + 2l')\, I' + i[\tfrac{1}{2}(1 - r) + 2l]\, I \,, \qquad (7.8.25)$$

where l and l' are integers.

If $r = +1$, the ϕ_js occur in pairs $(\phi, -\phi)$. If $r = -1$, they all do so except for $\phi_p = 0$ and $\phi_{2p} = I$. Using (7.8.23), the imaginary part of (7.8.25) is therefore satisfied, while the real part gives, for $r = \pm 1$,

$$\gamma_1 + \ldots + \gamma_{2p} = 2p - 4 \times \text{integer} \,, \qquad (7.8.26)$$

as in (7.7.20). Thus r and all but one γ_j can be chosen independently, giving $2^{2p} = 2^n$ eigenvalues. This is the expected number, since V and W are 2^n by 2^n matrices.

Substituting these results for $u_1, \ldots, u_{2p}$ into (7.8.15), λ can be chosen to ensure (7.8.13) and (7.8.14), giving

$$\Lambda(u) = \rho[h(iu)]^{-p} \prod_{j=1}^{2p} e^{-\pi \gamma_j u/4I}\, H(iu - \phi_j + \tfrac{1}{2} i\, \gamma_j\, I') \,. \qquad (7.8.27)$$

This result can be slightly simplified by squaring both sides and using the relation (15.2.4) between $H(u + iI')$ and $\Theta(u)$, giving

$$\Lambda^2(u) = \rho' \prod_{j=1}^{2p} \frac{H(iu - \phi_j + \tfrac{1}{2} i\, \gamma_j\, I')\, \Theta(iu - \phi_j - \tfrac{1}{2} i\, \gamma_j\, I')}{H(iu)\, \Theta(iu)} \,, \qquad (7.8.28)$$

where ρ' is another constant.

Using (15.1.6), this can in turn be written as

$$\Lambda^2(u) = D \prod_{j=1}^{2p} k^{\frac{1}{2}}\, \text{sn}\, (iu - \phi_j + \tfrac{1}{2} i\, \gamma_j\, I') \,, \qquad (7.8.29)$$

where D is independent of $\gamma_1, \ldots, \gamma_{2p}$, being given by

$$D = \rho' \prod_{j=1}^{2p} \frac{\Theta(iu - \phi_j + \tfrac{1}{2} iI')\, \Theta(iu - \phi_j - \tfrac{1}{2} iI')}{H(iu)\, \Theta(iu)} \,. \qquad (7.8.30)$$

From (15.2.4) and (15.2.3), D is a doubly periodic function of iu, with poles of order $2p$ at $iu = 0$ and I', and $4p$ simple zeros, at $iu = \phi_j \pm \tfrac{1}{2} iI'$, $j = 1, \ldots, 2p$. From (7.8.18)–(7.8.22), one such function is

$$\frac{(k\, \text{sn}^2 iu)^{2p} + r}{(k^{\frac{1}{2}}\, \text{sn}\, iu)^{2p}} \,. \qquad (7.8.31)$$

The ratio of D to the expression (7.8.31) is therefore an entire doubly periodic function of iu. From theorem 15a, it is therefore a constant.

To within a normalization constant, therefore, the term D in (7.8.29) can be replaced by the expression (7.8.31). The normalization constant can now easily be obtained from (7.8.16), giving

$$\Lambda^2(u) = \tau\left[\left(\frac{2}{k \operatorname{sn} iu}\right)^{2p} + r(2 \operatorname{sn} iu)^{2p}\right]$$

$$\times \prod_{j=1}^{2p} k^{\frac{1}{2}} \operatorname{sn}(iu - \phi_j + \tfrac{1}{2} i \gamma_j I'), \tag{7.8.32}$$

where

$$\begin{aligned}\tau &= +1 \quad \text{if } r = +1, \\ &= -i \quad \text{if } r = -1.\end{aligned} \tag{7.8.33}$$

7.9 General Expressions for the Eigenvalues

Having used elliptic functions, and in particular their factorisation theorem 15c, to evaluate the eigenvalues Λ, we can now eliminate them. From (7.8.21), (15.4.12), (15.4.4) and (15.4.5),

$$k^{\frac{1}{2}} \operatorname{sn}(\phi_j - \tfrac{1}{2} i I') = -\exp(i\theta_j), \tag{7.9.1}$$

$$\operatorname{cn}(\phi_j - \tfrac{1}{2} i I') \operatorname{dn}(\phi_j - \tfrac{1}{2} i I') = -i k^{\frac{1}{2}} \exp(i\theta_j)c_j, \tag{7.9.2}$$

where

$$c_j = k^{-1}(1 + k^2 - 2k \cos 2\theta_j)^{\frac{1}{2}}. \tag{7.9.3}$$

Using the addition formula (15.4.21), it follows that

$$k^{\frac{1}{2}} \operatorname{sn}(iu - \phi_j + \tfrac{1}{2} i I')$$

$$= \frac{\operatorname{cn} iu \ \operatorname{dn} iu \ - i k c_j \operatorname{sn} iu}{\exp(-i\theta_j) - k \exp(i\theta_j) \operatorname{sn}^2(iu)}. \tag{7.9.4}$$

Also, from (7.8.5),

$$\sinh 2K = -i \operatorname{sn} iu$$

$$\cosh 2K = \operatorname{cn} iu$$

$$\sinh 2L = i/(k \operatorname{sn} iu)$$

$$\cosh 2L = i \operatorname{dn} iu/(k \operatorname{sn} iu), \tag{7.9.5}$$

while from (15.2.6)

$$k^{\frac{1}{2}} \operatorname{sn}(iu - \phi_j - \tfrac{1}{2} i I') = [k^{\frac{1}{2}} \operatorname{sn}(iu - \phi_j + \tfrac{1}{2} i I')]^{-1} \tag{7.9.6}$$

Using these relations, (7.8.32) can be written (dropping the explicit dependence of Λ on u) as

$$\Lambda^2 = \tau(-4)^p[(\sinh 2L)^{2p} + r\,(\sinh 2K)^{2p}]\prod_{j=1}^{2p}(\mu_j)^{\gamma_j}, \qquad (7.9.7)$$

where

$$\mu_j = \frac{\cosh 2K\,\cosh 2L + c_j}{\exp(i\theta_j)\,\sinh 2K + \exp(-i\theta_j)\,\sinh 2L}. \qquad (7.9.8)$$

Analytic Continuation to $T \geqslant T_c$

This result has only been obtained for $k < 1$, since only then can the elliptic function definitions of Chapter 15 be used. However, for finite p each eigenvalue must be an algebraic function of $\exp(2K)$ and $\exp(2L)$, so (7.9.7) can be analytically continued to $k \geqslant 1$, i.e. to $T \geqslant T_c$.

In doing this, the only difficulty is the sign of each c_j. Provided $0 < \theta_j < \pi$ there is no problem: (7.9.3) is positive for $k < 1$ and tends to a strictly positive limit as $k \to 1$, so the analytic continuation of (7.9.3) is the positive square root.

On the other hand, if $r = -1$ and $j = 2p$, then $\theta_j = \pi$ and, for $k < 1$, (7.9.3) gives

$$c_{2p} = (1 - k)/k \quad \text{if } r = -1. \qquad (7.9.9)$$

This tends to zero as $k \to 1$, and its analytic continuation is clearly negative for $k > 1$.

Thus the formulae (7.9.7), (7.9.8) and (7.9.3) apply not only for $k < 1$, but also for $k \geqslant 1$, *provided that the positive sign is chosen in* (7.9.3) *except when $r = -1$, $j = 2p$ and $k \geqslant 1$.*

Counting of the Eigenvalues

One disadvantage of this method, as with any method that does not depend on an explicit representation of the transfer matrix, is that it only proves that any eigenvalue of VW must be of the form (7.9.7), with an appropriate choice of $\gamma_1, \ldots, \gamma_{2p}$. It does *not* tell us how many eigenvalues there are for a particular choice of $\gamma_1, \ldots, \gamma_{2p}$, if indeed there are any.

There are two ways round this problem: one can consider a low- or high-temperature limit, when at least some of the eigenvalues (notably the largest) can easily be uniquely identified, or one can compare with a direct calculation using spinor operators (Kaufman, 1949). One does in fact find that for each choice of r and $\gamma_1, \ldots, \gamma_{2p}$ satisfying (7.8.26), there is one and only one eigenvalue given by (7.9.7).

Maximum Eigenvalue and the Free Energy

In the thermodynamic limit the partition function Z is given by (7.2.5). Since the lattice has m rows of $2p$ sites, the total number N of sites is $2mp$. From (1.7.6) and (7.2.5), the free energy f per site is therefore given by

$$-f/k_BT = (2p)^{-1} \ln \Lambda_{max} ,\qquad (7.9.10)$$

where k_B is Boltzmann's constant and Λ_{max} is the eigenvalue of greatest modulus.

From (7.6.1), (7.9.3) and (7.9.8), for K and L real,

$$\mu_j\mu_j^* = \frac{\cosh 2K \cosh 2L + c_j}{\cosh 2K \cosh 2L - c_j},\qquad (7.9.11)$$

while for K and L positive,

$$0 \le c_j \le \cosh 2K \cosh 2L .\qquad (7.9.12)$$

It follows that $|\mu_j| \ge 1$, so the RHS of (7.9.7) is maximised by choosing $\gamma_1 = \ldots = \gamma_p = +1$, which is allowed by (7.8.26). The product of the denominators of $\mu_1, \ldots, \mu_{2p}$ in (7.9.7), as given by (7.9.8), can then be calculated using (7.7.14). It exactly cancels with the leading factors in (7.9.7), except 4^p, giving

$$\Lambda_{max}^2 = \prod_{j=1}^{2p} 2(\cosh 2K \cosh 2L + c_j) .\qquad (7.9.13)$$

This is true for either $r = +1$ or $r = -1$. However, from the Perron–Frobenius theorem (Gantmacher, 1959, p. 53) the maximum eigenvalue of a matrix with all positive entries corresponds to an eigenvector with all positive entries. From (7.6.2), this can only happen if $r = +1$.
Define

$$F(\theta) = \ln \{2 [\cosh 2K \cosh 2L + k^{-1}(1 + k^2 - 2k \cos 2\theta)^{\frac{1}{2}}]\} .\qquad (7.9.14)$$

Then from (7.7.14), (7.9.3) and (7.9.13), setting $r = +1$, it follows that

$$\ln \Lambda_{max} = \tfrac{1}{2} \sum_{j=1}^{2p} F[\pi(j - \tfrac{1}{2})/2p] .\qquad (7.9.15)$$

This is a sum over $\frac{1}{2}F(\theta_j)$, where $\theta_1, \ldots, \theta_{2p}$ are uniformly distributed over the interval $(0, \pi)$. In the limit of p large it therefore becomes the usual definition of the integral of $\frac{1}{2}F(\theta)$, divided by the sub-interval length $\pi/2p$. Thus (7.9.10) becomes

$$-f/k_B T = (2\pi)^{-1} \int_0^\pi F(\theta)\, d\theta. \qquad (7.9.16)$$

This is the principal result of this chapter: the free energy of the square lattice Ising model in the thermodynamic limit.

7.10   Next-Largest Eigenvalues: Interfacial Tension, Correlation Length and Magnetization for $T < T_c$

In this section it will usually be supposed that $0 < k < 1$.

Asymptotic Degeneracy and Interfacial Tension

What is the next-largest eigenvalue of the transfer matrix? Clearly one candidate is Λ_1, the eigenvalue obtained by setting $r = -1, \gamma_1 = \ldots = \gamma_{2p} = +1$. From (7.9.13), (7.9.14), (7.9.3) and (7.7.14), this is given by

$$\ln \Lambda_1 = \frac{1}{2} \sum_{j=1}^{2p} F(\pi j/2p) \quad \text{if } k < 1. \qquad (7.10.1)$$

The two sums in (7.9.15) and (7.10.1) differ only by terms that are exponentially small when p is large. To see this, Fourier analyse $F(\theta)$:

$$F(\theta) = \sum_{m=0}^{\infty} a_m \cos 2m\theta. \qquad (7.10.2)$$

Now substitute this expression for $F(\theta)$ into (7.9.15) and (7.10.1) and interchange the j and m summations. The j summation is then easily performed, giving

$$\ln \Lambda_{\max} = p(a_0 - a_{2p} + a_{4p} - a_{6p} + \ldots)$$

$$\ln \Lambda_1 = p(a_0 + a_{2p} + a_{4p} + a_{6p} + \ldots). \qquad (7.10.3)$$

This transformation is a special case of the Poisson summation formula (Courant and Hilbert, 1953, Vol. 1, p. 76). It is ideally suited to evaluating $\Lambda_{\max}$ and Λ_1 for large p, since the a_m usually tend exponentially to zero with increasing $|m|$.

To see this, set

$$z = \exp(2i\theta) \tag{7.10.4}$$

and consider $F(\theta)$, given by (7.9.14), as a function of z. It has branch points at $z = 0$, k, k^{-1} and ∞, and is analytic in the annulus $k < |z| < k^{-1}$.

The Fourier expansion (7.10.2) can be written

$$F(\theta) = \tfrac{1}{2} \sum_{m=0}^{\infty} a_m(z^m + z^{-m}), \tag{7.10.5}$$

which is plainly a Laurent series. Since F is analytic on $|z| = 1$, this series converges. More strongly, since F is analytic in the annulus $k < |z| < k^{-1}$, and singular at $z = k$, k^{-1}, the series (7.10.5) must converge for $k < |z| < k^{-1}$ and diverge when $z = k$ or k^{-1}. From the ratio test for series convergence, it follows at once that

$$a_m \sim k^m \quad \text{as } m \to \infty. \tag{7.10.6}$$

Since $\Lambda_1 < \Lambda_{max}$, it now follows from (7.10.3) that when p is large

$$\Lambda_1/\Lambda_{max} = 1 - \mathcal{O}(k^{2p}). \tag{7.10.7}$$

For $k < 1$, the two largest eigenvalues Λ_{max} and Λ_1 are therefore *asymptotically degenerate*, in that their ratio differs from unity by terms that vanish exponentially with the width of the lattice.

The rate of this exponential decay is a measure of the interfacial tension s, as can be seen by the following argument (Fisher, 1969).

Consider the quantity

$$Z'_N = \text{Trace}\,(VW)^{m/2}R, \tag{7.10.8}$$

where V, W are the row-to-row transfer matrices and R is the spin reversal operator defined by (7.3.14). As in (7.2.3), this is the partition function of a lattice of m rows, but with the anti-cyclic condition that the spins in the top row are the reverse of those in the bottom row.

In an ordered ferromagnetic state the spins are either all mostly up, or all mostly down, within a region. Suppose that near the bottom of the lattice they are mostly up. Then from the anti-cyclic boundary condition, near the top of the lattice they must be mostly down, as in Fig. 7.4.

Somewhere in between, there must be a line running across the width of the lattice separating the domains of mostly up and mostly down spins. There are n sites per row, so this line will give an extra contribution ns to the free energy, where s is the interfacial tension per unit length. Thus

$$-k_B T \ln Z'_N = Nf + ns, \tag{7.10.9}$$

where f is the usual free energy per site of the lattice, given by (1.7.6). Thus

$$Z'_N/Z_N = \exp(-ns/k_B T) \, . \qquad (7.10.10a)$$

This is not quite right: it is the correct contribution to the partition function of a separation line but there are many such lines, as can be seen by considering the zero-temperature limit. In this, all spins above the separation line must be down and spins below it must be up, and the line must be of length $2n$ lattice spaces, which is the minimum possible length.

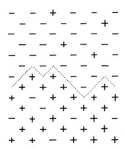

Fig. 7.4. The two domains induced at low temperatures by requiring that the spins in the top row be the reverse of those in the bottom row. Below the separation line (shown dotted), there is a 'sea' of up-spins containing 'islands' of down spins. Above the separation line the reverse is true.

All such arrangements minimize the energy under the given boundary conditions, so are equally likely. Given the face in which the left-hand end of the line lies, there are $(2n)!/(n!)^2$ such lines: this is the number of walks of equal length that one may take in a rectangular-grid city to get from 0th Street and 0th Avenue to nth Street and nth Avenue. The RHS of (7.10.10a) should therefore be multiplied by this factor, but since for n large it is effectively an exponential, it can be absorbed into the definition of s.

On the other hand, since there are m rows, there are $\tfrac{1}{2}m$ faces in which the left-hand end of the separation line may lie. Thus the RHS of (7.10.10a) should also be multiplied by this factor. This cannot be absorbed into s, and clearly persists for non-zero temperatures, so (7.10.10a) should be replaced by

$$Z'_N/Z_N = \tfrac{1}{2}m \exp(-ns/k_B T) \, . \qquad (7.10.10b)$$

From (7.2.3) and (7.10.8), it follows that

$$\tfrac{1}{2}m \exp(-ns/k_B T) = \frac{\text{Trace } (VW)^{m/2}R}{\text{Trace } (VW)^{m/2}} \, . \qquad (7.10.11)$$

Since R commutes with V and W, this result can in turn be written as

$$\tfrac{1}{2}m \exp(-ns/k_BT) = \sum_j r_j\Lambda_j^m \Big/ \sum_j \Lambda_j^m , \qquad (7.10.12)$$

where the Λ_j^2 are the eigenvalues of VW, and the r_j the corresponding eigenvalues of R.

When m is large only the two largest eigenvalues Λ_{max} and Λ_1 contribute to the sums in (7.10.12). The corresponding eigenvalues of R are $+1$ and -1, so

$$\tfrac{1}{2}m \exp(-ns/k_BT) = \frac{\Lambda_{max}^m - \Lambda_1^m}{\Lambda_{max}^m + \Lambda_1^m}. \qquad (7.10.13)$$

Set

$$\varepsilon = 1 - \Lambda_1/\Lambda_{max} , \qquad (7.10.14)$$

then (7.10.13) becomes

$$\tfrac{1}{2}m \exp(-ns/k_BT) = \frac{1 - (1 - \varepsilon)^m}{1 + (1 - \varepsilon)^m}. \qquad (7.10.15)$$

From (7.10.7), ε vanishes exponentially with n. The definition (7.10.9) of s is sensible only if m and n are large and of the same order, in which case ε is effectively small in (7.10.15), so

$$\tfrac{1}{2}m \exp(-ns/k_BT) = \tfrac{1}{2}m\varepsilon. \qquad (7.10.16)$$

From this and (7.10.14), it follows that

$$\Lambda_1/\Lambda_{max} = 1 - \mathbb{O}\{\exp(-ns/k_BT)\}. \qquad (7.10.17)$$

This is a general result, applicable to any two-dimensional ferromagnetic system with $T < T_c$. Comparing it with (7.10.7), we see that for the Ising model

$$\exp(-s/k_BT) = k. \qquad (7.10.18)$$

Thus the interfacial tension s is large and positive at low temperatures $(k \ll 1)$, decreases with increasing temperature (increasing k), and vanishes at the critical temperature $(k = 1)$.

Correlation Length

After Λ_{max} and Λ_1, what is the next-largest eigenvalue of $(VW)^{\frac{1}{2}}$? From (7.9.7) and (7.8.26), this is obtained by negating the two γ_js corresponding to the smallest $|\mu_j|$s. For $r = +1$, from (7.9.11), (7.9.3) and (7.7.14), the next-largest eigenvalues Λ_2 therefore corresponds to

$$\gamma_1 = \gamma_{2p} = -1 , \quad \gamma_2 = \ldots = \gamma_{2p-1} = +1 , \qquad (7.10.19)$$

and

$$\Lambda_2/\Lambda_{\max} = \pm(\mu_1\mu_{2p})^{-1}. \qquad (7.10.20)$$

Since $\theta_{2p} = \pi - \theta_1$, it follows from (7.9.8) that $\mu_{2p} = -\mu_1^*$, so from (7.9.11),

$$\frac{\Lambda_2}{\Lambda_{\max}} = \frac{\cosh 2K \cosh 2L - c_1}{\cosh 2K \cosh 2L + c_1}, \qquad (7.10.21)$$

choosing for convenience the lower sign in (7.10.20).

In the limit of p large, θ_1 tends to zero and c_1 to $|1 - k|/k$, so

$$\Lambda_2/\Lambda_{\max} = A, \qquad (7.10.22)$$

where

$$A = \frac{\cosh 2K \cosh 2L - |1 - k|/k}{\cosh 2K \cosh 2L + |1 - k|/k}. \qquad (7.10.23)$$

A similar argument applies for $r = -1$. If Λ_2 is now taken to be the next-largest eigenvalue for $r = -1$, then when p is large we again obtain the result (7.10.22), to within an irrelevant sign.

Thus all eigenvalues Λ_j other then $\Lambda_{\max}$ and Λ_1 satisfy the inequality (for $p \to \infty$)

$$|\Lambda_j| \le A\,\Lambda_{\max}, \qquad (7.10.24)$$

where $0 < A < 1$ provided $k \ne 1$. This result justifies the simplification of (7.10.12) to (7.10.13).

Provided all the eigenvalues Λ_j are real, the ratio $\Lambda_2/\Lambda_{\max}$ is related to the correlation length ξ. To see this, let P and Q be two sites on the lattice, and σ_P, σ_Q the corresponding spins. Then from (1.4.4) and (1.8.1) the expectation value of the product $\sigma_P\sigma_Q$ is

$$\langle\sigma_P\sigma_Q\rangle = Z_N^{-1} \sum_\sigma \sigma_P\sigma_Q \exp\left[K \sum_{(i,j)} \sigma_i\sigma_j + L \sum_{(i,k)} \sigma_i\sigma_k \right], \quad (7.10.25)$$

where the outer sum is over all values of all spins, and the inner sums have the same meanings as in (6.2.1).

Using the same argument that led to (7.2.1), (7.10.25) can be written

$$\langle\sigma_P\sigma_Q\rangle = Z_N^{-1} \sum_{\phi_1} \sum_{\phi_2} \ldots \sum_{\phi_m} \sigma_P\sigma_Q V_{\phi_1,\phi_2}$$

$$\times W_{\phi_2,\phi_3} V_{\phi_3,\phi_4} \ldots W_{\phi_m,\phi_1}. \qquad (7.10.26)$$

Let ϕ be a set of n spins $\{\sigma_1, \ldots, \sigma_n\}$ and let s_1 be the 2^n by 2^n diagonal matrix with entries

$$(s_1)_{\phi,\phi'} = 0 \quad \text{if } \phi' \ne \phi \qquad (7.10.27)$$

$$= \sigma_1 \quad \text{if } \phi' = \phi.$$

Then if P, Q are the first sites in rows x and y respectively, where x and y are odd, (7.10.26) can be written as

$$\langle \sigma_P \sigma_Q \rangle = Z_N^{-1} \text{ Trace } VW \ldots VW \, s_1 \, VW \ldots$$

$$\times VW \, s_1 \, VW \ldots VW \,, \tag{7.10.28}$$

where s_1 occurs before the xth and yth matrices V. Thus

$$\langle \sigma_P \sigma_Q \rangle = Z_N^{-1} \text{ Trace } (VW)^{\frac{1}{2}(x-1)} s_1 \, (VW)^{\frac{1}{2}(y-x)}$$

$$\times s_1 \, (VW)^{\frac{1}{2}(m-y+1)} \,. \tag{7.10.29}$$

The argument now closely parallels that of Section 2.2 for the one-dimensional Ising model. Let U be the matrix of eigenvectors of VW, and D the corresponding diagonal matrix with diagonal elements $D_{jj} = \Lambda_j$, $j = 1, 2, 3, \ldots$. Then for all integers x,

$$(VW)^{x/2} = U D^x U^{-1} \,. \tag{7.10.30}$$

Using this result, (7.10.29) can be written

$$\langle \sigma_P \sigma_Q \rangle = Z_N^{-1} \sum_i \sum_j t_{ij} \Lambda_j^{y-x} t_{ji} \Lambda_i^{m+x-y} \,, \tag{7.10.31}$$

where

$$t_{ij} = (U^{-1} s_1 U)_{ij} \,. \tag{7.10.32}$$

Now let $m \to \infty$. The i summation in (7.10.31) is then dominated by the value for which $\Lambda_i = \Lambda_{\max}$. Call this value 0. Using (7.2.5) it follows that

$$\langle \sigma_P \sigma_Q \rangle = \sum_j t_{0j} \, (\Lambda_j / \Lambda_{\max})^{y-x} t_{j0} \,. \tag{7.10.33}$$

From (7.3.14) and (7.10.27) it is apparent that

$$s_1 R = -R s_1 \,, \tag{7.10.34}$$

so whereas the transfer matrix VW commutes with R, the spin operator s_1 anti-commutes with it.

Since $R^2 = I$, there is a representation in which

$$R = \begin{pmatrix} I & 0 \\ 0 & -I \end{pmatrix} \,. \tag{7.10.35}$$

Using the commutation and anti-commutation properties mentioned above, it follows that V, W and U are all block-diagonal, i.e. of the form

$$\begin{pmatrix} \diagbox & 0 \\ 0 & \diagbox \end{pmatrix}$$

while s_1, and hence $U^{-1} s_1 U$, is of the form

It follows at once that

$$t_{ij} = 0 \quad \text{unless} \quad r_i = -r_j, \tag{7.10.36}$$

where r_i and r_j are the eigenvalues of R corresponding to Λ_i and Λ_j, respectively.

Since $\Lambda_0 = \Lambda_{max}$ corresponds to $r_0 = +1$, this implies that the t_{0j} and t_{j0} in (7.10.33) vanish unless $r_j = -1$. The summation can therefore be restricted to such values of j, giving

$$\langle \sigma_P \sigma_Q \rangle = t_{01} t_{10} (\Lambda_1 / \Lambda_{max})^{y-x} + t_{02} t_{20} (\Lambda_2 / \Lambda_{max})^{y-x} + \ldots . \tag{7.10.37}$$

In the limit $n \to \infty$, $\Lambda_1 = \Lambda_{max}$. Provided $\Lambda_2, \Lambda_3, \ldots$ are all real, it follows that for $y - x$ large

$$\langle \sigma_P \sigma_Q \rangle = t_{01} t_{10} + \mathcal{O}[(\Lambda_2 / \Lambda_{max})^{y-x}] . \tag{7.10.38}$$

The correlation function g_{PQ} is defined in (1.7.21), which can normally be written in the form

$$g_{PQ} = \langle \sigma_P \sigma_Q \rangle - \lim_{y-x \to \infty} \langle \sigma_P \sigma_Q \rangle , \tag{7.10.39}$$

i.e. g_{PQ} is the difference between $\langle \sigma_P \sigma_Q \rangle$ and its limiting value for P, Q far apart.

With this definition, (7.10.38) implies that for $y - x$ large

$$g_{PQ} \sim (\Lambda_2 / \Lambda_{max})^{y-x} . \tag{7.10.40}$$

Since $y - x$ is the distance between sites P and Q, the definition (1.7.24) of the correlation length ξ gives

$$\xi^{-1} = \ln(\Lambda_{max} / \Lambda_2) . \tag{7.10.41}$$

This is a quite general result, but it is not immediately applicable to the present problem. To see this, note that Q is vertically above P, so g_{PQ} is the vertical correlation function on the diagonal square lattice. If K and L are interchanged, this must become the same as the horizontal correlation function. However, if P and Q lie in the same row, at positions 1 and j, then repeating the argument of (7.10.26)–(7.10.33) gives

$$\langle \sigma_P \sigma_Q \rangle = (U^{-1} s_1 s_j U)_{00} . \tag{7.10.42}$$

Here U is the matrix of eigenvectors of VW, which we have seen depend on K and L only via k. Thus g_{PQ} and ξ must also depend only on k, whereas from (7.10.22, 23) this is not true of Λ_{max}/Λ_2.

The reason for this apparent contradiction is that the transfer matrix VW is not in general symmetric, so its eigenvalues are not all real. The eigenvalue Λ_2 is merely the largest of a band of complex eigenvalues, with different arguments. In the limit of n large this band becomes continuous and the $j = 2$ contribution to (7.10.33) can be cancelled by the contributions of eigenvalues arbitrarily close in modulus to Λ_2, but with different arguments. Johnson et al. (1972a, 1973) have explicitly found such a phenomenon for the eight-vertex model discussed in Chapter 10.

Fortunately in this case it is easy to retrieve the situation, since ξ can depend only on k. For a given value of k, consider the isotropic case $K = L$. From (7.4.2), the transfer matrix is then symmetric, so its eigenvalues are real and the formula (7.10.41) is valid. Using (7.6.1) and (7.10.22, 23), it follows that, for $0 < k < 1$,

$$\xi^{-1} = - \ln k . \qquad (7.10.43)$$

Comparing this result with (7.10.18), we see that the interfacial tension s and the correlation length ξ satisfy the simple exact relation

$$s\xi = k_B T . \qquad (7.10.44)$$

Spontaneous Magnetization

From (1.7.22), the magnetization M is given by

$$M = \langle \sigma_P \rangle . \qquad (7.10.45)$$

Care has to be taken in evaluating this average for $T < T_c$ and $H = 0$. For the finite zero-field system of this chapter, $\langle \sigma_P \rangle$ must be zero, since for every state in which $\sigma_P = +1$, there is an equally likely state (obtained by reversing all spins) in which $\sigma_P = -1$.

What should be done is clear from Fig. 1.1: when $H = 0$ the magnetization can take any value between M_0 and $-M_0$, where the spontaneous magnetization M_0 is defined by

$$M_0 = \lim_{H \to 0^+} \langle \sigma_P \rangle ; \qquad (7.10.46)$$

i.e. $\langle \sigma_P \rangle$ is to be evaluated for $H > 0$ in the thermodynamic limit, then H allowed to tend to zero.

Now for $H > 0$ it is certainly true that the correlation g_{PQ} defined by

(1.7.21) tends to zero as the P and Q become far apart. Since the system is translation invariant, $\langle \sigma_P \rangle = \langle \sigma_Q \rangle$, so it follows that

$$\langle \sigma_P \rangle^2 = \lim_{y-x \to \infty} \langle \sigma_P \sigma_Q \rangle . \qquad (7.10.47)$$

This can be taken as a definition of $\langle \sigma_P \rangle$ for $H > 0$. Letting $H \to 0$, it then provides a definition of M_0, so from (7.10.38), (7.10.46) and (7.10.47),

$$M_0 = (t_{01} t_{10})^{\frac{1}{2}} . \qquad (7.10.48)$$

The calculation of this quantity is quite technical and I refer the reader to the excellent book by McCoy and Wu (1973). One property, however, can readily be deduced from the above: the eigenvectors of VW, and hence the matrix U, depend on K and L only via k. From (7.10.32) therefore, so do the t_{ij}. Equation (7.10.48) therefore gives

$$M_0 = \text{function of } k \text{ only} . \qquad (7.10.49)$$

In fact, it was found by Onsager in 1949, and a proof published by Yang in 1952, that

$$M_0 = (1 - k^2)^{1/8} = k'^{1/4} . \qquad (7.10.50)$$

In view of the difficulty of the calculation, this is an amazingly simple result. It is curious that no simple way has been found to derive it. A derivation, using corner transfer matrices and applicable to the more general eight-vertex model, is given in Section 13.7.

7.11 Next-Largest Eigenvalue and Correlation Length for $T > T_c$

As in the previous section, let Λ_1 be the maximum eigenvalue for $r = -1$, but now suppose that $k > 1$.

From (7.9.9), c_{2p} is now negative. The formula (7.10.1) would still be true if $F(\pi)$ were defined by (7.9.14) with the square root negated, but it is more sensible to keep the square root positive for all θ, in which case the term $j = 2p$ in (7.10.1) must be corrected to give

$$\ln \Lambda_1 = \tfrac{1}{2} \ln A + \tfrac{1}{2} \sum_{j=1}^{2p} F(\pi j / 2p) , \qquad (7.11.1)$$

using the definition (7.10.23) of A.

The argument of equations (7.10.2)–(7.10.6) can again be used to show that for large p the summations in (7.9.15) and (7.11.1) differ by exponentially small terms. The only difference is that now the annulus of analyticity of $F(\theta)$ is $k^{-1} < |z| < k$, so $a_m \sim k^{-m}$.

Subtracting (7.9.15) from (7.11.1), it follows that for p large

$$\Lambda_1/\Lambda_{\max} = A^{\frac{1}{2}}. \qquad (7.11.2)$$

Since $A < 1$, Λ_1 is therefore no longer asymptotically degenerate with $\Lambda_{\max}$, and from (7.10.13) there is no interfacial tension.

It is still true that all other eigenvalues satisfy (7.10.24), so they are less than Λ_1. If the eigenvalues of VW are all real, then the formula (7.10.33) gives, for $y - x$ large,

$$\langle \sigma_P \sigma_Q \rangle \sim A^{\frac{1}{2}(y-x)}. \qquad (7.11.3)$$

As in the previous section, this result can only be true for $K = L$, since $\langle \sigma_P \sigma_Q \rangle$ is a function only of k. From (1.7.21) and (1.7.24) the correlation length ξ is therefore given by

$$\xi = 2/\ln k \qquad (7.11.4)$$

for all K, L such that $k > 1$.

This is small at high temperatures (k large), increases with decreasing temperature (decreasing k), and becomes infinite at the critical temperature ($k = 1$). Note that the high-temperature formula (7.11.4) differs from the low-temperature one (7.10.43) in a factor of -2. There is no spontaneous magnetization.

7.12 Critical Behaviour

From (7.6.1) and (6.2.2),

$$k = [\sinh(2J/k_BT) \sinh(2J'/k_BT)]^{-1}, \qquad (7.12.1)$$

where J and J' are the interaction energies of the Ising model in the two directions. Normally, J and J' are regarded as fixed, and the temperature T as a variable.

As T increases monotonically from 0 to ∞, so does k. Thus k is itself a measure of the temperature.

Free Energy and the Exponent α

The free energy f is given by (7.9.14) and (7.9.16). For positive k and real θ, $F(\theta)$ is an analytic function not only of θ, but also of $\cosh 2K \cosh 2L$

and k, except only when $\theta = 0$ and $k = 1$. Thus f is an analytic function of K and L, except possibly when $k = 1$.

Since the square root in (7.9.14) vanishes when $\theta = 0$ and $k = 1$, the dominant singular behaviour of f is given by expanding $F(\theta)$ in powers of this square root and retaining only the first two terms, i.e. setting

$$F(\theta) = \ln(2\cosh 2K \cosh 2L)$$
$$+ k^{-1}\operatorname{sech} 2K \operatorname{sech} 2L\,(1 + k^2 - 2k\cos 2\theta)^{\frac{1}{2}}. \quad (7.12.2)$$

Substituting this into (7.9.16), the contribution of the first term to f is analytic even at $k = 1$, so the dominant singular part f_s of f is given only by the second term. Using the relation $\cos 2\theta = 2\cos^2\theta - 1$ and changing the integration variable from θ to $\frac{1}{2}\pi - \theta$, it is found that

$$-f_s/k_B T = \frac{1 + k}{\pi k \cosh 2K \cosh 2L}\,E(k_1)\,. \quad (7.12.3)$$

where

$$k_1 = 2k^{\frac{1}{2}}/(1 + k) \quad (7.12.4)$$

and $E(k)$ is the complete elliptic integral of the second kind of modulus k:

$$E(k) = \int_0^{\pi/2} (1 - k^2\sin^2\theta)^{\frac{1}{2}}\,d\theta\,. \quad (7.12.5)$$

Near $k = 1$ this integral satisfies the approximate formula (Gradshteyn and Ryzhik, 1965, Paragraph 8.114.3)

$$E(k) \sim 1 + \tfrac{1}{4}(1 - k^2)\ln[16/(1 - k^2)]\,, \quad (7.12.6)$$

so from (7.12.3), again neglecting analytic contributions to f,

$$-f_s/k_B T = \frac{(1 + k)(1 - k)^2}{2\pi k \cosh 2K \cosh 2L}\ln\left|\frac{1 + k}{1 - k}\right|\,. \quad (7.12.7)$$

Clearly, f is in fact singular at $k = 1$.

A critical temperature can be defined either as a value of T for which f is a singular function, or one at which the spontaneous magnetization or interfacial tension vanishes, or one at which the correlation length ξ becomes infinite. By any of these criteria, it is now evident that the square lattice Ising model has one and only one critical temperature T_c, given by $k = 1$, i.e.

$$\sinh(2J/k_B T_c)\,\sinh(2J'/k_B T_c) = 1\,. \quad (7.12.8)$$

Near $T = T_c$, $k - 1$ is proportional to $T - T_c$. Thus the definition (1.1.3) can be replaced by

$$t = k - 1\,, \quad (7.12.9)$$

and (7.12.7) gives

$$f_s \propto t^2 \ln |t| . \tag{7.12.10}$$

This result can be written as

$$f_s \propto t^2 \lim_{\alpha \to 0} \frac{1 - |t|^{-\alpha}}{\alpha} . \tag{7.12.11}$$

Comparing this with the definitions (1.7.7)–(1.7.9) of u, C and the critical exponents α and α', we see that in this limiting sense

$$\alpha = \alpha' = 0 . \tag{7.12.12}$$

Other Exponents

From (7.10.18), (7.10.50), (7.10.43) and (7.11.4), near $T = T_c$ the interfacial tension s, the spontaneous magnetization M_0, and the correlation length ξ behave as

$$s \sim -t , \quad M_0 \sim (-t)^{1/8} \quad \text{as } t \to 0^- , \tag{7.12.13}$$
$$\xi \sim |t|^{-1} \quad \quad \text{as } t \to 0^\pm .$$

Comparing these results with (1.7.34), (1.1.4) and (1.7.25), we see that the corresponding exponents μ, β, ν, ν' exist and are given by

$$\mu = 1 , \quad \beta = \tfrac{1}{8} , \quad \nu = \nu' = 1 . \tag{7.12.14}$$

The scaling relations (1.2.15) and (1.2.16) are therefore satisfied.

Since the two-dimensional Ising model has only been solved in zero field $(H = 0)$, a complete test of scaling is not possible. Even so, there is a wealth of numerical results (e.g. Sykes *et al.*, 1973b; Domb, 1974; Baxter and Enting, 1979) and of mathematical theorems applicable to the model in a field. For instance, Abraham (1973) has rigorously proved that

$$\gamma = 7/4 , \tag{7.12.15}$$

in agreement with (1.2.14). There is no reason to suppose that the scaling hypothesis is not satisfied. In particular the exponent δ defined by (1.1.5) is presumably given by

$$\delta = 15 . \tag{7.12.16}$$

7.13 Parametrized Star – Triangle Relation

In the above working I have delayed introducing elliptic functions for as long as possible: until Section 7.8. There they were needed in order to express $\exp(2K)$, $\exp(2L)$ as meromorphic functions of some variable u, while satisfying (7.6.1) for k independent of u.

This equation (7.6.1) is the 'commutation condition': two transfer matrices with the same value of k, but different values of u, commute. This was established in the first part of Section 7.3, using the star – triangle relation (6.4.4), (6.4.5). In fact (7.6.1) is merely a re-interpretation of (6.4.13).

Thus it would have been perfectly natural to have introduced elliptic functions as early as Section 6.4, so as to obtain a parametrization of (6.4.13), and indeed of the full star – triangle relations (6.4.8).

Onsager (1944, pp. 135 and 144) noted that this was an obvious thing to do: in Section 6.4 the K_1, K_2, K_3, L_1, L_2, L_3 satisfy relations similar to those of hyperbolic trigonometry (Coxeter, 1947). It is well known that these can be simplified by using elliptic functions (Greenhill, 1892, Paragraph 129): Onsager calls this a 'uniformizing substitution'. The resulting identities are very simple and have analogues in other models: let us therefore not leave the Ising model without noting them.

For $j = 1, 2, 3$, let K_j, L_j in Section 6.4 be given by (7.8.5), with K, L, u replaced by K_j, L_j, u_j. Then (6.4.13) is automatically satisfied.

Substituting these expressions for $K_1, \ldots, L_3$ into (6.4.14) and (6.4.15) gives

$$R^2 = -2i/(k^2 \operatorname{sn} iu_1 \operatorname{sn} iu_2 \operatorname{sn} iu_3) , \qquad (7.13.1)$$

$$-i \operatorname{sn} iu_1 \operatorname{cn} iu_2 \operatorname{cn} iu_3 - \operatorname{cn} iu_1 \operatorname{sn} iu_2 \operatorname{sn} iu_3 = k^{-1} \operatorname{dn} iu_1 . \qquad (7.13.2)$$

From (15.2.5), the functions $\operatorname{sn} u$, $\operatorname{cn} u$, $\operatorname{dn} u$ are all strictly periodic, of periods $4I$ and $4iI'$. Comparing (7.13.2) with the formula (15.4.22), it follows that one set of solutions of (7.13.2) is

$$u_1 = (4n + 1) I' - u_2 - u_3 + 4imI , \qquad (7.13.3a)$$

for all integers m, n.

From (15.2.5), (7.13.2) is unchanged by negating u_2 and u_3, or by negating u_2 and incrementing u_3 by $2I' + 2iI$, or by interchanging u_2 and u_3. From (7.13.3a), (7.13.2) is therefore also satisfied by

$$u_1 = (4n + 1) I' + u_2 + u_3 + 4imI ,$$

$$= (4n - 1) I' + u_2 - u_3 + (4m - 2) i I , \qquad (7.13.3b)$$

$$= (4n - 1) I' + u_3 - u_2 + (4m - 2) i I .$$

The difference between the RHS and LHS of (7.13.2) is a periodic function of iu_1, with periods $4I$, $4iI'$. Within each period rectangle it has four poles, at $iu_1 = \pm iI'$, $\pm iI' + 2I$. From the theorem 15b it therefore has just four zeros per such period rectangle. These are all accounted for by (7.13.3), so this is the complete set of solutions of (7.13.2).

In addition to (6.4.15), there are two other relations obtained by permuting the suffixes 1, 2, 3, i.e. permuting u_1, u_2, u_3 in (7.13.2). The solution (7.13.3a) is unchanged by this, but those in (7.13.3b) are not. The correct solution is therefore (7.13.3a).

From (7.8.5), incrementing u_1 by $4I'$ or $4iI$ leaves K_1, L_1 unchanged. Without loss of generality one can therefore take the solution of (7.13.2) to be

$$u_1 + u_2 + u_3 = I' . \tag{7.13.4}$$

The relations (6.4.8) are now satisfied.

Operator Form

Now let us look at the operator form (6.4.25) of the star–triangle relation. The operators U_i are functions of K, L, and, from (6.4.13), every operator has the same value k^{-1} of $\sinh K \ \sinh L$. Regarding k as constant, they are therefore functions of the single variable u in (7.8.5).

The middle operator has arguments L_2, K_2, rather than K_2, L_2. From (15.2.6) and (15.2.5), interchanging K and L is equivalent to replacing u by $I' - u$. Thus u_2 should be replaced by $I' - u_2$, which from (7.13.4) is $u_1 + u_3$. Writing U_i as a function of u, rather than K and L, the equation (6.4.25) therefore takes the simple form

$$U_{i+1}(u_1) \, U_i(u_1 + u_3) \, U_{i+1}(u_3)$$
$$= U_i(u_3) \, U_{i+1}(u_1 + u_3) \, U_i(u_1) . \tag{7.13.5}$$

This is an operator identity, true for $i = 1, \ldots, 2N - 2$ and all complex numbers u_1, u_3. In particular, it is true when u_1, u_3 take their 'physical' values $0 < u_1 < I'$, $0 < u_3 < I'$, corresponding to K_1, L_1, K_3, L_3 being real.

7.14   The Dimer Problem

Before moving on to the next chapter, it is appropriate to mention the planar dimer problem. This is because its solution by Kasteleyn (1961) and by Temperley and Fisher (1961) was the next major advance in exact statistical mechanics after Onsager's solution of the Ising model; and because the zero-field Ising model partition function can itself be expressed as a dimer problem.

A 'dimer' is an object that occupies two adjacent lattice sites, e.g. a dumb-bell shaped molecule. The 'dimer problem' is to determine the number of ways of covering a given lattice with dimers, so that all sites are occupied and no two dimers overlap. If there are N sites, then N must be even and there must be $N/2$ dimers.

A simple illustration is to ask the number of ways of covering a chessboard with dominoes, each domino filling two squares. Fisher (1961) used his result to work this number out: it is $12\,988\,816$.

For any lattice, the number of dimer coverings is clearly

$$Z = (m!\,2^m)^{-1} \sum_P b(p_1, p_2)\, b(p_3, p_4)\, b(p_5, p_6) \ldots b(p_{N-1}, p_N)\,, \quad (7.14.1)$$

where $m = \tfrac{1}{2}N$, the sum is over all permutations $P = \{p_1, \ldots, p_N\}$ of the integers $1, \ldots, N$, and

$$b(i, j) = 1 \quad \text{if sites } i \text{ and } j \text{ are adjacent}\,, \qquad (7.14.2)$$
$$= 0 \quad \text{otherwise}\,.$$

This expression counts the number of ways of grouping the N sites in m nearest-neighbour pairs, which is the same thing as covering them with dimers. The factor $1/m!$ allows for the fact that no distinction is made between pairs, and the factor 2^{-m} is because no distinction is made between a pair (i, j) and a pair (j, i).

Unfortunately there is in general no easy way to calculate the sum in (7.14.1).

However, what one can make progress with is the expression

$$Pf(A) = (m!\,2^m)^{-1} \sum_P \varepsilon_p\, a(p_1, p_2)\, a(p_3, p_4) \ldots a(p_{N-1}, p_N)\,, \quad (7.14.3)$$

where

$$a(i, j) = -a(j, i)\,, \qquad (7.14.4)$$

and ε_p is the signature of the permutation P, being $+1$ for even permutations and -1 for odd ones.

If A is the N by N matrix with elements $a(i, j)$ (i.e. a_{ij}), then (7.14.3) is known as the 'Pfaffian' of A (Muir, 1882). It is simply the square root of the determinant of A:

$$Pf(A) = (\det A)^{\frac{1}{2}}\,, \qquad (7.14.5)$$

and determinants are comparitively simple to calculate, mainly because the determinant of a product of matrices is the product of the determinants.

Kasteleyn (1961) and Temperley and Fisher (1961), therefore asked the question: can (7.14.1) be put into the form (7.14.3) by a judicious choice of the signs of the $a(i, j)$? In general the answer is no, but for any planar lattice (i.e. ones with no crossing edges) it turns out to be yes. Further, for a regular lattice the resulting matrix A is effectively cyclic, so its determinant can be calculated.

Following this solution of the planar dimer problem, Kasteleyn (1963) showed that the square-lattice zero-field Ising model partition function can be expressed as a dimer problem on a decorated lattice, and was therefore able to re-derive Onsager's solution. As was mentioned in Section 7.1, this Pfaffian method has proved very useful for calculating Ising model properties (Montroll *et al.*, 1963; McCoy and Wu, 1973; Thompson, 1972).

8

ICE-TYPE MODELS

8.1 Introduction

Following the solution of the Ising model and the dimer problem, the next class of statistical mechanical models to prove tractable was that of the 'ice-type' models, which was solved by Lieb (1967a, b, c) for three archetypal cases, then more generally by Sutherland (1967).

There exist in nature a number of crystals with hydrogen bonding. The most familiar example is ice, where the oxygen atoms form a lattice of coordination number four, and between each adjacent pair of atoms is an hydrogen ion. Each ion is located near one or other end of the bond in which it lies. Slater (1941) proposed (on the basis of local electric neutrality) that the ions should satisfy the *ice rule*:

Of the four ions surrounding each atom, two are close to it, and two are removed from it, on their respective bonds.

This means that the partition function is given by (1.4.1), i.e.

$$Z = \sum \exp(-\mathscr{E}/k_B T) , \qquad (8.1.1)$$

where the sum is now over all arrangements of the hydrogen ions that are allowed by the ice rule, and $\mathscr{E}$ is the energy of the arrangement.

For ice itself, $\mathscr{E}$ is the same for all allowed arrangements. With a suitable choice of the zero of the energy scale, $\mathscr{E}$ can therefore be taken to be zero. Z then becomes simply the number of allowed arrangements, and the residual entropy is

$$S = k_B \ln Z. \qquad (8.1.2)$$

This is non-zero, since there are many arrangements allowed by the ice rule. One of them is shown in Fig. 8.1(a) for the square lattice.

127

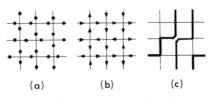

(a)   (b)   (c)

Fig. 8.1. An arrangement of hydrogen ions on a 3 by 3 square lattice (with cyclic boundary conditions), satisfying the ice rule: (a) the positions of the hydrogen ions on the bonds, (b) the corresponding electric dipoles, (c) the corresponding line representation.

Of course real ice, and other crystals, are three-dimensional, but unfortunately the only exact solutions we have for three-dimensional ice-type models are for very special 'frozen' states (Nagle, 1969b).

In this chapter only ice-type models on the square lattice will be considered. They exhibit similar behaviour to three-dimensional reality, and have the enormous advantage of being solvable! (In particular, square ice is really quite a good approximation to real ice, since the residual entropy is only weakly sensitive to the structure of the lattice.)

The hydrogen-ion bonds between atoms form electric dipoles, so can conveniently be represented by arrows placed on the bonds pointing toward the end occupied by the ion, as in Fig. 8.1(b). The ice rule is then equivalent to stating that at each site (or vertex) of the lattice there are two arrows in, and two arrows out. There are just six such ways of arranging the arrows, as shown in Fig. 8.2. (For this reason the ice-type models are sometimes known as 'six-vertex' models, as opposed to the 'eight-vertex' model of Chapter 10.)

In general, each of these six local arrangements will have a distinct energy: let us call them $\varepsilon_1, \ldots, \varepsilon_6$, using the ordering of Fig. 8.2. Then the partition function is given by (8.1.1), where

$$\mathscr{E} = n_1\varepsilon_1 + n_2\varepsilon_2 + \ldots + n_6\varepsilon_6 \qquad (8.1.3)$$

and n_j is the number of vertices in the lattice of type j.

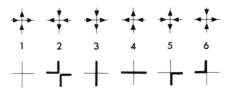

Fig. 8.2. The six arrow configurations allowed at a vertex, and the corresponding line configurations.

We now have a very general model that includes three important models as special cases.

Ice

As was remarked above, the ice model is obtained by taking all energies to be zero, i.e.

$$\varepsilon_1 = \varepsilon_2 = \ldots = \varepsilon_6 = 0 \,. \tag{8.1.4}$$

KDP

Potassium dihydrogen phosphate, KH_2PO_4 (referred to hereafter as KDP), forms a hydrogen-bonded crystal of coordination number four, and orders ferroelectrically at low temperatures (i.e. all dipoles tend to point in the same general direction). Slater (1941) argued that it could be represented by an ice-type model with an appropriate choice of $\varepsilon_1, \ldots, \varepsilon_6$. For the square lattice such a choice is

$$\varepsilon_1 = \varepsilon_2 = 0, \varepsilon_3 = \varepsilon_4 = \varepsilon_5 = \varepsilon_6 > 0 \,. \tag{8.1.5}$$

The ground state is then either the one with all arrows pointing up and to the right, or all pointing down and to the left. Either state is typical of an ordered ferroelectric.

F Model

Rys (1963) suggested that a model of anti-ferroelectrics could be obtained by choosing

$$\varepsilon_1 = \varepsilon_2 = \varepsilon_3 = \varepsilon_4 > 0, \varepsilon_5 = \varepsilon_6 = 0 \,. \tag{8.1.6}$$

The ground state is then one in which only vertex arrangements 5 and 6 occur. There are only two ways of doing this. One is shown in Fig. 8.3,

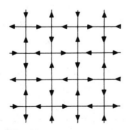

Fig. 8.3. One of the two ground-state energy configurations of the anti-ferroelectric ice-type model. Only vertex configurations 5 and 6 occur.

and the other is obtained by reversing all arrows. Note that arrows alternate in direction, as would be expected in an ordered antiferroelectric (Nagle, 1969a).

Restrictions

In this chapter the following restrictions will be imposed on $\varepsilon_1, \ldots, \varepsilon_6$:

$$\varepsilon_1 = \varepsilon_2, \quad \varepsilon_3 = \varepsilon_4, \quad \varepsilon_5 = \varepsilon_6 . \tag{8.1.7}$$

These ensure that the model is unchanged by reversing all dipole arrows, which one would expect to be the situation for a model in zero external electric field. Thus this is a 'zero-field' model which includes the ice, KDP and F models as special cases.

In fact the third condition $\varepsilon_5 = \varepsilon_6$ is no restriction at all. From Fig. 8.2 it is obvious that vertex arrangement 5 is a 'sink' of horizontal arrows, whereas 6 is a 'source'. If cylindrical or toroidal boundary conditions are imposed, then there must be as many sinks as sources, so $n_5 = n_6$. From (8.1.3) it follows that ε_5 and ε_6 only enter the partition function in the combination $\varepsilon_5 + \varepsilon_6$, so there is no loss of generality in choosing $\varepsilon_5 = \varepsilon_6$.

The other two conditions ($\varepsilon_1 = \varepsilon_2$ and $\varepsilon_3 = \varepsilon_4$) are more ones of convenience than necessity, since the working of Sections 8.2–8.7 can easily be generalized to the unrestricted case (so long as each of the six energies, e.g. ε_1, is the same for all sites of the square lattice). The effect of relaxing them (i.e. introducing electric fields) will be discussed in Section 8.12.

8.2 The Transfer Matrix

Yet another way of representing the hydrogen-ion dipoles is to draw a line on an edge if the corresponding arrow points down or to the left, otherwise to leave the edge empty. A typical arrangement of lines is shown in Fig. 8.1(c), and the six allowed line arrangements at a vertex are shown in Fig. 8.2.

Suppose the lattice has M rows and N columns, and impose cyclic (i.e. toroidal) boundary conditions. Consider a row of N vertical edges (between two adjacent rows of sites). There are M such rows: label them $r = 1, 2,$ $\ldots, M$ sequentially upwards. Let φ_r denote the 'state' of row r: i.e. the arrangement of lines on the N vertical edges. Since each edge may or may not be occupied by a line, φ_r has 2^N possible values. Then as usual we can write the partition function as

$$Z = \sum_{\varphi_1} \sum_{\varphi_2} \ldots \sum_{\varphi_M} V(\varphi_1, \varphi_2) \, V(\varphi_2, \varphi_3) \ldots V(\varphi_{M-1}, \varphi_M) \, V(\varphi_M, \varphi_1)$$

$$= \text{Trace } V^M, \tag{8.2.1}$$

where V is the 2^N by 2^N transfer matrix, with elements

$$V(\varphi, \varphi') = \sum \exp[-(m_1\varepsilon_1 + m_2\varepsilon_2 + \ldots + m_6\varepsilon_6)/k_BT]. \tag{8.2.2}$$

In (8.2.2), φ is the arrangement of lines on one row of vertical edges, and φ' is the arrangement on the row above, as in Fig. 8.4. The summation is over all allowed arrangements of lines on the intervening horizontal edges.

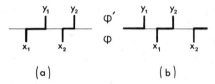

Fig. 8.4. The two typical arrangements of lines in adjacent rows (for $n = 2$). The $y_1, \ldots, y_n$ must interlace $x_1, \ldots, x_n$.

These arrangements must satisfy the ice rule at each vertex: if there is no such arrangement, then $V(\varphi, \varphi')$ is zero. There are at most two such arrangements. The $m_1, \ldots, m_6$ are the numbers of intervening vertices of types $1, \ldots, 6$.

Let Λ be an eigenvalue of V, and g the corresponding eigenvector. Then

$$\Lambda g = Vg. \tag{8.2.3}$$

As in Sections 2.1 and 7.2, when M is large it follows from (8.2.3) that

$$Z \sim \Lambda_{\text{max}}^M, \tag{8.2.4}$$

where Λ_{max} is the largest of the 2^N eigenvalues of V.

8.3 Line-Conservation

In Fig. 8.2 and 8.1(c) the four lines in vertex arrangement 2 are divided into two pairs. This makes it clear that the lines link together to form continuous non-crossing paths through the lattice. If one starts by following a path upwards, or to the right, then one will always be travelling in one or other of these two directions, never down or to the left. The cyclic boundary conditions ensure that a path never ends.

Suppose there are n such paths from the bottom of the lattice to the top. Each path will go through a row of vertical edges once and only once. It follows that:

> if there are n lines on the bottom row of vertical edges, then there are n lines on every row.

In particular, there must be n lines on the second row, which means that $V(\varphi, \varphi')$ is zero unless φ and φ' contain the same number of lines.

The matrix V therefore breaks up into $N + 1$ diagonal blocks, one between the state with no lines, another between states with one line, and so on up to the state with N lines. Thus n, the number of lines per row, is a 'quantum number' of the matrix V. We can restrict our attention to states with a given value of n.

The obvious way to identify such a state is to specify the positions $x_1, \ldots, x_n$ of the lines, ordered so that

$$1 \leqslant x_1 < x_2 < \ldots < x_n \leqslant N. \tag{8.3.1}$$

Let $X = \{x_1, \ldots, x_n\}$ be such a specification, and let $g(X)$ be the corresponding element of the eigenvector g. Then (8.2.3) can be written

$$\Lambda g(X) = \sum_Y V(X, Y) g(Y), \tag{8.3.2}$$

where $V(X, Y)$ is the element of V between states X and Y, and is still given by (8.2.2). Using (8.1.7), it is convenient to set

$$\omega_j = \exp(-\varepsilon_j/k_B T), \quad j = 1, \ldots, 6, \tag{8.3.3a}$$

$$a = \omega_1 = \omega_2, \quad b = \omega_3 = \omega_4, \quad c = \omega_5 = \omega_6. \tag{8.3.3b}$$

(Thus $\omega_1, \ldots, \omega_6$ are the Boltzmann weights of vertex arrangements $1, \ldots, 6$.) Then (8.2.2) becomes

$$V(X, Y) = \sum a^{m_1 + m_2} b^{m_3 + m_4} c^{m_5 + m_6}, \tag{8.3.4}$$

where X, Y replace ϕ, ϕ'; again the summation is over the allowed arrangements of lines on the intervening row of horizontal edges; and $m_1, \ldots, m_6$ are the numbers of intervening vertices of types $1, \ldots, 6$. Two typical cases are shown in Fig. 8.4 (with $n = 2$).

The problem now is to solve the eigenvalue equation (8.3.1) for a given value of n. It is very helpful to begin by considering the simple cases $n = 0, 1$ and 2.

The Case $n = 0$

If $n = 0$, then there are no vertical lines in the two successive rows. There are two possible arrangements of lines on the intervening horizontal row of edges: either all the edges are empty, or they all contain a line. In the first instance, all vertices are of type 1; in the second they are all of type 4. Thus the $n = 0$ block of V is a one-by-one matrix, with value

$$\Lambda = a^N + b^N. \tag{8.3.5}$$

The Case $n = 1$

If $n = 1$, we can write $g(X)$ as $g(x)$, where x is the position of the vertical line in the row. This x can take the values $1, \ldots, N$, so this block of V is an N by N matrix, with elements $V(x, y)$.

If x is less than y, then all horizontal edges between x and y must contain a line, all others must be empty (as in the first half of Fig. 8.4a). If it is greater than y, the reverse is true. If $x = y$, then either all horizontal edges are empty, or all are full. Counting $m_1, \ldots, m_6$ for the various cases, the equation (8.3.2) becomes

$$\Lambda g(x) = a^{N-1} b \, g(x) + \sum_{y=x+1}^{N} a^{N+x-y-1} b^{y-x-1} c^2 g(y)$$

$$+ a \, b^{N-1} g(x) + \sum_{y=1}^{x-1} a^{x-y-1} b^{N+y-x-1} c^2 g(y). \tag{8.3.6}$$

We look for a solution of the form

$$g(x) = z^x, \tag{8.3.7}$$

where z is a complex number. Substituting this form for $g(x)$ into (8.3.6), and summing some elementary geometric series, the equation becomes

$$\Lambda z^x = a^N L(z) \, z^x - a^{x-1} b^{N-x} c^2 z^{N+1}/(a - bz)$$

$$+ b^N M(z) \, z^x + a^{x-1} b^{N-x} c^2 z/(a - bz), \tag{8.3.8}$$

where

$$L(z) = [ab + (c^2 - b^2)z]/(a^2 - abz), \tag{8.3.9}$$

$$M(z) = [a^2 - c^2 - abz]/(ab - b^2 z).$$

The second and fourth terms on the RHS of (8.3.8) are 'boundary terms', coming from the $y = N$ and $y = 1$ summation limits in (8.3.6), respectively.

They differ only by a factor $(-z^N)$, so their sum can be made to cancel by choosing

$$z^N = 1 .\qquad (8.3.10)$$

The remaining first and third terms on the RHS are 'wanted terms', in that they have the same form as the LHS (constant $\times z^x$). Thus (8.3.8) is now satisfied if

$$\Lambda = a^N L(z) + b^N M(z) . \qquad (8.3.11)$$

There are N solutions of (8.3.10) for the complex number z. With (8.3.7), these give the N expected eigenvectors of this block of the matrix V. The corresponding eigenvalues are given by (8.3.11).

The equations (8.3.7) and (8.3.10) could have been predicted on translation invariance grounds, but it turns out to be a mistake in this problem to introduce this consideration too early: for $n > 1$ it obscures the structure of $g(X)$.

The Case $n = 2$

When $n = 2$, $g(X)$ becomes $g(x_1 , x_2)$, where x_1 and x_2 are the positions of the two lines. The summation in (8.3.1) is over all allowed line-positions y_1 and y_2 in the upper row of vertical edges, given that there are lines in positions x_1 and x_2 in the lower row.

The two archetypal cases are shown in Fig. 8.4. There are special cases when y_1 or y_2 equals x_1 or x_2, but the ice rule ensures that y_1 and y_2 must always satisfy either

$$x_1 \leqslant y_1 \leqslant x_2 \leqslant y_2 \qquad \text{or} \qquad y_1 \leqslant x_1 \leqslant y_2 \leqslant x_2 .$$

Thus y_1 and y_2 must interlace x_1 and x_2.

Counting $m_1 , \ldots , m_6$, allowing for the special cases, (8.3.2) becomes

$$\Lambda g(x_1 , x_2) = \sum_{y_1 = x_1}^{x_2} \sum_{y_2 = x_2}^{N}{}^{*} a^{x_1 - 1} E(x_1 , y_1) D(y_1 , x_2) E(x_2 , y_2)$$

$$\times c\, a^{n - y_2} g(y_1 , y_2) + \sum_{y_1 = 1}^{x_1} \sum_{y_2 = x_1}^{x_2}{}^{*} b^{y_1 - 1} D(y_1 , x_1)$$

$$\times E(x_1 , y_2) D(y_2 , x_2)\, c\, b^{N - x_2} g(y_1 , y_2) , \qquad (8.3.12)$$

where

$$D(y,x) = a/c \qquad \text{if } x = y,$$
$$= c\,a^{x-y-1} \qquad \text{if } x > y, \qquad (8.3.13)$$
$$E(x,y) = b/c \qquad \text{if } y = x,$$
$$= c\,b^{y-x-1} \qquad \text{if } y > x.$$

The * in the summations means that any terms with $y_1 = y_2$ are to be excluded. In each case there is only one such term: $y_1 = x_2 = y_2$ in the first sum, $y_1 = x_1 = y_2$ in the second.

The first step in solving (8.3.12) is an obvious generalisation of (8.3.7): try

$$g(x_1, x_2) = A_{12}\, z_1^{x_1} z_2^{x_2}, \qquad (8.3.14)$$

where A_{12}, z_1, z_2 are complex numbers.

The summations in (8.3.12) are now straightforward, if rather tedious, to perform. The easiest way is to first ignore the *, then subtract off the contribution of the terms spuriously included. The first double summation in (8.3.12) then gives

$$A_{12}\{a^{x_1}(L_1 a^{x_2-x_1} z_1^{x_1} + M_1 b^{x_2-x_1} z_1^{x_2})$$
$$\times (L_2 a^{N-x_2} z_2^{x_2} - \rho_2 b^{N-x_2} z_2^{N})$$
$$- a^{N+x_1-x_2} b^{x_2-x_1} (z_1 z_2)^{x_2}\}, \qquad (8.3.15a)$$

and the second gives

$$A_{12}\{(\rho_1 a^{x_1} + M_1 b^{x_1} z_1^{x_1})$$
$$\times (L_2 a^{x_2-x_1} z_2^{x_1} + M_2 b^{x_2-x_1} z_2^{x_2}) b^{N-x_2}$$
$$- a^{x_2-x_1} b^{N+x_1-x_2} (z_1 z_2)^{x_1}\}. \qquad (8.3.15b)$$

Here $L_j \equiv L(z_j)$, $M_j \equiv M(z_j)$, and

$$\rho_j \equiv \rho(z_j) = c^2 z_j/(a^2 - ab z_j). \qquad (8.3.16)$$

Expanding the products in (8.3.15a), or (8.3.15b), gives a total of five terms. These can be grouped into three classes.

Wanted terms

These are terms that have the same form as $g(x_1, x_2)$ itself, i.e. they are proportional to $z_1^{x_1} z_2^{x_2}$. There is one each in (8.3.15a) and (8.3.15b), and their sum is

$$A_{12}\,(a^N L_1 L_2 + b^N M_1 M_2)\, z_1^{x_1} z_2^{x_2}. \qquad (8.3.17)$$

Using (8.3.14), they cancel with the LHS of (8.3.12) if

$$\Lambda = a^N L_1 L_2 + b^N M_1 M_2 . \qquad (8.3.18)$$

Unwanted internal terms

These have the form $(z_1 z_2)^{x_2}$, or $(z_1 z_2)^{x_1}$, and include the final correction terms in (8.3.15a and b). Their sum in (8.3.15a) is

$$A_{12}\, a^{N+x_1-x_2}\, b^{x_2-x_1}\, (M_1 L_2 - 1)\, (z_1 z_2)^{x_2} , \qquad (8.3.19a)$$

and in (8.3.15b) it is

$$A_{12}\, a^{x_2-x_1}\, b^{N+x_1-x_2}\, (M_1 L_2 - 1)\, (z_1 z_2)^{x_1} . \qquad (8.3.19b)$$

Using (8.3.9), one can verify that

$$M_1 L_2 - 1 = -c^2 s_{12}/[(a - b z_1)\, (a - b z_2)] , \qquad (8.3.20)$$

where if

$$\Delta = (a^2 + b^2 - c^2)/2ab , \qquad (8.3.21)$$

then

$$s_{12} = 1 - 2\Delta\, z_2 + z_1 z_2 . \qquad (8.3.22)$$

Boundary terms

These come from either the $y_2 = N$ or the $y_1 = 1$ summation limits in (8.3.12), and are characterized in (8.3.15) by the fact that they contain a factor ρ_2 or ρ_1. Define

$$R_j(x , x') = L_j a^{x'-x} z_j^x + M_j b^{x'-x} z_j^{x'} . \qquad (8.3.23)$$

Then the sum of the boundary terms in (8.3.15a) is

$$- A_{12} a^{x_1} b^{N-x_2} R_1(x_1 , x_2)\, \rho_2 z_2^N , \qquad (8.3.24a)$$

and in (8.3.15b) it is

$$A_{12} a^{x_1} b^{N-x_2} R_2(x_1 , x_2)\, \rho_1 . \qquad (8.3.24b)$$

Elimination of unwanted terms

To satisfy (8.3.12), the unwanted terms (8.3.19) and (8.3.24) must be eliminated. How can this be done? A fairly obvious idea is to generalize the ansatz (8.3.14) and to try a linear superposition of such terms, i.e.

$$g(x_1, x_2) = \sum_r A_{12}^{(r)} z_{1,r}^{x_1} z_{2,r}^{x_2}. \tag{8.3.25}$$

Put another way, we try summing over various choices of z_1 and z_2, with appropriate coefficients A_{12}.

The wanted terms will certainly cancel if they do so for each value of r, i.e. if (8.3.18) is satisfied for every choice of z_1 and z_2. Since Λ is independent of r, the RHS of (8.3.18) must therefore be the same for all choices of z_1 and z_2.

Also, it may be possible to cancel the unwanted internal terms (8.3.19) if for every choice of z_1 and z_2 there is another choice z_1' and z_2' with the same value of $z_1 z_2$. Together with the previous remark, this means that z_1' and z_2' must satisfy

$$z_1' z_2' = z_1 z_2$$

$$a^N L(z_1') L(z_2') + b^N M(z_1') M(z_2')$$
$$= a^N L(z_1) L(z_2) + b^N M(z_1) M(z_2). \tag{8.3.26}$$

Eliminating z_2' and using (8.3.9), this gives a quadratic equation for z_1'. There are therefore just two solutions for z_1' and z_2', and it is obvious that they are:

$$z_1' = z_1, \quad z_2' = z_2 \quad \text{and} \quad z_1' = z_2, \quad z_2' = z_1, \tag{8.3.27}$$

since interchanging z_1' and z_2' leaves (8.3.26) unchanged.

(For more complicated problems, notably staggered ice-type models with different weights on the two sub-lattices, there are additional solutions for z_1' and z_2'. Regarding (8.3.14) as a 'plane wave' trial function, these z_1' and z_2' can be regarded as 'scattered waves', the two equations (8.3.26) playing the role of total momentum and total energy conservation. For $n = 2$, such problems can be solved by using these scattered waves, but unfortunately the working does not then generalize in any apparently useful way to $n > 2$.)

From (8.3.27), it follows that in addition to the choice (z_1, z_2), we should also include the choice (z_2, z_1). Thus there are just two terms in (8.3.25), and the resulting ansatz for $g(x_1, x_2)$ can be written

$$g(x_1, x_2) = A_{12} z_1^{x_1} z_2^{x_2} + A_{21} z_2^{x_1} z_1^{x_2}. \tag{8.3.28}$$

Summing (8.3.19) over these two choices (the second choice is obtained by interchanging the suffixes 1 and 2, except in x_1 and x_2), it is obvious that the unwanted terms cancel if

$$(M_1 L_2 - 1) A_{12} + (M_2 L_1 - 1) A_{21} = 0. \tag{8.3.29}$$

Using (8.3.20), this condition simplifies to

$$s_{12}A_{12} + s_{21}A_{21} = 0 .$$ (8.3.30)

Finally, summing the boundary terms (8.3.24a) and (8.3.24b) together over the two choices, we obtain

$$a^{x_1} b^{N-x_2} \{\rho_2 R_1(x_1, x_2) (A_{21} - z_2^N A_{12})$$

$$+ \rho_1 R_2(x_1, x_2) (A_{12} - z_1^N A_{21})\}.$$ (8.3.31)

Clearly this will vanish for all x_1 and x_2 iff

$$z_1^N = A_{12}/A_{21} , \quad z_2^N = A_{21}/A_{12} .$$ (8.3.32)

Solving (8.3.30) for A_{12}/A_{21}, (8.3.32) then gives two equations for z_1 and z_2. These can in principle be solved (there are many solutions, corresponding to the different eigenvectors of V). To within a normalization constant, the elements of the eigenvector g are then given by (8.3.28), and the eigenvalue Λ by (8.3.18).

8.4   Eigenvalues for Arbitrary n

The solution of the eigenvalue problem for arbitrary n is a straightforward generalization of that for $n = 2$. The appropriate generalizations will be briefly indicated in this section. A fuller description (for the ice model) is given by Lieb (1967a).

The eigenvalue equation (8.3.12) becomes

$$\Lambda g(x_1, \ldots, x_n) = \sum_{y_1=x_1}^{x_2} \ldots \sum_{y_{n-1}=x_{n-1}}^{x_n} \sum_{y_n=x_n}^{N} {}^* a^{x_1-1}$$

$$E_{11}D_{12}E_{22}D_{23} \ldots E_{nn} c \, a^{N-y_n} g(y_1, \ldots, y_n)$$

$$+ \sum_{y_1=1}^{x_1} \sum_{y_2=x_1}^{x_2} \ldots \sum_{y_n=x_{n-1}}^{x_n} {}^* b^{y_1-1} D_{11}E_{12}D_{22}E_{23} \ldots$$

$$D_{nn} c \, b^{N-x_n} g(y_1, \ldots, y_n) ,$$ (8.4.1)

where $1 \leqslant x_1 < x_2 < \ldots < x_n \leqslant N$, the $*$ means that no two of $y_1, \ldots, y_n$ can be equal, and $D_{ij} = D(y_i , x_j)$, $E_{ij} = E(x_i , y_j)$.

One first tries taking

$$g(x_1, \ldots, x_n) = A_{1 \ldots n} z_1^{x_1} \ldots z_n^{x_n} ,$$ (8.4.2)

where $A_{1 \ldots n}$ is a constant coefficient.

The first n-fold summation on the RHS of (8.4.1) then gives

$$A_{1...n}\{a^{x_1}R_1(x_1, x_2) ... R_{n-1}(x_{n-1}, x_n)$$

$$\times (L_n a^{N-x_n} z_n^{x_n} - \rho_n b^{N-x_n} z_n^N) - \text{correction terms}\}, \qquad (8.4.3a)$$

and the second gives

$$A_{1...n}\{(\rho_1 a^{x_1} + M_1 b^{x_1} z_1^{x_1})R_2(x_1, x_2) ... R_n(x_{n-1}, x_n)b^{N-x_n}$$

$$- \text{correction terms}\}. \qquad (8.4.3b)$$

Here the 'correction terms' arise from explicitly subtracting off spurious contributions from $y_1 = y_2$, $y_2 = y_3$, ..., or $y_{n-1} = y_n$.

From (8.3.23), each $R_j(x, x')$ is a sum of two terms. Expanding the products of the $n - 1$ Rs therefore gives 2^{n-1} terms. In each of (8.4.3a) and (8.4.3b), only one of these is 'wanted' (i.e. has the same form as $f(x_1, ..., x_n)$. Equating these wanted terms on both sides of (8.4.1) gives

$$\Lambda = a^N L_1 ... L_n + b^N M_1 ... M_n. \qquad (8.4.4)$$

Apart from the 'boundary terms' containing a factor ρ_n or ρ_1, all other terms contain at least one factor of the form

$$(z_j z_{j+1})^{x_{j+1}} \quad \text{or} \quad (z_j z_{j+1})^{x_j}. \qquad (8.4.5)$$

They can be made to cancel by adding terms in (8.4.2) with z_j and z_{j+1} interchanged. Doing this for all j, and all initial choices of $z_1, ..., z_n$ thereby generated, one is led to replacing (8.4.2) by

$$g(x_1, ..., x_n) = \sum_P A_{p_1, ..., p_n} z_{p_1}^{x_1} ... z_{p_n}^{x_n}, \qquad (8.4.6)$$

where the sum is over all $n!$ permutations $P = \{p_1, ..., p_n\}$ of the integers $1, ..., n$.

This trial form for the eigenfunction is the same as that used by Bethe (1931) for diagonalizing the quantum-mechanical Hamiltonian of the one-dimensional Heisenberg model. For that reason it is known as the Bethe ansatz.

Evaluating the internal unwanted terms containing the factors (8.4.5) (these include contributions from the 'correction terms'), one finds they cancel provided the following generalization of (8.3.30) is satisfied, for all permutations P and $j = 1, ..., n - 1$:

$$s_{p_j, p_{j+1}} A_{p_1, ..., p_n} + s_{p_{j+1}, p_j} A_{p_1, ..., p_{j+1}, p_j, ..., p_n} = 0. \qquad (8.4.7)$$

This leaves only the boundary terms, containing a factor ρ_j for some value of j. Replacing $z_1, ..., z_n$ in (8.4.3a) by $z_2, ..., z_n, z_1$, it becomes obvious that the boundary terms therein will cancel with those in (8.4.3b) if

$$-z_1^N A_{2\ldots n1} + A_{1\ldots n} = 0.\qquad(8.4.8)$$

Making all possible permutations of $z_1, \ldots, z_n$, all boundary terms will therefore cancel if

$$z_{p1}^N = A_{p_1,\ldots,p_n}/A_{p_2,\ldots,p_n,p_1},\qquad(8.4.9)$$

for all permutations P.

These conditions (8.4.4), (8.4.7) and (8.4.9) do in fact ensure that the eigenvalue equation (8.4.1) is satisfied (see Lieb, 1967a, for a full treatment of the sufficiency of these conditions for the ice model). It is not immediately obvious that they can all be satisfied, since there are many more equations than unknowns. However, it is easy to verify that (8.4.7) has the solution (to within a normalization factor):

$$A_{p_1,\ldots,p_n} = \varepsilon_P \prod_{1 \leqslant i < j \leqslant n} s_{p_j,p_i},\qquad(8.4.10)$$

where ε_P is the signature ($+1$ for even permutations, -1 for odd ones) of the permutation P. Substituting this into (8.4.9) then gives, for all P,

$$z_{p1}^N = (-)^{n-1} \prod_{l=2}^{n} s_{p_l,p_1}/s_{p_1,p_l}.\qquad(8.4.11)$$

The RHS of this equation is symmetric in $p_2, \ldots, p_n$, so there are only n such distinct equations, namely

$$z_j^N = (-)^{n-1} \prod_{\substack{l=1 \\ \neq j}}^{n} s_{l,j}/s_{j,l}\qquad(8.4.12)$$

for $j = 1, \ldots, n$.

Thus we have n equations for $z_1, \ldots, z_n$. These can in principle be solved, and the coefficients A_P calculated from (8.4.10). The eigenfunction g is then given by (8.4.6), the eigenvalue Λ by (8.4.4).

8.5　Maximum Eigenvalue; Location of $z_1, \ldots, z_n$

Unfortunately the equations (8.4.12) have not in general been solved for finite n and N. (This contrasts with the Ising model, where all eigenvalues can be explicitly obtained for finite N.) It turns out that they can be solved for the maximum eigenvalue in the thermodynamic limit (N large), but reasonable care is necessary to ensure it is the maximum eigenvalue that is obtained.

A remarkable feature of the equations (8.4.2), (8.4.10), (8.4.12) and (8.3.22) is that they are not merely of the same form as Bethe's ansatz for the Heisenberg model: they are exactly the same! Thus the eigenvectors of this model are those of our transfer matrix V. This meant that Lieb (1967a, b, c) was able to use the known properties of the Heisenberg model, in particular the work of Yang and Yang (1966) to identify and evaluate the maximum eigenvalue in the limit $N \to \infty$. This work is quite rigorous, and the interested reader is referred to it. Here I shall merely give some plausible arguments to locate the solution of (8.4.12) corresponding to the maximum eigenvalue, and to evaluate it for N large.

The Case $n = 2$

Again it is instructive to consider the case $n = 2$, when (8.4.12) becomes (8.3.32). Multiplying the two equations (8.3.32) together gives

$$(z_1 z_2)^N = 1 , \tag{8.5.1}$$

which implies that

$$z_1 z_2 = \tau , \tag{8.5.2}$$

where τ is an Nth root of unity.

This relation is a simple consequence of the translation invariance of V, since from (8.3.28) it implies that

$$g(x_1 + 1 , x_2 + 1) = \tau g(x_1 , x_2) . \tag{8.5.3}$$

From the Perron – Frobenius theorem (Frobenius, 1908), the eigenvector corresponding to the maximum eigenvalue must have all its entries non-negative. From (8.5.3), this can only be so if $\tau = 1$, so we must choose the solution

$$z_1 z_2 = 1 \tag{8.5.4}$$

of (8.5.1) (i.e. $g(x_1 , x_2)$ is itself translation invariant, as we would expect). Also, from (8.3.28) and (8.3.32),

$$g(x_1 , x_2) = A_{21}(z_1^{N+x_1} z_2^{x_2} + z_2^{x_1} z_1^{x_2}) , \tag{8.5.5}$$

so, using (8.5.4) and setting

$$z_1 = \exp(ik) , \quad r = \tfrac{1}{2}N - 1 , \tag{8.5.6}$$

we have

$$g(x_1, x_2) \propto \cos k(x_1 - x_2 + \tfrac{1}{2}N) . \qquad (8.5.7)$$

Now, $x_1 - x_2 + \tfrac{1}{2}N$ can take all integer (or half integer) values from $-r$ to r. To ensure that all values of $g(x_1, x_2)$ be non-negative, it is therefore sufficient that k should either be real and lie in the interval $[-\pi/2r, \pi/2r]$, or that k should be pure imaginary.

Further, negating k merely interchanges z_1 and z_2, leaving the eigenvector g unchanged. Thus we can just as well limit our search to real values of k in the interval $[0, \pi/2r]$, or positive pure imaginary values.

Now use (8.3.22), (8.3.30) and (8.3.32) to write z_1^N as a rational function of z_1 and z_2, and use (8.5.4) to eliminate z_2. The resulting equation for z_1 is

$$\Delta(z_1^{N-1} + z_1) = z_1^N + 1 , \qquad (8.5.8)$$

or using (8.5.6),

$$\Delta = \cos(r+1)k/\cos rk . \qquad (8.5.9)$$

Plotting the RHS of (8.5.9) as a function of k for k real, and for k pure imaginary, it is easily seen that:

if $\Delta < 1$, (8.5.9) has one real solution in the interval $(0, \pi/2r)$, and no pure imaginary solution;
if $\Delta > 1$, (8.5.9) has no real solution in $(0, \pi/2r)$, but has a single positive imaginary solution.

In both cases it is evident from (8.5.7) that all values of $g(x_1, x_2)$ are strictly positive so, from the Perron – Frobenius theorem, we have located the solution corresponding to the maximum eigenvalue of V in the $n = 2$ sub-block.

When $n = 1$, it is obvious from (8.3.7) and (8.3.10) that the eigenvector with all positive entries is given by $z = 1$.

Admittedly these $n = 2$ and $n = 1$ results provide very slender evidence, but they do in fact point in the right direction: when $\Delta < 1$ (and $n \leq \tfrac{1}{2}N$) the solution of (8.4.12) that maximizes Λ is such that $z_1, \ldots, z_n$ are distinct, lie on the unit circle, are distributed symmetrically about unity, and are packed as closely as possible. The equation (8.4.12) does admit solutions with two or more of $z_1, \ldots, z_n$ equal, but these must be discarded since from (8.4.6) and (8.4.10) all elements of g then vanish identically.

The $n = 2$ and $n = 1$ results also suggest that for $\Delta > 1$ the $z_1, \ldots, z_n$ are all positive real, but we shall not need this hypothesis.

8.6 The Case $\Delta > 1$

The case $\Delta > 1$ is trivial. Let Λ_n be the maximum eigenvalue for a given value of n. Then from the above results it can be verified that, for N sufficiently large,

$$\Lambda_0 > \Lambda_1 \quad \text{and} \quad \Lambda_0 > \Lambda_2 . \tag{8.6.1}$$

In fact, it can be shown that (Lieb, 1967c)

$$\Lambda_0 \geqslant \Lambda_n , \quad n = 0, \ldots, N . \tag{8.6.2}$$

Thus the maximum eigenvalue is simply Λ_0, i.e. from (8.3.5),

$$\Lambda_{\max} = a^N + b^N . \tag{8.6.3}$$

From (1.7.6), (8.2.4) and (8.3.3), remembering that the lattice has MN sites, the free energy is therefore given for N large by

$$f = \min(\varepsilon_1 , \varepsilon_3) . \tag{8.6.4}$$

The system is 'frozen' in the sense that if one vertical arrow is fixed to be up, then the probability of any other arrow being up is unity, no matter how far it is from the first. (Providing of course that the proper thermodynamic limit is taken; both arrows must be deep within an infinite lattice.) This is complete ferroelectric order.

Such frozen solutions exist for three dimensional ice-type models (Nagle, 1969b): one of the very few exact results in three dimensional statistical mechanics.

8.7 Thermodynamic Limit for $\Delta < 1$

If $z_1, \ldots, z_n$ lie on the unit circle, then the equations (8.4.12) involve complex numbers. They can be reduced to a set of real equations as follows.

Define the 'wave numbers' $k_1, \ldots, k_n$, and the function $\Theta(p, q)$ by

$$z_j = \exp(ik_j), \quad s_{l,j}/s_{j,l} = \exp[-i\Theta(k_j , k_l)] . \tag{8.7.1}$$

From (8.3.22) it follows that

$$e^{-i\Theta(p,q)} = \frac{1 - 2\Delta\, e^{ip} + e^{ip+iq}}{1 - 2\Delta\, e^{iq} + e^{ip+iq}} , \tag{8.7.2}$$

and hence that

$$\Theta(p,q) = 2 \tan^{-1}\{\Delta \sin \tfrac{1}{2}(p-q)/[\cos \tfrac{1}{2}(p+q) - \Delta \cos \tfrac{1}{2}(p-q)]\}, \quad (8.7.3)$$

so $\Theta(p,q)$ is a real function.

The product in (8.4.12) is unchanged by including the term $l = j$, so (8.4.12) can now be written

$$\exp(iNk_j) = (-)^{n-1} \prod_{l=1}^{n} \exp[i\Theta(k_j, k_l)]. \quad (8.7.4)$$

Both sides of this equation are unimodular, i.e. of the form $\exp(i\theta)$, so it is natural to take logarithms and divide by i, giving (for $j = 1, \ldots, n$)

$$Nk_j = 2\pi I_j - \sum_{l=1}^{n} \Theta(k_j, k_l), \quad (8.7.5)$$

where each I_j is an integer if n is odd, and half an odd integer if n is even.

The equation (8.7.5) is consistent with the hypothesis that $k_1, \ldots, k_n$ are real, since so then are both sides of (8.7.5).

We want $k_1, \ldots, k_n$ to be distinct, symmetrically distributed about the origin, and packed as closely as possible. This suggests choosing

$$I_j = j - \tfrac{1}{2}(n+1), \quad j = 1, \ldots, n. \quad (8.7.6)$$

Yang and Yang (1966) proved that (8.7.5) then has a unique real solution for $k_1, \ldots, k_n$.

The ratio n/N is the proportion of up arrows in each row of the lattice, so it is the probability of finding a vertical arrow to be up. When N is large we expect this probability to tend to its appropriate thermodynamic limit. This means that we are interested in solving (8.7.5) in the limit of N and n large, n/N remaining fixed.

In this limit $k_1, \ldots, k_n$ become densely packed in some fixed interval $(-Q, Q)$, so they effectively form a continuous distribution. Let the number of k_js lying between k and $k + dk$ be $N\rho(k)\,dk$. Then in the limit of N large, $\rho(k)$ is the distribution function. Since the total number of k_js is n, $\rho(k)$ must satisfy

$$\int_{-Q}^{Q} \rho(k)\,dk = n/N. \quad (8.7.7)$$

For a given value k of k_j, $I_j + \tfrac{1}{2}(n+1)$ is the number of k_ls with $l < j$. Thus (8.7.5) becomes

$$Nk = -\pi(n+1) + 2\pi N \int_{-Q}^{k} \rho(k')\,dk' - N \int_{-Q}^{Q} \Theta(k,k')\,\rho(k')\,dk'. \quad (8.7.8)$$

Differentiating with respect to k, dividing by N and rearranging, this gives

$$2\pi \rho(k) = 1 + \int_{-Q}^{Q} \frac{\partial \Theta(k,k')}{\partial k} \rho(k') \, dk' . \tag{8.7.9}$$

This is a linear integral equation for $\rho(k)$. For a given ratio n/N, Q is determined by (8.7.7).

The eigenvalue Λ is given by (8.4.4). In the limit of N large both terms on the RHS grow exponentially, the larger completely dominating the smaller. From (1.7.6), (8.2.4) and (8.3.3), the free energy f is therefore given by

$$f = \min\left\{ \varepsilon_1 - k_B T \sum_{j=1}^{n} [\ln L(z_j)]/N \, , \, \varepsilon_3 - k_B T \sum_{j=1}^{n} [\ln M(z_j)]/N \right\}. \tag{8.7.10}$$

In the limit of N, n large, these sums become integrals, giving

$$f = \min\left\{ \varepsilon_1 - k_B T \int_{-Q}^{Q} [\ln L(e^{ik})] \, \rho(k) \, dk \, , \right.$$
$$\left. \varepsilon_3 - k_B T \int_{-Q}^{Q} [\ln M(e^{ik})] \, \rho(k) \, dk \right\}. \tag{8.7.11}$$

Since $\rho(k)$ is an even function, these integrals are real.

8.8 Free Energy for $-1 < \Delta < 1$

The problem now is to solve the linear integral equation (8.7.9). For the case $\Delta = -1$ Hulthén (1938) noted that by making an appropriate change of the variable k, the equation can be transformed to one with a difference kernel. Walker (1959) generalized this to $\Delta < -1$, and Yang and Yang (1966) to $\Delta < 1$. There are more complicated models that can be solved by the Bethe ansatz method (Lieb and Wu, 1968; Baxter, 1969, 1970b, c, 1973a; Baxter and Wu, 1974; Kelland, 1974a). In every case such a transformation to a difference kernel exists. (See also the remarks following (8.13.77) and (10.4.31), remembering that trigonometric functions are special cases of elliptic functions.)

For $-1 < \Delta < 1$ the appropriate transformation is to replace k by α, where if

$$\Delta = -\cos\mu, \quad 0 < \mu < \pi, \tag{8.8.1}$$

then

$$\exp(ik) = [\exp(i\mu) - \exp(\alpha)]/[\exp(i\mu + \alpha) - 1]. \tag{8.8.2}$$

Differentiating logarithmically gives

$$\frac{dk}{d\alpha} = \frac{\sin \mu}{\cosh \alpha - \cos \mu}, \tag{8.8.3}$$

so k is a real monotonic increasing function of α, odd, going from $\mu - \pi$ to $\pi - \mu$ as α increases from $-\infty$ to ∞.

In (8.7.2), let $p = k(\alpha)$ and $q = k(\beta)$. Then the equation simplifies to

$$\exp[-i\Theta(p, q)] = \frac{\exp(\alpha - \beta) - \exp(2i\mu)}{\exp(\beta - \alpha) - \exp(2i\mu)} \tag{8.8.4}$$

so $\Theta(p, q)$ is a function only of $\alpha - \beta$ (and the constant μ).

Let $(2\pi)^{-1} R(\alpha)$ be the transformed distribution function, defined by

$$R(\alpha) \, d\alpha = 2\pi \rho(k) \, dk. \tag{8.8.5}$$

Making the substitutions (8.8.2), (8.8.4) and (8.8.5) in (8.7.8), differentiating with respect to α and using (8.8.3), the integral equation becomes

$$R(\alpha) = \frac{\sin \mu}{\cosh \alpha - \cos \mu} - \frac{1}{2\pi} \int_{-Q_1}^{Q_1} \frac{\sin 2\mu}{\cosh(\alpha - \beta) - \cos 2\mu} R(\beta) \, d\beta, \tag{8.8.6}$$

where $(-Q_1, Q_1)$ is the interval on the α line corresponding to $(-Q, Q)$ on the k line. The side condition (8.7.7) becomes simply

$$(2\pi)^{-1} \int_{-Q_1}^{Q_1} R(\alpha) \, d\alpha = n/N. \tag{8.8.7}$$

The free energy is given by (8.7.11) and (8.3.9). On making the substitutions (8.8.2) and (8.8.5) it becomes natural to define another constant, w, by

$$a/b = [\exp(i\mu) - \exp(iw)]/[\exp(i\mu + iw) - 1], \quad -\mu < w < \mu. \tag{8.8.8}$$

From (8.3.21) and (8.8.1) it then follows that

$$a : b : c = \sin \tfrac{1}{2}(\mu - w) : \sin \tfrac{1}{2}(\mu + w) : \sin \mu, \tag{8.8.9}$$

and from (8.8.2) that

$$L(e^{ik}) = \frac{\exp(iw + i\mu) - \exp(\alpha - i\mu)}{\exp(\alpha) - \exp(iw)},$$

$$M(e^{ik}) = \frac{\exp(iw - i\mu) - \exp(\alpha + i\mu)}{\exp(\alpha) - \exp(iw)}. \tag{8.8.10}$$

Using (8.8.5), the formula (8.7.11) for the free energy now becomes

$$f = \min\{\varepsilon_1 - (k_B T/2\pi) \int_{-Q_1}^{Q_1} [\ln L(e^{ik})] R(\alpha) \, d\alpha,$$

$$\varepsilon_3 - (k_B T/2\pi) \int_{-Q_1}^{Q_1} [\ln M(e^{ik})] R(\alpha) \, d\alpha\}. \tag{8.8.11}$$

Solution by Fourier Integrals

Since (8.8.6) is a linear integral equation with a difference kernel, it can be solved by Fourier integrals if $Q_1 = \infty$.
Suppose this is so. Let

$$\overline{R}(x) = (2\pi)^{-1} \int_{-\infty}^{\infty} R(\alpha) \exp(ix\alpha) \, d\alpha. \tag{8.8.12}$$

Multiplying both sides of (8.8.6) by $\exp(ix\alpha)$ and integrating over α, we obtain

$$\overline{R}(x) = \frac{\sinh(\pi - \mu)x}{\sinh \pi x} - \frac{\sinh(\pi - 2\mu)x}{\sinh \pi x} \overline{R}(x). \tag{8.8.13}$$

It immediately follows that

$$\overline{R}(x) = \tfrac{1}{2} \operatorname{sech} \mu x. \tag{8.8.14}$$

From (8.8.12), the LHS of (8.8.7) is simply $\overline{R}(0)$, which from (8.8.14) is $\tfrac{1}{2}$. Thus if $Q_1 = \infty$, then

$$n = \tfrac{1}{2}N. \tag{8.8.15}$$

We want to choose n so as to maximize Λ, or equivalently to minimize the expression (8.8.11) for f. It makes very good sense to assume that (8.8.15) is the correct value of n, since it corresponds to the symmetric situation when there are as many down arrows as up ones in each row of the lattice. Further justification of this argument is given by Lieb (1967).

We could also have predicted that $Q_1 = \infty$ by arguing that if the first term dominates in (8.4.4), then Λ is maximized by choosing n as large as possible so that $L_1, \ldots, L_n$ all have modulus greater than unity. (If the second term, then $M_1, \ldots, M_n$.) Thus if there is a real value of k for which $L [\exp(ik)]$, or $M [\exp(ik)]$, is unimodular, then this should correspond to $k = \pm Q$. For $-1 < \Delta < 1$ there are two such values, and from (8.8.10) they obviously correspond to $\alpha = \pm \infty$, i.e. $Q_1 = \infty$, $Q = \pi - \mu$.

Thus we are in the fortunate position that we can calculate $R(\alpha)$ analytically when, and usually only when, $n = \tfrac{1}{2}N$, and this is precisely the desired value to obtain Λ_{max}. Nature can be kind!

If $w = 0$, then $a = b$, $\varepsilon_1 = \varepsilon_3$ and $|L| = |M|$, so the two terms in (8.8.11) are equal. If $w < 0$, then the first term is the smaller; if $w > 0$, the second term.

Consider first the case $w < 0$. Since $R(\alpha)$ is even, the function $\ln L[\exp(ik)]$ in (8.8.11) can be replaced by its even part, which is also its real part. The Fourier transform of this is

$$(2\pi)^{-1} \int_{-\infty}^{\infty} \exp(ix\alpha) \ln |L(e^{ik})| \, d\alpha$$
$$= \frac{\sinh (\mu + w)x \, \sinh (\pi - \mu)x}{x \sinh \pi x}. \quad (8.8.16)$$

The Fourier transform of $R(\alpha)$ is given by (8.8.12) and (8.8.14). Using these results, (8.8.11) becomes, for $w < 0$,

$$f = \varepsilon_1 - k_B T \int_{-\infty}^{\infty} \frac{\sinh (\mu + w)x \, \sinh (\pi - \mu)x}{2x \sinh \pi x \cosh \mu x} \, dx. \quad (8.8.17)$$

Using (8.8.3), (8.8.9) and the formula

$$\sinh (\mu + w)x = \sinh (\mu - w)x + 2 \cosh \mu x \sinh w x, \quad (8.8.18)$$

this result can be written as

$$f = \varepsilon_3 - k_B T \int_{-\infty}^{\infty} \frac{\sinh (\mu - w)x \, \sinh (\pi - \mu)x}{2x \sinh \pi x \cosh \mu x} \, dx. \quad (8.8.19)$$

However, this is precisely the result obtained for $w > 0$. Thus both (8.8.17) and (8.8.19) are valid throughout the interval $-\mu < w < \mu$. There is no singularity at $w = 0$.

In general these integral expressions for the free energy cannot be analytically evaluated. An exception is when μ is a rational fraction of π. For instance, for the ice model (8.1.4), $a = b = c = 1$, $w = 0$ and $\mu = 2\pi/3$. The integral in (8.8.17) can then be evaluated by summing over residues in the upper half-plane, giving

$$Z^{1/MN} = \exp(-f/k_B T) = (4/3)^{3/2}, \quad (8.8.20)$$

which is Lieb's (1967a) result for the residual entropy of square ice.

8.9 Free Energy for $\Delta < -1$

If $\Delta < -1$, the parameters μ, α defined by (8.8.2) are purely imaginary, so can be replaced by $-i\lambda$, $-i\alpha$, where λ and the new α are real. Then

(8.8.1) and (8.8.2) become

$$\Delta = -\cosh\lambda, \quad \lambda > 0, \tag{8.9.1}$$

$$e^{ik} = (e^{\lambda} - e^{-i\alpha})/(e^{\lambda - i\alpha} - 1). \tag{8.9.2}$$

As α increases from $-\pi$ to π, k also increases from $-\pi$ to π.
The integral equation (8.8.6) now becomes

$$R(\alpha) = \frac{\sinh\lambda}{\cosh\lambda - \cos\alpha} - \frac{1}{2\pi} \int_{-Q_1}^{Q_1} \frac{\sinh 2\lambda}{\cosh 2\lambda - \cos(\alpha - \beta)} R(\beta)\, d\beta, \tag{8.9.3}$$

and the side condition (8.8.7) remains unchanged.

Whereas (8.8.6) was solvable by Fourier integrals if $Q_1 = \infty$, the corresponding equation (8.9.3) is solvable by Fourier series if $Q_1 = \pi$. Set, for all integers m,

$$\overline{R}_m = (2\pi)^{-1} \int_{-\pi}^{\pi} R(\alpha) \exp(im\alpha)\, d\alpha. \tag{8.9.4}$$

Multiplying (8.9.3) by $\exp(im\alpha)$ and integrating, we obtain

$$\overline{R}_m = \exp(-\lambda|m|) - \exp(-2\lambda|m|)\, \overline{R}_m, \tag{8.9.5}$$

so

$$\overline{R}_m = \tfrac{1}{2} \operatorname{sech}\lambda m. \tag{8.9.6}$$

From (8.9.4), the LHS of (8.9.7) is simply $\overline{R}_0$, so again we have $n = \tfrac{1}{2}N$. This is the value of n for which we expect Λ to obtain its maximum value.

The free energy is still given by (8.8.10) and (8.8.11), but now w (like μ and the old α) is pure imaginary, so we replace it by $-iv$, where $-\lambda < v < \lambda$. Then (8.8.9) and (8.8.10) become

$$a:b:c = \sinh\tfrac{1}{2}(\lambda - v) : \sinh\tfrac{1}{2}(\lambda + v) : \sinh\lambda, \tag{8.9.7}$$

$$L\,(e^{ik}) = \frac{\exp(v + \lambda) - \exp(-\lambda - i\alpha)}{\exp(-i\alpha) - \exp(v)},$$

$$M\,(e^{ik}) = \frac{\exp(v - \lambda) - \exp(\lambda - i\alpha)}{\exp(-i\alpha) - \exp(v)}. \tag{8.9.8}$$

If $v < 0$, then the first term in (8.8.11) is the lesser, and $\ln L[\exp(ik)]$ can easily be Taylor expanded in powers of $\exp(i\alpha)$. Doing this, then using (8.9.6), gives

$$f = \varepsilon_1 - k_B T \left\{ \tfrac{1}{2}(\lambda + v) + \sum_{m=1}^{\infty} \frac{\exp(-m\lambda) \sinh m(\lambda + v)}{m \cosh m\lambda} \right\}. \tag{8.9.9}$$

This expression can be re-arranged as

$$f = \varepsilon_3 - k_B T \left\{ \tfrac{1}{2}(\lambda - v) + \sum_{m=1}^{\infty} \frac{\exp(-m\lambda)\sinh m(\lambda - v)}{m\cosh m\lambda} \right\}, \quad (8.9.10)$$

but this is precisely the result obtained when $v > 0$. Thus both (8.9.9) and (8.9.10) are valid throughout the interval $-\lambda < v < \lambda$.

Equations for Finite n

Instead of taking the limit $N, n \to \infty$ and then making the change of variables in (8.9.1), we could equally well have made the change of variables first. The intermediate equations will be needed in the next two chapters, so it is convenient to give them here.

Let α_j be the value of α when $k = k_j$. Then from (8.7.1) and (8.9.1)

$$z_j = \exp(ik_j) = \sinh \tfrac{1}{2}(\lambda + i\alpha_j)/\sinh \tfrac{1}{2}(\lambda - i\alpha_j), \quad (8.9.11)$$

for $j = 1, \ldots, n$. Using this, (8.7.1) and (8.7.2), the original finite-n equations (8.4.12), (8.4.4) can be written

$$\left[\frac{\sinh \tfrac{1}{2}(\lambda + i\alpha_j)}{\sinh \tfrac{1}{2}(\lambda - i\alpha_j)} \right]^N = -\prod_{l=1}^{n} \frac{\sinh[\tfrac{1}{2}i(\alpha_l - \alpha_j) - \lambda']}{\sinh[\tfrac{1}{2}i(\alpha_l - \alpha_j) + \lambda']}, \quad (8.9.12)$$

$$\Lambda = a^N \prod_{j=1}^{n} \frac{\sinh \tfrac{1}{2}(v + i\alpha_j + 2\lambda')}{\sinh \tfrac{1}{2}(v + i\alpha_j)} + b^N \prod_{j=1}^{n} \frac{\sinh \tfrac{1}{2}(v + i\alpha_j - 2\lambda')}{\sinh \tfrac{1}{2}(v + i\alpha_j)},$$

$$(8.9.13)$$

where

$$\lambda' = \lambda - i\pi. \quad (8.9.14)$$

As $n \to \infty$, $\alpha_1, \ldots, \alpha_n$ tend to a continuous distribution on the line interval $(-\pi, \pi)$. The number of α_js in the interval $(\alpha, \alpha + d\alpha)$ is then $(2\pi)^{-1} NR(\alpha)\, d\alpha$. In this limit the equations (8.9.12) for $\alpha_1, \ldots, \alpha_n$ reduce to the integral equation (8.9.3) for $R(\alpha)$.

8.10   Classification of Phases

We have seen that the free energy takes a different analytic form depending on whether $\Delta > 1$, $1 > \Delta > -1$, or $-1 > \Delta$. In terms of the Boltzmann weights a, b, c, it follows from (8.3.21) that there are four cases to consider, the four regimes being shown in Fig. 8.5.

I. Ferroelectric Phase: $a > b + c$

In this case $\Delta > 1$ and, from (8.3.2) and (8.3.3), $\varepsilon_1 < \varepsilon_3, \varepsilon_5$. Thus the lowest energy state is one in which all vertices are of type 1, or all of type 2. Either all arrows point up or to the right, or all point down or to the left.

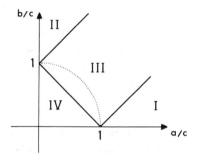

Fig. 8.5. The phase diagram of the zero field ice-type model, in terms of the Boltzmann weights a, b, c. The dotted circular quadrant corresponds to the free-fermion case, when $\Delta = 0$ and the model can be solved by Pfaffians.

Thus at very low temperatures the system is ferroelectrically ordered (all parallel arrows point the same way), and the free energy f is equal to ε_1.

However, from Section 8.6, this is the value of f throughout the regime I. This means that excited states give a negligible contribution to the partition function and throughout the regime I the system is frozen in one or other of the two ground states. As explained in Section 8.6, there is complete ferroelectric order.

II. Ferroelectric Phase: $b > a + c$

This is the same as case I, except that now it is vertex types 3 and 4 that are dominant. There is complete ferroelectric order: effectively all arrows either point up and to the left, or they all point down and to the right.

III. Disordered Phase: $a, b, c < \frac{1}{2}(a + b + c)$

This is the case when $-1 < \Delta < 1$. It includes the infinite temperature case $a = b = c = 1$, so one might expect the system to be disordered. This is true in the sense that all correlations decay to zero with increasing distance r.

However, if $a^2 + b^2 = c^2$ (when the weights must lie in this regime III), then from $(8.3.21, 22)$ it follows that $\Delta = 0$ and $s_{ij} = s_{ji}$. The equations $(8.4.12)$ then simplify dramatically, A_P is proportional to ε_P, and the eigenfunction $(8.4.6)$ is simply a determinant.

In this case the problem can be solved by the Pfaffian method mentioned at the end of the last chapter (Fan and Wu, 1970; Wu and Lin, 1975) and the correlations calculated. It is found (Baxter, 1970a) that they decay as an inverse power law in r, rather than an exponential. From $(1.7.24)$, the correlation length ξ is therefore infinite, so the system is not disordered in the usual sense.

As will be shown in Chapter 10, the ice-type model is a special case of the eight-vertex model, which can also be solved. In this regime III, the ice-type model corresponds to the eight-vertex model being at a critical temperature. There are infinitely many eigenvalues of the transfer matrix which are degenerate with the maximum one. There is no spontaneous order or interfacial tension, but the correlation length is infinite.

The ice-type model therefore has a very unusual property: it is critical for all a, b, c in the regime III.

IV. Anti-Ferroelectric Phase $c > a + b$

In this case $\Delta < -1$ and $\varepsilon_5 < \varepsilon_1, \varepsilon_3$. The lowest energy state is either that shown in Fig. 8.3, or the one obtained from it by reversing all arrows. In either case the arrows alternate in direction.

At sufficiently low temperatures we therefore expect the system to be in an ordered state with this anti-ferroelectric ordering. Since the free energy is analytic throughout the regime IV, we expect this to be the anti-ferroelectric ordered regime. This is confirmed by the following results for $\Delta < -1$, i.e. for $c > a + b$.

Interfacial Tension

For $\Delta < -1$, $Q_1 = Q = \pi$, so the maximum eigenvalue corresponds to $z_1, \ldots, z_n$ being distributed round the whole of the unit circle. A more careful analysis (Baxter, 1973b) reveals that for N even there are actually two such solutions in the $n = \frac{1}{2}N$ sub-space. The numerically larger of the two eigenvalues (Λ_0) corresponds to an eigenvector which is symmetric with respect to reversing all arrows; the smaller (Λ_1) is negative, and corresponds to an anti-symmetric eigenvector. The $z_1, \ldots, z_n$ of one solution interlace the $z_1, \ldots, z_n$ of the other, and for N large

$$\Lambda_1/\Lambda_0 = -1 + \mathcal{O}\{\exp(-Ns/k_BT)\}, \qquad (8.10.1)$$

where if

$$x = \exp(-\lambda), \qquad (8.10.2)$$

then

$$\exp(-s/k_BT) = 2x^{\frac{1}{2}} \prod_{m=1}^{\infty} \left(\frac{1 + x^{4m}}{1 + x^{4m-2}}\right)^2. \qquad (8.10.3)$$

(This result is derived in Section 10.10 for the more general eight-vertex model.)

Thus Λ_0 and Λ_1 are asymptotically degenerate. From an argument parallel to that of Section 7.10, s is the interfacial tension between the two ordered anti-ferroelectric phases.

Spontaneous Staggered Polarization

Regard the ground state arrow arrangement shown in Fig. 8.3 as a 'standard' configuration. For any arrow configuration, assign a parameter τ_i to each i according to the rule:

$\tau_i = +1$ if the arrow on edge i points in the same

direction as the arrow in edge i in Fig. 8.3,

$\tau_i = -1$ if the arrow points in the opposite direction.

Then $\langle \tau_i \rangle$ is the mean 'polarization' of the electric dipole on edge i, normalized to lie between -1 and 1. It is defined with respect to the alternating arrow pattern of Fig. 8.3, so is a 'staggered' polarization.

We have exactly the same problem defining it that we had for the Ising model spontaneous magnetization in Section 7.10. If we define it as in (1.4.4), then it must be zero, since for every state with an up (or right) arrow on edge i, there is another state (obtained by reversing all arrows) with the same energy and a down (or left) arrow on edge i.

However, by using only symmetries which leave the standard configuration of Fig. 8.3 unchanged (arrow reversal plus translation, and mirror reversal plus rotation plus arrow reversal) one can show that $\langle \tau_i \rangle$ must have the same value for all edges i, horizontal or vertical. If P_0 is this common value, then by analogy with (7.10.47) we can define it by

$$P_0^2 = \lim \langle \tau_i \tau_j \rangle, \qquad (8.10.4)$$

where the limit is that in which edges i and j are infinitely far apart. This P_0 is the *spontaneous* staggered polarization. (Just as M_0 is the limit of M

when $H \to 0^+$, so is P_0 the limit of $\langle \tau_i \rangle$ obtained by applying a staggered electric field to the ice-type model, then turning it off.)

We now use an argument similar to that of (7.10.25)–(7.10.37). Consider a particular column C of the lattice. Let $\tau_i' = +1$ if the vertical arrow in row i of this column points up, -1 if it points down. Let s be the 2^N by 2^N diagonal matrix with entries $+1$ (-1) for row-states with an up (down) arrow in column C. Then, for $j > 1$,

$$\langle \tau_i' \tau_j' \rangle = t_{01} t_{10} (\Lambda_1/\Lambda_0)^{j-i} + t_{02} t_{20} (\Lambda_2/\Lambda_0)^{j-i} + \ldots , \qquad (8.10.5)$$

where if U is the matrix of eigenvectors of the transfer matrix V, then t_{01} is the element $(0, 1)$ of $U^{-1} s U$, i.e.

$$t_{01} = (U^{-1} s U)_{01} , \qquad (8.10.6)$$

and similarly for t_{10}, t_{02}, etc. The summation in (8.10.5) is over all eigenvalues $\Lambda_1, \Lambda_2, \ldots$ that correspond to eigenvectors which are anti-symmetric with respect to reversing all arrows. Thus Λ_0 $(=\Lambda_{max})$ is not included, but Λ_1 is.

First take the limit $N \to \infty$ in (8.10.5). From (8.10.1), $\Lambda_1/\Lambda_0 \to -1$. All the other eigenvalues remain strictly less than Λ_0 in modulus. Thus if we now let $j - i$ become large, we obtain

$$\langle \tau_i' \tau_j' \rangle \sim t_{01} t_{10} (-1)^{j-i} . \qquad (8.10.7)$$

The τ_i' are defined relative to a regular configuration (all arrows up), while the τ_i are defined relative to the staggered configuration of Fig. 8.3. It follows that $\tau_i' \tau_j' = (-1)^{i+j} \tau_i \tau_j$, so from (8.10.4) and (8.10.7)

$$P_0^2 = t_{01} t_{10} . \qquad (8.10.8)$$

The matrix elements t_{01} and t_{10} have been calculated (Baxter, 1973c: the calculation is quite intricate and complicated). The result is

$$P_0 = \prod_{m=1}^{\infty} \left(\frac{1 - x^{2m}}{1 + x^{2m}} \right)^2 . \qquad (8.10.9)$$

Correlation Length

After Λ_0 and Λ_1, the next-largest eigenvalue Λ_2 is the maximum eigenvalue in the $n = \frac{1}{2} N - 1$ (or $n = \frac{1}{2} N + 1$) subspace, for N even. Again $z_1, \ldots, z_n$ are distributed round the unit circle, but there is a hole in the distribution at $z = -1$. Such incomplete distributions can be handled (Yang and Yang, 1968; Gaudin, 1971; Takahshi and Suzuki, 1972; Johnson and

McCoy, 1972). Specializing the more general eight-vertex result of Johnson *et al.* (1972a, 1973), we obtain that for N infinite

$$\ln(\Lambda_0/\Lambda_2) = \lambda + v + 2 \sum_{m=1}^{\infty} \frac{(-1)^m \sinh m(\lambda + v)}{m \cosh m\lambda}. \qquad (8.10.10)$$

This expression is valid only for $-\lambda < v < 0$, since for $v \ge 0$ the summation diverges. However, expanding the summand in powers of $\exp(-2\lambda)$ and summing term-by-term gives

$$\left(\frac{\Lambda_2}{\Lambda_0}\right)^{\frac{1}{2}} = x^{\frac{1}{4}}(z^{\frac{1}{2}} + z^{-\frac{1}{2}}) \prod_{m=1}^{\infty} \frac{(1 + x^{4m}z)(1 + x^{4m}z^{-1})}{(1 + x^{4m-2}z)(1 + x^{4m-2}z^{-1})}, \qquad (8.10.11)$$

where $z = \exp(v)$. This product formula is convergent and valid throughout the allowed interval $-\lambda < v < \lambda$.

The correlation length ξ can now be obtained by reasoning similar to that of Section 7.10. The formula (7.10.41) only necessarily holds if the transfer matrix V is symmetric, which is true only if $a = b$ and $v = 0$. Indeed, Johnson, Krinsky and McCoy argued that ξ must be the same as the decay length of the correlation between two vertical arrows in the same row (instead of the same column). Like $\langle \sigma_P \sigma_Q \rangle$ in (7.10.42), this correlation depends on the Boltzmann weights a, b, c only via the eigenvector matrix U. From (8.4.6), (8.4.10), (8.4.12) and (8.3.22), these eigenvectors depend only on Δ. From (8.9.1) and (8.9.7), this means that U is a function of λ, but not of v. (This point will be taken up in the next chapter.) Thus ξ must also be independent of v, in contradiction to (7.10.41) and (8.10.11).

As with the Ising model, this argument demolishes one derivation, but provides another. The equations (7.10.41) and (8.10.11) are valid for $v = 0$, when V is symmetric and its eigenvalues are real. Since ξ is independent of v, the resulting expression for ξ must be valid throughout the allowed range $-\lambda < v < \lambda$. It is

$$\xi^{-1} = -\ln\left\{ 2 x^{\frac{1}{4}} \prod_{m=1}^{\infty} \left(\frac{1 + x^{4m}}{1 + x^{4m-2}} \right)^2 \right\}. \qquad (8.10.12)$$

Johnson *et al*, (1973) verified this explicitly by properly summing (8.10.5) over all relevant eigenvalues. Comparing this result with (8.10.3), we see that the interfacial tension s and the correlation length ξ satisfy the exact relation

$$s \, \xi = k_B T. \qquad (8.10.13)$$

This is the same as the corresponding Ising model relation (7.10.44).

8.11   Critical Singularities

Consider a given set of values of the interaction energies $\varepsilon_1, \ldots, \varepsilon_6$, satisfying (8.1.7). For a temperature T, the Boltzmann weights a, b, c are given by (8.3.3). they correspond to a point in the $(a/c, b/c)$ plane of Fig. 8.5.

As T increases from 0 to ∞, this point traces out a path in the plane, always ending at the point $(1, 1)$ in regime III. Depending on the values of $\varepsilon_1, \varepsilon_3, \varepsilon_5$, this path may or may not cross from one regime into another.

If the lowest two of $\varepsilon_1, \varepsilon_3, \varepsilon_5$ are equal, then the path always lies inside regime III. The free energy is analytic for all temperatures T.

If one of $\varepsilon_1, \varepsilon_3, \varepsilon_5$ is less than both the others, then at sufficiently low temperatures the path will be in regime I, II or IV. As T increases it will cross into regime III, at a 'transition temperature' T_c. There can only be one such transition temperature.

The free energy has a singularity at $T = T_c$, so in this sense this is a critical point. It is, however, a very unusual critical point since, as was remarked in the previous section, the correlation length is infinite throughout regime III.

There are three cases to consider.

Ferroelectric: $\varepsilon_1 < \varepsilon_3, \varepsilon_5$

In this case, at sufficiently low temperatures, the weights a, b, c lie in regime I of Fig. 8.5. The model then has complete ferroelectric order. A typical example is the KDP model discussed in Section 8.1.

The transition temperature T_c is given by the condition

$$a = b + c \tag{8.11.1}$$

For $T > T_c$, f is given by (8.8.9) and (8.8.17); for $T < T_c$, f is simply equal to ε_1.

As $T \rightarrow T_c^+$, it follows from (8.3.21), (8.8.1) and (8.8.8) that $\mu \rightarrow \pi$, $w \rightarrow -\pi$. Thus it is useful to define δ, ε by

$$\mu = \pi - \delta, \quad w = -\pi + \varepsilon. \tag{8.11.2}$$

Then (8.8.9) becomes

$$a : b : c = \sin \tfrac{1}{2}(\delta + \varepsilon) : \sin \tfrac{1}{2}(\varepsilon - \delta) : \sin \delta. \tag{8.11.3}$$

As $T \rightarrow T_c^+$, δ and ε tend through positive values to zero, their ratio remaining non-zero and finite. The temperature difference $T - T_c$ is proportional to

$$t = (b + c - a)/a, \tag{8.11.4}$$

provided t is small. We can therefore use this as our definition of the deviation of T from T_c, instead of (1.1.3). From (8.11.3), t is related to δ and ε by

$$t = 2 \sin \tfrac{1}{2}\delta \, (\cos \tfrac{1}{2}\delta - \cos \tfrac{1}{2}\varepsilon)/\sin \tfrac{1}{2}(\delta + \varepsilon) \,, \qquad (8.11.5)$$

and for δ and ε small this gives

$$t \sim \tfrac{1}{4}\delta(\varepsilon - \delta) \,. \qquad (8.11.6)$$

Thus both δ and ε vanish as $t^{\frac{1}{2}}$ when $t \to 0^+$.

Now make the substitutions (8.11.2) into (8.8.17) and let t become small. We obtain, using (8.11.6),

$$f = \varepsilon_1 - k_B T_c \delta(\varepsilon - \delta) \int_{-\infty}^{\infty} \frac{x \, dx}{\sinh 2\pi x} + \mathcal{O}(t^{\frac{3}{2}}) = \varepsilon_1 - \tfrac{1}{2}k_B T_c t + \mathcal{O}(t^{\frac{3}{2}}) \,. \tag{8.11.7}$$

This is the result for $t > 0$. For $t < 0$, f is simply ε_1. Clearly f is continuous at $t = 0$, its first derivative (the internal energy) has a step-discontinuity, and its second derivative (the specific heat) diverges as $t^{-\frac{1}{2}}$ for $t > 0$. Using the definition (1.7.10) of the critical exponent α, it follows (because of the step-discontinuity) that

$$\alpha = 1 \,, \qquad (8.11.8)$$

corresponding to a first-order transition.

For $T < T_c$ the system is completely ordered, so the spontaneous polarization P_0 is

$$P_0 = 1 \,. \qquad (8.11.9)$$

Just as we defined in (1.1.4) a critical exponent β for a magnetization M_0, so can we define an exponent β_e for an electrical polarization P_0. In this case it follows that

$$\beta_e = 0 \,. \qquad (8.11.10)$$

Above T_c the correlation length is always infinite, whereas below T_c it is zero and the interfacial tension is infinite. The exponents ν, ν' and μ cannot therefore be sensibly defined. Despite this pathological behaviour, the model is interesting in that it is one of the very few that can be solved in the presence of a symmetry-breaking field (in this case a direct electric field). This calculation will be outlined in the next section and the critical equation of state obtained.

Ferroelectric: $\varepsilon_3 < \varepsilon_1, \varepsilon_5$

Exactly the same results hold for this case as for the previous case, provided ε_1 is interchanged with ε_3, a with b, and regime I with regime II. Indeed, this corresponds merely to mirror-reversing the lattice, or rotating it through 90°.

Anti-Ferroelectric: $\varepsilon_5 < \varepsilon_1, \varepsilon_3$

At sufficiently low temperatures the weights lie in regime IV of Fig. 8.5. The transition from regime IV to regime III occurs at a critical temperature T_c given by

$$c = a + b \tag{8.11.11}$$

For $T > T_c$, f is given by (8.8.9) and (8.8.17); for $T < T_c$ it is given by (8.9.7) and (8.9.9).

This case is quite different from the previous two, because the ordered state is one of partial anti-ferroelectric order, rather than complete ferroelectric order.

The 'singular part' f_{sing} of the free energy can be defined by (1.7.10a). Comparing (8.8.1) with (8.9.1), and (8.8.9) with (8.9.7), we see that the analytic continuations from $T > T_c$ to $T < T_c$ of μ and w are $-i\lambda$ and $-iv$, respectively. In both cases w/μ is real and $-1 < w/\mu < 1$. Near T_c, $\Delta \simeq -1$ and μ is small.

We therefore want to continue analytically the integral in (8.8.17) from small real positive values of μ and $\mu + w$ to small negative imaginary values. To do this, we first use the evenness of the integrand to write (8.8.17) as

$$f = \varepsilon_1 - k_B T \mathscr{P} \int_{-\infty}^{\infty} \frac{\sinh{(\mu + w)x}\ \exp[(\pi - \mu)x]}{2x \sinh \pi x \ \cosh \mu x}\, dx\,, \tag{8.11.12}$$

where $\mathscr{P}$ denotes the principal-value integral. If μ, w have negative imaginary parts, this integral can be closed round the upper half of the complex x-plane. Summing over residues then gives, setting $\mu = -i\lambda$ and $w = -iv$:

$$f = \varepsilon_1 - k_B T \left\{ \frac{1}{2}(\lambda + v) + \sum_{m=1}^{\infty} \frac{\exp(-m\lambda)\sinh m(\lambda + v)}{m \cosh m\lambda} \right.$$

$$\left. - i\sum_{m=1}^{\infty} \frac{(-)^m \exp[-(m - \frac{1}{2})\pi^2/\lambda]\cosh[(m - \frac{1}{2})\pi v/\lambda]}{(m - \frac{1}{2})\sinh[(m - \frac{1}{2})\pi^2/\lambda]} \right\}. \tag{8.11.13}$$

(for the F-model, when $v = 0$, this result is given in eqn. A13 of Glasser et al. (1972).)

Clearly the real part (for λ, v real) of this expression is the same as (8.9.9). Thus f_{sing} is the imaginary part of (8.11.13). Near T_c, λ is small, so

$$f_{\text{sing}} \simeq -4ik_B T_c \exp(-\pi^2/\lambda) \cosh(\pi v/2\lambda)\,. \tag{8.11.14}$$

Analogously to (8.11.4), let us define the deviation of T from T_c to be

$$t = (a + b - c)/c. \tag{8.11.15}$$

Then, from (8.9.7), v/λ remains finite and non-zero as $t \to 0$, while for λ and v small

$$t \simeq -\tfrac{1}{2}(\lambda^2 - v^2). \tag{8.11.16}$$

Thus near T_c both λ and v are proportional to $(-t)^{\frac{1}{2}}$.

It follows that the free energy has an unusual singularity at $T = T_c$, namely

$$f_{\text{sing}} \propto \exp\left[-\text{constant}/(-t)^{\frac{1}{2}}\right]. \tag{8.11.17}$$

This is a very weak singularity. It and all its derivatives tend to zero as $t \to 0^-$. In fact, all temperature derivatives of the free energy exist and are the same on both sides of the transition (Glasser *et al.*, 1972; Lieb and Wu, 1972, pp. 392–407). The transition is of *infinite order*.

Clearly (8.11.17) is not of the usually postulated form (1.7.10b), so the exponent α does not properly exist. If one insists on giving it a value, the only sensible choice is

$$\alpha = -\infty. \tag{8.11.18}$$

For $T < T_c$, the correlation length ξ, the interfacial tension s, and the spontaneous staggered polarization P_0 are given by (8.10.3), (8.10.9) and (8.10.12). Their critical behaviour is most easily obtained by noting that these infinite product expressions are precisely those that relate elliptic moduli and integrals to their corresponding nome. In fact, from (15.1.1)–(15.1.4):

$$\exp(-1/\xi) = \exp(-s/k_B T) = k^{\frac{1}{4}}, \tag{8.11.19}$$

$$P_0 = 2k'I/\pi,$$

where k, k', I are the modulus, conjugate modulus and elliptic integral corresponding to the nome

$$q = x^2 = \exp(-2\lambda). \tag{8.11.20}$$

Near T_c, λ becomes small so x and k approach one. Then $I' \to \tfrac{1}{2}\pi$, so from (15.1.3)

$$I \simeq -\tfrac{1}{2}\pi^2/\ln q = \pi^2/4\lambda. \tag{8.11.21}$$

Also, replacing k and q in (15.1.4) by their conjugates k' and q', where $q' = \exp(-\pi I/I')$, we obtain

$$k' \simeq 4q'^{\frac{1}{4}} = 4\exp(-\pi^2/4\lambda), \tag{8.11.22}$$

$$\ln k \simeq -8q' = -8\exp(-\pi^2/2\lambda). \tag{8.11.23}$$

Using these formulae in (8.11.19), it follows that near T_c

$$\xi^{-1} = s/k_B T \simeq 4 \exp(-\pi^2/2\lambda) , \; P_0 \simeq (2\pi/\lambda) \exp(-\pi^2/4\lambda) . \quad (8.11.24)$$

Thus ξ^{-1}, s and P_0 all tend rapidly to zero as $\lambda \to 0$, i.e. as $T \to T_c$. They each have an essential singularity similar to that of f_{sing}, i.e. of the form (8.11.17). They do *not* vanish as simple power laws. The definitions (1.7.9), (1.7.34) and (1.1.4) of their critical exponents ν, μ and β_e therefore fail.

On the other hand, from (8.11.14) and (8.11.24) it is apparent that the proportionality relations

$$\xi^{-1} \propto s \propto (-t)^{\frac{1}{2}} P_0^2 \propto f_{\text{sing}}^{\frac{1}{2}} \quad (8.11.25)$$

are satisfied. If these quantities did vanish as power laws, then (8.11.25) would imply the exponent relations

$$\nu = \mu = \tfrac{1}{2} + 2\beta_e = \tfrac{1}{2}(2 - \alpha) . \quad (8.11.26)$$

In this sense we can therefore say that these exponent relations are satisfied. In particular, the scaling relations (1.2.15) and (1.2.16) hold true.

Unfortunately, applying a direct electric field does not break the degeneracy of the anti-ferroelectric ground states. To do this it is necessary to apply a *staggered* electric field, alternating in direction on successive edges. The model has not been solved in the presence of such a field, so we are unable to apply any further tests of the scaling hypothesis in this case.

8.12 Ferroelectric Model in a Field

In the absence of fields the partition function Z is given by (8.1.1) and (8.1.3), and the vertex energies $\varepsilon_1, \ldots , \varepsilon_6$ satisfy the arrow-reversal symmetry relations (8.1.7).

The arrow-reversal symmetry can be broken by applying vertical and horizontal fields E and E', respectively. These give each vertical up-pointing (down-pointing) arrow an extra energy $-E$ $(+E)$, and horizontal right-pointing (left-pointing) arrows an energy $-E'$ $(+E')$.

If desired, these energies can be incorporated into the vertex energies by sharing out the energy of each arrow between its end-point vertices. If $\varepsilon_1, \ldots , \varepsilon_6$ are the original zero-field vertex energies, satisfying (8.1.7), then from Fig. 8.2(a) the six resulting vertex energies are

$$\varepsilon_1 - E - E', \quad \varepsilon_2 + E + E', \quad \varepsilon_3 + E - E', \quad \varepsilon_4 - E + E', \quad \varepsilon_5, \quad \varepsilon_6 . \quad (8.12.1)$$

As was remarked in Section 8.1, there is no loss of generality in choosing $\varepsilon_5 = \varepsilon_6$, so any six energies can be fitted to (8.12.1), using (8.1.7). This is therefore the general six-vertex model.

This can be solved (Yang, 1967; Lieb and Wu, 1972): the working of Sections 8.2–8.7 can be appropriately generalized, leading to a linear integral equation of the form (8.7.9). In general this equation can no longer be solved analytically, but its properties can be studied and it can of course be solved numerically.

The generalization is particularly simple if $E' = 0$, i.e. only the vertical electric field E is applied, so from now on let us consider this case. Rather than incorporating the vertical field into the vertex energies, let us keep $\varepsilon_1, \ldots, \varepsilon_6$ as the vertex energies, still satisfying (8.1.7). Then (8.1.3) must be replaced by

$$\mathscr{E} = n_1\varepsilon_1 + \ldots + n_6\varepsilon_6 - E(N_t - 2N_d) , \qquad (8.12.2)$$

where N_t is the total number of vertical edges and N_d the number of down-pointing arrows. (Thus $N_t - 2N_d$ is the number of up arrows minus the number of down ones.)

The vital point to remember is that of Section 8.3: there are exactly n down arrows in each row. Since there are M rows of N columns, it follows that

$$N_t - 2N_d = M(N - 2n) . \qquad (8.12.3)$$

Replacing (8.1.3) by (8.12.2), and noting that the transfer matrix V breaks up into $N + 1$ diagonal blocks, each with its own value of n, the equation (8.2.1) therefore becomes

$$Z = \sum_{n=0}^{N} \exp[EM(N - 2n)/k_BT] \text{ Trace } V_n^M , \qquad (8.12.4)$$

where V_n is the nth diagonal block of the original transfer matrix V. If Λ_n is the maximum eigenvalue in this block, then when M is large

$$Z \sim \sum_{n=0}^{N} \Lambda_n^M \exp[EM(N - 2n)/k_BT] . \qquad (8.12.5)$$

Further, for M large the summation in (8.12.5) will be dominated by the value of n which maximizes the summand. From (1.7.6) the free energy per site is therefore

$$f = f_n - E(1 - 2n/N) , \qquad (8.12.6)$$

where

$$f_n = -N^{-1} k_BT \ln \Lambda_n \qquad (8.12.7)$$

and n must be chosen to minimize the RHS of (8.12.6).

Since V is the original zero-field transfer matrix, for a given value of n this f_n is precisely the f given by (8.7.7), (8.7.9) and (8.7.11). The only difference between the previous working and that of this section is that originally we chose n to minimize f_n itself (the appropriate value being $n = \frac{1}{2}N$). Now we must minimize (8.12.6).

More explicitly, for each value of n we must solve (8.7.7) and (8.7.9) for Q and $\rho(k)$, using the definition (8.7.3) of Θ. Then we must calculate f_n from (8.7.11), and finally choose n to minimize (8.12.6).

The polarization P is the expectation value of $(N_t - 2 Nd)/N_t$, and since the summation in (8.12.5) is dominated by the appropriate value of n, this is simply given by

$$P = 1 - 2n/N. \qquad (8.12.8)$$

As E varies, so may n; but since n is chosen so that (8.12.6) is stationary with respect to variations in n (for given E), it follows at once that

$$-\frac{\partial f}{\partial E} = 1 - 2n/N = P. \qquad (8.12.9)$$

This equation is the expected analogue for electrical systems of (1.7.14).

Critical Equation of State

Let us suppose that ε_1 is less than ε_3 and ε_5. Then in zero field there is a transition at a temperature T_c given by (8.11.1). Close to this temperature, the above programme can be carried out to first order in the temperature variable t of (8.11.4).

The easiest way to do this is to go back to the equations (8.4.12), (8.3.22) for $z_1, \ldots, z_n$. Defining $k_1, \ldots, k_n$ by (8.7.1) these equations give

$$e^{iNk_j} = (-)^{n-1} \prod_{\substack{l=1 \\ \neq j}}^{n} \frac{1 - 2\Delta\, e^{ik_j} + e^{i(k_j + k_l)}}{1 - 2\Delta\, e^{ik_l} + e^{i(k_j + k_l)}} \qquad (8.12.10)$$

for $j = 1, \ldots, n$.

Define t, δ as in Section 8.11. Then from (8.8.1) and (8.11.2),

$$\Delta = \cos \delta, \qquad (8.12.11)$$

and $t, \delta \to 0$ as $T \to T_c$.

In Section 8.8 it was shown that if $n = \frac{1}{2} N$, then $Q = \pi - \mu = \delta$. Thus $k_1, \ldots, k_n$ are distributed over the interval $(-\delta, \delta)$. For $n < \frac{1}{2} N$ we expect them to be distributed over some smaller interval centred on the origin.

If δ is small, it follows that $k_1, \ldots, k_n$ are of order δ. Expanding both sides of (8.12.10) to order δ, and taking logarithms, it follows that

$$Nk_j = 2 \sum_{\substack{l=1 \\ \neq j}}^{n} (\delta^2 - k_j k_l)/(k_j - k_l) . \tag{8.12.12}$$

Now let n, N tend to infinity, keeping n/N fixed. As in Section 8.7, $k_1, \ldots, k_n$ effectively form a continuous distribution over some interval $(-Q, Q)$. Again let $N\rho(k) \, dk$ be the number of k_js between k and $k + dk$. Then (8.12.12) becomes the integral equation

$$k = 2\mathscr{P} \int_Q^Q \frac{\delta^2 - kk'}{k - k'} \rho(k') \, dk' , \tag{8.12.13}$$

where $-Q < k < Q$ and $\mathscr{P}$ means that the principal-value integral must be used. Again Q is related to n/N by the condition (8.7.7). Using the definition (8.12.8) of the polarization P, this condition is

$$\int_{-Q}^{Q} \rho(k) \, dk = \tfrac{1}{2}(1 - P) . \tag{8.12.14}$$

Writing kk' in (8.12.13) as $k^2 - k(k - k')$, and using (8.12.14), the integral equation becomes

$$\frac{Pk}{2(\delta^2 - k^2)} = \mathscr{P} \int_{-Q}^{Q} \frac{\rho(k') \, dk'}{k - k'} . \tag{8.12.15}$$

This is a singular integral equation with a Cauchy kernel (Muskhelishvili, 1953), and can be solved exactly. (One brute-force way is to transform from k, k' to α, α', where $k = Q \tanh \alpha$, and then use Fourier integrals.) The solution is, for $Q \leq \delta$,

$$\rho(k) = \frac{P\delta(Q^2 - k^2)^{\frac{1}{2}}}{2\pi(\delta^2 - Q^2)^{\frac{1}{2}}(\delta^2 - k^2)} . \tag{8.12.16}$$

Substituting this into (8.12.14) and using the formula

$$\int_{-Q}^{Q} \frac{(Q^2 - k^2)^{\frac{1}{2}}}{\delta^2 - k^2} \, dk = \pi[1 - (1 - Q^2/\delta^2)^{\frac{1}{2}}] , \tag{8.12.17}$$

we find that Q is given by

$$Q = \delta(1 - P^2)^{\frac{1}{2}} . \tag{8.12.18}$$

Since $0 \leq n \leq N$, P lies in the interval $(-1, 1)$; so Q is always less than δ and the above solution is valid for all allowed values of n/N.

For a given value of n, the free energy f_n is given by (8.7.11). Since

$\varepsilon_1 < \varepsilon_3$, the first term is the smaller, so we must expand $L\,[\exp(ik)]$ about $\delta = 0$ and $k = 0$.

Using (8.3.9) and (8.11.4), noting that $k \sim \delta$ and $t \sim \delta^2$, we obtain

$$\ln L\,(e^{ik}) = 2t + ik + \left(1 - \frac{a}{c}\right)k^2 + \mathcal{O}(\delta^3)\,. \tag{8.12.19}$$

Substituting this expression into (8.7.11), and using (8.12.16) and (8.12.18), then gives

$$f_n = \varepsilon_1 - k_B T\left[t(1 - P) + \left(1 - \frac{a}{c}\right)\delta^2(1 - P)^2/4 + \mathcal{O}(\delta^3)\right]\,. \tag{8.12.20}$$

However, from (8.11.3) and (8.11.6), in the limit of δ small,

$$\left(1 - \frac{a}{c}\right)\delta^2 = -2t\,, \tag{8.12.21}$$

so, neglecting terms small compared with t, (8.12.20) simplifies to

$$f_n = \varepsilon_1 - \tfrac{1}{2}k_B T_c\, t(1 - P^2)\,. \tag{8.12.22}$$

Now we choose n to minimize (8.12.6), remembering that n must lie between 0 and N, and is related to P by (8.12.8). This gives

$$P = E/(k_B T_c\, t) \quad \text{if } |E| < k_B T_c\, t\,, \tag{8.12.23a}$$

$$= \text{sign}(E) \quad \text{otherwise}\,. \tag{8.12.23b}$$

The resulting division of the $(t\,, E)$ plane is shown in Fig. 8.6. In Region A the system is disordered, with polarization P given by (8.12.23a). In regions B, C it is completely ordered, with $P = +1, -1$, respectively.

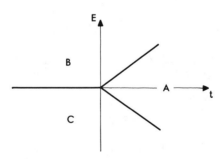

Fig. 8.6. Phase diagram of the ferroelectric model near its critical point $t = E = 0$. The boundaries of the disordered phase A are the lines $E = \pm k_B T_c\, t$.

Scaling Hypothesis

Equation (8.12.23) is the critical equation of state, valid for all negative t, and for small positive t. In view of its quite complicated derivation, it is amazingly simple.

Remembering that P is the electrical analogue of M, and E of H, we can compare (8.12.23) with the form (1.2.1) predicted by the scaling hypothesis, namely

$$E/k_B T_c = P \, |P|^{\delta-1} h_s(t|P|^{-1/\beta}) \,. \tag{8.12.24}$$

A little thought shows that if $h_s(x)$ is of the general form shown in Fig. 1.4, and if

$$\lim_{x \to +\infty} x^{-1} h_s(x) = 1 \,, \quad \delta = 1 + \beta^{-1} \,, \tag{8.12.25}$$

then (8.12.24) reduces to (8.12.23) in the limit $\beta \to 0^+$. Thus the scaling hypothesis is satisfied in this limiting sense, and the critical exponents are

$$\alpha = 1, \, \beta_e = 0, \, \gamma_e = 1, \, \delta_e = \infty \,, \tag{8.12.26}$$

using the suffix e to denote 'electrical' exponents. These results of course agree with our previous observations (8.11.8) and (8.11.10).

Apart from the restriction in (8.12.25), $h_s(x)$ is undetermined. This is a pity, since of the two-dimensional Ising, ice-type and eight-vertex models, only this ferroelectric model has been solved in a symmetry-breaking field. It would be extremely interesting to obtain an exact two-dimensional scaling function.

8.13 Three-Colourings of the Square Lattice

The ice model is a special case of the 'six-vertex' or 'ice-type' model in which $\varepsilon_1, \dots, \varepsilon_6$ are all zero, as in (8.1.4). Lenard (Lieb, 1967a) has pointed out that the model is equivalent to counting the number of ways of colouring the faces of the square lattice with three colours, so that no two adjacent faces are coloured alike.

To see this, consider some such colouring of the lattice, and label the colours 1, 2, 3. Place arrows on the edges of the lattice according to the rule:

if an observer in one face, with colour σ, looks across an edge to a neighbouring face which has colour $\sigma + 1$ (mod 3), then place an arrow in the intervening edge pointing to the observer's left; if the neighbouring face has colour $\sigma - 1$ (mod 3), point the arrow to the right.

Now imagine the observer walking once round a site. Let I be the number of increases in colour (left-pointing arrows) that he sees, and D the number of decreases (right-pointing arrows). Since he returns to the original colour, it must be true that $I - D = 0$ (mod 3). Since there are four faces round each site, it is also true that $I + D = 4$. The only non-negative solution of these equations is $I = D = 2$, so there are two arrows into the site, and two out. The ice rule (Section 8.1) is therefore satisfied at this and every site of the lattice.

To every three-colouring of the lattice there therefore corresponds an arrow covering of the edges that satisfies the ice rule. Conversely, to every such arrow covering there correspond three allowed colourings of the lattice (one square can be coloured arbitrarily; the colours of the rest are then uniquely determined). Thus the number of ways of colouring the lattice is $3Z_{ice}$, where Z_{ice} is the ice-model partition function. It is also equal to

$$\Sigma \, G(N_1 , N_2 , N_3) \, , \qquad (8.13.1)$$

where $G(N_1 , N_2 , N_3)$ is the number of allowed ways of colouring the faces so that N_1 have colour 1, N_2 have colour 2, and N_3 have colour 3. If there are N_t faces altogether, then plainly

$$N_1 + N_2 + N_3 = N_t \, . \qquad (8.13.2)$$

The summation in (8.13.1) is over all non-negative integers N_1, N_2, N_3 satisfying (8.13.2).

An obvious generalization of the colouring problem is to calculate the $G(N_1 , N_2 , N_3)$ individually, instead of just their sum. Equivalently, we can attempt to calculate the generating function

$$Z_G = \Sigma \, z_1^{N_1} z_2^{N_2} z_3^{N_3} \, G(N_1 , N_2 , N_3) \, , \qquad (8.13.3)$$

for arbitrary values of z_1, z_2, z_3.

A more obviously statistical-mechanical way of looking at this problem is to regard the colours 1, 2, 3 as three species of particles. Each face of the lattice contains just one particle, adjacent particles must be of different species. Then Z_G in (8.13.3) is the grand-partition function of this close-packed lattice gas; z_1, z_2, z_3 are the three activities.

It turns out that this problem can be solved (Baxter, 1970c, 1972a). More precisely, one can calculate the limiting 'partition function per site'

$$\kappa = \lim_{N_t \to \infty} Z_G^{1/N_t} \, , \qquad (8.13.4)$$

where as usual the limit $N_t \to \infty$ means the thermodynamic limit in which both the height and the width of the lattice become large.

The method is again that of the Bethe ansatz. Here I shall give the required modifications of Sections 8.2 to 8.8.

Another problem that can be solved is that of four-colouring the sites of the triangular lattice (Baxter, 1970b). As I point out in Sections 12.1 and 12.2, both these colouring problems (with unit activities) are special cases of the Potts model.

Transfer Matrix

Let M be the number of rows of the lattice, and N the number of columns. Impose cyclic (i.e. toroidal) boundary conditions. Then

$$N_t = MN. \qquad (8.13.5)$$

Consider a row of the lattice. Let $\sigma_1, \ldots, \sigma_N$ be the colours of the N faces, as in Fig. 8.7. Place arrows on the intervening vertical edges according to the above rule. Then there is an up arrow in position j if $\sigma_{j+1} = \sigma_j + 1$, a down arrow if $\sigma_{j-1} = \sigma_j - 1$.

Fig. 8.7. A row of faces of the square lattice, coloured $\sigma_1, \ldots, \sigma_N$. Arrows are placed on the intervening edges according to the rule given in Section 8.13. The particular configuration shown corresponds to $\sigma_2 = \sigma_1 + 1$, $\sigma_3 = \sigma_2 - 1, \ldots, \sigma_1 = \sigma_N - 1$.

Let there be n down arrows in the row, in positions $x_1, x_2, \ldots, x_n$, where $1 \leq x_1 < x_2 < \ldots < x_n \leq N$. Then the colours $\sigma_1, \ldots, \sigma_N$ are uniquely determined by specifying σ_1 and $x_1, \ldots, x_n$. Let us refer to σ_1 simply as σ. Then the product of the activities for this row is

$$D_\sigma(X) = z_{\sigma_1} z_{\sigma_2} \ldots z_{\sigma_N}$$

$$= (z_1 z_2 z_3)^{N/3} \prod_{j=1}^{n} \zeta(x_j + j + \sigma), \qquad (8.13.6)$$

where

$$\zeta(\sigma) = z_\sigma z_{\sigma+1}/(z_1 z_2 z_3)^{2/3}. \qquad (8.13.7)$$

Here X denotes the set $\{x_1, \ldots, x_n\}$ and we use the modulo 3 conventions

$$z_{\sigma+3} = z_\sigma, \quad \zeta(\sigma + 3) = \zeta(\sigma). \qquad (8.13.8)$$

We also note that for the colouring to be consistent with the cyclic boundary condition $\sigma_{N+1} = \sigma_1$, N and n must be such that

$$N - 2n = 0, \quad \text{modulo } 3. \tag{8.13.9}$$

We now need both σ and X to specify the state of a row. Let ϕ denote both σ and X. Then again we have (8.2.1), where $Z = Z_G$ and V is the transfer matrix. The elements of V can now be taken to be

$$V(\phi, \phi') = D_\sigma(X) \text{ if } \phi, \phi' \text{ consistent},$$

$$= 0 \text{ otherwise}. \tag{8.13.10}$$

Here $\phi = \{\sigma, X\}$ is the colouring on one row of faces, and $\phi' = \{\sigma', Y\}$ is the colouring on the row above. Thus $D_\sigma(X)$ is the activity product for the lower row.

X denotes the positions of the down arrows in the lower row, Y the positions of those in the upper row. We still have the ice rule (two arrows into each site, and two out), so it is still true that all rows of the lattice have the same number of down arrows, and we can regard this number (n) as fixed. Further if $X = \{x_1, \ldots, x_n\}$ and $Y = \{y_1, \ldots, y_n\}$ then $y_1, \ldots, y_n$ must interlace $x_1, \ldots, x_n$. Altogether, it follows that ϕ and ϕ' are consistent if either:

$$\sigma' = \sigma + 2 \text{ and } 1 \leqslant x_1 \leqslant y_1 \leqslant x_2 \leqslant \ldots \leqslant y_n \leqslant N; \tag{8.13.10a}$$

or

$$\sigma' = \sigma + 1 \text{ and } 1 \leqslant y_1 \leqslant x_1 \leqslant y_2 \leqslant \ldots \leqslant x_n \leqslant N. \tag{8.13.10b}$$

We still have (8.2.3) and (8.2.4), where g is an eigenvector of the transfer matrix V. Let $g_\sigma(X)$ be the element of g corresponding to the row-state $\phi = \{\sigma, X\}$. Then the eigenvalue equation (8.3.2) becomes

$$\Lambda g_\sigma(X) = D_\sigma(X) \left\{ \sum_L g_{\sigma+2}(Y) + \sum_R g_{\sigma+1}(Y) \right\}. \tag{8.13.11}$$

Here L denotes a summation over $y_1, \ldots, y_n$ subject to the restrictions (8.13.10a); R denotes a summation subject to (8.13.10b). We adopt the modulo 3 convention

$$g_{\sigma+3}(X) = g_\sigma(X). \tag{8.13.12}$$

Bethe Ansatz

As in Section 8.3, we can successively consider the cases $n = 0, 1, 2, \ldots$ This leads us to modify (8.4.6), and instead to try the Bethe-type ansatz

$$g_\sigma(X) = \sum_P A'_P \prod_{j=1}^{n} \phi_{p_j}(x_j + j + \sigma). \tag{8.13.13}$$

Here $P = \{p_1, \ldots, p_n\}$ is a permutation of the integers $\{1, \ldots, n\}$, the sum is over all $n!$ such permutations, the coefficients A'_P and the functions $\phi_j(x)$ are at our disposal.

We can ensure that the condition (8.13.12) is satisfied by requiring that there exist wave numbers $k_1, \ldots, k_n$ such that

$$\phi_j(x + 3) = \phi_j(x) \exp(3ik_j) \qquad (8.13.14)$$

and

$$k_1 + \ldots + k_n = 0. \qquad (8.13.15)$$

Thus $\phi_j(x)$ can be regarded as a plane wave modulo 3. The condition (8.13.15) implies that we are seeking a translational invariant eigenvector of the transfer matrix: this must include the eigenvector corresponding to the maximum eigenvalue $\Lambda_{\max}$.

When $z_1 = z_2 = z_3 = 1$ we regain the ice model and expect the functions $\phi_j(x)$ to be pure plane waves. To maintain the analogy with Section 8.4, we must use not the coefficients A'_P, but the related set

$$A_P = A'_P \exp[i(k_{p_1} + 2k_{p_2} + \ldots + nk_{p_n})]. \qquad (8.13.16)$$

When $z_1 = z_2 = z_3$, these should reduce to the coefficients A_P of the ice model.

Substituting the form (8.13.13) of $g_\sigma(X)$ into the eigenvalue equation (8.13.11), we find as in Section 8.4 that there are 'wanted terms', 'internal unwanted terms' and 'boundary terms'. The wanted terms give (for n even)

$$\Lambda = 2 (z_1 z_2 z_3)^{N/3} \gamma_1 \ldots \gamma_n, \qquad (8.13.17)$$

where

$$\gamma_j \left\{ \frac{\phi_j(x)}{\zeta_j(x)} - \frac{\phi_j(x + 1)}{\zeta_j(x + 1)} \right\} = \phi_j(x + 2) \qquad (8.13.18)$$

for $j = 1, \ldots, n$ and all integers x.

The equations (8.13.14) and (8.13.18) form a cubic eigenvalue equation for γ in terms of k_j, the solution of which can be written as $\gamma_j = \gamma(k_j)$, where

$$\gamma(k) = i \exp(3ik/2)/g(k) \qquad (8.13.19)$$

and the function $g = g(k)$ (not to be confused with the eigenvector g) is defined by

$$g^3 - 3Bg + 2 \sin(3k/2) = 0. \qquad (8.13.20)$$

The constant B is given by

$$B = [\zeta(1) + \zeta(2) + \zeta(3)]/3$$

$$= (z_2 z_3 + z_3 z_1 + z_1 z_2)/[3(z_1 z_2 z_3)^{2/3}]. \qquad (8.13.21)$$

We still get the equations (8.4.7) from the vanishing of the internal unwanted terms. In fact at first sight we appear to get three such equations, each with its own function s_{pq}. However, closer examination shows that all three are in fact the same. (If they were not, then the Bethe ansatz would fail.) It turns out that

$$s_{jl} = g(k_j) \exp[i(k_l + \tfrac{1}{2}k_j)] + g(k_l) \exp[-i(k_j + \tfrac{1}{2}k_l)]. \quad (8.13.22)$$

The coefficients A_P are therefore again given by (8.4.10). Further, (8.4.8) and (8.4.9), with z_j therein replaced by $\exp(ik_j)$, are still the conditions for the boundary terms to vanish. Thus again we obtain (8.4.12). i.e.

$$\exp(iNk_j) = \prod_{\substack{l=1 \\ \neq j}}^{n} [-s_{lj}/s_{jl}] \quad (8.13.23)$$

for $j = 1, \ldots, n$.

These equations (8.13.23), together with (8.13.20) and (8.13.22), determine $k_1, \ldots, k_n$; $\gamma_1, \ldots, \gamma_n$ are then given by (8.13.19). Note that these equations involve the activities z_1, z_2, z_3 only via the single dimensionless parameter B. Why this should be so is not clear.

When $z_1 = z_2 = z_3$, then $B = 1$ and (8.13.20) has the solution $g(k) = 2 \sin(k/2)$. Substituting this into the above equations, the ice model results of Section 8.4 are regained.

Of course (8.13.23) has many solutions for $k_1, \ldots, k_n$, corresponding to the various eigenvalues of V. We are interested in the solution corresponding to the maximum eigenvalue. From (8.2.4), with $Z = Z_G$, and from (8.13.4) and (8.13.5), we then have

$$\kappa = \lim_{N \to \infty} \Lambda_{max}^{1/N} ; \quad (8.13.24)$$

or, using (8.13.19) and (8.13.15), and writing g_j for $g(k_j)$,

$$\kappa = (z_1 z_2 z_3)^{1/3} \lim_{N \to \infty} [(-)^{n/2} g_1 \ldots g_n]^{-1/N}. \quad (8.13.25)$$

The Limit $N \to \infty$

I expect the analysis of Section 8.7 to apply also to the three-colouring problem: in the limit of n and N large, $k_1, \ldots, k_n$ form a continuous distribution over some interval $(-Q, Q)$, with distribution function $\rho(k)$ satisfying

$$2\pi \rho(k) = 1 + \int_{-Q}^{Q} \frac{\partial \Theta(k, k')}{\partial k} \rho(k') \, dk', \quad (8.13.26)$$

$$\int_{-Q}^{Q} \rho(k)\, dk = n/N\,,\qquad(8.13.27)$$

the function $\Theta(k\,,k')$ being defined by (8.7.1). From (8.13.22) it follows that

$$\exp[-i\,\Theta(p\,,q)]$$
$$= \frac{g(q)\,\exp[i(p+\tfrac{1}{2}q)] + g(p)\,\exp[-i(q+\tfrac{1}{2}p)]}{g(p)\,\exp[i(q+\tfrac{1}{2}p)] + g(q)\,\exp[-i(p+\tfrac{1}{2}q)]}\,.\qquad(8.13.28)$$

The $g_1,\dots,g_n$ occur in pairs of opposite sign, so from (8.13.25)

$$\ln\kappa = \tfrac{1}{3}\ln(z_1 z_2 z_3) - \int_{-Q}^{Q}\ln|g(k)|\,\rho(k)\,dk\,.\qquad(8.13.29)$$

Transformation to an Integral Equation with a Difference Kernel

An important step in the solution of the ice-type models is the transformation (8.8.2) from the variable k to a new variable α. Using this transformation to go from the variables p, q to new variables α, β, we found in (8.8.4) that $\exp[-i\,\Theta(p\,,q)]$ becomes a function only of $\alpha - \beta$. The integral equation (8.13.26) then has a difference kernel: for the required value of Q it can be solved by Fourier transformation.

Can this procedure be repeated for the present case, i.e. does there exist a function $k(\alpha)$ such that if

$$p = k(\alpha),\, q = k(\beta)\,,\qquad(8.13.30)$$

then

$$\Theta(p\,,q) = \text{function only of } \alpha - \beta\,?\qquad(8.13.31)$$

If so, then

$$\frac{dp}{d\alpha}\frac{\partial\,\Theta(p\,,q)}{\partial p} + \frac{dq}{d\beta}\frac{\partial\,\Theta(p\,,q)}{\partial q} = 0\,.\qquad(8.13.32)$$

From (8.13.20) and (8.13.28) we can verify that

$$\Theta(p\,,q) = p - q[1 - g^2(p)/B] + \mathcal{O}(q^2)\,.\qquad(8.13.33)$$

Taking the limit $q \to 0$ in (8.13.32), and choosing $\alpha = 0$ to be a zero of $k(\alpha)$, it follows that

$$k'(\alpha) = k'(0)\,[1 - B^{-1}g^2(k)]\,.\qquad(8.13.34)$$

Substituting this result back into (8.13.32), we obtain

$$[B - g^2(p)]\frac{\partial\,\Theta(p\,,q)}{\partial p} + [B - g^2(q)]\frac{\partial\,\Theta(p\,,q)}{\partial q} = 0\,.\qquad(8.13.35)$$

Using (8.13.28) and (8.13.20), we can verify directly that this needed identity is indeed satisfied, for all complex numbers p and q. This in turn means that (8.13.31) is correct; there is indeed a transformation that reduces (8.13.26) to an integral equation with a difference kernel!

The function on the RHS of (8.13.31) is readily evaluated by setting $\beta = 0$. Then $q = 0$, so from (8.13.33) and (8.13.30) we obtain

$$\Theta(p, q) = k(\alpha - \beta). \tag{8.13.36}$$

Change variables in (8.13.26) from k, k' to α, β, where $k = k(\alpha)$ and $k' = k(\beta)$. Define $R(\alpha)$ as in (8.8.5). Then we obtain the integral equation

$$R(\alpha) = k'(\alpha) + (2\pi)^{-1} \int_{-Q_1}^{Q_1} k'(\alpha - \beta) R(\beta) \, d\beta, \tag{8.13.37}$$

where $Q = k(Q_1)$. (The function $k(\alpha)$ is monotonic increasing and odd.) The side condition (8.13.27) and the equation (8.13.29) for κ become

$$(2\pi)^{-1} \int_{-Q_1}^{Q_1} R(\alpha) \, d\alpha = n/N, \tag{8.13.38}$$

$$\ln \kappa = \tfrac{1}{3}\ln(z_1 z_2 z_3) - (2\pi)^{-1} \int_{-Q_1}^{Q_1} \ln|g(\alpha)| R(\alpha) \, d\alpha \tag{8.13.39}$$

(regarding g now as a function of α, rather than k.)

The Functions $g(\alpha)$, $k(\alpha)$

The functions $g(\alpha)$, $k(\alpha)$ are defined by (8.13.20) and (8.13.34). We want to solve (8.13.37) for $R(\alpha)$, and then evaluate κ from (8.13.39).

Eliminating k between (8.13.20) and (8.13.34), we obtain the relation

$$\frac{d\alpha}{dg} = [2B/k'(0)] \{4 - g^2(3B - g^2)^2\}^{-\frac{1}{2}} \tag{8.13.40}$$

between α and g.

This equation can be integrated using elliptic integrals (Gradshteyn and Ryzhik, 1965, Paragraph 3.147.2). Let u, v, w be the three values of x which satisfy the cubic equation

$$x(3B - x)^2 = 4, \tag{8.13.41}$$

and let k_m, τ be the constants

$$k_m = [(u - v)w/(u - w)v]^{\frac{1}{2}}, \tag{8.13.42}$$

$$\tau = [uw/(u - w)]^{\frac{1}{2}} B/k'(0). \tag{8.13.43}$$

From (8.13.21), $B \geqslant 1$. It follows that u, v, w are real and positive. Let us choose them so that

$$u > v \geqslant w .$$ (8.13.44)

Then k_m is real, satisfying

$$0 < k_m \leqslant 1 .$$ (8.13.45)

Now introduce a new variable s, related to g by

$$g^2 = uw s^2/(u - w + w s^2) .$$ (8.13.46)

Substituting this expression for g into (8.13.40), the differential equation becomes

$$\frac{d\alpha}{ds} = \tau[(1 - s^2)(1 - k_m^2 s^2)]^{-\frac{1}{2}}.$$ (8.13.47)

Integrating, remembering that $k = g = s = 0$ when $\alpha = 0$, we obtain

$$\alpha = \tau \int_0^s [(1 - t^2)(1 - k_m^2 t^2)]^{-\frac{1}{2}} dt .$$ (8.13.48)

This gives α as a function of s (it is actually an elliptic integral). We are interested in s as a function of α. From (15.5.6) we see that

$$s = \text{sn}(\tau^{-1}\alpha, k_m) ,$$ (8.13.49)

where $\text{sn}(u, k)$ is the elliptic sn function of argument u and modulus k, defined by (15.1.1)–(15.1.6).

From now on let us regard the elliptic modulus k_m (not to be confused with the wave numbers $k_1, \ldots, k_n$ above) as understood. Also, we are free to choose the scale of α in any convenient way: let us do so so that

$$\tau = 1 .$$ (8.13.50)

Then $k'(0)$ is defined by (8.13.43), and s is simply sn α.

As is shown in Section 15.2, the function sn α is meromorphic (i.e. its only singularities are poles). Since u, v, w are constants, it follows from (8.13.46) that g^2 is also a meromorphic function of α.

Let η be one of the poles of $g^2(\alpha)$. Then from (8.13.46),

$$\text{sn}^2 \eta = -(u - w)/w .$$ (8.13.51)

As in Chapter 15, let I and I' be the complete elliptic integrals of the first kind of moduli $k_m, k_m' = (1 - k_m^2)^{\frac{1}{2}}$, respectively. From now on let q be the 'nome', defined by (15.1.1)–(15.1.4), i.e.

$$q = \exp(-\pi I'/I) .$$ (8.13.52)

As α moves along the imaginary axis in the complex plane, from 0 to iI', sn α is also pure imaginary, and goes from 0 to $+i\infty$. The RHS of (8.13.51) is negative, so we can choose η to lie in the interval $(0, iI')$ of the imaginary axis.

Using (15.4.4), (15.4.5), (8.13.42) and (8.13.51), we have that

$$\text{cn } \eta = (u/w)^{\frac{1}{2}}, \quad \text{dn } \eta = (u/v)^{\frac{1}{2}}, \tag{8.13.53}$$

$$1 - k_m^2 \, \text{sn}^4 \, \eta = u(v + w - u)/(vw). \tag{8.13.54}$$

Taking square roots of (8.13.41), we obtain a cubic equation for $x^{\frac{1}{2}}$. By considering the sum of the roots of this equation (taking proper account of their sign), we obtain the homogeneous relation

$$u^{\frac{1}{2}} = v^{\frac{1}{2}} + w^{\frac{1}{2}}. \tag{8.13.55}$$

Squaring, this implies that

$$v + w - u = -2(vw)^{\frac{1}{2}}. \tag{8.13.56}$$

Multiplying both sides by $u/(vw)$ and using (8.13.53) and (8.13.54), we obtain

$$1 - k_m^2 \, \text{sn}^4 \, \eta = -2 \, \text{cn } \eta \, \text{ dn } \eta. \tag{8.13.57}$$

Using (15.4.21) with $u = -v = \eta$, it follows that

$$\text{sn } 2\eta = -\text{sn } \eta, \tag{8.13.58}$$

i.e., using (15.2.5),

$$\text{sn } 2\eta = \text{sn}(2iI' - \eta). \tag{8.13.59}$$

This equation has just one solution in the interval $(0, iI')$, namely

$$\eta = 2iI'/3 \tag{8.13.60}$$

Thus η is this simple fraction of iI'.

Using (8.13.51), remembering that we now regard g as a function of α, we can write (8.13.46) as

$$g^2(\alpha) = u \, \text{sn}^2\alpha/(\text{sn}^2\alpha - \text{sn}^2\eta). \tag{8.13.61}$$

We can express the constant u in terms of η. From (8.13.41), the product of the three roots is 4, so $uvw = 4$. Using (8.13.53), it follows that

$$\text{cn } \eta \, \text{dn } \eta = \tfrac{1}{2}u^{3/2}. \tag{8.13.62}$$

Using (15.1.6) and (15.4.30), this can be written as

$$\tfrac{1}{2}u^{3/2} = H_1(\eta) \, \Theta_1(\eta) \, \Theta^2(0)/[H_1(0) \, \Theta_1(0) \, \Theta^2(\eta)], \tag{8.13.63}$$

where $H(u)$, $H_1(u)$, $\Theta(u)$, $\Theta_1(u)$ are the elliptic theta functions defined in (15.1.5).

From (15.2.3b) (with u therein replaced by $-\eta$),

$$H(2\eta) = q^{-1/3} H(\eta) \; ; \tag{8.13.64}$$

and from (15.4.17) (with u, v replaced by η, $-\eta$),

$$2 H(\eta) \, \Theta(\eta) \, H_1(\eta) \, \Theta_1(\eta) = H(2\eta) \, \Theta(0) \, H_1(0) \, \Theta_1(0) . \tag{8.13.65}$$

Eliminating $H(2\eta)/H(\eta)$ and $H_1(\eta) \, \Theta_1(\eta)$ between these last three equations, we obtain

$$u = q^{-2/9} \, \Theta^2(0)/\Theta^2(\eta) . \tag{8.13.66}$$

Using this in (8.13.61), together with (15.1.6) and (15.4.19), we find that

$$g^2(\alpha) = q^{-2/9} H^2(\alpha)/[H(\alpha - \eta) \, H(\alpha + \eta)] . \tag{8.13.67}$$

Now consider the function $k(\alpha)$. From (8.13.34) and (8.13.43), using $\tau = 1$,

$$k'(\alpha) = [uw/(u - w)]^{\frac{1}{2}}[B - g^2(\alpha)] . \tag{8.13.68}$$

The RHS of this equation is a meromorphic function of α; like $g^2(\alpha)$, it has simple poles when the denominator in (8.13.61) vanishes, i.e. when

$$\alpha = \pm \, \eta + 2mI + 2inI' , \tag{8.13.69}$$

for all integers m, n. The residue at such a pole can be obtained in the usual way by differentiating the denominator. Using (15.5.1a) and (15.5.5), this gives

$$\mathrm{Res}[k'(\alpha)] = - [uw/(u - w)]^{\frac{1}{2}}u \, \mathrm{sn} \, \alpha /(2 \, \mathrm{cn} \, \alpha \, \mathrm{dn} \, \alpha) , \tag{8.13.70}$$

Substituting the values (8.13.69) of α, using the periodicity relations (15.2.5), together with (8.13.51) and (8.13.62), we find that

$$\mathrm{Res}[k'(\alpha)] = \mp i , \tag{8.13.71}$$

the upper (lower) sign being used if the upper (lower) one is used in (8.13.69).

A function of α that has precisely these poles and residues is

$$i\left\{\frac{H'(\alpha + \eta)}{H(\alpha + \eta)} - \frac{H'(\alpha - \eta)}{H(\alpha - \eta)}\right\} . \tag{8.13.72}$$

The difference between this and $k'(\alpha)$ is therefore a meromorphic function with no poles, i.e. an entire function. Further, it is doubly periodic, with periods $2I$ and $2iI'$, so it must be bounded. By Liouville's theorem it is therefore a constant. (This is an example of the use of theorem 15a in Section 15.3.) Integrating (8.13.68) and using $k(0) = 0$, we therefore obtain

$$k(\alpha) = C\alpha + i \ln[H(\eta + \alpha)/H(\eta - \alpha)],\qquad(8.13.73a)$$

where C is some constant.

Since $g^2(\alpha)$ is periodic, with periods $2I$ and $2iI'$, so from (8.13.20) is $\sin^2(3k/2)$. Using (15.2.3), this fixes C to be

$$C = -2\pi/(3I).\qquad(8.13.73b)$$

This completes the derivation of the functions $g(\alpha)$ and $k(\alpha)$. We shall need the Fourier expansion of $k(\alpha)$, and the Fourier integrals

$$G_m = (2I)^{-1} \int_{-I}^{I} \exp(im\pi\alpha/I) \ln[g^2(\alpha)] \, d\alpha,\qquad(8.13.74)$$

where m is an integer. From (8.13.61), (8.13.73), (8.13.60) and the product expansion (15.1.5) of the elliptic theta function $H(u)$, it is straightforward to establish that

$$G_0 = -4\pi I'/(9I),$$

$$G_m = (r^{2m} - 1)/[m(1 + r^m + r^{2m})], \; m \neq 0,\qquad(8.13.75)$$

and that

$$k(\alpha) = \frac{\pi\alpha}{3I} + 2 \sum_{m=1}^{\infty} \frac{r^m \sin(m\pi\alpha/I)}{m(1 + r^m + r^{2m})}.\qquad(8.13.76)$$

Here α is real and

$$r = q^{2/3} = \exp(-2\pi I'/3I).\qquad(8.13.77)$$

Elliptic functions occur very frequently in the exactly solved two-dimensional models in statistical mechanics. This model is interesting in that they are needed to transform (8.13.26) to an integral equation with a difference kernel. They also occur in this way in the original method of solving the three-spin model (Baxter and Wu, 1973, 1974). As I remark at the end of Section 10.4, I suspect that the 'difference kernel' transformation is closely related to the elliptic function parametrization of the generalized star – triangle relation.

Solution of the Integral Equation

We can solve the integral equation (8.13.37) by Fourier series, provided $2Q_1$ is a period of the elliptic functions, i.e. if

$$Q_1 = I.\qquad(8.13.78)$$

Substituting the form (8.13.76) for $k(\alpha)$, and setting

$$R(\alpha) = (\pi/I) \left\{ R_0 + 2 \sum_{m=1}^{\infty} R_m \cos(m\pi\alpha/I) \right\}, \qquad (8.13.79)$$

it is easy to find that

$$R_m = r^m/(1 + r^{2m}). \qquad (8.13.80)$$

From (8.13.38), it follows that

$$n = \tfrac{1}{2}N, \qquad (8.13.81)$$

so there are as many up arrows as down ones. As in Section 8.8, we expect this case to give the maximum eigenvalue of the transfer matrix V.

From (8.13.78) and (8.13.76), $k(\alpha)$ increases monotonically from $-\pi/3$ to $\pi/3$ as α increases from $-Q_1$ to Q_1. Thus $Q = \pi/3$ in (8.13.26): the wave-numbers $k_1, \ldots, k_n$ fill the interval $(-\pi/3, \pi/3)$. This is the same interval as for the original ice model, though of course the distribution is in general different.

Substituting into (8.13.39) the Fourier series (8.13.79) for $R(\alpha)$, we obtain

$$\ln \kappa = \tfrac{1}{3}\ln(z_1z_2z_3) - \tfrac{1}{2} \sum_{m=-\infty}^{\infty} G_m R_m \qquad (8.13.82)$$

(taking $R_{-m} = R_m$). Using (8.13.75) and (8.13.80), it follows that

$$\ln \kappa = \tfrac{1}{3}\ln(z_1z_2z_3) - \tfrac{1}{6}\ln r$$

$$+ \sum_{m=1}^{\infty} \frac{r^m - r^{3m}}{m(1 + r^{2m})(1 + r^m + r^{2m})}. \qquad (8.13.83)$$

This gives $\kappa/(z_1z_2z_3)^{1/3}$ as a function of r. We can regard r as defined by (8.13.77) and (8.13.66). Using (15.1.5), these give

$$u^{\frac{1}{2}} = r^{-1/6} \prod_{m=1}^{\infty} \frac{(1 - r^{3m - 1\frac{1}{2}})^2}{(1 - r^{3m - 2\frac{1}{2}})(1 - r^{3m - \frac{1}{2}})}. \qquad (8.13.84)$$

To summarize: given z_1, z_2, z_3, define B by (8.13.21) and let u be the largest root of (8.13.41). Define r $(0 < r < 1)$ by (8.13.84). Then κ, the partition-function-per-site of the weighted three-colouring problem, is given by (8.13.83). Note that $\kappa/(z_1z_2z_3)^{1/3}$ depends on z_1, z_2, z_3 only via B.

This form of the result is convenient when B is large, which is when one of z_1, z_2, z_3 is large, or small, compared with the others. Then r is small and the infinite series and product are rapidly convergent.

Critical Behaviour

Considered as a function of the positive real variables z_1, z_2, z_3, the partition-function-per-site κ is analytic except when $z_1 = z_2 = z_3$. In this case $B = 1$, $u = 2$ and $r = 1$.

It follows that r is just less than one if z_1, z_2, z_3 are nearly equal. The expressions (8.13.83), (8.13.84) are no longer convenient, since the series and product converge only slowly. It is then useful to apply the Poisson summation formula of Section 15.8 to (8.13.83), and the conjugate modulus formula (15.7.2b) to (8.13.84). This converts the equations to the form:

$$\kappa = (4/3)^{3/2} \, (z_1 z_2 z_3)^{1/3} \prod_{n=1}^{\infty} \frac{(1 + s^n)^2 (1 - s^{4n/3})^3}{(1 - s^n)^2 (1 - s^{4n})}, \qquad (8.13.85)$$

$$u^{\frac{1}{2}} = 2 \prod_{n=1}^{\infty} \frac{(1 + s^{4n/3})^3}{1 + s^{4n}}. \qquad (8.13.86)$$

Here $s = \exp(-3\pi I/2I')$, but we can regard it as defined by (8.13.86). It is small when B is close to one. In particular, when $z_1 = z_2 = z_3 = 1$, then $B = 1$, $u = 2$, $s = 0$ and we regain the ice-model result (8.8.20), namely $\kappa = (4/3)^{3/2}$ (Lieb, 1976a).

We can also examine the way in which the ice-model limit is approached. Let z_1, z_2, z_3 differ from unity by terms of order ε. Then B exceeds unity by terms of order ε^2. By scaling ε appropriately, we can choose

$$B = 1 + \varepsilon^2, \qquad (8.13.87)$$

where $|\varepsilon| \ll 1$. From (8.13.41) it follows that

$$u^{\frac{1}{2}} = 2\{1 + \tfrac{1}{3}\varepsilon^2 + \mathcal{O}(\varepsilon^4)\}, \qquad (8.13.88)$$

so from (8.13.86)

$$s \sim |\varepsilon/3|^{3/2}, \qquad (8.13.89)$$

and from (8.13.85)

$$\kappa = (4/3)^{3/2} \, (z_1 z_2 z_3)^{1/3} \{1 + 4 |\varepsilon/3|^{3/2} + \mathcal{O}(\varepsilon^2)\}. \qquad (8.13.90)$$

Thus κ has a 'singular part' proportional to $\varepsilon^{3/2}$. In this sense we can say that the three-colouring problem has a critical point at $z_1 = z_2 = z_3$. Defining the critical exponent α analogously to (1.7.10), we have

$$\alpha = \tfrac{1}{2}. \qquad (8.13.91)$$

It is possible to explicitly eliminate the variable r between (8.13.83) and (8.13.84) (Baxter, 1970c), giving

$$\kappa = (z_1 z_2 z_3)^{\frac{1}{3}} \frac{4uvw^{\frac{1}{2}}}{[(u-w)^3 u]^{\frac{1}{2}} - [(v-w)^3 v]^{\frac{1}{2}}}. \tag{8.13.92}$$

This makes it clear that $\kappa/(z_1 z_2 z_3)^{1/3}$ is an algebraic function of B.

9

ALTERNATIVE WAY OF SOLVING THE ICE-TYPE MODELS

9.1 Introduction

In Chapter 8 the ice-type models have been solved by using a Bethe ansatz for the eigenvectors of the transfer matrix. This method depends heavily on the fact that the number of 'lines', or down arrows, is conserved from row to row. It is not clear how to generalize the method to models without such conservation.

The purpose of this chapter is to examine the results of Chapter 8, and to show how they suggest an alternative route by which they can be derived. This alternative route can be called the 'commuting transfer matrices' method: it will be used in Chapter 10 to solve the eight-vertex model.

9.2 Commuting Transfer Matrices

Let V be the row-to-row transfer matrix of an ice-type model. From (8.3.3) and (8.3.4), it is a function of the Boltzmann weights a, b, c.

Let g be an eigenvector of V, as in (8.2.3). Then the elements of g are given by (8.4.6), (8.4.10) and (8.3.22), where $z_1, \ldots , z_n$ are solutions of the equations (8.4.12).

However, these equations for g involve a, b, c only via the combination

$$\Delta = (a^2 + b^2 - c^2)/2ab . \qquad (9.2.1)$$

Thus if we consider two transfer matrices, with different values of a, b, c but the same value of Δ, then they have common eigenvectors.

If all eigenvectors are given by the Bethe ansatz and span the 2^N dimensional vector space (which is the case), and if P is the matrix of eigenvectors, then it follows that

$$V = P V_d P^{-1} , \qquad (9.2.2)$$

where V_d is diagonal, V and V_d are functions of a, b, c, but P is a function only of Δ.

In Chapter 8 we were led to the parametrization (8.9.7) (or (8.8.9)), namely

$$a, b, c = \rho \sinh \tfrac{1}{2}(\lambda - v), \rho \sinh \tfrac{1}{2}(\lambda + v), \rho \sinh \lambda , \qquad (9.2.3)$$

where ρ is a normalization factor.

Regard ρ, λ and v as variables, not necessarily real. Then (9.2.3) defines a, b, c. The matrices V, V_d and P are now functions of ρ, λ and v.

However, from (8.9.1)

$$\Delta = -\cosh \lambda . \qquad (9.2.4)$$

Thus P is a function only of λ: it is independent of v and ρ.

We can regard λ and ρ as fixed constants, and v as a (complex) variable, and exhibit the dependence of V on v by writing it as $V(v)$. Then (9.2.2) implies that

$$V(v) V(u) = V(u) V(v) , \qquad (9.2.5)$$

for all complex numbers u, v; i.e. the transfer matrices $V(u)$ and $V(v)$ commute.

9.3   Equations for the Eigenvalues

Now consider the equations (8.4.4) and (8.4.12), which exactly define the eigenvalues Λ of V for finite n and N. By analogy with (8.7.1) and (8.9.1) (replacing α_j by iv_j), let us transform from $z_1, \dots , z_n$ to

$$z_j = (e^\lambda - e^{v_j})/(e^{\lambda + v_j} - 1) . \qquad (9.3.1)$$

Then from (8.3.22), (8.3.9) and (9.2.3),

$$s_{jk} = \frac{\sinh \lambda \, \sinh \tfrac{1}{2}(2\lambda + v_j - v_k)}{\sinh \tfrac{1}{2}(\lambda - v_j) \, \sinh \tfrac{1}{2}(\lambda - v_k)} , \qquad (9.3.2)$$

$$L(z_j) = - \frac{\sinh \tfrac{1}{2}(v - v_j + 2\lambda)}{\sinh \tfrac{1}{2}(v - v_j)} ,$$

$$M(z_j) = - \frac{\sinh \tfrac{1}{2}(v - v_j - 2\lambda)}{\sinh \tfrac{1}{2}(v - v_j)} . \qquad (9.3.3)$$

For given values of ρ and $v_1, \ldots, v_n$ let us define functions $\phi(v)$, $q(v)$ by

$$\phi(v) = \rho^N \sinh^N(v/2) \qquad (9.3.4)$$

$$q(v) = \prod_{l=1}^{n} \sinh \tfrac{1}{2}(v - v_l). \qquad (9.3.5)$$

Then (8.4.4) can be written

$$\Lambda = [\phi(\lambda - v) \, q(v + 2\lambda') + \phi(\lambda + v) \, q(v - 2\lambda')]/q(v), \qquad (9.3.6)$$

where

$$\lambda' = \lambda - i\pi. \qquad (9.3.7)$$

From (8.4.12), (9.3.1) and (9.3.2) the $v_1, \ldots, v_n$ are given by the n equations

$$\frac{\phi(\lambda - v_j)}{\phi(\lambda + v_j)} = - \frac{q(-2\lambda' + v_j)}{q(2\lambda' + v_j)}, \quad j = 1, \ldots, n. \qquad (9.3.8)$$

9.4 Matrix Function Relation that Defines the Eigenvalues

In the Bethe ansatz method, a considerable amount of work is needed to establish the equations (8.4.12), i.e. (9.3.8). We can now observe that they are a simple corollary of the commutation relations and (9.3.6).

To do this, we use a similar argument to that of Section 7.7 for the Ising model. From (9.2.2), if Λ is the eigenvalue of V corresponding to column r of P, then

$$\Lambda = (P^{-1} V P)_{rr}. \qquad (9.4.1)$$

Regard ρ and λ as fixed, v as a variable. The the RHS of (9.4.1) is a sum over elements of V, with coefficients from P that are independent of v. From (8.3.4) and (9.2.3), each element of V is an entire function of v. Thus

$$\Lambda = \Lambda(v), \qquad (9.4.2)$$

is also an entire function.

Now look at (9.3.6). The RHS is the ratio of two entire functions, and the denominator $q(v)$ vanishes when $v = v_1, \ldots, v_n$. Since the ratio must be entire, the numerator must also vanish at these values. The equations (9.3.8) follow immediately.

Thus (9.3.6), considered as a relation between the functions $\Lambda(v)$ and $q(v)$, defines $\Lambda(v)$.

Every allowed solution of (9.3.8) defines $v_1, \ldots, v_n$, and hence an eigenvalue $\Lambda(v)$ and a function $q(v)$. This is true for each value of n. Altogether there must be 2^N such eigenvalues $\Lambda(v)$ and associated functions $q(v)$.

Let us label these $\Lambda_r(v)$, $q_r(v)$, $r = 1, \ldots, 2^N$. The matrix V_d in (9.2.2) is a diagonal matrix with entries $\Lambda_1, \ldots, \Lambda_{2^N}$. Similarly, let Q_d be the diagonal matrix with entries $q_1, \ldots, q_{2^N}$. Then the full set of equations (9.3.6) (one for each eigenvalue) can be written as the single matrix equation

$$V_d(v)\, Q_d(v) = \phi(\lambda - v)\, Q_d(v + 2\lambda') + \phi(\lambda + v)\, Q_d(v - 2\lambda'). \quad (9.4.3)$$

(The factors $\phi(\lambda - v)$, $\phi(\lambda + v)$ are the same for each eigenvalue, so are simple scalar coefficients).

Now define the non-diagonal matrix function

$$Q(v) = P\, Q_d(v)\, P^{-1}. \quad (9.4.4)$$

and again exhibit the dependence of the transfer matrix on v by writing it as $V(v)$. Pre-multiplying (9.4.3) by P, post-multiplying by P^{-1}, using (9.2.2), (9.4.4) and the fact that P is independent of v, (9.4.3) becomes

$$V(v)\, Q(v) = \phi(\lambda - v)\, Q(v + 2\lambda') + \phi(\lambda + v)\, Q(v - 2\lambda'), \quad (9.4.5)$$

which is a relation between the matrix functions $V(v)$, $Q(v)$.

Since $V_d(v)$, $Q_d(v)$ are diagonal for all v, $Q(v)$ commutes with $V(u)$ and $Q(u)$ for all complex numbers u and v.

It was shown in Section 8.3 that $V(v)$ breaks up into $N + 1$ diagonal blocks, one for each value of n. It therefore certainly breaks up into two blocks, one with n even and the other with n odd. This is simply a consequence of the commutation relation

$$V(v)\, S = S\, V(v), \quad (9.4.6)$$

where S is the diagonal operator that has entries $+1$ (-1) for row-states with an even (odd) number of down arrows.

The matrix P can therefore also be chosen to commute with S.

From (9.3.5), since $n < N$, all the diagonal elements of $Q_d(v)$ are of the form

$$\sum_r d_r \exp(rv/2), \quad (9.4.7)$$

where for n even (odd) the sum is over all even (odd) values of r in the interval $-N < r < N$. The coefficients d_r are independent of v; some may be zero.

From (9.4.4), each element of $Q(v)$ is a sum of expressions of the form (9.4.7), either all with n even, or all with n odd. Thus each element of $Q(v)$ is itself of the form (9.4.7).

9.5 Summary of the Relevant Matrix Properties

To summarize: we have used the results of the Bethe ansatz calculation of Chapter 8 to establish the following properties:

(i) Given a transfer matrix V for a particular set of values of a, b, c; there are infinitely many other transfer matrices (with different a, b, c but the same Δ) that commute with V.

(ii) If a, b, c are defined in terms of ρ, λ, v by (9.2.3), and if ρ, λ are regarded as constants and v as a complex variable, then matrices $V(u)$, $V(v)$ commute for all values of u, v.

(iii) All elements of $V(v)$ are entire functions of v.

(iv) There exists a matrix function $Q(v)$ such that the matrix relation (9.4.5) is satisfied for all complex numbers v.

(v) The determinant of $Q(v)$ is not identically zero, and matrices $Q(u)$, $Q(v)$, $V(v)$ commute for all values of u, v.

(vi) The matrices $Q(v)$, $V(v)$ commute with the diagonal operator S that has entries $+1$ (-1) for row-states with an even (odd) number of down arrows. They therefore break up into two diagonal blocks. Within each block all elements of $Q(v)$ are of the form (9.4.7), where $-N < r < N$ and r takes even (odd) values.

Sufficiency

These properties (i)–(vi) are in fact sufficient to define the eigenvalues of $V(v)$. All we have to do is to reason backwards to (9.3.6) and (9.3.8) as follows.

From the commutation properties (ii) and (v), there exists a matrix P (independent of v) such that

$$P^{-1}V(v)\,P = V_d(v), \quad P^{-1}Q(v)\,P = Q_d(v), \tag{9.5.1}$$

where $V_d(v)$ and $Q_d(v)$ are diagonal. From (iv), (9.4.5) is satisfied and can therefore be put into the diagonal form (9.4.3). Let $\Lambda(v)$ be a particular eigenvalue of $V(v)$, and $q(v)$ the corresponding eigenvalue of $Q(v)$. Then the corresponding entry in (9.4.3) is the function relation (9.3.6).

From (9.5.1), $\Lambda(v)$ and $q(v)$ are sums over elements of $V(v)$, $Q(v)$, respectively, weighted by coefficients (from P^{-1} and P) that are independent of v. From (iii) it follows that $\Lambda(v)$ is entire. From (vi) it follows that $q(v)$ is entire and of the form (9.4.7), where r takes either all even or all odd values.

If $q(v)$ were indentically zero for all v, then so would be the determinant of $Q(v)$. Provided this does not occur, it must be possible to write $q(v)$ in the form (9.3.5) (where $0 \leqslant n \leqslant N$), together with a non-zero factor that cancels out of (9.3.6). As shown at the beginning of Section 9.4, the relation (9.3.6) now implies the equations (9.3.8). These define $v_1, \ldots,$ v_n. From (9.3.5), $q(v)$ is now known, so (9.3.6) gives $\Lambda(v)$.

These equations are exact for finite n and N. Of course it still remains to solve (9.3.8), and in general this can only be done analytically in the limit $n, N \to \infty$, using methods such as those of Sections 8.6–8.9. Even so, the equations (9.3.8) are an enormous simplification of the original eigenvalue problem: they could for instance be solved rapidly on a computer even for moderately large values of n and N. In this sense they are a 'solution' of the eigenvalue problem.

Note that n does not occur in the properties (i)–(vi). Thus one may hope to generalize them to models where there is no line conservation. This will be done in Chapter 10.

9.6 Direct Derivation of the Matrix Properties: Commutation

Can the properties (i)–(vi) be established without using the Bethe ansatz, hence giving an alternative way of diagonalizing $V(v)$? They can, as will be shown in this and the next two sections.

Consider a horizontal row of the lattice and the adjacent vertical edges. With each edge i associate a 'spin' μ_i such that $\mu_i = +1$ if the corresponding arrow points up or to the right, and $\mu_i = -1$ if the arrow points down or to the left.

Let $\alpha_1, \ldots, \alpha_N$ be the spins on the lower row of vertical edges: $\beta_1, \ldots, \beta_n$ the spins on the upper row; and $\mu_1, \ldots, \mu_N$ the spins on the horizontal edges; as indicated in Fig. 9.1. Denoting the set $\{\alpha_1, \ldots, \alpha_N\}$ by $\boldsymbol{\alpha}$; and $\{\beta_1, \ldots, \beta_N\}$ by $\boldsymbol{\beta}$; it is obvious that $\boldsymbol{\alpha}$ ($\boldsymbol{\beta}$) specifies the spins in the lower (upper) row. Thus the transfer matrix V has elements $V_{\alpha\beta}$, and these are given by

$$V_{\alpha\beta} = \sum_{\mu_1} \ldots \sum_{\mu_N} w(\mu_1, \alpha_1 | \beta_1, \mu_2) \, w(\mu_2, \alpha_2 | \beta_2, \mu_3)$$

$$\ldots w(\mu_N, \alpha_N | \beta_N, \mu_1). \tag{9.6.1}$$

Here $w(\mu, \alpha|\beta, \mu')$ is the Boltzmann weight of the vertex configuration specified by the spins μ, α, β, μ'. From Fig. 8.2 and (8.3.3) it follows that

$$w(+, +|+, +) = w(-, -|-, -) = a,$$

$$w(+, -|-, +) = w(-, +|+, -) = b, \qquad (9.6.2)$$

$$w(+, -|+, -) = w(-, +|-, +) = c,$$

and $w(\mu, \alpha|\beta, \mu')$ is zero for all other values of μ, α, β, μ'.

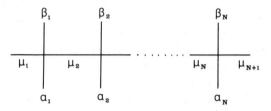

Fig. 9.1. A row of the square lattice, showing the 'spins' associated with the various edges. The cyclic boundary condition is that $\mu_{N+1} = \mu_1$.

Let V' be another transfer matrix, defined by (9.6.1) and (9.6.2), but with a, b, c replaced by a', b', c'. Denote the corresponding vertex weight function w by w'. Then from (9.6.1)

$$(VV')_{\alpha\beta} = \sum_\gamma V_{\alpha\gamma} V'_{\gamma b}$$

$$= \sum_{\mu_1,\ldots,\mu_N} \sum_{\nu_1,\ldots,\nu_N} \prod_{i=1}^N S(\mu_i, \nu_i|\mu_{i+1}, \nu_{i+1}|\alpha_i, \beta_i) \qquad (9.6.3)$$

where

$$S(\mu, \nu|\mu', \nu'|\alpha, \beta) = \sum_\gamma w(\mu, \alpha|\gamma, \mu') w'(\nu, \gamma|\beta, \nu'). \qquad (9.6.4)$$

This S is simply the Boltzmann weight of a pair of sites, one above the other, as indicated in Fig. 9.2, summed over the possible arrow configurations on the intervening edge.

Let $S(\alpha, \beta)$ be the four-by-four matrix with rows labelled by (μ, ν), columns labelled by (μ', ν'), and elements $S(\mu, \nu|\mu', \nu'|\alpha, \beta)$. Then (9.6.3) can be written more compactly as

$$(VV')_{\alpha\beta} = \text{Tr } S(\alpha_1, \beta_1) S(\alpha_2, \beta_2) \ldots S(\alpha_N, \beta_N). \qquad (9.6.5)$$

Similarly,

$$(V'V)_{\alpha\beta} = \text{Tr } S'(\alpha_1, \beta_1) S'(\alpha_2, \beta_2) \ldots S'(\alpha_N, \beta_N), \qquad (9.6.6)$$

where S' is defined in the same way as S, but with w, w' interchanged in (9.6.4).

To establish property (i), we want to find a V' that commutes with V, i.e. the right hand sides of (9.6.5) and (9.6.6) are the same. Clearly this will be so if there exists a four-by-four non-singular matrix M such that

$$S(\alpha, \beta) = M\, S'(\alpha, \beta)\, M^{-1}, \qquad (9.6.7)$$

for $\alpha = \pm 1$ and $\beta = \pm 1$.

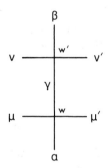

Fig. 9.2. The lattice segment whose weight (summed over the internal edge spin γ) is the $S(\mu, \nu | \mu', \nu' | \alpha, \beta)$ of eq. (9.6.4).

Star – Triangle Relation

The matrix M has rows labelled by (μ, ν); columns by (μ', ν'). If we write the elements as $w''(\nu, \mu | \nu', \mu')$, post-multiply (9.6.7) by M and write the matrix products explicitly, using (9.6.4), we obtain

$$\sum_{\gamma, \mu'', \nu''} w(\mu, \alpha | \gamma, \mu'')\, w'(\nu, \gamma | \beta, \nu'')\, w''(\nu'', \mu'' | \nu', \mu')$$

$$= \sum_{\gamma, \mu'', \nu''} w''(\nu, \mu | \nu'', \mu'')\, w'(\mu'', \alpha | \gamma, \mu')\, w(\nu'', \gamma | \beta, \nu'), \qquad (9.6.8)$$

for $\alpha, \beta, \mu, \nu, \mu', \nu' = \pm 1$.

We can regard $w''(\nu, \mu | \nu', \mu')$ as a 'Boltzmann weight function' for a vertex with surrounding edge spins ν, μ, ν', μ'. Then (9.6.8) can be given the simple graphical interpretation indicated in Fig. 9.3: the combined weight of the left-hand trilateral (summed over spins on internal edges) must be the same as that of the right-hand trilateral. This must be true for all values of the six exterior spins.

One figure can be obtained from the other by shifting a line across the intersection of the other two. In both figures the lines (μ, ν'), (α, β)

intersect at the vertex with weight function w. Similarly (α, β), (ν, μ') intersect with function w'; (ν, μ'), (μ, ν') with w''.

This picture of the equation (9.6.8) can be very illuminating, as will be shown in Chapter 11.

The equation (9.6.8) can also be written in terms of operators: let U_i be the matrix with elements

$$(U_i)_{\alpha\beta} = \delta(\alpha_1, \beta_1) \ldots \delta(\alpha_{i-1}, \beta_{i-1})$$

$$w(\alpha_i, \alpha_{i+1} | \beta_i, \beta_{i+1}) \, \delta(\alpha_{i+2}, \beta_{i+2}) \ldots \delta(\alpha_N, \beta_N). \qquad (9.6.9)$$

Thus U_i acts on the spins in position i and $i + 1$, leaving the rest unchanged. It can be interpreted as a vertex operator.

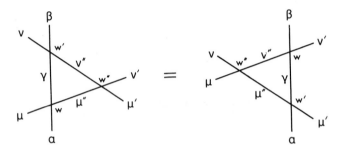

Fig. 9.3. The lattice segments whose weights (summed over the internal edge spins γ, μ'', ν'') are the left- and right-hand sides of eq. (9.6.8). This relation is the 'star–triangle' relation of the vertex models.

Similar, define U_i' and U_i'' by (9.6.9), with w replaced by w, w'', respectively. Then (9.6.8) implies that

$$U_{i+1} U_i' U_{i+1}'' = U_i'' U_{i+1}' U_i, \qquad (9.6.10)$$

and it is obvious from (9.6.9.) that

$$U_i U_j' = U_j' U_i, \qquad (9.6.11)$$

if $|i - j| \geq 2$.

This equation (9.6.10) is the same as (6.4.27). Thus the present operators U_i satisfy the same star–triangle property as the corresponding Ising model operators of Section 6.4. Since (9.6.10) is a direct corollary of (9.6.8), I shall therefore call (9.6.8) the 'star–triangle' relation for the ice-type models. (Strictly speaking, a more accurate name would be 'trilateral-to-trilateral'.)

From (9.6.11), the U_i operators also satisfy (6.4.28). Further, if N is replaced by $N + 1$ in (9.6.9), then the element

$$(\beta_1, \ldots, \beta_{N-1}, \mu_N, \alpha_N | \beta_1, \ldots, \beta_N, \mu_{N+1})$$

of U_N is the weight of the extreme-right vertex in Fig. 9.1. Thus U_N can be thought of as the operator which adds this vertex, going from edge spins β_N, μ_{N+1} to spins μ_N, α_N.

Similarly, the product $U_1 \ldots U_N$ adds the row of vertices in Fig. 9.1, going from $\beta_1, \ldots, \beta_N, \mu_{N+1}$ to $\mu_1, \alpha_1, \ldots, \alpha_N$. Apart from boundary conditions and a shift of spin indices, it is therefore the transfer matrix and the partial commutation argument of (6.4.30)–(6.4.31) applies. In all these respects the U_i operators of this chapter therefore correspond to those of Section 6.4.

To summarize: the transfer matrix V commutes with another transfer matrix V' if w'' can be chosen to satisfy (9.6.8). This is analogous to the star – triangle relation of the Ising model.

Solution of the Star – Triangle Relation

Given w, we want to find w', w'' so that (9.6.8) is satisfied. One trivial solution is $w' \propto w, w''(\nu, \mu | \nu', \mu') = \delta(\nu, \nu') \delta(\mu, \mu')$; but this is not interesting since it implies only that V commutes with a scalar multiple of itself. We want solutions in which w' is not simply proportional to w.

From Fig. 9.3 it is obvious that w'' plays a very similar role to w and w'. Thus it is natural to take w'' also to be given by (9.6.2), but with a, b, c replaced by a'', b'', c''.

If a, b, c are given, then a', b', c' and a'', b'', c'' are at our disposal. Since (9.6.8) is homogeneous in w' and w'', this leaves us four disposable parameters.

On the other hand, $\alpha, \beta, \mu, \nu, \mu', \nu'$ in (9.6.8) can each independently take the values ± 1, so (9.6.8) represents 64 scalar equations. At first sight the task of satisfying all of them seems hopeless!

Fortunately there are many simplifications. From (9.6.2), $w(\mu, \alpha | \beta, \nu) = 0$ unless $\mu + \alpha = \beta + \nu$, and similarly for w', w''. It follows that both sides of (9.6.8) are zero unless $\nu + \mu + \alpha = \beta + \nu' + \mu'$. This leaves only 20 non-trivial equations.

Negating all spins leaves w, w', w'' and (9.6.8) unchanged, so these 20 occur in 10 identical pairs.

Further, interchanging the pairs, $(\alpha, \beta), (\mu, \nu'), (\mu', \nu), (\mu'', \nu'')$ merely interchanges the two sides of (9.6.8). This implies that 4 of the remaining

10 equations are satisfied identically, while the rest occur in three equivalent pairs. Thus (9.6.8) finally reduces to just three equations, namely

$$ac'a'' = bc'b'' + ca'c''\,,$$
$$ab'c'' = ba'c'' + cc'b''\,, \tag{9.6.12}$$
$$cb'a'' = c\underset{\sim}{a}'b'' + bc'c''\,.$$

Eliminating a'', b'', c'' leaves the single equation

$$(a^2 + b^2 - c^2)/(ab) = (a'^2 + b'^2 - c'^2)/(a'b')\,. \tag{9.6.13}$$

Defining Δ as in (8.3.21) and (9.2.1), it follows that w'' can be chosen to satisfy the star – triangle relation (9.6.8) provided that

$$\Delta = \Delta'\,. \tag{9.6.14}$$

Thus if V and V' have different values of a, b, c, but the same value of Δ; then they commute. We have therefore directly established the commutation property observed in Section 9.2. This completes step (i) of Section 9.5.

9.7   Parametrization in Terms of Entire Functions

To establish the properties (ii) and (iii) of Section 9.5, we need to parametrize a, b, c in terms of three other variables, say ρ, λ and v, so that a, b, c are entire functions of v, but Δ is independent of v.

An obvious parametrization is to regard a, Δ as 'constants' and to introduce a variable $x = b/a$. Then from (9.21),

$$a = a\,, \quad b = ax\,, \quad c = a(1 + x^2 - 2\Delta x)^{\frac{1}{2}}. \tag{9.7.1}$$

However, c is not an entire function of x: it is the square root of a quadratic polynomial in x.

There is a simple way of parametrizing a function $F = [(x - x_1)(x - x_2)]^{\frac{1}{2}}$, namely to define

$$t^2 = (x - x_1)/(x - x_2)\,, \tag{9.7.2}$$

i.e. to set

$$x = (x_1 - t^2 x_2)/(1 - t^2)\,. \tag{9.7.3}$$

Then the sign of t can be chosen so that

$$F = (x_1 - x_2)t/(1 - t^2)\,, \tag{9.7.4}$$

so both x and F are rational functions of t.

In our case x_1 and x_2 are the zeros of $1 + x^2 - 2\Delta x$. Thus

$$\Delta = \tfrac{1}{2}(x_1 + x_1^{-1}) , \quad x_2 = x_1^{-1} , \tag{9.7.5}$$

and (9.7.1) becomes

$$a = a, \quad b = a(x_1 - t^2 x_1^{-1})/(1 - t^2), \tag{9.7.6}$$

$$c = a(x_1 - x_1^{-1})t/(1 - t^2) .$$

We can 're-normalize' to remove the denominators by setting $a = \rho' \, x_1(1 - t^2)$. Then

$$a = \rho' \, x_1(1 - t^2), \quad b = \rho' \, (x_1^2 - t^2), \tag{9.7.7}$$

$$c = \rho'(x_1^2 - 1) \, t .$$

With this parametrization, a, b, c are entire functions of ρ', x_1 and t, but Δ depends only on x_1. Varying t changes $a:b:c$, but leaves Δ unaltered. This completes the derivation of (ii) and (iii), and we could continue to use this parametrization, regarding ρ', x_1 as constants and t as a variable. However, to regain contact with the results of the Bethe ansatz (and to make the subsequent generalization to the eight-vertex model more straightforward) it is useful to finally transform from ρ', x_1, t to ρ, λ, v by setting

$$x_1 = -\exp(-\lambda), \quad t = \exp[\tfrac{1}{2}(v - \lambda)], \quad \rho' = \tfrac{1}{2}\rho t^{-1} x_1^{-1}. \tag{9.7.8}$$

We then regain (9.2.1), (9.2.3) and the properties (ii) and (iii) of Section 9.5.

Parametrized Star – Triangle Operator Relation

So far we have used only the corollary (9.6.14) of the star – triangle relations (9.6.12). Since (9.6.12) is unchanged by interchanging the twice-primed and unprimed weights a, b, c; another obvious corollary is $\Delta'' = \Delta'$. Thus (a', b', c') and (a'', b'', c'') can also be parametrized in the form (9.2.3), all sets having the same value of λ.

They have different values of v; let us call them v' and v'', respectively [and similarly for ρ, but these normalization factors cancel trivially out of (9.6.12)]. Substituting the resulting expressions (9.2.3) for $a, \ldots, c''$ into (9.6.12), all three equations are satisfied if

$$\sinh \tfrac{1}{4}(\lambda + v - v' + v'') = 0 . \tag{9.7.9}$$

Incrementing v' by $4\pi i$ leaves a', b', c' unchanged, so without loss of generality we can take the solution of (9.7.9) to be

$$v' = \lambda + v + v'' . \tag{9.7.10}$$

For some purposes it is convenient to use

$$u = \tfrac{1}{2}(\lambda + v) , \tag{9.7.11}$$

as a variable, instead of v. Then (9.2.3) becomes

$$a , b , c = \rho \sinh (\lambda - u) , \rho \sinh u , \rho \sinh \lambda , \tag{9.7.12}$$

and if $u' = \tfrac{1}{2}(\lambda + v')$, $u'' = \tfrac{1}{2}(\lambda + v'')$, then (9.7.10) becomes

$$u' = u + u'' . \tag{9.7.13}$$

The vertex operators U_i defined by (9.6.9) and (9.6.2) are functions of a, b, c. From (9.7.12) they therefore depend on ρ, λ, u. Regarding ρ, λ as constants, we can regard U_i as a function of u and write it as $U_i(u)$. Then the U_i' and U_i'' in (9.6.10) are $U_i(u')$ and $U_i(u'')$, respectively. Using (9.7.13), the star – triangle operator relation (9.6.10) becomes

$$U_{i+1}(u) \, U_i(u + u'') \, U_{i+1}(u'') = U_i(u'') \, U_{i+1}(u + u'') \, U_i(u) . \tag{9.7.14}$$

This is an identity, true for $i = 1, \ldots, N - 2$, and for all complex numbers u and u''. In particular, it is true for $0 < u/\lambda < 1, 0 < u''/\lambda < 1$: for $\Delta < 1$ these are the 'physical' values of u, u''; corresponding to positive Boltzmann weights a, b, c, a'', b'', c''.

Comparing (9.7.14) with (7.13.5), we again see a very close analogy between the 'star – triangle' relations of the ice-type and Ising models.

9.8   The Matrix $Q(v)$

Column Vectors y

The next step is to obtain property (iv), i.e. to construct the matrix $Q(v)$ that satisfies (9.4.5).

Let y be a particular column of $Q(v)$. Then (9.4.5) implies that

$$V(v) y = y' + y'' , \tag{9.8.1}$$

where y' and y'' are proportional to y, with v replaced by $v + 2\lambda'$ and $v - 2\lambda'$, respectively.

Let us try to construct y, y', y'' directly, and let $y(\alpha_1, \ldots, \alpha_N)$ be the element $(\alpha_1, \ldots, \alpha_N)$ of the vector y. Then the product $V(v) y$ simplifies if $y(\alpha_1, \ldots, \alpha_N)$ has the product form

$$y(\alpha_1, \ldots, \alpha_N) = g_1(\alpha_1) g_2(\alpha_2) \ldots g_N(\alpha_N), \qquad (9.8.2)$$

i.e. y is a direct product of the two-dimensional vectors $g_1, \ldots, g_N$. In fact, from (9.6.1), the element $(\alpha_1, \ldots, \alpha_N)$ of $V(v) y$ is

$$[V(v) y]_\alpha = \mathrm{Tr}\, G_1(\alpha_1) G_2(\alpha_2) \ldots G_N(\alpha_N), \qquad (9.8.3a)$$

where $G_i(+)$ and $G_i(-)$ are two-by-two matrices with (μ, μ') elements

$$[G_i(\alpha)]_{\mu\mu'} = \sum_\beta w(\mu, \alpha | \beta, \mu') g_i(\beta). \qquad (9.8.3b)$$

Explicitly, using (9.6.2),

$$G_i(+) = \begin{pmatrix} ag_i(+) & 0 \\ cg_i(-) & bg_i(+) \end{pmatrix}$$

$$G_i(-) = \begin{pmatrix} bg_i(-) & cg_i(+) \\ 0 & ag_i(-) \end{pmatrix} \qquad (9.8.4)$$

We want the RHS of (9.8.3a) to decompose into the sum of two terms, each like (9.8.2). This will be so if there exist two-by-two matrices $P_1, \ldots, P_N$ such that

$$G_i(\alpha) = P_i H_i(\alpha) P_{i+1}^{-1}, \qquad (9.8.5)$$

where each $H_i(\alpha)$ is upper-right triangular, and $P_{N+1} = P_1$.

To see this, substitute the form (9.8.5) of each $G_i(\alpha)$ into (9.8.3a). The P_is cancel, so the effect is to replace each $G_i(\alpha)$ by $H_i(\alpha)$. If $H_i(\alpha)$ has the form

$$H_i(\alpha) = \begin{pmatrix} g_i'(\alpha) & g_i'''(\alpha) \\ 0 & g_i''(\alpha) \end{pmatrix}, \qquad (9.8.6)$$

then (9.8.3a) gives

$$[V(v)y]_\alpha = g_1'(\alpha_1) \ldots g_N'(\alpha_N) + g_1''(\alpha_1) \ldots g_N''(\alpha_N). \qquad (9.8.7)$$

Pair Propagation Through a Vertex

Can (9.8.5) and (9.8.6) be satisfied, i.e. can we choose the P_i so that the the bottom-left element of $P_i^{-1} G_i(\alpha) P_{i+1}$ vanishes for both $\alpha = +1$ and $\alpha = -1$? If p_i is the first column of P_i, this is equivalent to requiring that

$$G_i(\alpha) p_{i+1} = g_i'(\alpha) p_i, \qquad (9.8.8)$$

for $\alpha = \pm 1$. Here $G_i(+)$ and $G_i(-)$ are two-by-two matrices; the p_i are two-dimensional vectors; $g_i'(+)$ and $g_i'(-)$ are scalars.

Let the elements of p_i be $p_i(+)$ and $p_i(-)$. Using (9.8.3b), the condition (9.8.8) can be written explicitly as

$$\sum_{\beta,\mu'} w(\mu, \alpha | \beta, \mu') g_i(\beta) p_{i+1}(\mu') = g_i'(\alpha) p_i(\mu), \qquad (9.8.9)$$

for $\alpha, \mu = \pm 1$.

This equation can be interpreted graphically as in Fig. 9.4. Let μ, α, β, μ' be the edge spins round a vertex, as shown. With the upper and right-hand edges associate weights $g_i(\beta)$, $p_{i+1}(\mu')$. Sum over all values of β, μ', weighted by the vertex weight w. This gives a function of μ and α.

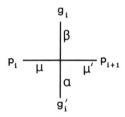

Fig. 9.4. Pair-propagation through a vertex: graphical representation of eq. (9.8.9).

The condition (9.8.9) implies that this function factors into a weight $g_i'(\alpha)$ for the lower edge, $p_i(\mu)$ for the left-hand one.

Thus we can think of (9.8.9) as saying that the vertical and horizontal functions (or vectors) g_i, p_{i+1} 'propagate' through the vertex to become g_i', p_i. Indeed, using (9.6.9), (9.8.9) can be written in fairly obvious operator notation as

$$U_i(g_i \otimes p_{i+1}) = p_i \otimes g_i'. \qquad (9.8.10)$$

Since $\mu, \alpha = \pm 1$, (9.8.9) represents four scalar equations. Explicitly they are

$$a \, g_i(+) \, p_{i+1}(+) = g_i'(+) \, p_i(+)$$

$$b \, g_i(-) \, p_{i+1}(+) + c \, g_i(+) \, p_{i+1}(-) = g_i'(-) \, p_i(+)$$

$$c \, g_i(-) \, p_{i+1}(+) + b \, g_i(+) \, p_{i+1}(-) = g_i'(+) \, p_i(-) \qquad (9.8.11)$$

$$a \, g_i(-) \, p_{i+1}(-) = g_i'(-) \, p_i(-).$$

These equations are homogeneous and linear in $g_i(+)$, $g_i(-)$, $g_i'(+)$, $g_i'(-)$, so these variables can be eliminated (by taking the determinant of coefficients), leaving

$$\frac{a^2 + b^2 - c^2}{ab} = \frac{r_{i+1}}{r_i} + \frac{r_i}{r_{i+1}}, \tag{9.8.12}$$

where

$$r_i = p_i(-)/p_i(+) \quad \text{for } i = 1, \ldots, N. \tag{9.8.13}$$

This is a quadratic recurrence relation between r_i and r_{i+1}. There are two interesting things about it: firstly, it involves a, b, c only via the Δ defined by (9.2.1); secondly, from (9.2.4) it explicitly factors into the simple form

$$r_{i+1} = -r_i \exp(\pm\lambda). \tag{9.8.14}$$

Column Vectors $y(v)$ and the Matrix $Q_R(v)$

The relation (9.8.14) must hold for $i = 1, \ldots, N$; but the choice of sign can be made independently for each i. Thus the most general solution for $r_1, \ldots, r_{N+1}$ is

$$r_i = (-)^i r \exp[\lambda(\sigma_1 + \ldots + \sigma_{i-1})], \tag{9.8.15}$$

where r is arbitrary and each σ_i has value ± 1. The cyclic boundary condition $r_{N+1} = r_1$ is satisfied if

$$\sigma_1 + \ldots + \sigma_N = 0, \tag{9.8.16}$$

which implies that N must be even.

(If $|\Delta| < 1$, then λ is pure imaginary. If λ equals $2\pi i m/n$, where m and n are integers, then it is sufficient that $\sigma_1 + \ldots + \sigma_N$ be a multiple of n. Such cases are often of particular interest, e.g. the pure ice model has $\lambda = 2\pi i/3$).

We can choose all $p_i(+)$, $g_i(+)$ to be unity, so $p_i(-) = r_i$. Solving (9.8.11), using (9.8.15) and (9.2.3), then gives

$$g_i(+) = 1, \quad g_i(-) = r_i \exp[\tfrac{1}{2}(\lambda + v)\sigma_i],$$

$$g_i'(+) = a, \quad g_i'(-) = -ar_i \exp[\tfrac{1}{2}(3\lambda + v)\sigma_i]. \tag{9.8.17}$$

The equation (9.8.8) is now satisfied, where p_i is the first column of P_i. It follows that the matrix $H_i(\alpha)$ defined by (9.8.5) must be of the form (9.8.6), whatever the choice of the second column of P_i (so long as no P_i is singular). The elements $g_i''(\alpha)$ can be obtained by taking the determinant of both sides of (9.8.5) and using (9.8.4) and (9.8.6). This gives

$$g_i''(\alpha) = \frac{ab\, g_i^2(\alpha)\, \det(P_{i+1})}{g_i'(\alpha)\, \det(P_i)}. \tag{9.8.18}$$

Substituting this expression into the last term in (9.8.7), the P_i-determinants cancel out, so we can ignore them in (9.8.18) (or we could require them to be unity). From (9.8.17) it then follows that

$$g_i''(+) = b, \quad g_i''(-) = -br_i \exp[\tfrac{1}{2}(v - \lambda)\sigma_i] . \qquad (9.8.19)$$

For a given i, let us define a two-dimensional vector function $h_i(v)$ of v by

$$h_i(v) = \begin{pmatrix} 1 \\ r_i \exp[\tfrac{1}{2}(\lambda + v)\sigma_i] \end{pmatrix} , \qquad (9.8.20)$$

where we take r in (9.8.15), and hence all $r_1, \ldots, r_N$, to be independent of v. Let g_i be the two-dimensional vector

$$g_i = \begin{pmatrix} g_i(+) \\ g_i(-) \end{pmatrix} . \qquad (9.8.21)$$

and similarly for g_i' , g_i'' .

Then the equations (9.8.17), (9.8.19) can be written very neatly as

$$g_i = h_i(v) , \quad g_i' = a h_i(v + 2\lambda') , \quad g_i'' = b h_i(v - 2\lambda') , \qquad (9.8.22)$$

where $\lambda' = \lambda + i\pi$, as in (9.3.7). If we also define a 2^N-dimensional vector function $y(v)$ of v by

$$y(v) = h_1(v) \otimes h_2(v) \otimes \ldots \otimes h_N(v) , \qquad (9.8.23)$$

then, using (9.8.2), the equation (9.8.7) can be written

$$V(v) y(v) = a^N y(v + 2\lambda') + b^N y(v - 2\lambda') . \qquad (9.8.24)$$

From (9.2.3) and (9.3.4), $a^N = \phi(\lambda - v)$ and $b^N = \phi(\lambda + v)$. There are many choices of the $y(v)$, corresponding to different choices of r, $\sigma_1, \ldots, \sigma_N$ in (9.8.15), subject only to the restriction (9.8.16). Let $Q_R(v)$ be a 2^N by 2^N matrix whose columns are linear combinations (with coefficients that are independent of v) of such vectors $y(v)$. Then it follows immediately from (9.8.24) that

$$V(v) Q_R(v) = \phi(\lambda - v) Q_R(v + 2\lambda') + \phi(\lambda + v) Q_R(v - 2\lambda') , \qquad (9.8.25)$$

which is basically the equation (9.4.5) required by property (iv) of Section 9.5.

Row Vectors $y^T(-v)$ and Matrix $Q_L(v)$

We still have to satisfy (v) and (vi). From (9.6.1) and (9.6.2) (by interchanging α_i and β_i, and negating all μ_i) it can be seen that interchanging

a with b is equivalent to transposing the transfer matrix V. Thus, from (9.2.3),

$$V(-v) = V^T(v) \, . \tag{9.8.26}$$

If we define

$$Q_L(v) = Q_R^T(-v) \, , \tag{9.8.27}$$

then transposing (9.8.25) and negating v gives

$$Q_L(v) \, V(v) = \phi(\lambda - v) \, Q_L(v + 2\lambda') + \phi(\lambda + v) \, Q_L(v - 2\lambda') \, , \tag{9.8.28}$$

so $Q_L(v)$ plays a similar role to $Q_R(v)$, except that it pre-multiplies the transfer matrix, instead of post-multiplying.

The vector $y(v)$ is defined by (9.8.23), (9.8.20) and (9.8.15), so depends on r and $\sigma_1, \ldots, \sigma_N$, as well as v. This can be exhibited by writing it as $y(v|r, \sigma)$. Consider the scalar product

$$y^T(-u|r', \sigma') \, y(v|r, \sigma) \, , \tag{9.8.29}$$

of two such vectors. This is readily evaluated as

$$\prod_{i=1}^{N} \{1 + rr' \exp[\lambda(\sigma_1 + \ldots + \sigma_{i-1} + \sigma_1' + \ldots + \sigma_{i-1}')$$

$$+ \tfrac{1}{2}(\lambda + v)\sigma_i + \tfrac{1}{2}(\lambda - u)\sigma_i']\} \, . \tag{9.8.30}$$

In particular, this expression depends on $\sigma_1, \ldots, \sigma_N$, u and v so let us call it $J(u, v|\sigma_1, \ldots, \sigma_N)$, and consider the ratio

$$J(u, v|\ldots, \sigma_{j+1}, \sigma_j, \ldots)/J(u, v|\ldots, \sigma_j, \sigma_{j+1}, \ldots) \, , \tag{9.8.31}$$

the numerator differing from the denominator only in the interchange of σ_j and σ_{j+1}. Since all but the $i = j$ and $i = j + 1$ terms in (9.8.30) are symmetric in σ_j, σ_{j+1}, this ratio simplifies, leaving only these terms in the numerator and denominator. A simple direct calculation (using the fact that σ_j, σ_{j+1}, σ_j', σ_{j+1}' only take the values ± 1) then reveals that the ratio (9.8.31) is a symmetric function of u and v.

However, it is obvious from (9.8.30) that $J(u, v|\sigma_1, \ldots, \sigma_N)$ is symmetric in u and v if $\sigma_i = \sigma_i'$ for $i = 1, \ldots, N$. Since all values of $\sigma_1, \ldots, \sigma_N$ allowed by (9.8.16) are permutations of this particular set of values, and since all such permutations can be obtained by successive interchanges of pairs (σ_j, σ_{j+1}), it follows that (9.8.30) is always a symmetric function of u and v. Thus

$$y^T(-u|r', \sigma') \, y(v|r, \sigma) = y^T(-v|r', \sigma') \, y(u|r, \sigma) \, . \tag{9.8.32}$$

Now consider the matrix product $Q_L(u) Q_R(v)$. Since any column of $Q_R(v)$ is a linear combination of vectors $y(v|r, \sigma)$, and any row of $Q_L(u)$ is a linear combination of vectors $y^T(-u|r', \sigma')$, it follows immediately from (9.8.32) that

$$Q_L(u) Q_R(v) = Q_L(v) Q_R(u) , \qquad (9.8.33)$$

for all complex numbers u, v.

$Q_R(v)$ a Non-Singular Matrix

Consider now the set of vectors $y(v|r, \sigma)$ formed by letting r take all possible complex number values, and $\sigma = \{\sigma_1, \ldots, \sigma_N\}$ taking all the $\binom{N}{\frac{1}{2}N}$ values allowed by (9.8.16). I want to assert that there are values of v for which these vectors span all 2^N-dimensional space, so that $Q_R(v)$ and $Q_L(-v)$ can then be chosen non-singular. Unfortunately I know of no simple way to completely prove this, but it is almost certainly correct and the following argument supports the assertion.

From (9.8.23), (9.8.20) and (9.8.15), the element $(\alpha_1, \ldots, \alpha_N)$ of $y(v|r, \sigma)$ contains a factor

$$r^{\frac{1}{2}(N - \alpha_1 - \ldots - \alpha_N)} , \qquad (9.8.34)$$

the other terms being independent of r. Thus

$$y(v|r, \sigma) = \sum_{n=0}^{N} r^n y_n(v|\sigma) , \qquad (9.8.35)$$

where each $y_n(v|\sigma)$ has non-zero elements only when $\alpha_1 + \ldots + \alpha_N = N - 2n$, i.e. when there are n down arrow spins.

Let $\mathcal{V}_n$ be the $\binom{N}{n}$-dimensional space of vectors whose elements are zero unless $\alpha_1 + \ldots + \alpha_N = N - 2n$. Then it would be sufficient to show (for $n = 0, \ldots, N$) that $\mathcal{V}_n$ is spanned by the vectors $y_n(v|\sigma)$ obtained by letting σ take all possible values.

Since there are $\binom{N}{\frac{1}{2}N}$ such values of σ permitted by (9.8.16), there are at least as many vectors $y_n(v|\sigma)$ as the dimensionality of $\mathcal{V}_n$. The most delicate case is $n = \frac{1}{2}N$, when there are just enough vectors.

Each element $(\alpha_1, \ldots, \alpha_N)$ of $y_n(v|\sigma)$ contains a factor

$$\exp\left[\frac{1}{4}v \sum_{i=1}^{N} \sigma_i(1 - \alpha_i)\right] , \qquad (9.8.36)$$

all other terms being independent of v and non-zero. From (9.8.16), this factor simplifies to

$$\exp\left[-\tfrac{1}{4} v \sum_{i=1}^{N} \sigma_i \alpha_i \right]. \tag{9.8.37}$$

If $n = \tfrac{1}{2}N$ and v is large and negative, then there is a single dominant element of $y_n(v|\sigma)$ given by $\alpha_1, \ldots, \alpha_N = \sigma_1, \ldots, \sigma_N$: this maximizes (9.8.37) and is consistent with (9.8.16). Thus there are $\binom{N}{n}$ column vectors, each with its dominant element in a different row. These vectors clearly form a basis of $\mathcal{V}_n$.

The assertion is therefore certainly true for $n = \tfrac{1}{2}N$. Since this subspace contains the maximum eigenvalue of V, this eigenvalue can certainly be obtained by the present methods. More generally, for $n \neq \tfrac{1}{2}N$, there are more vectors $y_n(v|\sigma)$ than necessary and there is no reason to suppose they do not (for general values of v) span $\mathcal{V}_n$.

$Q(v)$ and its Commutation Relations

From now on let us therefore suppose that the determinant of $Q_R(v)$, and hence $Q_L(v)$, does not vanish identically (it may of course vanish for a finite number of complex values of v). Let v_0 be a value for which it is non-zero and define

$$Q(v) = Q_R(v)\, Q_R^{-1}(v_0). \tag{9.8.38}$$

Taking $u = v_0$ in (9.8.33), it follows that

$$Q(v) = Q_L^{-1}(v_0)\, Q_L(v). \tag{9.8.39}$$

Post-multiplying (9.8.25) by $Q_R^{-1}(v_0)$, and pre-multiplying (9.8.28) by $Q_L^{-1}(v_0)$, therefore gives
$$V(v)\, Q(v) = Q(v)\, V(v)$$
$$= \phi(\lambda - v)\, Q(v + 2\lambda') + \phi(\lambda + v)\, Q(v - 2\lambda'). \tag{9.8.40}$$

Also, from (9.8.39) and (9.8.38),

$$Q(u)\, Q(v) = Q_L^{-1}(v_0)\, Q_L(u)\, Q_R(v)\, Q_R^{-1}(v_0). \tag{9.8.41}$$

From (9.8.33), this is unaltered by interchanging u with v, so

$$Q(u)\, Q(v) = Q(v)\, Q(u). \tag{9.8.42}$$

Thus this matrix function $Q(v)$ satisfies (9.4.5) and all the properties (iv) and (v) of Section 9.5.

Commutation Relations Involving S

Finally, the diagonal operator S in (vi) is

$$S = \begin{pmatrix} 1 & 0 \\ 0 & -1 \end{pmatrix} \otimes \cdots \otimes \begin{pmatrix} 1 & 0 \\ 0 & -1 \end{pmatrix}. \tag{9.8.43}$$

From (9.8.20) and (9.8.23) it follows that

$$S\, y(v) = y(v + 2\pi i)\,, \tag{9.8.44}$$

and, since all columns of $Q_R(v)$ are linear combinations of vectors $y(v)$,

$$S\, Q_R(v) = Q_R(v + 2\pi i)\,. \tag{9.8.45}$$

Transposing, negating v, using (9.8.27) and the fact that $Q_R(v + 4\pi i) = Q_R(v)$, gives

$$Q_L(v)\, S = Q_L(v + 2\pi i)\,. \tag{9.8.46}$$

Post-multiplying (9.8.45) by $Q_R^{-1}(v_0)$, pre-multiplying (9.8.46) by $Q_L^{-1}(v_0)$, and using (9.8.38) and (9.8.39), it follows that

$$S\, Q(v) = Q(v)\, S = Q(v + 2\pi i)\,. \tag{9.8.47a}$$

Also, since $w(\mu, \alpha|\beta, \mu')$ is unchanged by multiplication by $\mu\alpha\beta\mu'$, it follows from (9.6.1) that

$$S\, V(v) = V(v)\, S\,. \tag{9.8.47b}$$

From (9.8.20) and (9.8.23), as $v \to \pm\infty$ any element of $y(v)$, and hence $Q(v)$, grows at most as fast as $\exp(\frac{1}{2}Nv)$. The properties (vi) of Section 9.5 now follow immediately from (9.8.47).

As was shown in Section 9.5, the properties (i)–(vi) imply the equations (9.3.6), (9.3.8) for the eigenvalues Λ of $V(v)$. Thus we have derived these equations without using the Bethe ansatz. There are two key steps in the working: the star triangle relation (9.6.8) and the vertex propagation relation (9.8.9). It is worth noting that both of these are local properties, the first of a triangle of three vertices, the second of a single vertex.

9.9   Values of ρ, λ, v

All the equations of this chapter are algebraic identities, so they are true for all values of a, b, c and ρ, λ, v, real or complex. It is not necessary to locate their values in the complex plane until one starts the analysis of the solution of (9.3.8), letting $N \to \infty$ and choosing the solution corresponding to the maximum eigenvalue Λ. (This analysis was performed in Sections 8.5–8.9.)

If the vertex interaction energies $\varepsilon_1, \ldots, \varepsilon_6$ are real [and satisfy (8.1.7)], then the Boltzmann weights a, b, c given by (8.3.3) are real and positive. When locating ρ, λ, v there are four cases to consider, being the four phases shown in Fig. 8.5. The ρ, λ, v can be chosen so that:

(I) $\Delta > 1$, $a > b + c$:

$\rho = -\rho'$, $\lambda = i\pi + \lambda'$, $v = -i\pi - v'$,

where ρ', λ', v' are real and $\rho' > 0$, $v' > \lambda' > 0$.

(II) $\Delta > 1$, $b > a + c$:

$\rho = -\rho'$, $\lambda = i\pi + \lambda'$, $v = i\pi + v'$,

where ρ', λ', v' are real and $\rho' > 0$, $v' > \lambda' > 0$.

(III) $-1 < \Delta < 1$, $a + b > c > |a - b|$:

$\rho = -i\rho'$, $\lambda = i\mu$, $v = iw$,

where ρ', μ, w are real and $\rho' > 0$, $\pi > \mu > |w|$.

(These are the μ, w of Section 8.8.)

(IV) $\Delta < -1$, $c > a + b$:

ρ, λ, v are real, $\rho > 0$, $\lambda > |v|$.

10

SQUARE LATTICE EIGHT-VERTEX MODEL

10.1 Introduction

Lieb's (1967a, b, c) solution of the ice-type, or six-vertex, models was the most significant new exact result since the work of Berlin and Kac (1952) on the spherical model, and the pioneering work of Onsager (1944) on the Ising model.

Even so, as models of critical phenomena the ice-type models have some unsatisfactory pathological behaviour: the ferroelectric ordered state is 'frozen' (i.e. the ordering is complete even at non-zero temperatures), and the anti-ferroelectric critical properties do not diverge or vanish as simple powers of $T - T_c$ (see Section 8.11).

The first of these unusual properties is certainly connected with the ice-rule: starting from a configuration with all arrows pointing up or to the right, the simplest deformation that can be made is to draw a line right through the lattice (going generally in the SW − NE direction) and reverse *all* arrows on this line. For an infinite lattice with ferroelectric ordering, this costs an infinite amount of energy, so gives an infinitesimal contribution to the partition function.

Sutherland (1970), and Fan and Wu (1970), therefore suggested generalizing the ice-type models as follows:

On every edge of the square lattice place an arrow;

Allow only configurations such that there are an even number of arrows into (and out of) each site;

There are eight possible arrangements of arrows at a site, or 'vertex', as shown in Fig. 10.1 (hence the name of the model). To arrangement j assign an energy $\varepsilon_j (j = 1, \ldots, 8)$. Then the partition function is

$$Z = \sum_C \exp[-(n_1\varepsilon_1 + \ldots + n_8\varepsilon_8)/k_BT] \,, \qquad (10.1.1)$$

where the sum is over all allowed configurations C of arrows on the lattice, n_j is the number of vertex arrangements of type j in configuration C, k_B is Boltzmann's constant, T is the temperature.

The first six vertex arrow arrangements in Fig. 10.1 are those permitted by the ice rule (Fig. 8.2). The last two (all arrows in, or all out) are new. Starting from the lattice state with all arrows pointing up or to the right, one can now make local deformations (e.g. reverse all arrows round a square) that cost only a finite energy, so one no longer expects the ferro electric state to be completely ordered, and may hope that the model will be in other respects also less pathological.

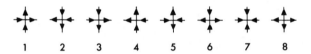

Fig. 10.1. The eight arrow configurations allowed at a vertex.

It is clear from (10.1.1) that Z is a function of the eight Boltzmann weights

$$\omega_j = \exp(-\varepsilon_j/k_BT), j = 1, \ldots, 8 \,. \qquad (10.1.2)$$

From Fig. 10.1, vertex 7 is a sink of arrows, 8 is a source. If toroidal boundary conditions are imposed on the lattice, it follows that

$$n_7 = n_8 \,. \qquad (10.1.3)$$

Similarly, reversing all vertical arrows gives vertex 5 to be a sink, 6 a source, so

$$n_5 = n_6 \,. \qquad (10.1.4)$$

Thus $\varepsilon_5, \ldots, \varepsilon_8$ in (10.1.1) occur only in the combinations $\varepsilon_5 + \varepsilon_6$, $\varepsilon_7 + \varepsilon_8$, so without loss of generality we can choose

$$\varepsilon_5 = \varepsilon_6, \quad \varepsilon_7 = \varepsilon_8 \,. \qquad (10.1.5a)$$

A particularly interesting situation is when we also have

$$\varepsilon_1 = \varepsilon_2, \quad \varepsilon_3 = \varepsilon_4 \,. \qquad (10.1.5b)$$

The model is then unchanged by reversing all arrows. Regarding the arrows as electric dipoles, this means that no external electric fields are applied, so this specialized model is known as the 'zero-field' eight-vertex model.

The solution of the zero-field eight-vertex model will be given in this chapter. The full model has not been solved. In this respect the six- and eight-vertex models differ: the former can be solved even in electric fields (Section 8.12).

10.2 Symmetries

Consider the zero-field model and set

$$a = \omega_1 = \omega_2, \quad b = \omega_3 = \omega_4,$$
$$c = \omega_5 = \omega_6, \quad d = \omega_7 = \omega_8.$$
(10.2.1)

Then from (10.1.1) and (10.1.2)

$$Z = \Sigma a^{n_1 + n_2} b^{n_3 + n_4} c^{n_5 + n_6} d^{n_7 + n_8},$$
(10.2.2)

so clearly Z is a function $Z(a, b; c, d)$ of a, b, c, d.

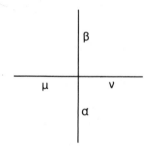

Fig. 10.2. The four arrow spins μ, α, β, ν on the edges at a vertex. The vertex configuration $(\mu, \alpha, \beta, \nu)$ has weight $w(\mu, \alpha | \beta, \nu)$ given by (10.2.3).

Fan and Wu (1970) showed that this function has several symmetries. Let μ, α, ν, β be the 'arrow-spins' associated with the four edges round a vertex, as in Fig. 10.2. They have value $+1$ (-1) if the corresponding arrow points up or to the right (down or to the left). Then the Boltzmann weight of the vertex in Fig. 10.2 is $w(\mu, \alpha | \beta, \nu)$, where

$$w(+, + | +, +) = w(-, - | -, -) = a,$$
$$w(+, - | -, +) = w(-, + | +, -) = b,$$
$$w(+, - | +, -) = w(-, + | -, +) = c,$$
$$w(+, + | -, -) = w(-, - | +, +) = d,$$
(10.2.3)

and $w(\mu, \alpha|\beta, \nu)$ is zero for all other values of μ, α, β, ν. (This is the generalization of (9.6.2).) This definition can be written more neatly as

$$w(\mu, \alpha|\beta, \nu) = \tfrac{1}{4}\{a'(1 + \alpha\beta\mu\nu) + b'(\alpha\beta + \mu\nu) + c'(\alpha\nu + \beta\mu)$$
$$+ d'(\beta\nu + \alpha\mu)\}, \qquad (10.2.4)$$

for all μ, α, β, ν, where

$$a' = \tfrac{1}{2}(a + b + c + d), \quad b' = \tfrac{1}{2}(a + b - c - d), \qquad (10.2.5)$$
$$c' = \tfrac{1}{2}(a - b + c - d), \quad d' = \tfrac{1}{2}(a - b - c + d).$$

Suppose the lattice has M rows (labelled $i = 1, \dots, M$) and N columns ($j = 1, \dots, N$). Then with this definition of w,

$$Z = \sum_{\alpha, \mu} w_{11} w_{12} \dots w_{MN}, \qquad (10.2.6)$$

where

$$w_{ij} = w(\mu_{ij}, \alpha_{ij}|\alpha_{i+1,j}, \mu_{i,j+1}), \qquad (10.2.7)$$

and the summation in (10.2.6) can be extended over all values (± 1) of the edge arrow spins $\alpha_{11}, \dots, \alpha_{MN}, \mu_{11}, \dots, \mu_{MN}$.

Using the expression (10.2.4) for the function w, (10.2.7) becomes

$$w_{ij} = w_{ij}^{(1)} + \dots + w_{ij}^{(8)}, \qquad (10.2.8)$$

where $w_{ij}^{(k)}$ corresponds to the kth additive term on the RHS of (10.2.4), and is a simple product of a weight and arrow spins, e.g. $w_{ij}^{(3)} = \tfrac{1}{4}b'\alpha_{ij}\alpha_{i+1,j}$.

Substituting the form (10.2.8) of w_{ij} into (10.2.6), the summand can be expanded into 8^{MN} terms of the form

$$w_{11}^{(k_{11})} \dots w_{MN}^{(k_{MN})}, \qquad (10.2.9)$$

where each k_{ij} is an integer between 1 and 8. Each such term (10.2.9) can be represented by an arrow graph G on the original lattice: at each site (i, j) draw the k_{ij}th vertex arrow configuration of Fig. 10.1. There are then two arrows on every edge, one from each of the end-sites.

Consider a particular edge, say the vertical one between sites $(i - 1, j)$ and (i, j), with arrow spin α_{ij}. Only two factors in (10.2.9) can contain α_{ij}, namely those for sites $(i - 1, j)$ and (i, j). Comparing (10.2.4) and Fig. 10.1, we find that α_{ij} is absent from (present in) either factor if the corresponding arrow in G is up (down).

But (10.2.9) must be summed over $\alpha_{11}, \dots, \mu_{MN}$, in particular over α_{ij}. If (10.2.9) contains an odd power of α_{ij}, it will give zero contribution to the sum, so can be ignored. This leaves only terms with an even power of α_{ij}: this power can be either zero (α_{ij} absent from both factors), or two (α_{ij} present in each factor). In either case the corresponding two arrows in G point the same way.

This applies to all edges, both vertical and horizontal, so a term (10.2.9) contributes to (10.2.6) only if all edges in G contain a pair of parallel arrows. Replace each such pair by a single arrow pointing in the common direction. Summing (10.2.9) over α, μ now merely gives a factor 4^{MN}, which cancels the factors 1/4 in each $w_{ij}^{(k)}$. Thus (10.2.6) and (10.2.9) give

$$Z = \sum a'^{m_1 + m_2} \, b'^{m_3 + m_4} \, c'^{m_5 + m_6} \, d'^{m_7 + m_8} \, , \qquad (10.2.10)$$

where m_k is the number of vertices in G of type k ($k = 1, \ldots, 8$), and the sum is over all arrow coverings G such that each vertex is one of the eight shown in Fig. 10.1.

But (10.2.10) is precisely (10.2.2), with a, b, c, d replaced by a', b', c', d'. Thus

$$Z(a, b; c, d) = Z(a', b'; c', d'). \qquad (10.2.11)$$

The method used in deriving this result is basically that of the 'weak-graph expansion' (Nagle, 1968; Nagle and Temperley, 1968; Wegner, 1973). Like the Ising model duality relation (Section 6.2), (10.2.11) relates a high-temperature model (a, b, c, d almost equal) to a low-temperature one ($a \gg b, c, d$). Indeed, it will be shown in the next section that the Ising model is a special case of the eight-vertex model: the duality relation (6.2.14) can in fact be deduced from (10.2.11).

Some other simple symmetries (which relate high-temperature models to high-temperature ones, and low to low) are readily deduced from (10.1.1)–(10.2.2). Reversing all horizontal arrows, it is obvious from Fig. 10.1 that

$$Z(a, b; c, d) = Z(b, a; d, c), \qquad (10.2.12)$$

while rotating through 90° gives

$$Z(a, b; c, d) = Z(b, a; c, d). \qquad (10.2.13)$$

Suppose M, N are even. Then the lattice can be divided into two sub-lattices A and B such that every site in A has neighbours only in B, and vice-versa. Reverse all arrows on horizontal (vertical) edges that have an A site on the left (top) end. The new model is still a zero-field eight vertex model, but with a, b, c, d replaced by c, d, a, b; so

$$Z(a, b; c, d) = Z(c, d; a, b). \qquad (10.2.14)$$

From (10.1.3) and (10.1.4), (10.2.2) contains only even powers of c and d. From (10.2.14), Z must also be an even function of a and b. Thus

$$Z(a, b; c, d) = Z(\pm a, \pm b; \pm c, \pm d), \qquad (10.2.15)$$

where each sign can be chosen independently.

All these symmetry relations (10.2.11)–(10.2.15) can be summarized by introducing

$$w_1 = \tfrac{1}{2}(a + b), \quad w_2 = \tfrac{1}{2}(a - b), \qquad (10.2.16)$$

$$w_3 = \tfrac{1}{2}(c + d), \quad w_4 = \tfrac{1}{2}(c - d),$$

and regarding Z as a function $Z[w_1, \ldots, w_4]$ of $w_1, \ldots, w_4$, instead of a, b, c, d. The symmetries then become

$$Z[w_1, w_2, w_3, w_4] = Z[\pm w_i, \pm w_j, \pm w_k, \pm w_l], \qquad (10.2.17)$$

for any choices of the signs, and any permutations (i, j, k, l) of $(1, 2, 3, 4)$. Thus Z is unaltered by negating or interchanging any of $w_1, \ldots w_4$.

10.3 Formulation as an Ising Model with Two- and Four-Spin Interactions

When thinking of the eight-vertex model as a generalization of the six-vertex, it is natural to describe it in terms of arrows on lattice edges, and view it as a model of a ferroelectric, the arrows being electric dipoles.

However, the eight-vertex model can also be formulated in terms of spins, and viewed as a generalization of the Ising model of a magnet (Wu, 1971; Kadanoff and Wegner, 1971).

To see this, associate spins σ_{ij} with the *faces* of the square lattice, as in Fig. 10.3. Each spin can either have value $+1$, or -1. Allow interactions between nearest and next-nearest neighbour spins. Then the most general translation-invariant Hamiltonian satisfying (1.7.4) is

$$\mathscr{E} = -\sum_{i=1}^{M}\sum_{j=1}^{N} \{J_v \sigma_{ij}\sigma_{i,j+1} + J_h \sigma_{ij}\sigma_{i+1,j} + J\sigma_{i,j+1}\sigma_{i+1,j}$$

$$+ J'\sigma_{ij}\sigma_{i+1,j+1} + J''\sigma_{ij}\sigma_{i,j+1}\sigma_{i+1,j}\sigma_{i+1,j+1}\}. \qquad (10.3.1)$$

Thus this model contains a four-spin interaction between the spins round a site. The partition function is given by (1.7.5) with $H = 0$: denote it by Z_I.

Now define, for all i, j,

$$\alpha_{ij} = \sigma_{ij}\sigma_{i,j+1} \qquad (10.3.2)$$

$$\mu_{ij} = \sigma_{ij}\sigma_{i+1,j}.$$

Then (10.3.1) can be written

$$\mathscr{E} = - \sum_{i=1}^{M} \sum_{j=1}^{N} \{J_v \, \alpha_{ij} + J_h \mu_{ij}$$

$$+ J\alpha_{ij}\mu_{ij} + J'\alpha_{i+1,j}\mu_{ij} + J''\alpha_{ij}\alpha_{i+1,j}\} , \qquad (10.3.3)$$

and, for all i and j,

$$\mu_{ij}\alpha_{ij}\alpha_{i+1,j}\mu_{i,j+1} = 1 . \qquad (10.3.4)$$

To any σ-spin configuration there corresponds an α, μ-spin configuration satisfying (10.3.4). Conversely, to any α, μ-spin configuration satisfying (10.3.4), there correspond two σ-spin configurations satisfying (10.3.2).

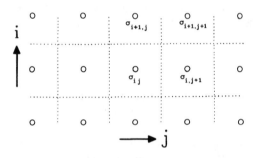

Fig. 10.3. The eight-vertex model square lattice, shown by dotted lines; and the sites of the dual lattice, shown as open circles.

(To see this converse, fix one face spin, say σ_{11} arbitrarily. Then (10.3.2) defines the neighbouring face spins, and so on; (10.3.4) ensures that the definitions are consistent. Thus there are just two solutions of (10.3.2), depending on the choice of the first spin.)

It follows that

$$Z_I = 2 \sum_{\alpha,\mu} \exp(-\mathscr{E}/k_B T) , \qquad (10.3.5)$$

where the sum is over all values (± 1) of $\alpha_{11} , \ldots , \mu_{MN}$ satisfying (10.3.4).

However, since $w(\mu , \alpha | \beta , \nu)$ in (10.2.3) vanishes unless $\mu\alpha\beta\nu = 1$, (10.2.6) is unchanged by imposing the condition (10.3.4). Thus the sums in (10.2.6) and (10.3.5) are the same, provided that (for $\mu\alpha\beta\nu = 1$)

$$w(\mu , \alpha | \beta , \nu) = \exp\{[\tfrac{1}{2}J_v(\alpha + \beta) + \tfrac{1}{2}J_h(\mu + \nu)$$

$$+ J\alpha\mu + J'\beta\mu + J''\alpha\beta]/k_B T\} , \qquad (10.3.6)$$

(sharing out the $J_v \alpha_{ij}$ energies between sites (i, j) and $(i - 1, j)$, and similarly for $J_h \mu_{ij}$. Hence

$$Z_I = 2Z_{8V}, \qquad (10.3.7)$$

where Z_{8V} is the partition function of the eight-vertex model defined above, with α_{ij}, μ_{ij} being the edge arrow spins, and with (using Fig. 10.1, (10.3.6)

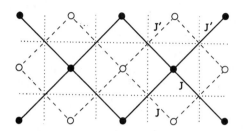

Fig. 10.4. The Ising spins of Fig. 10.3. The solid and broken lines link pairs of spins that interact via the diagonal terms (with coefficients J and J') in eq. (10.3.1). Note the automatic division into two sub-lattices: solid and open circles.

and (10.1.2))

$$\varepsilon_1 = -J_h - J_v - J - J' - J'', \quad \varepsilon_2 = J_h + J_v - J - J' - J'',$$

$$\varepsilon_3 = -J_h + J_v + J + J' + J'', \quad \varepsilon_4 = J_h - J_v + J + J' - J'', \qquad (10.3.8)$$

$$\varepsilon_5 = \varepsilon_6 = J - J' + J'',$$

$$\varepsilon_7 = \varepsilon_8 = -J + J' + J''.$$

Thus this general Ising-type model is equivalent to the general eight-vertex model, and vice-versa. In particular, the zero-field eight-vertex model ($\omega_1 = \omega_2$, $\omega_3 = \omega_4$) corresponds to the Ising-type model with $J_h = J_v = 0$, i.e. with only diagonal and four-spin interactions. In this case, from (10.1.2) and (10.2.1),

$$a = \exp[(J + J' + J'')/k_B T], \quad b = \exp[(-J - J' + J'')/k_B T] \qquad (10.3.9)$$

$$c = \exp[(-J + J' - J'')/k_B T], \quad d = \exp[(J - J' - J'')/k_B T].$$

More particularly, if $J'' = 0$ then only the diagonal interactions remain and a, b, c, d satisfy the condition

$$ab = cd. \qquad (10.3.10)$$

As is evident from Fig. 10.4, the Ising-type model then factors into two independent nearest-neighbour square Ising models, one on the sub-lattice of solid circles, the other on the sub-lattice of open circles. These two models are identical: both have interaction strength J in one direction, J' in the other. If f_{8V} is the free energy per site of this eight-vertex model, then in the thermodynamic limit it follows that

$$f_{8V} = f_{\text{Ising}} , \qquad (10.3.11)$$

where f_{Ising} is the free energy per site of the usual square-lattice nearest-neighbour Ising model.

The zero-field eight-vertex model therefore contains as special cases both the zero-field ice-type model of Chapters 8 and 9, and the Ising model of Chapter 7. In general it can be regarded as two identical Ising models, one on each sub-lattice of faces, coupled via a four-spin interaction round each site.

10.4   Star – Triangle Relation

Here I shall show how the zero-field eight-vertex model can be solved by generalizing the method of Sections 9.6–9.8: this is the way it was originally done (Baxter, 1971a, 1972b).

The Bethe ansatz method of Sections 8.3 and 8.4 can in fact also be appropriately generalized (Baxter, 1973a), but is very cumbersome: it does have the merit of providing formulae for the eigenvectors of the transfer matrix, as well as the eigenvalues, but no use has yet been made of these.

To obtain the results in a form analogous to those of Section 9.6–9.8, it is necessary to use elliptic functions. I shall introduce them at an early stage, though Kumar (1974) has shown that they can be deferred at least until Section 10.7.

Again we try to satisfy the 'star – triangle' relation (9.6.8), only now $w(\mu, \alpha | \beta, \nu)$ is given by (10.2.3) rather than (9.6.2). The three equations (9.6.12) are thereby replaced by the six equations

$$ac'a'' + da'd'' = bc'b'' + ca'c''$$

$$ab'c'' + dd'b'' = ba'c'' + cc'b''$$

$$cb'a'' + bd'd'' = ca'b'' + bc'c'' \qquad (10.4.1)$$

$$ad'b'' + db'c'' = bd'a'' + cb'd''$$

$$aa'd'' + dc'a'' = bb'd'' + cd'a''$$

$$da'a'' + ac'd'' = db'b'' + ad'c'' .$$

These equations are homogeneous and linear in a'', b'', c'', d''. The determinant of coefficients of the first, third, fourth and sixth equations is

$$(cda'b' - abc'd')[(a^2 - b^2)(c'^2 - d'^2) + (c^2 - d^2)(a'^2 - b'^2)]. \quad (10.4.2)$$

For a'', b'', c'', d'' not to be all zero, this determinant must vanish.

The aim here is to construct a class of transfer matrices (with weights a', b', c', d') that all commute with the original matrix V (with weights a, b, c, d). If V itself is to be a member of this class (this seems desirable, but it may not be essential), then we want (10.4.2) to vanish when a', b', c', d' equal a, b, c, d. In this case the first factor vanishes, but the second does not.

In general, therefore, we require the first factor in (10.4.2) to vanish, i.e.

$$\frac{cd}{ab} = \frac{c'd'}{a'b'}. \quad (10.4.3)$$

The first, third, fourth and sixth equations in (10.4.1) can now be solved for $a'':b'':c'':d''$. Using (10.4.3), they give (to within a common factor)

$$a'' = a(cc' - dd')(b^2c'^2 - c^2a'^2)/c$$

$$b'' = b(dc' - cd')(a^2c'^2 - d^2a'^2)/d$$

$$c'' = c(bb' - aa')(a^2c'^2 - d^2a'^2)/a$$

$$d'' = d(ab' - ba')(b^2c'^2 - c^2a'^2)/b. \quad (10.4.4)$$

Substituting these into either the second or fifth equation in (10.4.1), using (10.4.3), gives

$$\frac{a^2 + b^2 - c^2 - d^2}{ab} = \frac{a'^2 + b'^2 - c'^2 - d'^2}{a'b'}. \quad (10.4.5)$$

Define

$$\Delta = (a^2 + b^2 - c^2 - d^2)/2(ab + cd)$$

$$\Gamma = (ab - cd)/(ab + cd), \quad (10.4.6)$$

Similarly, define Δ', Γ' by (10.4.6) with a, b, c, d replaced by a', b', c', d'. Then (10.4.3) and (10.4.5) are equivalent to

$$\Delta = \Delta', \quad \Gamma = \Gamma'. \quad (10.4.7)$$

It follows that any two transfer matrices commute provided they have the same values of Δ and Γ. Apart from a trivial normalization factor, this leaves one degree of freedom in choosing a, b, c, d, so a non-trivial class of transfer matrices can be constructed, each member commuting with every other.

Parametrization in Terms of Entire Functions

The next step is to generalize Section 9.7, i.e. to parametrize a, b, c, d in terms of four other variables, say ρ, k, λ and v, so that a, b, c, d are entire functions of v, but Δ, Γ are independent of v (and of the normalization factor ρ).

First eliminate d between the two equations (10.4.6). This gives

$$2\Delta(1 + \gamma)\,ab = a^2 + b^2 - c^2 - a^2b^2\gamma^2c^{-2}\,, \qquad (10.4.8)$$

where

$$\gamma = (1 - \Gamma)/(1 + \Gamma) = cd/ab\,. \qquad (10.4.9)$$

Eq. (10.4.8) is a symmetric biquadratic relation between a/c and b/c. If b/c is given, then it is a quadratic equation for a/c, with discriminant

$$\Delta^2(1 + \gamma)^2(b/c)^2 - [(b/c)^2 - 1][1 - \gamma^2(b/c)^2]\,. \qquad (10.4.10)$$

This is a quadratic form in $(b/c)^2$, and can be written as

$$(1 - y^2b^2/c^2)(1 - k^2y^2b^2/c^2)\,, \qquad (10.4.11)$$

where k, y depend only on Δ, γ, being given by

$$k^2y^4 = \gamma^2$$
$$(1 + k^2)y^2 = 1 + \gamma^2 - \Delta^2(1 + \gamma)^2\,. \qquad (10.4.12)$$

We want to parametrize b/c as a function of some variable u (say), so that the square root of (10.4.11) is meromorphic. As is shown in Section 15.4, this can be done by taking

$$b/c = y^{-1}\,\mathrm{sn}\,iu\,, \qquad (10.4.13)$$

where sn u is the Jacobian elliptic sn function of argument u and modulus k, and the factor i in the argument is introduced for later convenience. The square root of (10.4.11) is then cn iu dn iu, so the solution of (10.4.8) is

$$\frac{a}{c} = \frac{y[\Delta(1 + \gamma)\,\mathrm{sn}\,iu + y\,\mathrm{cn}\,iu\,\mathrm{dn}\,iu]}{y^2 - \gamma^2\,\mathrm{sn}^2\,iu}\,. \qquad (10.4.14)$$

This is a meromorphic function of u. It can be simplified by defining λ by

$$k\,\mathrm{sn}\,i\lambda = -\gamma/y\,. \qquad (10.4.15)$$

Then (10.4.12) and (10.4.9) give

$$y = \operatorname{sn} i\lambda, \quad \gamma = - k \operatorname{sn}^2 i\lambda, \tag{10.4.16}$$

$$\Gamma = (1 + k \operatorname{sn}^2 i\lambda)/(1 - k \operatorname{sn}^2 i\lambda)$$

$$\Delta = - \operatorname{cn} i\lambda \operatorname{dn} i\lambda/(1 - k \operatorname{sn}^2 i\lambda). \tag{10.4.17}$$

Using the elliptic function addition formula (15.4.21), (10.4.14) gives

$$a/c = \operatorname{sn} i(\lambda - u) / \operatorname{sn} i\lambda, \tag{10.4.18}$$

so from (10.4.9) and (10.4.16),

$$d/c = -k \operatorname{sn} iu \ \operatorname{sn} i(\lambda - u). \tag{10.4.19}$$

The function sn u is a generalization of the trigonometric sine function: from (15.1.4)–(15.1.6) it reduces to sin u when $k = 0$. Just as it was convenient to use the hyperbolic sine function sinh u in Chapter 9, so it is convenient here to use the function snh u, defined by

$$\operatorname{snh} u = -i \operatorname{sn} iu = i \operatorname{sn}(-iu). \tag{10.4.20}$$

It is a meromorphic function of u, real if u is real (and $0 < k < 1$).
Using this, from (10.4.13, 16, 18, 19) we have

$$a : b : c : d = \operatorname{snh}(\lambda - u) : \operatorname{snh} u : \tag{10.4.21}$$

$$\operatorname{snh} \lambda : k \operatorname{snh} \lambda \operatorname{snh} u \ \operatorname{snh}(\lambda - \mu).$$

From (15.1.6) and (10.4.20),

$$\operatorname{snh} u = -ik^{-\frac{1}{2}} H(iu)/\Theta(iu), \tag{10.4.22}$$

where the theta functions $H(u)$, $\Theta(u)$ are entire. Define v by

$$u = \tfrac{1}{2}(\lambda + v). \tag{10.4.23}$$

Using (10.4.22) in (10.4.21), the Θ function denominators can be multiplied out, giving

$$a = -i\rho \, \Theta(i\lambda) \, H[\tfrac{1}{2}i(\lambda - v)] \, \Theta[\tfrac{1}{2}i(\lambda + v)],$$

$$b = -i\rho \, \Theta(i\lambda) \, \Theta[\tfrac{1}{2}i(\lambda - v)] \, H[\tfrac{1}{2}i(\lambda + v)],$$

$$\tag{10.4.24}$$

$$c = -i\rho \, H(i\lambda) \, \Theta[\tfrac{1}{2}i(\lambda - v)] \, \Theta[\tfrac{1}{2}i(\lambda + v)],$$

$$d = \ \ i\rho \, H(i\lambda) \, H[\tfrac{1}{2}i(\lambda - v)] \, H[\tfrac{1}{2}i(\lambda + v)],$$

where ρ is some normalization factor. If ρ, λ, v are real, then so are a, b, c, d.

This completes the generalization of steps (i), (ii), (iii) of Section 9.5: a, b, c, d are defined in terms of ρ, k, λ, v by (10.4.24); from (10.4.17), Γ and Δ depend only on k and λ.

Keep ρ, k and λ fixed; regard the transfer matrix V, given by (9.6.1) and (10.2.3), as a function $V(v)$ of v. Then

$$V(v)\, V(v') = V(v')\, V(v)\,, \qquad (10.4.25)$$

for all complex numbers v, v'. From (10.4.24), a, b, c, d are entire functions of v: so therefore are all elements of $V(v)$.

If λ, u are held fixed and k allowed to tend to zero, then snh $u \to$ sinh u. From (10.4.21), $d \to 0$ (relative to a, b, c), so we regain the six-vertex model of Chapters 8 and 9 (Δ, λ, u and v having the same meaning as therein). In particular, (10.4.21) becomes (9.7.12), and (10.4.17) gives (9.2.4).

Relation between u, u', u''

The equations (10.4.1) are unaltered by interchanging the unprimed and double-primed variables. Thus (10.4.7) further implies that

$$\Delta = \Delta' = \Delta'', \quad \Gamma = \Gamma' = \Gamma''\,. \qquad (10.4.26)$$

The weights a', b', c', d' [and a'', b'', c'', d''] can therefore also be put into the form (10.4.22), with the same values of k and λ. The values of ρ, u and v will be different: let us call them ρ', u' and v' [ρ'', u'' and v''].

The first of the equations (10.4.1) can be written

$$c'(aa'' - bb'') = a'(cc'' - dd'')\,. \qquad (10.4.27)$$

Substituting the expressions (10.4.19) and using the identity (15.4.23), this becomes

$$\text{sn}\, i(\lambda - u - u'') = \text{sn}\, i(\lambda - u)\,. \qquad (10.4.28a)$$

Proceeding similarly [but using (15.4.24)], the fourth of the equations (10.4.1) also gives (10.4.27). The second and fifth give

$$\text{sn}\, i(u' - u) = \text{sn}(iu'')\,, \qquad (10.4.28b)$$

while the third and sixth give

$$\text{sn}\, i(u' - u'') = \text{sn}(iu)\,. \qquad (10.4.28c)$$

The general solution of these equations (10.4.28) is

$$u' = u + u'' + 4miI + 2nI'\,, \qquad (10.4.29)$$

where I, I' are the complete elliptic integrals defined in Chapter 15, and m, n are any integers. However, incrementing u' by $4iI$ or $2I'$ does not affect (10.4.19), so without loss of generality we can choose

$$u' = u + u'' . \qquad (10.4.30)$$

This is exactly the same as the six-vertex relation (9.7.13), so the eight-vertex operators U_i defined by (9.6.9), (10.2.3) and (10.4.22) also satisfy the star – triangle operator relation (9.7.14). i.e.

$$U_{i+1}(u) \, U_i(u + u'') \, U_{i+1}(u'') = U_i(u'') \, U_{i+1}(u + u'') \, U_i(u) . \qquad (10.4.31)$$

Note that u'' is just the difference of u' and u. I suspect that this is closely related to the 'transformation to a difference kernel' that occurs in the Bethe ansatz, as in equations (8.8.2) – (8.8.4) and (8.13.30) – (8.13.38). The elliptic function parametrization has been introduced here simply on the grounds of mathematical convenience, but suppose we had originally required that a, b, c, d be functions of some variable u (and a', b', c', d' the same functions of u': and a'', b'', c'', d'' of u'') so that Δ and Γ be constants and that u'' be a function only of $u' - u$. We would then have been led inexorably from the star – triangle relations (10.4.1) to the elliptic function parametrization (10.4.21), just as in Section 8.13 we were led from (8.13.31) to (8.13.67) and (8.13.73).

10.5 The Matrix $Q(v)$

Pair Propagation through a Vertex

Now we seek to generalize Section 9.8. The first ten equations generalize trivially (we still want $H_i(\alpha)$ to be upper-right triangular). The 'pair propagation' conditions (9.8.11) become (replacing the integer i by j)

$$\begin{aligned}
ag_j(+) \, p_{j+1}(+) + dg_j(-) \, p_{j+1}(-) &= g_j'(+) \, p_j(+) \\
bg_j(-) \, p_{j+1}(+) + cg_j(+) \, p_{j+1}(-) &= g_j'(-) \, p_j(+) \\
cg_j(-) \, p_{j+1}(+) + bg_j(+) \, p_{j+1}(-) &= g_j'(+) \, p_j(-) \\
dg_j(+) \, p_{j+1}(+) + ag_j(-) \, p_{j+1}(-) &= g_j'(-) \, p_j(-) .
\end{aligned} \qquad (10.5.1)$$

They are still homogeneous and linear in $g_j(+)$, $g_j(-)$, $g_j'(+)$, $g_j'(-)$. Equating to zero the determinant of coefficients, and using (10.4.6) and (10.4.9), we obtain

$$2\Delta(1 + \gamma) \, r_j r_{j+1} = r_j^2 + r_{j+1}^2 - \gamma(1 + r_j^2 r_{j+1}^2) , \qquad (10.5.2)$$

where r_j is again given by (9.8.13).

This is a symmetric biquadratic relation between r_j and r_{j+1}. It involves a, b, c, d only via the 'constants' Δ and Γ. Further, it is exactly the same as (10.4.8) with a, b, c replaced by $r_{j+1}, r_j, \gamma^{\frac{1}{2}}$, respectively.

We can therefore apply the solution (10.4.21) of (10.4.8) to the relation (10.5.2). Replacing u by t, this gives (for a particular value of j)

$$\gamma^{-\frac{1}{2}} r_{j+1} = \text{snh}(\lambda - t)/\text{snh}\,\lambda\,, \qquad (10.5.3)$$

$$\gamma^{-\frac{1}{2}} r_j = \text{snh}\, t/\text{snh}\,\lambda\,,$$

i.e., using (10.4.16) and (10.4.20),

$$r_j = k^{\frac{1}{2}}\,\text{snh}\, t, \quad r_{j+1} = -k^{\frac{1}{2}}\,\text{snh}(t - \lambda)\,. \qquad (10.5.4)$$

From (15.2.5) and (10.4.20), r_j is unchanged by replacing t by $2iI - t$, while r_{j+1} becomes $-k^{\frac{1}{2}}\,\text{snh}(t + \lambda)$. Thus if

$$r_j = k^{\frac{1}{2}}\,\text{snh}\, t\,, \qquad (10.5.5)$$

then two solutions of (10.5.2) are

$$r_{j+1} = -k^{\frac{1}{2}}\,\text{snh}(t \pm \lambda)\,. \qquad (10.5.6)$$

Since (10.5.2) is a quadratic equation for r_{j+1}, these are all the solutions.

Now consider these equations sequentially for $j = 1, \ldots, N$, determining t for each value. The choice of sign in (10.5.6) can be made independently for each j, so the most general solution for $r_1, \ldots, r_{N+1}$ is

$$r_j = (-)^j k^{\frac{1}{2}}\,\text{snh}\, s_j\,, \qquad (10.5.7)$$

where

$$s_j = s + \lambda(\sigma_1 + \ldots + \sigma_{j+1})\,, \qquad (10.5.8)$$

s is an arbitrary constant and each σ_j has value ± 1. (These $\sigma_1, \ldots, \sigma_N$ have no connection with the Ising spins of Section 10.3). The cyclic boundary condition $r_{N+1} = r_1$ is satisfied if N is even and

$$\sigma_1 + \ldots + \sigma_N = 0\,. \qquad (10.5.9)$$

(As in the six-vertex models, this condition can be relaxed for the special values $(4imI + 2rI')/n$ of λ, where m, r, n are integers: it is then sufficient that $\sigma_1 + \ldots + \sigma_N$ be an integer multiple of n. Such cases are often of particular interest: the Ising model ($K'' = 0$) has $\lambda = \frac{1}{2}I'$.)

Clearly this solution for $r_1, \ldots, r_N$ is similar to that of the six-vertex model in (9.8.15). If we set $r = k^{\frac{1}{2}} e^s$ and let $k \to 0$ while keeping r fixed, then (10.5.7) reduces to (9.8.15).

Eliminating $g_j'(+)$ between the first and third of the equations (10.5.1), we obtain

$$\frac{g_j(-)}{g_j(+)} = \frac{ar_j - br_{j+1}}{c - dr_j r_{j+1}}. \tag{10.5.10}$$

Using (10.4.21), (10.5.7), (10.5.8) and the identity (15.4.23), this becomes

$$\frac{g_j(-)}{g_j(+)} = (-)^j k^{\frac{1}{2}} \operatorname{snh}(s_j + \sigma_j u). \tag{10.5.11}$$

Taking the ratios of the first and second equations (10.5.1) now gives [again using (15.4.23)]

$$\frac{g_j'(-)}{g_j'(+)} = (-)^{j+1} k^{\frac{1}{2}} \operatorname{snh}[s_j + \sigma_j(u + \lambda)]. \tag{10.5.12}$$

We are still free to choose $p_j(+)$, $g_j(+)$ arbitrarily. An important property in Chapter 9 was that the elements of $Q(v)$ were entire functions of v (or u). In Section 9.8 this came about because each $g_j(+)$ and $g_j(-)$ was entire. From (10.5.11) and (10.4.22), this can be ensured in this more general situation by choosing

$$g_j(+) = \Theta[i(s_j + \sigma_j u)], \tag{10.5.13a}$$

for then

$$g_j(-) = (-)^{j+1} i H[i(s_j + \sigma_j u)]. \tag{10.5.13b}$$

Similarly, from (10.5.7) and (10.4.22), we can choose $p_j(+)$ so that

$$p_j(+) = \Theta(is_j), \quad p_j(-) = (-)^{j+1} i H(is_j). \tag{10.5.14}$$

Using (10.4.24), (10.4.23) and (15.4.25), the first of the equations (10.5.1) now gives

$$g_j'(+) = \rho h(\lambda - u)\, \Theta[is_j + i\sigma_j(u + \lambda)], \tag{10.5.15a}$$

so from (10.5.12),

$$g_j'(-) = \rho h(\lambda - u)(-)^j i H[is_j + i\sigma_j(u + \lambda)], \tag{10.5.15b}$$

where the function $h(u)$ is defined by

$$h(u) = -i\,\Theta(0)\, H(iu)\, \Theta(iu). \tag{10.5.16}$$

The matrices $G_j(\pm)$ now contain non-zero entries $dg_j(\mp)$ instead of the zeros in (9.8.4). Using (10.4.24), (10.5.13) and (15.4.25 or 26), their determinants are

$$\det G_j(+) = \rho^2 h(u) \, h(\lambda - u) \, \Theta[is_j + i\sigma_j(u + \lambda)] \, \Theta[is_j + i\sigma_j(u - \lambda)]$$

$$\det G_j(-) = -\rho^2 h(u) \, h(\lambda - u) \, H[is_j + i\sigma_j(u + \lambda)] \, H[is_j + i\sigma_j(u - \lambda)].$$

(10.5.17)

As in Chapter 9, we can calculate $g_j''(\alpha)$ in (9.8.6) by taking determinants in (9.8.5). The determinants of P_i and P_{i+1} can again be ignored, since their contribution to $g_j''(\alpha)$ cancels out of (9.8.7) (or we can require that $\det P_i = 1$). Using (10.5.15) and (10.5.17), we are left with

$$g_j''(+) = \rho \, h(u) \, \Theta[is_j + i\sigma_j(u - \lambda)]$$

$$g_j''(-) = \rho \, h(u) \, (-)^j i H[is_j + i\sigma_j(u - \lambda)].$$

(10.5.18)

Column Vectors $y(v)$

The 2^N-dimensional vector y has elements given by (9.8.2). Thus it is a direct product of the two-dimensional vectors $g_1, \ldots, g_N$:

$$y = g_1 \otimes g_2 \otimes \ldots \otimes g_N,$$

(10.5.19)

where

$$g_j = \begin{pmatrix} g_j(+) \\ g_j(-) \end{pmatrix},$$

(10.5.20)

i.e., using (10.4.23) and (10.5.13),

$$g_j = \begin{pmatrix} \Theta[is_j + \frac{1}{2}i(\lambda + v)\sigma_j] \\ (-)^{j+1} i H[is_j + \frac{1}{2}i(\lambda + v)\sigma_j] \end{pmatrix}.$$

(10.5.21)

Comparing (9.8.1) and (9.8.7), the vector y' (y'') is also defined by (10.5.19), but with each g_j replaced by g_j' (g_j''). From (10.5.15) and (15.2.5), g_j' can be obtained from g_j by multiplying by $\rho h(\lambda - u)$ and incrementing u by λ', where

$$\lambda' = \lambda - 2iI.$$

(10.5.22)

Regard y, defined by (10.5.19) and (10.5.21), as a function $y(v)$ of v (k, λ, s being kept constant). From (10.4.23), incrementing u by λ' is equivalent to incrementing v by $2\lambda'$, so

$$y' = \{\rho h[\tfrac{1}{2}(\lambda - v)]\}^N y(v + 2\lambda').$$

(10.5.23a)

Similarly, using (10.5.18),

$$y'' = \{\rho h[\tfrac{1}{2}(\lambda + v)]\}^N y(v - 2\lambda').$$

(10.5.23b)

It is obviously convenient to define a function

$$\phi(v) = [\rho\, h(v/2)]^N \; ; \qquad (10.5.24)$$

the equation (9.8.1) can now be written

$$V(v)\, y(v) = \phi(\lambda - v)\, y(v + 2\lambda') + \phi(\lambda + v)\, y(v - 2\lambda')\,, \qquad (10.5.25)$$

so we have generalized (9.8.24) to the eight-vertex model. Again there are many choices of $y(v)$, corresponding to different choices of s, $\sigma_1, \ldots, \sigma_N$ in (10.5.8), subject to (10.5.9). Let $Q_R(v)$ be a 2^N by 2^N matrix whose columns are linear combinations (with coefficients that are independent of v) of such vectors $y(v)$. Then, from (10.5.25),

$$V(v)\, Q_R(v) = \phi(\lambda - v)\, Q_R(v + 2\lambda') + \phi(\lambda + v)\, Q_R(v - 2\lambda')\,. \qquad (10.5.26)$$

Row Vectors $y^T(-v)$ and Matrix $Q_L(v)$

Equations (9.8.26)–(9.8.29) generalize to the eight-vertex model, the only explicit modification necessary being to change r in (9.8.29) to s. Let s_j' be defined by (10.5.8), with s, $\sigma_1, \ldots, \sigma_{j-1}$ replaced by s', $\sigma_1', \ldots, \sigma_{j-1}'$. Using (10.5.19), (10.5.21) and the identity (15.4.27), we find that

$$y^T(-u|s', \sigma')\, y(v|s, \sigma) = \prod_{j=1}^{N} \tau F[s_j - s_j' + \tfrac{1}{2}\lambda(\sigma_j - \sigma_j') + \tfrac{1}{2}(v\sigma_j + u\sigma_j')]$$

$$\times\, G[s_j + s_j' + \tfrac{1}{2}\lambda(\sigma_j + \sigma_j') + \tfrac{1}{2}(v\sigma_j - u\sigma_j')]\,, \qquad (10.5.27)$$

where

$$\tau = 2q^{\frac{1}{4}}/[H_1(0)\,\Theta_1(0)]\,,$$

$$F(u) = -\,H[\tfrac{1}{2}i(I' + u)]\,H[\tfrac{1}{2}i(I' - u)]\,, \qquad (10.5.28)$$

$$G(u) = H_1[\tfrac{1}{2}i(I' + u)]\,H_1[\tfrac{1}{2}i(I' - u)]\,.$$

We can now use the same inductive argument as that following (9.8.31) to show that the RHS of (10.5.27) is a symmetric function of u and v. (We need only (10.5.27) and (10.5.8): the definitions (10.5.28) are irrelevant. It is necessary to split (10.5.27) into two factors, one containing only F functions, the other containing only G functions. The inductive argument applies to each, but one appeals initially to the case $\sigma_j = \sigma_j'$ for the F-factor, $\sigma_j = -\,\sigma_j'$ for the G-factor.)

The relation (9.8.33) therefore also generalizes to the eight-vertex model, i.e.

$$Q_L(u)\, Q_R(v) = Q_L(v)\, Q_R(u)\,, \qquad (10.5.29)$$

where

$$Q_L(v) = Q_R^T(-v).$$ (10.5.30)

$Q_R(v)$ a Non-Singular Matrix

Each vector $y(v)$ is given by (10.5.19), (10.5.21), (10.5.8) and (10.5.9). There are many such vectors, since s can be any complex number, and $\sigma_1, \ldots, \sigma_N$ any set of integers ± 1 satisfying (10.5.9). We want the set of all such vectors to span all 2^N-dimensional space (except possibly for special values of v).

As in Chapter 9, I am not able to give a full proof of this, but it is almost certainly so (it is for $N = 2$ and 4).

It would not be generally true only if all determinants of all possible matrices $Q_R(v)$ vanished identically for all k, λ, v. If this were so, they would vanish for $k = 0$, which is the six-vertex model: in this case we know that the eigenvalues of the transfer matrix are correctly given by assuming $Q_R(v)$ to be non-singular, and we have strong direct evidence that it is.

Let us therefore assume that $Q_R(v)$ is non-singular for some value v_0. Defining $Q(v)$ by (9.8.38), i.e.

$$Q(v) = Q_R(v)\, Q_R^{-1}(v_0),$$ (10.5.31)

the relations (9.8.39) – (9.8.42) follow, in particular

$$V(v)\, Q(v) = Q(v)\, V(v)$$
$$= \phi(\lambda - v)\, Q(v + 2\lambda') + \phi(\lambda + v)\, Q(v - 2\lambda'),$$ (10.5.32)

$$Q(u)\, Q(v) = Q(v)\, Q(u),$$ (10.5.33)

for all complex numbers v, u.

Commutation Relations Involving S and R

From (15.2.3a) and (15.2.4), the theta functions $\Theta(u)$, $H(u)$ satisfy

$$\Theta(u + 2I) = \Theta(u), \quad H(u + 2I) = -H(u).$$ (10.5.34)

The effect of incrementing v by $4iI$ in (10.5.21) is therefore to negate the function H.

Define the diagonal operator S by (9.8.43): it has entries $+1$ (-1) for row-states with an even (odd) number of down arrows. From (10.5.19) and (10.5.21), pre-multiplying y by S is equivalent to negating every H, so

$$S\,y(v) = y(v + 4iI)\,. \tag{10.5.35}$$

This is the generalization to the eight-vertex model of (9.8.44). The equations (9.8.45) – (9.8.47) can at once be similarly generalized (merely replace π by $2I$). In particular, they give

$$S\,Q(v) = Q(v)\,S = Q(v + 4iI)\,, \tag{10.5.36a}$$

$$S\,V(v) = V(v)\,S\,. \tag{10.5.36b}$$

From (15.2.3b) and (15.2.4), the theta functions $H(u)$, $\Theta(u)$ satisfy the relations

$$H(u + iI') = iq^{-\frac{1}{4}} \exp(-\tfrac{1}{2}i\pi u/I)\,\Theta(u)\,, \tag{10.5.37}$$

$$\Theta(u + iI') = iq^{-\frac{1}{4}} \exp(-\tfrac{1}{2}i\pi u/I)\,H(u)\,.$$

Define a 2^N by 2^N matrix R by

$$R = \begin{pmatrix} 0 & 1 \\ 1 & 0 \end{pmatrix} \otimes \begin{pmatrix} 0 & 1 \\ 1 & 0 \end{pmatrix} \otimes \ldots \otimes \begin{pmatrix} 0 & 1 \\ 1 & 0 \end{pmatrix}. \tag{10.5.38}$$

(Multiplication by R has the effect of reversing all arrows.) Then from (10.5.19) and (10.5.21) it follows that

$$y(v + 2I') = q^{-N/4} \exp\left\{ \tfrac{1}{2}\pi \sum_{j=1}^{N} [s_j\sigma_j + \tfrac{1}{2}(\lambda + v)]/I \right\} RS\,y(v)\,. \tag{10.5.39}$$

From (10.5.8) and (10.5.9),

$$\sum_{j=1}^{N} s_j\sigma_j = \lambda \sum_{1 \leq i < j \leq N} \sigma_i\sigma_j$$

$$= \tfrac{1}{2}\lambda\{(\sigma_1 + \ldots + \sigma_N)^2 - \sigma_1^2 - \ldots - \sigma_N^2\} = -\tfrac{1}{2}N\lambda\,, \tag{10.5.40}$$

so

$$y(v + 2I') = q^{-N/4} \exp(N\pi v/4I)\,RS\,y(v)\,. \tag{10.5.41}$$

This relation is independent of s and $\sigma_1, \ldots, \sigma_N$, so is satisfied by all columns of $Q_R(v)$. Using also (10.5.30), it can readily be verified that

$$Q_R(v + 2I') = q^{-N/4} \exp(N\pi v/4I)\,RS\,Q_R(v)\,, \tag{10.5.42}$$

$$Q_L(v + 2I') = q^{-N/4} \exp(N\pi v/4I)\,Q_L(v)\,RS\,,$$

so, from (9.8.38) and (9.8.39),

$$RS\,Q(v) = Q(v)\,RS = q^{N/4} \exp(-N\pi v/4I)\,Q(v + 2I')\,. \tag{10.5.43a}$$

From (10.2.4), $w(\mu, \alpha | \beta, \nu)$ is unchanged by negating μ, α, β, ν. From (9.6.1) and (10.5.38) it follows that

$$R\, V(v) = V(v)\, R\,. \qquad (10.5.43b)$$

The matrices $Q(v)$, $Q(u)$, $V(v)$, $V(u)$, R, S therefore commute, for all complex numbers u and v.

From (9.6.1), (10.2.3) and (10.4.24), all elements of $V(v)$ are entire functions of v. From (10.5.19) and (10.5.21), so are all elements of $Q(v)$.

This completes the generalization to the eight-vertex model of the six-vertex model properties (i) – (vi) given in Section 9.5. The derivation has closely followed that given in Sections 9.6 – 9.8 for the six-vertex case.

10.6 Equations for the Eigenvalues of $V(v)$

The vital results of the previous two sections are (10.4.25), (10.5.24), (10.5.32), (10.5.33), (10.5.36) and (10.5.43), together with the fact that all elements of $V(v)$ and $Q(v)$ are entire functions of v.

We now generalize the 'sufficiency' argument of Section 9.5. Since all matrices commute, there exists a matrix P (independent of v) such that $V_d(v)$, $Q_d(v)$ in (9.5.1) are diagonal matrices. Equation (10.5.23) gives (9.4.3). Let $\Lambda(v)$ be a particular eigenvalue of $V(v)$, and $q(v)$ the corresponding eigenvalue of $Q(v)$. (This function $q(v)$ is not to be confused with the nome q of the elliptic functions.) Then the corresponding entry in the matrix equation (9.4.3) is the scalar equation (9.3.6), i.e.

$$\Lambda(v)\, q(v) = \phi(\lambda - v)\, q(v + 2\lambda') + \phi(\lambda + v)\, q(v - 2\lambda')\,, \qquad (10.6.1)$$

but now $\phi(v)$ is defined by (10.5.24), λ' by (10.5.22).

Since all elements of $V(v)$, $Q(v)$ are entire, so are $\Lambda(v)$, $q(v)$. Let $r(= \pm 1)$ be the eigenvalue of R corresponding to $\Lambda(v)$, $q(v)$; and $s(= \pm 1)$ the eigenvalue of S. Then from (10.5.36a) and (10.5.43a),

$$q(v + 4iI) = s\, q(v)\,, \qquad (10.6.2)$$

$$q(v + 2I') = rs\, q^{-N/4} \exp(N\pi v/4I)\, q(v)\,.$$

Integrating $q'(v)/q(v)$ round a period rectangle of width $2I'$ and height $4I$, then using Cauchy's integral formula (15.3.4), it is readily found that $q(v)$ has $\frac{1}{2}N$ zeros per period rectangle. Set

$$n = N/2\,, \qquad (10.6.3)$$

and let $v_1, \ldots, v_n$ be these zeros.

Consider the function

$$f(v) = q(v) \Big/ \prod_{j=1}^{n} h\left(\frac{v - v_j}{2}\right), \tag{10.6.4}$$

where $h(u)$ is defined by (10.5.16). From (10.6.2), (15.2.3) and (15.2.4),

$$f(v + 4iI) = (-)^n s\, f(v), \tag{10.6.5}$$

$$f(v + 2I') = (-)^n r s\, \exp[\pi(v_1 + \ldots + v_n)/2I]\, f(v).$$

From (10.6.4), $f(v)$ is entire and non-zero. From (10.6.5), $f'(v)/f(v)$ is therefore entire and doubly-periodic. From theorem 15a, it is therefore a constant, so $f(v)$ is of the form

$$f(v) = \text{constant} \times \exp(\tau v). \tag{10.6.6}$$

Substituting this into (10.6.5) gives

$$\tau = \pi(s - 1 + 2n + 4p')/8I, \tag{10.6.7a}$$

$$v_1 + \ldots + v_n = \tfrac{1}{2}(s-1+2n)I' + i(rs-1+2n)I + 2p'I' + 4ipI, \tag{10.6.7b}$$

where p, p' are integers.

Combining (10.6.4) and (10.6.6), to within a multiplicative factor that cancels out of all our subsequent calculations:

$$q(v) = \exp(\tau v) \prod_{j=1}^{n} h\left(\frac{v - v_j}{2}\right). \tag{10.6.8}$$

This is the eight-vertex generalization of (9.3.5). The function $h(u)$ has a simple zero at $u = 0$, so setting $v = v_j$ in (10.6.1) causes the LHS to vanish, leaving

$$\frac{\phi(\lambda - v_j)}{\phi(\lambda + v_j)} = -\frac{q(v_j - 2\lambda')}{q(v_j + 2\lambda')}, \quad j = 1, \ldots, n, \tag{10.6.9}$$

or, using (10.5.24) and (10.6.8),

$$\left\{\frac{h[\tfrac{1}{2}(\lambda - v_j)]}{h[\tfrac{1}{2}(\lambda + v_j)]}\right\}^N = -\exp(-4\tau\lambda') \prod_{l=1}^{n} \frac{h[\tfrac{1}{2}(v_j - v_l - 2\lambda')]}{h[\tfrac{1}{2}(v_j - v_l + 2\lambda')]}, \tag{10.6.10}$$

for $j = 1, \ldots, n$.

These are the eight-vertex generalizations of (8.4.12). They determine $v_1, \ldots, v_n$; $q(v)$ is then given by (10.6.8) and $\Lambda(v)$ by (10.6.1). There are many solutions of (10.6.10), corresponding to the different eigenvalues.

If $v_1, \ldots, v_n$ are distinct, then (10.6.10) ensures that the ratio of the RHS of (10.6.1) to $q(v)$ is an entire function, so that $\Lambda(v)$ is entire, as required. However, if any two of $v_1, \ldots, v_n$ are equal, then (10.6.10) is not a sufficient condition for $\Lambda(v)$ to be entire: it must be supplemented by further equations obtained by differentiating (10.6.1) with respect to v and then setting v equal to the common v_j value.

For this reason, solutions of (10.6.10) are in general spurious if any two of $v_1, \ldots, v_n$ are equal. (Note that in Chapter 8 we also rejected such solutions, though in that case it was because they gave the eigenvector to be zero.)

10.7 Maximum Eigenvalue: Location of $v_1, \ldots, v_n$

Principal Regime

Consider the case when

$$0 < k < 1, \quad 0 < \lambda < I', \quad |v| < \lambda, \quad \rho > 0, \qquad (10.7.1a)$$

so, from (10.4.23),

$$0 < u < \lambda. \qquad (10.7.1b)$$

From (10.4.21), the weights a, b, c, d all have the same sign; and from (10.4.24) they are all positive, so the restrictions (10.7.1a) are physically allowable.

From (10.4.17) and (15.4.4)

$$1 - \Delta^2 = (1 - k)^2 \, \mathrm{sn}^2 \, i\lambda / (1 - k \, \mathrm{sn}^2 \, i\lambda)^2, \qquad (10.7.2)$$

so, since $\mathrm{sn}^2 \, i\lambda$ is negative real,

$$\Delta < -1. \qquad (10.7.3)$$

From (10.4.6) this implies that

$$(a + b)^2 < (c - d)^2. \qquad (10.7.4)$$

The ratio d/c is given by (10.4.21) to be $k \, \mathrm{snh} \, u \, \mathrm{snh}(\lambda - u)$; this has a maximum when $u = \tfrac{1}{2}\lambda$, and from (15.4.24) this maximum must be less than one, so $d < c$. Taking positive square roots of (10.7.4) and noting that each RHS in (10.4.24) is positive, it follows that

$$c > a + b + d, \quad a > 0, \quad b > 0, \quad d > 0. \qquad (10.7.5)$$

The restrictions (10.7.1) therefore imply (10.7.5); conversely, if a, b, c, d satisfy (10.7.5), then there are unique real values of k, λ, v, ρ, u satisfying (10.4.21), (10.4.24) and (10.7.1).

The inequality (10.7.5) specifies a domain, or regime, in (a, b, c, d) space. This is the generalization to the eight-vertex model of the six-vertex anti-ferroelectric regime (regime IV in Fig. 8.5). The dominant Boltzmann weight is c, and the ground-state energy configurations of arrows on the lattice are either that shown in Fig. 8.3, or the configuration obtained from it by reversing all arrows. By extending the six-vertex notation to the eight-vertex, we are led automatically to regard (10.7.5) as the archetypal regime.

This has its disadvantages: if we regard the eight-vertex model as a generalization of the Ising model, as in Section 10.3, then it is natural to focus attention on the ferromagnetic regime, when J and J' are large. The dominant Boltzmann weight is then a, rather than c. Fortunately this can be converted to the case (10.7.5) by using the symmetry relation (10.2.14).

In fact, it will be shown in Section 10.11 that any set of values of a, b, c, d can be mapped into (10.7.5) (or its boundaries: since most properties are continuous these present no problem) by using the symmetry relations (10.2.11) – (10.2.17). They can equally well be all mapped into other regimes, notably $a > b + c + d$, but from now on I shall single out the regime (10.7.5) (with a, b, c, d all positive) and call it the *principal regime*.

Low-Temperature Limit

The equations (10.6.10) are quite complicated and for finite n have not in general been solved. I find it helpful to first look at the following simple limiting case: it gives some useful insights into the large-n behaviour.

Suppose $\varepsilon_5 < \varepsilon_1, \varepsilon_3, \varepsilon_7$ and T is small. Then from (10.1.2) and (10.2.1)

$$c \gg a, b, d, \qquad (10.7.6)$$

so the weights are certainly in the principal regime. It follows that $k \ll 1$; while I', λ, v are large, their ratios being of order unity. From (15.1.4) the nome q is small, so from (15.1.5)

$$\Theta(iu) \simeq 1, \quad H(iu) \sim iq^{\frac{1}{4}}\exp(\pi u/2I), \qquad (10.7.7)$$

provided $0 < \text{Re}(u/I') < 1$. Equation (10.4.24) therefore gives

$$c \simeq \rho q^{\frac{1}{4}}x^{-1}, \qquad (10.7.8)$$

where

$$q = \exp(-\pi I'/I), \quad x = \exp(-\pi\lambda/2I). \qquad (10.7.9)$$

Suppose $v_1, \ldots, v_n$ are all of order unity (or less). Then in this limit (10.6.10) becomes, using (10.6.3),

$$z_j^n + (-)^n \exp(-4\tau\lambda')(z_1 \ldots z_n)^{-1} = 0, \tag{10.7.10}$$

where

$$z_j = \exp(-\pi v_j/2I), \tag{10.7.11}$$

and $j = 1, \ldots, n$.

Equation (10.7.10) is a polynomial equation for z_j of degree n, so has n distinct roots. We want $v_1, \ldots, v_n$ to be distinct, so $z_1, \ldots, z_n$ must be the n roots of (10.7.10). It follows that

$$z^n + (-1)^n \exp(-4\tau\lambda')(z_1 \ldots z_n)^{-1} \equiv \prod_{j=1}^{n}(z - z_j), \tag{10.7.12}$$

for all complex numbers z. Setting $z = 0$ and taking square roots gives

$$z_1 \ldots z_n = \pm\exp(-2\tau\lambda'), \tag{10.7.13}$$

while (10.6.7b) and (10.7.11) give

$$z_1 \ldots z_n = rs(-)^n \exp(-2\tau I'). \tag{10.7.14}$$

From (10.6.7a), τ is real; λ' is given by (10.5.22), where $\lambda \neq I'$. Since $r = \pm 1$ and $s = \pm 1$, it follows that

$$\tau = 0, \quad s = (-)^n, \quad z_1 \ldots z_n = r. \tag{10.7.15}$$

The asymptotic formulae (10.7.7) fail if $\mathrm{Re}(u)$ becomes zero or negative: in this case we must use

$$\Theta(iu) \simeq 1, \quad H(iu) \sim 2i \, q^{\frac{1}{4}} \sinh(\pi u/2I), \tag{10.7.16}$$

for $|\mathrm{Re}(u/I')| < 1$. Then (10.6.8) and (10.5.16) give

$$q(v) = \prod_{j=1}^{n} 2q^{\frac{1}{4}} \sinh[\pi(v - v_j)/4I], \tag{10.7.17}$$

for $|\mathrm{Re}(v/I')| < 2$. Setting

$$z = \exp(-\pi v/2I), \tag{10.7.18}$$

this can be written

$$q(v) = (-)^n q^{n/4} z^{-n/2}(z_1 \ldots z_n)^{-\frac{1}{2}} \prod_{j=1}^{n}(z - z_j). \tag{10.7.19}$$

Using (10.7.12) and (10.7.13), this can in turn be written

$$q(v) = q^{n/4} z^{-n/2} (z_1 \ldots z_n)^{-\frac{1}{2}} \{(-z)^n + (z_1 \ldots z_n)^{-1}\} . \quad (10.7.20)$$

Now determine the asymptotic form of (10.6.1) in the low-temperature limit. If $|\text{Re}(v)| \ll \min(\lambda, 2I' - 2\lambda)$, we obtain [using (10.5.24), (10.6.8), (10.5.22), (10.5.16) and (10.7.7)]

$$\Lambda(v) \, q(v) = \rho^N q^{N/4} x^{-N} q^{n/4} z^{-n/2} (z_1 \ldots z_n)^{\frac{1}{2}} (R_1 + R_2) , \quad (10.7.21)$$

where

$$R_1 = (-z)^n, \quad R_2 = (z_1 \ldots z_n)^{-1}. \quad (10.7.22)$$

The first term on the RHS of (10.6.1) gives the R_1 term in (10.7.21), the second gives the R_2 term. However, from (10.7.20), $q(v)$ also contains a factor $R_1 + R_2$, so this cancels out of (10.7.21), leaving

$$\Lambda(v) = \rho^N q^{N/4} x^{-N} z_1 \ldots z_n , \quad (10.7.23)$$

so, from (10.7.8) and (10.7.15),

$$\Lambda(v) = r \, c^N , \quad (10.7.24)$$

in the low-temperature limit.

This is indeed the correct maximum transfer matrix eigenvalue in this limit. In fact there are two such eigenvalues, corresponding to $r = \pm 1$. For $r = +1 \, (-1)$ the corresponding eigenvector is symmetric (anti-symmetric) with respect to reversing all arrows, and the eigenvalue is positive (negative). The two eigenvalues are asymptotically degenerate in that their numerical difference vanishes exponentially as N becomes large. We have therefore located $v_1, \ldots, v_n$ for these eigenvalues, in this low temperature limit. From (10.7.15), (10.7.10) and (10.7.11), their values are

$$v_j = 2iI[2j - n - \tfrac{1}{2}(r+1)]/n, \quad j = 1, \ldots, n . \quad (10.7.25)$$

In Section 8.8 and 8.9 we remarked that the free energy is analytic at w (or v) $= 0$, even though the working for w positive differs from the working for w negative. It is easy to see how this comes about in the above equations: for v positive, R_1 is exponentially smaller than R_2 in the limit $n \to \infty$, so the first term on the RHS of (10.6.1) dominates. If v is negative, the situation is reversed. Thus when taking the thermodynamic limit in (10.6.1), or similarly in (8.4.4), the cases $v > 0$ and $v < 0$ must be discussed separately.

However, since the factor $R_1 + R_2$ is contained in $q(v)$, it cancels out of (10.7.21), i.e. of (10.6.1), so the end result is independent of whether it is R_1 or R_2 that dominates. Of course we have as yet only considered the low-temperature limit, but this argument generalizes to all temperatures.

10.8   Calculation of the Free Energy

Let us return to non-zero temperatures, i.e. to I', λ, v finite, and consider how to solve (10.6.10) in the limit of n large.

These equations are the eight-vertex generalizations of (8.9.12), with $v_j = -i\alpha_j$. In Section 8.9 we showed that we expected $\alpha_1, \ldots, \alpha_n$ to be real and distributed over the interval $(-\pi, \pi)$, i.e. over a semi-period of the relevant function $\sinh i\alpha/2$. In the eight-vertex case the function is $h(i\alpha/2)$, and the corresponding interval is $(-2I, 2I)$. In the low-temperature limit we have just observed in (10.7.25) that the $iv_1, \ldots, iv_n$ are indeed distributed over this interval.

One obvious way to solve (10.6.10) is therefore to assume that in the limit $n \to \infty$ the $v_1, \ldots, v_n$ are densely distributed along the line interval $(-2iI, 2iI)$, and to proceed as in Sections 8.7 – 8.9, thereby obtaining from (10.6.10) a linear integral equation for the distribution function of $v_1, \ldots, v_n$.

Here I shall use another method: it is a refinement of the method used in Baxter (1972b) and has the advantage that it discriminates between the two numerically largest eigenvalues, so can be used to obtain the interfacial tension (Baxter, 1973b).

Assumed Properties

First note that (10.6.7) and (10.7.11) determine τ and $z_1 \ldots z_n$ to within choices of the integers r, s, p, p'. Assuming that there are no discontinuous changes within the principal regime, these integers must keep their limiting low-density values. Hence (10.7.15) must be exactly correct throughout the principal regime.

Set

$$p(v) = \frac{\phi(\lambda - v)\, q(v + 2\lambda')}{\phi(\lambda + v)\, q(v - 2\lambda')}. \tag{10.8.1}$$

This is the ratio of the first term on the RHS of (10.6.1) to the second. In the low-temperature limit this is R_1/R_2, where R_1 and R_2 are given by (10.7.22), so it is then true that

$$p(v) \sim r\,(-)^n \exp(-n\pi v/2I). \tag{10.8.2}$$

For $\mathrm{Re}(v) > 0$, this vanishes exponentially as $n \to \infty$; for $\mathrm{Re}(v) < 0$ it grows exponentially.

Also, from (10.7.23), $\Lambda(v)$ is constant (for finite v) in the low-temperature limit. It therefore seems reasonable to assume (throughout the principal regime) that:

(i) there exists a positive real number δ such that $\ln[1 + p(v)]$ is analytic for $0 < \mathrm{Re}(v) < \delta$; and $\ln[1 + 1/p(v)]$ is analytic for $0 > \mathrm{Re}(v) > -\delta$,

(ii) $v_1, \ldots, v_n$ are pure imaginary,

(iii) $\Lambda(v)$ is analytic and non-zero in a vertical strip containing the imaginary axis.

Wiener – Hopf Factorization

I shall now show that (10.6.1) does admit solutions with these properties. First use (i) to make a Wiener – Hopf factorization (Paley and Wiener, 1934; Noble, 1958) of $1 + p(v)$: define $X_+(v)$, $X_-(v)$ by

$$\ln X_+(v) = \frac{1}{4iI} \int_{\alpha - 2iI}^{\alpha + 2iI} \frac{\ln[1 + p(v')]}{\exp[\pi(v - v')/2I] - 1} \, dv', \ \mathrm{Re}(v) > \alpha, \quad (10.8.3a)$$

$$\ln X_-(v) = -\frac{1}{4iI} \int_{\alpha' - 2iI}^{\alpha' + 2iI} \frac{\ln[1 + p(v')]}{\exp[\pi(v - v')/2I] - 1} \, dv', \ \mathrm{Re}(v) < \alpha', \quad (10.8.3b)$$

where $0 < \alpha < \alpha' < \delta$. Adding these equations and using the fact that $p(v)$ is periodic of period $4\pi i$, the RHS can be written as an integral round the rectangle $\alpha' - 2iI$, $\alpha' + 2iI$, $\alpha + 2iI$, $\alpha - 2iI$. Cauchy's residue theorem then gives

$$X_+(v) \, X_-(v) = 1 + p(v) \qquad (10.8.4)$$

This result can be used to define $X_+(v)$ for $\mathrm{Re}(v) \leqslant \alpha$, and $X_-(v)$ for $\mathrm{Re}(v) \geqslant \alpha'$.

The equation (10.6.1) can now be written as

$$\Lambda(v) = \phi(\lambda + v) \, q(v - 2\lambda') \, X_+(v) \, X_-(v)/q(v). \qquad (10.8.5)$$

From (10.8.3), $X_+(v)$ is analytic and non-zero (ANZ) for $\mathrm{Re}(v) > 0$, while $X_-(v)$ is ANZ for $\mathrm{Re}(v) < \delta$. The other terms on the RHS of (10.8.5) can be factored into products of similarly analytic functions. To do this, note from (10.5.16) and (15.1.5) that

$$h(u) = - h(-u) = \gamma \exp(\pi u/2I) \prod_{m=0}^{\infty} \{1 - q^m \exp(-\pi u/I)\}$$

$$\times \{1 - q^m \exp[-\pi(I' - u)/I]\}, \qquad (10.8.6)$$

where

$$\gamma = q^{\frac{1}{4}}\Theta(0) \prod_{m=1}^{\infty} (1 - q^{2m})^2 \,. \tag{10.8.7}$$

Substituting the expression (10.8.6) for $h(u)$ into (10.5.24) and (10.6.8) (with $\tau = 0$) gives

$$\phi(v) = \rho^N \gamma^N \exp(N\pi v/4I)A(v)\,A(2I' - v)\,, \tag{10.8.8}$$

$$q(v) = \gamma^n \exp[\pi(nv - v_1 - \ldots - v_n)/4I]F(v)\,G(v - 2I')\,, \tag{10.8.9a}$$

$$= (-\gamma)^n \exp[\pi(v_1 + \ldots + v_n - nv)/4I]F(v + 2I')\,G(v)\,, \tag{10.8.9b}$$

where

$$A(v) = \prod_{m=0}^{\infty} [1 - q^m \exp(-\pi v/2I)]^N, \tag{10.8.10}$$

$$F(v) = \prod_{j=1}^{n} \prod_{m=0}^{\infty} [1 - q^m \exp\{-\pi(v - v_j)/2I\}]\,, \tag{10.8.11a}$$

$$G(v) = \prod_{j=1}^{n} \prod_{m=0}^{\infty} [1 - q^m \exp\{\pi(v - v_j)/2I\}]\,. \tag{10.8.11b}$$

Note that $A(v)$ is a known function; while $v_1, \ldots, v_n$, and hence $F(v)$ and $G(v)$ are unknown. The object of the following manipulations is to obtain useful expressions for $F(v)$ and $G(v)$.

Substitute the results (10.8.10), (10.8.11) into (10.8.5), using (10.8.9b) for $q(v - 2\lambda')$ and (10.8.9a) for the denominator $q(v)$. This gives, using (10.5.22), (10.7.9) and (10.7.15),

$$\Lambda(v) = r \rho^N \gamma^N x^{-N} L_+(v)\,L_-(v)\,, \tag{10.8.12}$$

where

$$L_+(v) = A(\lambda + v)\,F(v + 2I' - 2\lambda)\,X_+(v)/F(v)\,,$$
$$L_-(v) = A(2I' - \lambda - v)\,G(v - 2\lambda)\,X_-(v)/G(v - 2I')\,. \tag{10.8.13}$$

Let

$$\delta' = \min(\delta, 2\lambda, 2I' - \lambda)\,; \tag{10.8.14}$$

using (10.7.1a), (10.8.10), (10.8.11) and assumption (ii), it is readily observed from (10.8.13) that $L_+(v)$ is ANZ for $\text{Re}(v) > 0$, while $L_-(v)$ is ANZ for $\text{Re}(v) < \delta'$. However, this last property, together with (10.8.12) and assumption (iii), implies that $L_+(v)$ is ANZ for $\text{Re}(v) \geq 0$. Altogether we finally have

$$L_+(v) \text{ is ANZ for } \text{Re}(v) \geq 0\,,$$
$$L_-(v) \text{ is ANZ for } \text{Re}(v) < \delta'\,. \tag{10.8.15}$$

Now repeat the working, but start by Wiener–Hopf factorizing $1 + [1/p(v)]$:

$$\ln Y_+(v) = \frac{\pm 1}{4iI} \int_{-\alpha-2iI}^{-\alpha+2iI} \frac{\ln[1 + 1/p(v')]}{\exp[\pi(v - v')/2I] - 1} \, dv' \,, \qquad (10.8.16)$$

taking the upper choice of signs if $\text{Re}(v) > -\alpha$, the lower if $\text{Re}(v) < -\alpha$, and choosing $0 < \alpha < \delta$. Then

$$Y_+(v) \, Y_-(v) = 1 + [1/p(v)] \,, \qquad (10.8.17)$$

$Y_+(v)$ is ANZ for $\text{Re}(v) > -\delta$, $Y_-(v)$ is ANZ for $\text{Re}(v) < 0$. Equation (10.6.1) becomes

$$\Lambda(v) = \phi(\lambda - v) \, q(v + 2\lambda') \, Y_+(v) \, Y_-(v)/q(v) \,. \qquad (10.8.18)$$

Using (10.8.8), (10.8.9a), (10.8.9b) for $\phi(\lambda - v)$, $q(v + 2\lambda')$, $q(v)$, respectively, this gives

$$\Lambda(v) = r\rho^N\gamma^N x^{-N} M_+(v) \, M_-(v) \,, \qquad (10.8.19)$$

where

$$\begin{aligned}
M_+(v) &= A(2I' - \lambda + v) \, F(v + 2\lambda) \, Y_+(v)/F(v + 2I') \,, \\
M_-(v) &= A(\lambda - v) \, G(v - 2I' + 2\lambda) \, Y_-(v)/G(v) \,.
\end{aligned} \qquad (10.8.20)$$

Equations for $F(v)$, $G(v)$, $p(v)$

From (10.8.20), $M_+(v)$ is ANZ for $\text{Re}(v) > -\delta'$, while $M_-(v)$ is ANZ for $\text{Re}(v) < 0$. Using (10.8.19) and property (iii), it follows that

$$\begin{aligned}
M_+(v) \text{ is ANZ for } \text{Re}(v) &> -\delta' \,, \\
M_-(v) \text{ is ANZ for } \text{Re}(v) &\leq 0 \,.
\end{aligned} \qquad (10.8.21)$$

Comparing (10.8.12) and (10.8.19), it is evident that

$$L_+(v)/M_+(v) = M_-(v)/L_-(v) \,. \qquad (10.8.22)$$

The LHS of this equation is ANZ for $\text{Re}(v) \geq 0$, periodic of period $4iI$, and $\to 1$ as $\text{Re}(v) \to +\infty$ (this last property follows from the definitions). The RHS is ANZ for $\text{Re}(v) \leq 0$, periodic of period $4iI$, and $\to$ constant as $\text{Re}(v) \to -\infty$. Altogether therefore, both sides are entire (and nonzero) and bounded. From Liouville's theorem they are therefore constant. This constant must be one, so

$$M_+(v) = L_+(v), \quad M_-(v) = L_-(v) \,. \qquad (10.8.23)$$

Using (10.8.13) and (10.8.20), the first of these equations is

$$S_+(v) = S_+(v + 2I' - 2\lambda) X_+(v)/Y_+(v) , \qquad (10.8.24)$$

where

$$S_+(v) = F(v) F(v + 2\lambda)/A(v + \lambda) . \qquad (10.8.25)$$

Regard $X_\pm(v)$, $Y_\pm(v)$ as known functions. Then equation (10.8.24) can be regarded as a recursion relation for $S_+(v)$: solving it gives

$$S_+(v) = \prod_{m=0}^{\infty} X_+[v + 2m(I' - \lambda)]/Y_+[v + 2m(I' - \lambda)] , \qquad (10.8.26)$$

and (10.8.25) can now be solved for $F(v)$, giving

$$F(v) = \prod_{m=0}^{\infty} \frac{A[v + (4m + 1)\lambda] S_+(v + 4m\lambda)}{A[v + (4m + 3)\lambda] S_+[v + (4m + 2)\lambda]} . \qquad (10.8.27)$$

(From their definitions, $X_+(v)$, $Y_+(v)$, $A(v)$ all tend exponentially to 1 as $\mathrm{Re}(v) \to +\infty$, so these infinite products converge.)

Similarly, the second of the equations (10.8.23) gives

$$S_-(v) = S_-(v - 2I' + 2\lambda) Y_-(v)/X_-(v) , \qquad (10.8.28)$$

where

$$S_-(v) = G(v) G(v - 2\lambda)/A(\lambda - v) . \qquad (10.8.29)$$

Since $G(v)$, $A(-v)$ tend exponentially to one as $\mathrm{Re}(v) \to -\infty$, so do $S_-(v)$ and $Y_-(v)/X_-(v)$. Thus

$$S_-(v) = \prod_{m=0}^{\infty} Y_-[v - 2m(I' - \lambda)]/X_-[v - 2m(I' - \lambda)] , \qquad (10.8.30)$$

$$G(v) = \prod_{m=0}^{\infty} \frac{A[(4m + 1)\lambda - v] S_-(v - 4m\lambda)}{A[(4m + 3)\lambda - v] S_-[v - (4m + 2)\lambda]} . \qquad (10.8.31)$$

The definition (10.8.1) of $p(v)$ can be expressed in terms of $A(v)$, $F(v)$, $G(v)$ by using (10.8.8) and (10.8.9). Use (10.8.9a) for $q(v + 2\lambda')$, (10.8.9b) for $q(v - 2\lambda')$, and (10.7.16). This gives

$$p(v) = r(-)^n \exp(-n\pi v/2I) A(\lambda - v) A(2I' - \lambda + v)$$
$$\times F(v + 2\lambda) G(v + 2\lambda - 2I')/[A(\lambda + v)$$
$$\times A(2I' - \lambda - v) F(v + 2I' - 2\lambda) G(v - 2\lambda)] . \qquad (10.8.32)$$

Iterative Calculation of $p(v)$

These results are exact, even for finite n, provided the assumptions (i)–(iii) are satisfied. A possible iterative method of solution, starting with some initial guess at $p(v)$ (satisfying assumption (i)) is: calculate $X_{\pm}(v)$, $Y_{\pm}(v)$ from (10.8.3) and (10.8.16); calculate $S_{+}(v)$ from (10.8.26) and (10.8.30); calculate $F(v)$, $G(v)$ from (10.8.27) and (10.8.31); calculate $p(v)$ from (10.8.32) and repeat.

Provided assumption (i) is satisfied, this procedure gives $F(v)$ to be ANZ for $\text{Re}(v) > 0$, and $G(v)$ for $\text{Re}(v) < 0$. From (10.8.11), $v_1, \ldots, v_n$ can therefore only be pure imaginary, so (ii) is satisfied. Also, from (10.8.12) or (10.8.19), $\Lambda(v)$ is analytic for $-\delta' < \text{Re}(v) < \delta'$, so (iii) is satisfied.

Now consider the case when n is large. Suppose, as is suggested by (10.8.2), that $p(v)$ vanishes exponentially with n for $0 < \text{Re}(v) < \delta$; and grows exponentially for $0 > \text{Re}(v) > -\delta$. Then from (10.8.3) and (10.8.16), $X_{+}(v)$, $X_{-}(v)$, $Y_{+}(v)$, $Y_{-}(v)$ are exponentially close to one, provided that $\text{Re}(v) > 0$, $\text{Re}(v) < \delta$, $\text{Re}(v) > -\delta$, $\text{Re}(v) < 0$, respectively.

For $\text{Re}(v) > 0$, each function $S_{+}(v)$ in (10.8.27) is therefore exponentially close to one; for $\text{Re}(v) < 0$ so is each function $S_{-}(v)$ in (10.8.31). From (10.8.32) it follows that for $|\text{Re}(v)| < \min(2\lambda, 2I' - 2\lambda)$

$$p(v) = r(-)^n \exp(-n\pi v/2I) \prod_{m=0}^{\infty} \frac{A[(4m+3)\lambda + v]}{A[(4m+1)\lambda + v]}$$

$$\frac{A[(4m+1)\lambda + v + 2I']\, A[4m+1)\lambda - v]\, A[(4m+3)\lambda - v + 2I']}{A[(4m+3)\lambda + v + 2I']\, A[(4m+3)\lambda - v]\, A[(4m+1)\lambda - v + 2I']}$$

$$\times \{1 + (\text{terms that vanish exponentially as } n \to \infty)\}. \qquad (10.8.33)$$

Now use the definition (10.8.10) of $A(v)$; together with (10.7.9) this implies that

$$A(v)/A(v + 2I') = [1 - \exp(-\pi v/2I)]^N. \qquad (10.8.34)$$

A quite remarkable feature of (10.8.33) is that $A(v)$ only occurs in the combination (10.8.34), so

$$p(v) = r(-)^n \exp(-n\pi v/2I) \prod_{m=0}^{\infty} \left[\frac{1 - x^{4m+3}z}{1 - x^{4m+1}z} \cdot \frac{1 - x^{4m+1}z^{-1}}{1 - x^{4m+3}z^{-1}} \right]^N, \qquad (10.8.35)$$

for $|\text{Re}(v)| < \min(2\lambda, 2I' - 2\lambda)$, where x, z are defined by (10.7.9) and (10.7.19), and the exponentially small corrections have been neglected.

The nome q has therefore disappeared from this expression. On the other hand, comparing (10.8.34) with (15.4.13), we see that $p(v)$ is related to the elliptic functions of nome x^2 or x. In fact, since $\text{am}(u)$ is defined by (15.4.13) with q replaced by q^2,

$$p(v) = r(-)^n \exp[-iN \, \text{am}(i\hat{v}, \hat{k})] \,, \tag{10.8.36}$$

where $\hat{k}$ is the elliptic modulus corresponding to the nome

$$\hat{q} = x \,, \tag{10.8.37}$$

and

$$\hat{v}/\hat{I} = v/(2I) \,. \tag{10.8.38}$$

The function $\exp[-i \, \text{am}(i\hat{v}, \hat{k})]$ has modulus less than one for $0 < \text{Re}(v) < 2\lambda$, and greater than one for $0 > \text{Re}(v) > -2\lambda$. Also, it is a meromorphic function of v. The assumption (i) is therefore satisfied by (10.8.36) with $0 < \delta < \min(2\lambda, 2I' - 2\lambda)$; so therefore are (ii) and (iii). Further, it is also true that this function $p(v)$ vanishes exponentially as $n \to \infty$, provided $0 < \text{Re}(v) < \delta$; it grows exponentially if $0 > \text{Re}(v) > -\delta$.

If we now substitute this expression for $p(v)$ back into (10.8.3) and (10.8.16) and continue the iterative procedure, we should obtain the exponentially small corrections to (10.8.36), then the corrections to the corrections, and so on. I have not done so, but expect that it should be possible to prove, with full mathematical rigor, that this procedure converges to a solution of (10.6.1) satisfying assumptions (i) – (iii), and with (10.8.36) as the exact large-N solution for $p(v)$.

Note that $v_1, \ldots, v_n$ are the zeros of $1 + p(v)$, lying on the imaginary axis. Like $p(v)$ therefore, for large N the values of $v_1/I, \ldots, v_n/I$ depend only on x: not on q (or z). I find this intriguing: I have no simple explanation as to why it should be so.

The Functions $p(v)$, $\Lambda(v)$ in the Thermodynamic Limit

The large-n formulae (10.8.35) and (10.8.36) give $p(v)$ only when $|\text{Re}(v)| < \min(2\lambda, 2I' - 2\lambda)$. To obtain $p(v)$ for other values of v, note that (10.8.27) gives $F(v)$ for all v, but only when $\text{Re}(v) > 0$ do the functions S_+ give a factor which is exponentially close to unity for large n. Thus (10.8.27) is 'useful' when $\text{Re}(v) > 0$, since to leading order in a large-n expansion the unknown functions $X_+(v), Y_+(v), S_+(v)$ can then be replaced by unity. Similarly, the expression (10.8.31) for $G(v)$ is useful when $\text{Re}(v) < 0$.

Using the second periodicity relation in (10.6.2), the function $q(v)$, for general values of v, can be written as proportional to $q(v^*)$, where $0 < \text{Re}(v^*) < 2I'$. This function $q(v^*)$ can then be factorized by (10.8.9a), giving an expression for $q(v)$ involving $F(v)$ only for $\text{Re}(v) > 0$, and $G(v)$ only for $\text{Re}(v) < 0$.

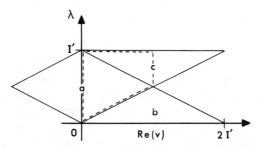

Fig. 10.5. Regions of applicability of the forms a, b, c of equation (10.8.39). Within the broken line $p(v)$ is exponentially small for large n.

Doing this in (10.8.1), using (10.8.27) and (10.8.31), we obtain (for n large)

$$p(v) = p^{(0)}(v) \qquad \text{for } |\text{Re}(v)| < \min(2\lambda, 2I' - 2\lambda), \qquad (10.8.39a)$$

$$p(v) = 1 \qquad \text{for } 2\lambda < \text{Re}(v) < 2I' - 2\lambda \qquad , \qquad (10.8.39b)$$

$$p(v) = p^{(0)}(v)\, p^{(0)}(v - 2I') \text{ for } 2I' - 2\lambda < \text{Re}(v) < 2\lambda \qquad , \qquad (10.8.39c)$$

where $p^{(0)}(v)$ is the function on the RHS of (10.8.35) and (10.8.36).

Together with the periodicity relations [a consequence of (10.6.2), (10.5.24) and (10.8.1)]

$$p(v + 4iI) = p(v + 2I') = p(v), \qquad (10.8.40)$$

these equations define $p(v)$ for general values of v. The regions of applicability of the three formulae (10.8.39) are shown in Fig. 10.5. Also shown is a broken line marking the boundary of the region in which $p(v)$ is exponentially small when N is large. From this it is apparent that in assumption (i) of this section the maximum value of δ is

$$\delta = \min(2\lambda, I'). \qquad (10.8.41)$$

Now calculate $L_-(v)$ from (10.8.13), $M_+(v)$ from (10.8.20), and calculate $\Lambda(v)$ from (10.8.12), using (10.8.23) to replace L_+ by M_+. (This route gives an expression which is manifestly analytic and non-zero for $-\delta' <$ $\mathrm{Re}(v) < \delta'$.) Neglecting exponentially small corrections, the result is

$$\Lambda(v) = r\rho^N \gamma^N x^{-N} \prod_{m=0}^{\infty} \frac{A[(4m+3)\lambda + v]}{A[(4m+5)\lambda + v]}$$

$$\times \frac{A[2I' + (4m-1)\lambda + v] A[(4m+3)\lambda - v] A[2I' + (4m-1)\lambda - v]}{A[2I' + (4m+1)\lambda + v] A[(4m+5)\lambda - v] A[2I' + (4m+1)\lambda - v]},$$

$$(10.8.42)$$

provided $|\mathrm{Re}(v)| < \delta$.

We have solved the functional equation (10.6.1) for the functions $q(v)$, $\Lambda(v)$. This is equivalent to solving (10.6.1) for $v_1, \ldots, v_n$: indeed $v_1, \ldots, v_n$ are the zeros of $1 + p(v)$, so can be obtained from (10.8.35).

These equations have many solutions, corresponding to the 2^N different eigenvalues $\Lambda(v)$ of the transfer matrix $V(v)$. In fact we have obtained just two such solutions, i.e. eigenvalues: one with $r = +1$ (arrow reversal symmetry), the other with $r = -1$ (anti-symmetry). From (10.8.42) they are equal in magnitude and opposite in sign (to within corrections that vanish exponentially as $n \to \infty$).

In the low temperature limit we have verified that these are the two numerically largest eigenvalues. From the Perron-Frobenius theorem (Frobenius, 1908), a matrix with positive entries has a unique maximum eigenvalue. It follows that these are the two numerically largest eigenvalues throughout the principal regime (10.7.1), within which the Boltzmann weights a, b, c, d are positive and our analysis is valid.

The result (10.8.42) can of course be analytically continued throughout the complex v-plane. Indeed, provided $|\mathrm{Re}(v)| < \delta$, it has a direct meaning: $\Lambda(v)$ is the eigenvalue associated with the eigenvector which is maximal in the principal regime (the eigenvector is independent of v). In particular, this analytic continuation satisfies

$$\Lambda(v) \Lambda(2\lambda - v) = \phi(\lambda + v) \phi(3\lambda - v), \qquad (10.8.43)$$

a relation to which I shall return in Chapter 13.

Free Energy

Taking logarithms of both sides of (10.8.42), using (10.8.10), the RHS becomes a double sum of terms like $\ln(1 - q^a x^b z^{\pm 1})$, where a and b are integers. Taylor expanding each such logarithm in powers of $q^a x^b z^{\pm 1}$, the summations can be performed term by term, giving

$$N^{-1}\ln[r\Lambda(v)] = \ln(\rho\gamma/x) - \sum_{m=1}^{\infty} \frac{(x^{3m} + q^m x^{-m})(z^m + z^{-m})}{m(1-q^m)(1+x^{2m})}.$$

(10.8.44)

Also, from (10.4.24), (10.8.7) and (15.1.5),

$$\ln c = \ln(\rho\gamma/x) + \sum_{m=1}^{\infty} \frac{2q^m - x^{2m} - q^{2m}x^{-2m} - q^m(x^m + x^{-m})(z^m + z^{-m})}{m(1-q^{2m})}.$$

(10.8.45)

As usual [equations (1.7.6) and (8.2.4)], the free energy f per site is related to the maximum eigenvalue Λ_{max} by

$$-f/k_B T = \lim_{N\to\infty} N^{-1}\ln\Lambda_{max},$$

(10.8.46)

N being the number of columns of the lattice.

From the Perron–Frobenius theorem (Frobenius, 1908), the maximum eigenvalue must have $r = +1$. Thus Λ_{max} is the $\Lambda(v)$ given by (10.8.44), with $r = +1$. Eliminating $\Lambda(v)$ and $\rho\gamma/x$ between these last three equations leaves

$$-f/k_B T = \ln c + \sum_{m=1}^{\infty} \frac{x^{-m}(x^{2m} - q^m)^2(x^m + x^{-m} - z^m - z^{-m})}{m(1-q^{2m})(1+x^{2m})}.$$

(10.8.47)

Alternatively, using (10.4.24), (10.8.7) and (15.4.27), one can establish that

$$\ln(c+d) = \ln(\rho\gamma/x) - \sum_{m=1}^{\infty} \frac{x^{2m} + q^m x^{-2m} - q^{m/2}(x^m + x^{-m} - z^m - z^{-m})}{m(1-q^m)},$$

(10.8.48)

and hence that

$$-f/k_B T = \ln(c+d)$$
$$+ \sum_{m=1}^{\infty} \frac{(x^{3m} - q^{m/2})(1 - q^{m/2}x^{-m})(x^m + x^{-m} - z^m - z^{-m})}{m(1-q^m)(1+x^{2m})}.$$

(10.8.49)

10.9 The Ising Case

It was shown in Section 10.3 that the eight-vertex model factors into two identical and independent Ising models when $J'' = 0$. The interaction coefficients are termed K, L in Chapter 7; $J/k_B T$, $J'/k_B T$ in this chapter.

Let k_I, u_I, q_I, . . . be the variables k, u, q, . . . of Chapter 7. Then from (7.6.1)

$$k_I^{-1} = \sinh(2J/k_B T) \sinh(2J'/k_B T) . \qquad (10.9.1)$$

Expanding the RHS as sums of exponentials and using (10.3.9), (10.3.10) and (10.4.6) gives

$$k_I^{-1} = (a^2 + b^2 - c^2 - d^2)/4ab = \Delta . \qquad (10.9.2)$$

The ferromagnetic ordered state of the Ising model ($J > 0$, $J' > 0$, $k_I < 1$) therefore lies in the regime $\Delta > 1$, $a > b + c + d$ of the eight-vertex model.

This regime can be mapped into the principal regime (10.7.5) by using the symmetry relation (10.2.14), i.e. by interchanging a with c, and b with d.

Do this, and then define ρ, k, λ, v by (10.4.24), or (10.4.21). We obtain

$$\text{snh } \lambda = k^{-\frac{1}{2}} , \qquad (10.9.3)$$

$$k^{\frac{1}{2}} \text{ snh } u = \exp(-2J'/k_B T) , \qquad (10.9.4)$$

and, from (10.9.2) and (10.4.17), noting that Δ is negated by interchanging a with c and b with d:

$$k_I^{-1} = \tfrac{1}{2} \text{ cn } i\lambda \text{ dn } i\lambda . \qquad (10.9.5)$$

Using (15.4.32), it follows from (10.9.3) that

$$\lambda = \tfrac{1}{2} I' , \qquad (10.9.6)$$

so λ is exactly in the middle of the interval $(0 , I')$ permitted by (10.7.1). From (10.7.9),

$$q = x^4 . \qquad (10.9.7)$$

Also, from (15.4.32) and (10.9.5),

$$k_I = 2k^{\frac{1}{2}}/(1 + k) . \qquad (10.9.8)$$

This relation between elliptic moduli is that of the Landen transformation (Section 15.6). If q_I is the nome corresponding to k_I, then from (15.6.2) and (10.9.7),

$$q_I = q^{\frac{1}{2}} = x^2 . \qquad (10.9.9)$$

From (7.10.50), using (15.1.4), the Ising model spontaneous magnetization can therefore be written as

$$M_0 = \prod_{n=1}^{\infty} \frac{1 - q_I^{2n-1}}{1 + q_I^{2n-1}} = \prod_{n=1}^{\infty} \frac{1 - x^{4n-2}}{1 + x^{4n-2}} . \qquad (10.9.10)$$

Also, using the transformation (15.6.4) in (10.9.4), comparing the result with (7.8.5) gives

$$u_I = (1 + k)u. \tag{10.9.11}$$

With these equivalences, it can be verified directly that (10.8.47) does indeed become the previous Ising model result (7.9.16) when $J'' = 0$.

10.10   Other Thermodynamic Properties

Interfacial Tension

In the principal regime the eight-vertex model has long range anti-ferro-electric order. The predominant pattern is either that of Fig. 8.3, or that obtained therefrom by reversing all arrows.

The interfacial tension between domains of these two types can be obtained in the same way as in Section 7.10. If Λ_0, Λ_1 are the two numerically largest eigenvalues, they are asymptotically degenerate in the sense that

$$\Lambda_1/\Lambda_0 = -1 + \mathcal{O}\left[\exp(-Ns/k_BT)\right], \tag{10.10.1}$$

where s is the interfacial tension.

These eigenvalues Λ_0, Λ_1 are the two eigenvalues discussed in the previous section, with $r = +1$, -1, respectively. We observed there that for large N they were equal in magnitude and opposite in sign. To obtain s we must keep some of the exponentially small corrections that we neglected in the previous section.

First re-derive (10.8.42), keeping all the $X_\pm(v)$, $Y_\pm(v)$ contributions, and using (10.8.3) and (10.8.16). Let $\Lambda^{(0)}(v)$ be the function on the RHS of (10.8.42). Then it is found that, for $|\text{Re}(v)| < \alpha$,

$$\ln[\Lambda(v)/\Lambda^{(0)}(v)] = \frac{1}{8il} \int_{\alpha-2il}^{\alpha+2il} \ln[1 + p(v')]\, D(v - v')\, dv'$$

$$- \frac{1}{8il} \int_{-\alpha-2il}^{-\alpha+2il} \ln[1 + 1/p(v')]\, D(v - v')\, dv' , \tag{10.10.2}$$

where if z, x are defined by (10.7.18) and (10.7.9),

$$D(v) = 1 + 2 \sum_{m=0}^{\infty} (-)^m \left\{ \frac{x^{2m}z^{-1}}{1 - x^{2m}z^{-1}} + \frac{x^{2m+2}z}{1 - x^{2m+2}z} \right\}. \tag{10.10.3}$$

This function $D(v)$ satisfies the relations

$$D(v) = -D(-v) = -D(v + 2\lambda) ; \tag{10.10.4}$$

in fact

$$D(v - \lambda) = \mathrm{dn}(i\hat{v}, \hat{k}), \qquad (10.10.5)$$

where $\mathrm{dn}(u, k)$ is the elliptic function defined in (15.1.6), and $\hat{k}$, $\hat{v}$ are defined by (10.8.37) and (10.8.38): again we see the occurrence of elliptic functions with nome x, rather than q.

Equation (10.10.2) is exact, even for finite n. The leading corrections (for large n) to $\Lambda(v)$ can be obtained by substituting into (10.10.2) the large-n expressions (10.8.39) for $p(v)$.

Provided $\alpha < \min(2\lambda, 2I' - 2\lambda)$, these are the expressions given in (10.8.35) and (10.8.36). The function $p(v)$ then has saddle points at $v = 2iI \pm \lambda$, i.e. $z = -x^{\pm 1}$. Taking $\alpha = \lambda$ in (10.10.2), integrating by steepest descents therefore gives

$$\ln[\Lambda(v)/\Lambda^{(0)}(v)] \sim p(2iI + \lambda). \qquad (10.10.6)$$

Setting $z = -x$ in (10.8.35):

$$p(2iI + \lambda) = r\, k_1^{N/2}, \qquad (10.10.7)$$

where

$$k_1^{\frac{1}{2}} = 2x^{\frac{1}{2}} \prod_{m=1}^{\infty} \left(\frac{1 + x^{4m}}{1 + x^{4m-2}} \right)^2. \qquad (10.10.8)$$

[This quantity k_1 is the elliptic modulus with nome x^2. It is related to $\hat{k}$ by $\hat{k} = 2k_1^{\frac{1}{2}}/(1 + k_1)$.]

For $r = +1$, $\Lambda(v)$ in (10.10.6) is Λ_0; for $r = -1$ it is Λ_1. From (10.8.42), $\Lambda^{(0)}(v)$ is the same for both cases except for a change in sign. Taking the difference, using (10.10.7), therefore gives

$$\Lambda_1/\Lambda_0 = -1 + \mathcal{O}(k_1^{N/2}). \qquad (10.10.9)$$

The two maximum eigenvalues are therefore asymptotically degenerate: they satisfy (10.10.1), the interfacial tension s being given by

$$\exp(-s/k_B T) = k_1^{\frac{1}{2}}. \qquad (10.10.10)$$

This argument fails if $\lambda > 2I'/3$, since then $p(v')$ in (10.10.2) is not given by (10.8.35) when $\alpha = \lambda$. Even so, (10.10.10) remains correct, as is shown in Appendix D of Baxter (1973d).

When $k \to 0$, then $q \to 0$, $I \to \pi/2$ and $\mathrm{snh}\, u \to \sinh u$. Using (10.4.23), we see that the relation (10.4.21) becomes the same as (8.9.7), with $d = 0$. Thus the eight-vertex model reduces to the six-vertex model, the principal regime (10.7.5) becomes the regime IV of Section 8.10, and λ, v then have the same meaning as in Section 8.10.

The interfacial tension of the six-vertex model is therefore given by (10.10.10) and (10.10.8), in the limit $q \to 0$. However, these equations are independent of q, so they already give the result (8.10.3) quoted above. The definition (10.7.9) of x reduces to (8.10.2).

Correlation Length

In addition to the largest two eigenvalues $\Lambda(v)$ of the transfer matrix $V(v)$, in the limit of N large it is also possible to calculate the next-largest, and so on. Indeed, a considerable amount of work has gone into doing such calculations, mainly because they enter the related problem of the partition function of the XYZ chain, which is discussed in Section 10.14 (Yang and Yang, 1969; Gaudin, 1971; Johnson and McCoy, 1972; Takahashi and Suzuki, 1972; Takahashi, 1973, 1974; Johnson and Bonner, 1980).

I shall not attempt to reproduce such calculations here. Let me merely remark that for any eigenvalue it is expected (for N large) that the zeros $v_1, \ldots, v_n$ of $q(v)$ are grouped in strings in the complex plane. All zeros in a string have the same imaginary part, the zeros are spaced uniformly at intervals of 2λ, and are symmetric about either the pure imaginary axis or about the vertical line $\mathrm{Re}(v) = I'$. Thus a string of m zeros consists of the complex numbers

$$iw + lI' + (2j - m - 1)\lambda, \quad j = 1, \ldots, m, \qquad (10.10.11)$$

where l is an integer (either 0 or 1).

For the two largest eigenvalues (i.e. those discussed in Section 10.8) each string has $l = 0$ and contains only one zero. After these, let Λ_2 be the next-largest eigenvalue. Johnson et al (1972a, 1972b, 1973) argued that this would correspond to either one of the strings having $l = 1$, or to two of them being replaced by a string of length two.

In fact there are $2N$ such eigenvalues, all corresponding to the diagonal operator S having eigenvalue $s = -(-1)^n$. In the limit $N \to \infty$, and provided $\lambda \leqslant \frac{1}{2}I'$, the largest of them is given by (8.10.10), or equivalently (8.10.11), i.e.

$$\left(\frac{\Lambda_2}{\Lambda_0}\right)^{\frac{1}{2}} = x^{\frac{1}{4}}(z^{\frac{1}{2}} + z^{-\frac{1}{2}}) \prod_{m=1}^{\infty} \frac{(1 + x^{4m} z)(1 + x^{4m} z^{-1})}{(1 + x^{4m-2} z)(1 + x^{4m-2} z^{-1})}, \qquad (10.10.12)$$

where λ, v, x, z are defined by (10.4.24), (10.7.9) and (10.7.18). (Like the results (10.10.8), (10.10.10) for the interfacial tension, this formula does not involve q, so has the same form for both the eight-vertex and six-vertex models.)

The argument following (8.10.11) still applies, so the arrow correlation length ξ is given by (8.10.12). Using (10.10.8), this can be written as

$$\xi^{-1} = \tfrac{1}{2} \ln(1/k_1) \, , \qquad (10.10.13)$$

provided as before that $\lambda \leqslant \tfrac{1}{2} I'$.

Eliminating k_1 between (10.10.10) and (10.10.13) therefore gives

$$s\xi = k_B T \, , \qquad (10.10.14)$$

s here being the interfacial tension. It can now be seen why this relation is satisfied by both the Ising model [eq. (7.10.44)] and the six-vertex model [eq. (8.10.13)]: both are special cases of the eight-vertex model, for which the relation holds generally, provided that $0 \leqslant \lambda/I' \leqslant \tfrac{1}{2}$. (The Ising model has $\lambda/I' = \tfrac{1}{2}$, the six-vertex model has $\lambda/I' = 0$.)

Yet more generally, the scaling hypothesis (Widom, 1965; Abraham, 1979) predicts that, for all systems near their critical point, s and ξ should satisfy a relation of the form (10.10.14).

Johnson *et al* (1973, equations (6.17)) also obtained ξ for $\tfrac{1}{2} I' < \lambda < I'$, and found that (10.10.12) and (10.10.13) fail for $\lambda > 2I'/3$. This is typical of the calculation of the lower eigenvalues of the eight-vertex model transfer matrix: the results differ for various sub-intervals of the line $0 < \lambda < I'$. In part this is due to strings becoming longer than the period $2I'$ of $q(v)$ and hence reappearing on the other side of a period rectangle. It greatly complicates the study of the lower eigenvalues.

It also means that many of the formulae, while appearing not to involve q (only x and z occur explicitly), do involve it in their domains of validity. This is a pity: if it were true for any eigenvalue Λ_j that Λ_j/Λ_0 was a function only of x and z [as in (10.10.9), (10.10.12) and equation 8 of Johnson *et al.* (1972a)], then these ratios could be obtained (for N large) from the explicit Ising model results.

It should be remarked that the results (10.10.10), (10.10.13) for the interfacial tension and correlation length do not quite reduce to (7.10.18), (7.10.43) in the pure Ising model case. In this case, from (10.9.8) and (10.9.9), k_I is the elliptic modulus with nome x^2, so $k_1 = k_I$ is the modulus used in Chapter 7 and we see that there are discrepancies of factors of 2. For the interfacial tension this is because of a change of length scale: the n in (7.10.17) corresponds to $N/2$ in (10.10.9). For the correlation length it is because in the principal regime (wherein the system is ordered) the Λ_2s discussed by Johnson *et al* (1972a, 1973) lie in a different diagonal block of the transfer matrix from Λ_0 (the corresponding eigenvalues of S have opposite sign), so the matrix elements t_{02}, t_{20} in (7.10.33) are zero: one has to go to the next-largest band of eigenvalues. The effect of this is to square Λ_2/Λ_0, and hence to remove the $\tfrac{1}{2}$ from (10.10.13). (For the disordered regimes the Λ_2 of Johnson *et al.* lies in the same diagonal block as Λ_0 and (10.10.13) is correct as written, providing k, λ, v are defined as in Section 10.11. The result then is to be compared with the low-temperature Ising result (7.11.4).)

Spontaneous Magnetization

We have seen that the eight-vertex model can be viewed either as a model of ferroelectricity (with dipoles represented by arrows on lattice edges), or of ferromagnetism (with Ising spins on lattice faces). One obtains a different order parameter depending on which viewpoint is adopted.

Let us first use the magnetic Ising picture of Section 10.3. Let σ_1 be a particular spin, and let

$$M_0 = \langle \sigma_1 \rangle , \qquad (10.10.15)$$

be the average value of this spin, calculated in the limit of an infinitesimally weak applied magnetic field, as in (1.1.1).

Suppose that the system is ferromagnetic, so J and J' are positive, and that $J'' > -\max(J, J')$. Then for sufficiently low temperatures T, the Boltzmann weights, given by (10.3.9), will satisfy

$$a > b + c + d . \qquad (10.10.16)$$

This regime can be obtained from the principal regime (10.7.5) by interchanging a with c, and b with d, using the symmetry argument of (10.2.14). It is the ferromagnetically ordered regime of the eight vertex model.

Define k, λ, v, ρ, satisfying (10.7.1), by interchanging a with c, and b with d, and then using (10.4.21)–(10.4.24). Define q, x, z by (10.7.9) and (10.7.19), and a parameter y by

$$q = x^2 y . \qquad (10.10.17)$$

Then M_0 can be regarded as a function of x, y and z.

Barber and Baxter (1973) expanded M_0 as a series in x, with coefficients that *a priori* are functions of y and z. (This is a partially summed low-temperature series.) To order x^4 they found that

$$M_0 = 1 + 0.x - 2x^2 + 0.x^3 + 2x^4 + \dots . \qquad (10.10.18)$$

All coefficients calculated were in fact constants.

It is not difficult to see that M_0 must be independent of z: as in (7.10.48) and (7.10.32), M_0 can be written solely in terms of a diagonal single-spin operator and the matrix P (U in Chapter 7) of eigenvectors of the transfer matrix. These are independent of v, and hence z; so therefore is M_0.

It is not obvious that M_0 should be independent of q. However, the interfacial tension is, the correlation length is (provided $\lambda < 2I'/3$), and so are the first five terms in the x-expansion of M_0. For these reasons, Barber and Baxter (1973) conjectured that M_0 is a function only of x.

However, for the pure Ising model, which from (10.9.7) is when $q = x^4$, M_0 is known from (10.9.10) to be

$$M_0 = \prod_{n=1}^{\infty} \frac{1 - x^{4n-2}}{1 + x^{4n-2}}, \tag{10.10.19}$$

so the conjecture implies that this formula is true for all q.

The conjecture has been verified: M_0 can be obtained from the corner transfer matrices, as will be shown in Chapter 13.

Remember that the k_1 defined by (10.10.8) is, from (15.1.4a), the elliptic modulus corresponding to the nome x^2, From (15.1.4b) it follows that

$$M_0 = k_1'^{1/4} = (1 - k_1^2)^{1/8}. \tag{10.10.20}$$

Spontaneous Polarization

Return now to the original arrow formulation and let

$$P_0 = \langle \alpha_1 \rangle, \tag{10.10.21}$$

where α_1 is the 'arrow spin' on some particular edge (vertical or horizontal), having values ± 1 depending on the arrow direction. From (10.3.2),

$$P_0 = \langle \sigma_1 \sigma_2 \rangle, \tag{10.10.22}$$

where σ_1, σ_2 are the Ising spins of the faces on either side of the lattice edge.

This P_0 is an 'order parameter' like M_0. From Fig. 10.1 and (10.1.5), the zero-field eight-vertex model is unchanged by reversing all arrows. This symmetry can be broken by adding a field-like contribution to the total energy of

$$-E \sum_i \alpha_i, \tag{10.10.23}$$

where E is the 'electric field' and the sum is over all vertical (or all horizontal) arrows.

For a ferroelectric model, where a or b is the largest of the Boltzmann weights a, b, c, d, this field breaks the degeneracy of the ground states.

If one now calculates (10.10.21) in the limit of a large lattice, then lets $E \rightarrow 0$ through positive values, the resulting expression for P_0 will be non-zero if a or b are sufficiently large.

For an anti-ferroelectric model, where c or d is the largest of the Boltzmann weights [as in the principal regime (10.7.5)], (10.10.23) does not break the degeneracy of the ground states (such as that in Fig. 8.3). It is necessary to 'stagger' E, alternating its sign on successive edges. Then the appropriate limiting value of P_0 is again non-zero for sufficiently large c or d.

This calculation has not been carried out, any more than the Ising model has been solved in a magnetic field. In fact P_0 itself has not been calculated, but it must, like M_0, be a function only of x and q, independent of z. Baxter and Kelland (1974) have conjectured that in the principal regime (10.7.5)

$$ P_0 = \prod_{n=1}^{\infty} \left(\frac{1 + q^n}{1 - q^n} \frac{1 - x^{2n}}{1 + x^{2n}} \right)^2 , \qquad (10.10.24) $$

This agrees with the six-vertex ($q = 0$) result (8.10.9). In the Ising case ($q = x^4$) it gives, using (10.9.10), $P_0 = M_0^2$: this is correct, since σ_1 and σ_2 in (10.10.22) lie on distinct sub-lattices and so are independent for the Ising case. The conjecture is also correct in the limit $q = x^2$, $\lambda = I'$, when $a = b = 0$, $c = d$ and the system is completely ordered.

Baxter and Kelland also set $q = x^2 y$, as in (10.10.17), and calculated P_0 to order x^4 in an expansion in powers of x, with coefficients that are functions of y. The result agreed with (10.10.24) so it seems very likely that (10.10.24) is exactly correct throughout the principal regime.

Noting that the nomes q, x^2 correspond to the moduli k, k_1, respectively, it follows from (15.1.7) that (10.10.24) can be written as

$$ P_0 = k_1' I_1 / (k' I) , \qquad (10.10.25) $$

where k', k_1' are the corresponding conjugate moduli, and I, I_1 are the complete elliptic integrals of the first kind.

10.11 Classification of Phases

Within the principal regime (10.7.5), the free energy is given by (10.8.47). In this case we see from (10.2.16) that

$$ w_3 > w_4 > w_1 > |w_2| . \qquad (10.11.1) $$

For general values of a, b, c, d, using the symmetry relation (10.2.17), the free energy is given by the following procedure:

(i) Calculate w_1 , . . . , w_4 from (10.2.16).

(ii) Negate and re-arrange the w_1 , . . . , w_4 as necessary to satisfy (10.11.1).

(iii) Calculate the mapped values of a, b, c, d from (10.2.16). These will lie in the principal regime.

(iv) Calculate ρ, k, λ, v from (10.4.16)–(10.4.24), q and x from (10.7.9), z from (10.7.18).

(v) Calculate the free energy f from (10.8.47) or (10.8.49).

The resulting function $f(a\,,b\,,c\,,d)$ is analytic except only when one of a, b, c, d is equal to the sum of the other three. Thus there are five regimes:

I. Ferroelectric: $a > b + c + d$, $\Delta > 1$,
II. Ferroelectric: $b > a + c + d$, $\Delta > 1$,
III. Disordered: $a, b, c, d < \frac{1}{2}(a + b + c + d)$, $-1 < \Delta < 1$,
IV. Anti-ferroelectric: $c > a + b + d$, $\Delta < -1$,   (principal regime) ,
V. Anti-ferroelectric: $d > a + b + c$, $\Delta < -1$.

In regimes I, II, IV, V the system is ordered: any such regime can be obtained from IV by using only the elementary symmetries (10.2.12)–(10.2.14). The interfacial tension s, correlation length ξ, magnetization M_0 and polarization P_0 are given by (10.10.10), (10.10.13) (without the $\frac{1}{2}$), (10.10.19), (10.10.23), respectively; q, x, z being defined as in (iv) above.

Note that the system is always ordered if Δ, as given by (10.4.6), is numerically greater than one. It is disordered if $|\Delta| < 1$.

This classification into regimes becomes more obvious if we explicitly solve (10.4.17) and (10.4.6) for the elliptic modulus k. Squaring the second equation (10.4.17), using (15.4.4) and (15.4.5), and eliminating $\mathrm{sn}^2\, i\lambda$ between this and (10.4.16), we obtain

$$\Delta^2 = (1 + k^{-1}\gamma)\,(1 + k\gamma)/(1 + \gamma)^2. \qquad (10.11.2)$$

Solving this for $k + k^{-1}$ and using (10.4.6) gives

$$k_0 + k_0^{-1} - 2 \qquad\qquad (10.11.3)$$

$$= \frac{(a - b - c - d)\,(a - b + c + d)\,(a + b - c + d)\,(a + b + c - d)}{4abcd}, \qquad (10.11.3)$$

$$k_0 + k_0^{-1} + 2 = \frac{4a'b'c'd'}{abcd}, \qquad (10.11.4)$$

where a', b', c', d' are defined by (10.2.5), and the suffix 0 means that k is to be evaluated directly from (10.11.3) or (10.11.4), the mapping procedure (i)–(iii) above being omitted. This means that k_0 is not necessarily in the interval $(0, 1)$.

In the ordered regimes I, II, IV, V, the RHS of (10.11.3) is positive. It is negative in the disordered regime III.

To map III into the principal regime it is necessary to use the duality relation (10.2.11). Various cases arise, depending on whether b', c', d' are positive or negative, but it is found that the free energy f is an analytic function of a, b, c, d throughout the regime III. This regime is disordered: there is no spontaneous magnetization or polarization. The correlation length is given precisely by (10.10.13).

Disorder Points

The system is disordered throughout the regime III, but it is particularly so when c' or d' vanish, i.e. when [using (10.2.5)] the point (a, b, c, d) lies on either of the surfaces

$$a + c = b + d \quad \text{or} \quad a + d = b + c. \tag{10.11.5}$$

The procedure (i)–(v) maps these cases to eight-vertex models in the principal regime IV, with either a or b zero. Since k_0 is then the same as k, and $0 \leqslant k \leqslant 1$, it is then apparent from (10.11.3) that $k = 0$. More precisely, as the mapped a (or b) tends to zero, then $k \to 0$ and, from (10.4.9)–(10.4.21), y and γ are proportional to k^{-1}, and $\lambda \to \infty$ while $I' - \lambda$ remains finite. From (10.7.9), q and x both tend to zero, x being proportional to $q^{\frac{1}{2}}$. From (10.10.8) and (10.10.13), k_1 and ξ therefore tend to zero. Thus the correlation length [as defined by (7.10.41), modified as in the argument following (8.10.11)] becomes zero: the system is completely disordered.

The free energy is given by (10.8.47), c therein being the mapped value of c and all terms in the summation being zero, so

$$-f/k_B T = \ln \tfrac{1}{2}(a + b + c + d). \tag{10.11.6}$$

[This simple result can be understood in at least two ways: one is to note that the mappings (10.2.11)–(10.2.15) can be used to map the model to a six-vertex model in the frozen ferroelectric regime I or II of Section 8.10; the other is to verify that $Vx = [\tfrac{1}{2}(a + b + c + d)]^N x$, where V is the transfer matrix and x is a vector with all elements unity—this calculation is particularly simple if V is replaced by the transfer matrix that builds the lattice up diagonally.]

Such points of complete disorder occur also in the anti-ferromagnetic triangular Ising model (Stephenson, 1970).

10.12   Critical Singularities

The free energy is an analytic function of a, b, c, d, and the correlation length ξ is finite, unless

$$a = b + c + d \quad \text{or} \quad b = a + c + d$$
$$\text{or} \quad c = a + b + d \quad \text{or} \quad d = a + b + c,$$

$$(10.12.1)$$

where a, b, c, d are all non-negative. These are the critical surfaces in (a, b, c, d) space.

If the energies $\varepsilon_1, \ldots, \varepsilon_8$ are held fixed, satisfying (10.1.5), and the temperature T varied, then the point (a, b, c, d) will trace out a path in this space. If one of ε_1, ε_3, ε_5, ε_7 (say ε_1) is less than the others, then for sufficiently small T this path lies inside one of the ordered regimes (regime I). On the other hand, for large T ($a = b = c = d = 1$) the path certainly lies in the disordered regime III. Thus it must cross a critical surface (and does so only once) at a critical point.

If two or more of ε_1, ε_3, ε_5, ε_7 are equal and less than the others, then the path always lies in the disordered regime III and there is no critical temperature.

Consider a point (a, b, c, d) close to one of the surfaces (10.2.1) and let

$$t = -\frac{(a - b - c - d)(a - b + c + d)(a + b - c + d)(a + b + c - d)}{16abcd}.$$

$$(10.12.2)$$

Then this t is zero on a critical surface and in general will vanish linearly with $T - T_c$, $(T - T_c)/t$ being positive. We can therefore regard this t as the 'deviation-from-critical-temperature' variable, and replace (1.1.3) by (10.12.2).

The four critical surfaces in (10.12.1) can all be mapped onto the surface $c = a + b + d$ by the trivial mappings (10.2.12)–(10.2.14). These merely re-arrange a, b, c, d and map order to order, disorder to disorder. There is therefore no loss of generality in focussing attention on critical points on the surface $c = a + b + d$, i.e. between regimes III and IV. I shall do this for the rest of this section.

Alternative Expressions for the Thermodynamic Properties

Consider first the principal regime IV, where t is negative and $k = k_0$. From (10.11.3) and (10.12.2)

$$k + k^{-1} = 2 - 4t, \qquad (10.12.3)$$

so k approaches one as $t \to 0_-$.

It follows that $q \to 1$ and the product definitions (15.1.5) of the theta functions become weakly convergent. This can be avoided by using (15.7.2) in (10.4.24), so as to express a, b, c, d in terms of theta functions of the modulus k' conjugate to k.

We shall also want to compare the correct free energy in regime III with its analytic continuation from regime IV. To do this it is convenient to work with the $w_1, \ldots, w_4$ in (10.2.16), rather than a, b, c, d. Substituting the conjugated expressions (10.4.24) into (10.2.16), using (15.4.29) to factor the RHS, and then using the product expansions (15.1.5), we finally obtain

$$w_1 = \tfrac{1}{2}(a + b) = \rho' \prod_{n=1}^{\infty} \frac{(1 - p^{2n-1}z')(1 - p^{2n-1}/z')}{(1 - p^{2n-1}x')(1 - p^{2n-1}/x')},$$

$$w_2 = \tfrac{1}{2}(a - b) = -\rho' \frac{\sin(V/2)}{\sin(\mu/2)} \prod_{n=1}^{\infty} \frac{(1 - p^{2n}z')(1 - p^{2n}/z')}{(1 - p^{2n}x')(1 - p^{2n}/x')},$$

$$w_3 = \tfrac{1}{2}(c + d) = \rho' \frac{\cos(V/2)}{\cos(\mu/2)} \prod_{n=1}^{\infty} \frac{(1 + p^{2n}z')(1 + p^{2n}/z')}{(1 + p^{2n}x')(1 + p^{2n}/x')},$$

$$w_4 = \tfrac{1}{2}(c - d) = \rho' \prod_{n=1}^{\infty} \frac{(1 + p^{2n-1}z')(1 + p^{2n-1}/z')}{(1 + p^{2n-1}x')(1 + p^{2n-1}/x')}, \qquad (10.12.4)$$

where

$$p = (q')^2 = \exp(-2\pi I/I'),$$

$$\mu = \pi\lambda/I', \qquad w = \pi v/I', \qquad (10.12.5)$$

$$x' = \exp(i\mu), \quad z' = \exp(iw),$$

and ρ' is some normalization factor (proportional to ρ) that we shall not need explicitly.

These equations, together with the restrictions (10.7.1a), i.e.

$$0 < p < 1, \quad |w| < \mu < \pi, \quad \xi > 0, \qquad (10.12.6)$$

define ρ', μ, w and p. As $t \to 0$, $p \to 0$ and it is apparent that ξ, μ, V tend to finite limits, these being their critical values. In particular, the critical value of μ is given by

$$\tan(\mu/2) = (cd/ab)^{\frac{1}{2}}. \qquad (10.12.7)$$

The free energy is given by (10.8.47) or, alternatively, (10.8.49). The latter is more convenient for the present purposes, since it involves $c + d$

(i.e. $2w_3$), rather than c: this makes it easier to compare regimes III and IV.

We have just noted that μ and w tend to finite limits as $t \to 0$ and $k \to 1$. Since I' then tends to $\pi/2$ and I to infinity, from (10.12.5), (10.7.9) and (10.7.18) this means that q, x, z all tend to one (from below). The sum in (10.8.49) therefore becomes an integral. Its behaviour near $t = 0$ can be studied by using the Poisson summation formula (15.8.1).

To do this, define

$$F(u) = \frac{[\cosh(\pi - 2\mu)u - \cosh \mu u][\cosh \mu u - \cosh wu]}{u \sinh \pi u \cosh \mu u}. \qquad (10.12.8)$$

Then, noting that $F(u)$ is an even function, that $F(0) = 0$, and using (10.7.9), (10.7.19) and (10.12.5), the equation (10.8.49) becomes

$$-f/k_B T = \ln(c + d) + \frac{I'}{4I} \sum_{m=-\infty}^{\infty} F\left(\frac{mI'}{2I}\right). \qquad (10.12.9)$$

From (15.8.1), this can be written

$$-f/k_B T = \ln(c + d) + \tfrac{1}{2} G(0) + \sum_{m=1}^{\infty} G\left(\frac{4\pi mI}{I'}\right), \qquad (10.12.10)$$

where

$$G(k) = \int_{-\infty}^{\infty} \exp(iku) F(u) \, du, \qquad (10.12.11)$$

and we have used the fact that $G(k)$, like $F(u)$, must be an even function.

For positive k, the integral in (10.12.11) can be closed round the upper half u-plane and evaluated as a sum of residues from the poles at $u = in$, $i(n - \tfrac{1}{2})\pi/\mu$, for $n = 1, 2, 3, \ldots$. Substituting the result into (10.12.10), the sum over m can then be performed to give

$$-f/k_B T = \ln(c + d) + \tfrac{1}{2} G(0)$$

$$+ 2 \sum_{n=1}^{\infty} \frac{(\cos 2n\mu - \cos n\pi \cos n\mu)(\cos n\mu - \cos nw) p^{2n}}{n \cos n\mu (1 - p^{2n})}$$

$$+ 2 \sum_{n=1}^{\infty} \frac{(-1)^n \cot[(n - \tfrac{1}{2})\pi^2/\mu] \cos[(n - \tfrac{1}{2})\pi w/\mu] p^{(2n-1)\pi/\mu}}{(n - \tfrac{1}{2})[1 - p^{(2n-1)\pi/\mu}]}. \qquad (10.12.12)$$

The equations (10.10.10), (10.10.13), (10.10.20) for the interfacial tension s, the correlation length ξ, and the spontaneous magnetization M_0, are

$$s/k_B T = \tfrac{1}{2} \ln(1/k_1) \,,$$

$$\xi = 2/\ln(1/k_1) \,, \qquad\qquad (10.12.13)$$

$$M_0 = (1 - k_1^2)^{1/8} \,,$$

where k_1, defined by (10.10.8), is the elliptic modulus with nome x^2.

To study the behaviour near $x = 1$, we simply go to the conjugate modulus and conjugate nome. The conjugate modulus is $k_1' = (1 - k_1^2)^{\frac{1}{2}}$, while from (15.1.3) (interchanging I and I'), the conjugate nome is $\exp[\pi^2/\ln(x^2)]$. From (10.7.9) and (10.12.5), the conjugate nome is therefore

$$q_1' = \exp(-\pi I/\lambda) = p^{\pi/2\mu}. \qquad\qquad (10.12.14)$$

Replacing k, q in (15.1.4a) by k_1', q_1', and squaring, we obtain

$$1 - k_1^2 = 16 \, p^{\pi/2\mu} \prod_{n=1}^{\infty} \left(\frac{1 + p^{\pi n/\mu}}{1 + p^{\pi(n - \frac{1}{2})/\mu}} \right)^8. \qquad\qquad (10.12.15)$$

The spontaneous polarization is believed to be given exactly by (10.10.25), i.e.

$$P_0 = k_1' I_1/(k' I). \qquad\qquad (10.12.16)$$

Again we want to work with conjugate nomes. Replacing k, I', q in (15.1.8) by k', I, q' and using (10.12.5), we obtain

$$k' I = p^{\frac{1}{4}} \ln(1/p) \prod_{n=1}^{\infty} \left(\frac{1 - p^{2n}}{1 - p^{2n-1}} \right)^2. \qquad\qquad (10.12.17)$$

Similarly, using k_1', I_1, q_1' and (10.12.14),

$$k_1' I_1 = (\pi/\mu) \, p^{\pi/4\mu} \ln(1/p) \prod_{n=1}^{\infty} \left(\frac{1 - p^{2n\pi/\mu}}{1 - p^{(2n-1)\pi/\mu}} \right)^2. \qquad\qquad (10.12.18)$$

Behaviour near Criticality

Given a, b, c, d in regime IV, the equations (10.12.4) and (10.12.5) define p, μ and V. The free energy, interfacial tension, etc. are then given by (10.12.12)–(10.12.18).

On the other side of the transition, in regime III, $w_1, \ldots, w_4$ must first be re-arranged as in the procedure (i)–(v) of the previous section. The effect of this is to interchange w_1 and w_4 before using (10.12.4) and (10.12.5). The free energy is again given by (10.10.12) and the correlation length ξ by (10.12.13); M_0 and P_0 are then zero, s is meaningless.

However, examining (10.12.4) it is apparent that p, μ, V are analytic functions of a, b, c, d even when $p = 0$, i.e. at the transition from regime IV to III. Also, interchanging w_1 and w_4 therein is equivalent to simply negating p, leaving μ and V unchanged.

It follows that p, μ, V can be analytically continued from regime IV into regime III, and that these analytic continuations differ from their correct values only in that p is negated.

Also, from (10.12.3), (10.12.5) and (15.1.4a), when t is small

$$p = -t/16 + \mathcal{O}(t^2), \qquad (10.12.19)$$

so p vanishes linearly with t.

Look at the expression (10.12.12) for the free energy. The first three terms involve only w_3, μ, V and p^2, all of which are analytic across the boundary between regimes III and IV, being the same functions of a, b, c, d on either side. Only the last term can therefore be in any sense singular, and for p small the dominant contribution to it is

$$-f_{\text{sing}}/k_B T = -4 \cot\left(\frac{\pi^2}{2\mu}\right) \cos\left(\frac{\pi w}{2\mu}\right) p^{\pi/\mu}. \qquad (10.12.20)$$

This f_{sing} is effectively the f_s defined by (1.7.10a). Since μ, w tend to finite limits as $t \to 0$ and μ is non-zero, the cot and cos terms in (10.12.20) are effectively constants. From (10.12.19) and the above comments, the correct value of p in either regime IV or regime III behaves for t small as

$$p \simeq |t|/16. \qquad (10.12.21)$$

It follows that

$$f_{\text{sing}} \sim |t|^{\pi/\mu}, \qquad (10.12.22a)$$

the critical value of μ being given by (10.12.7).

Exceptional cases occur if $\mu = \pi/m$, where m is an integer. If m is even the factor $\cot(\pi^2/2\mu)$ in (10.12.20) is infinite. This is due to the fact that two poles of $F(u)$ coincide. The residue of the resulting double pole should be calculated properly when evaluating (10.12.11). The effect of this is to introduce an extra factor $\ln|t|$, so

$$f_{\text{sing}} \sim |t|^{\pi/\mu} \ln|t|, \qquad (10.12.22b)$$

if π/μ is an even integer.

If the critical value of π/μ is an odd integer, then the factor $\cot(\pi^2/2\mu)$ vanishes in (10.12.20). To obtain the leading singularity in this case it is then necessary to consider the dependence of μ on the temperature variable t.

In the ordered regime IV it follows easily from (10.12.13)–(10.12.18) and (10.12.21) that for t small (and negative)

$$s, \xi^{-1} \sim (-t)^{\pi/2\mu}, \quad M_0 \sim (-t)^{\pi/16\mu}, \tag{10.12.23a}$$

$$P_0 \sim (-t)^{(\pi-\mu)/4\mu},$$

while in the disordered regime III (t positive)

$$\xi^{-1} \sim t^{\pi/2\mu}. \tag{10.12.23b}$$

(The formula (10.10.13) for the correlation length ξ is correct only for $\lambda \leq 2I'/3$, i.e. $\mu \leq 2\pi/3$; however, Johnson et al (1973) showed that it gives the correct critical behaviour even for $\mu > 2\pi/3$.)

Critical Exponents and Scaling

As in Section 8.11, define a critical exponent β_e for P_0, analogously to the definition (1.1.4) of β for M_0. To avoid confusion with the parameter μ above, denote the interfacial tension exponent in (1.7.34) by μ_s. Then from (1.1.4), (1.7.9), (1.7.25), (1.7.34), (10.12.22) and (10.12.23), the critical exponents α, α', β, β_e, ν, ν', μ_s are

$$\alpha = \alpha' = 2 - \pi/\mu, \quad \beta = \pi/16\mu, \tag{10.12.24}$$

$$\beta_e = (\pi - \mu)/4\mu, \quad \nu = \nu' = \mu_s = \pi/2\mu.$$

Since the eight-vertex model has only been solved in zero fields (both electric and magnetic), it is not possible to use it to fully test the scaling hypothesis (1.2.1). However, all the scaling predictions that can be tested, namely (1.2.15), (1.2.16), $\alpha = \alpha'$ and $\nu = \nu'$, are indeed satisfied.

If one accepts the other scaling predictions, then the other exponents can be calculated from (1.2.12)–(1.2.14). In particular, the 'magnetic' γ, δ, η are

$$\gamma = 7\pi/8\mu, \quad \delta = 15, \quad \eta = \tfrac{1}{4}. \tag{10.12.25}$$

Universality and Weak Universality

It is worth-while recapitulating the definition of μ at criticality. For given vertex energies ε_1, ε_3, ε_5, ε_7, the weights a, b, c, d are given by (10.1.2) and (10.2.1). The condition for criticality is (10.12.1). If $\varepsilon_5 < \varepsilon_1$, ε_3, ε_7, it follows that the critical temperature T_c is given by

$$\exp(-\varepsilon_5/k_B T_c) = \exp(-\varepsilon_1/k_B T_c) + \exp(-\varepsilon_3/k_B T_c)$$

$$+ \exp(-\varepsilon_7/k_B T_c). \tag{10.12.26}$$

From (10.12.17) and (10.12.6), the critical value of μ is given by

$$\tan(\mu/2) = \exp[(\varepsilon_1 + \varepsilon_3 - \varepsilon_5 - \varepsilon_7)/2k_B T_c], \quad 0 < \mu < \pi. \quad (10.12.27)$$

[Other cases occur when ε_1, ε_3 or ε_7 is the least energy, but they can be trivially mapped to this case by using the relations (10.2.12)–(10.2.14).]

By varying ε_1, ε_3, ε_5, ε_7, this μ can be given any value between 0 and π. Thus the exponents α, β, γ, ν, μ_s (but not δ and η) depend on the values of $\varepsilon_1, \ldots, \varepsilon_7$, and vary continuously with them.

This contradicts the universality hypothesis of Section 1.3: that critical exponents should not depend on the details of the interactions. Kadanoff and Wegner (1971) argued that this variation was due to the special symmetries and dimensionality of the zero-field eight-vertex model. For instance, in the magnetic picture of Section 10.3, suppose that J_h and J_v are not both zero (in the electric picture, this means a field is applied). Then Kadanoff and Wegner's argument suggests that the magnetic exponents should be exactly those of the Ising model. From this viewpoint, universality is expected to 'normally' hold, the eight-vertex model being a special exceptional case. This has been supported by approximate renormalization group calculations (van Leeuwen, 1975; Kadanoff and Brown, 1979; Knops, 1980).

There are two other models which are believed to have continuously variable exponents, though they have not been solved exactly. They are the Ashkin–Teller model discussed in Section 12.9 (Kadanoff, 1977; Zisook, 1980) and the square lattice Ising model with ferromagnetic nearest-neighbour interactions and anti-ferromagnetic next-nearest neighbour ones (Nightingale, 1977; Barber, 1979; Oitmaa, 1981).

Suzuki (1974) proposed what may be called 'weak universality'. Most exponents are defined as powers of the temperature difference $T - T_c$ (δ and η are exceptions). Suzuki suggested that it was more natural to use the inverse correlation length ξ^{-1} as the variable measuring departure from criticality. For instance, instead of (1.1.4) one should write

$$M_0 \sim \xi^{-\hat{\beta}}, \quad (10.12.28)$$

$\hat{\beta}$ being a critical exponent. (This idea is quite attractive: from a mathematical point of view the temperature is rather an uninteresting divisor of the Hamiltonian, while the correlation length gives valuable information on the near-critical behaviour of the system.)

From (1.1.4) and (1.7.25),

$$\hat{\beta} = \beta/\nu. \quad (10.12.29a)$$

Similarly, the reduced exponents for f_{sing}, s and χ are

$$\hat{\phi} = (2 - \alpha)/\nu, \quad \hat{\mu}_s = \mu_s/\nu, \quad \hat{\gamma} = \gamma/\nu, \qquad (10.12.29b)$$

while δ and η are not affected. From (10.12.24) we have for the eight-vertex model that

$$\beta = \frac{1}{8}, \quad \phi = 2, \quad \mu_s = 1, \quad \gamma = \frac{7}{4}, \quad \delta = 15, \quad \eta = \frac{1}{4}, \qquad (10.12.30)$$

all of which are fixed numbers, independent of μ.

Thus if one formulates 'weak universality' as the proposition that $\hat{\beta}, \hat{\phi}, \hat{\mu}_s, \hat{\gamma}, \delta, \eta$ should be independent of the details of the interactions, then the eight-vertex model is consistent with this hypothesis. Further, this hypothesis connects well with scaling, since the scaling relations (1.2.12)–(1.2.16) predict

$$\hat{\beta} = \frac{d}{\delta + 1}, \quad \hat{\phi} = d, \quad \hat{\mu}_s = d - 1,$$

$$\hat{\gamma} = d\frac{\delta - 1}{\delta + 1}, \quad \eta = 2 - d\frac{\delta - 1}{\delta + 1}. \qquad (10.12.31)$$

Thus scaling implies that if δ is universal, then weak universality must be satisfied.

10.13 An Equivalent Ising Model

We saw in Section 10.3 that the eight-vertex model can be regarded as two nearest-neighbour Ising models (one on each sub-lattice), linked by four-spin interactions. Some people are unhappy at the introduction of such four-spin interactions, feeling that they are somehow 'unphysical'. Jüngling (1975) has answered this objection by showing that the eight-vertex model, and in particular the zero-field eight-vertex model, is also equivalent to a square lattice Ising model with only two-spin interactions. These interactions are between nearest neighbours, and between next-next-nearest neighbours.

To see this, consider the square lattice in Fig. 10.6. It is drawn diagonally and the two sub-lattices are distinguished, their sites being shown as open circles and solid circles, respectively. Let N be the number of solid-circle sites.

Divide the lattice into N squares, as indicated by dotted lines, each containing one solid circle. Let the total energy be

$$\mathscr{E} = -\Sigma(J_1\sigma_i\sigma_m + J_2\sigma_j\sigma_m + J_3\sigma_k\sigma_m + J_4\sigma_l\sigma_m + J\sigma_i\sigma_k + J'\sigma_j\sigma_l),$$

$$(10.13.1)$$

where the sum is over all such squares; i, j, k, l, m are the sites within a square, arranged as in the example in Fig. 10.6; and $\sigma_i(= \pm 1)$ is the spin at any site i.

This energy contains only interactions between pairs of spins; either nearest-neighbours (e.g. $\sigma_i\sigma_m$), or next-next-nearest neighbours (e.g. $\sigma_i\sigma_k$). As usual, the partition function is

$$Z = \sum_\sigma \exp(-\mathscr{E}/k_B T) \,, \tag{10.13.2}$$

the sum being over the values of all the spins.

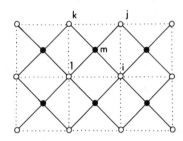

Fig. 10.6. Jüngling's formulation of the eight-vertex model: an Ising model on the lattice of solid lines, with only two-spin interactions, is equivalent to an Ising model on the lattice of dotted lines, with two- and four-spin interactions.

The summand in (10.13.2) factors into a product of N terms, one for each square. Each solid-circle spin enters just one such term, so the summation over its values (± 1) is easily performed. Doing this for each solid-circle spin gives

$$Z = \sum_\sigma^{(\circ)} \prod W(\sigma_i, \sigma_j, \sigma_k, \sigma_l) \,, \tag{10.13.3}$$

where the product is over all squares, the superfix ($\circ$) means that the sum is over all values of all the open-circle spins, and if

$$K_i = J_i/k_B T, \quad K = J/k_B T, \quad K' = J'/k_B T \,, \tag{10.13.4}$$

then

$$W(\sigma_1, \sigma_2, \sigma_3, \sigma_4) = 2 \exp(K\sigma_1\sigma_3 + K'\sigma_2\sigma_4)$$
$$\times \cosh(K_1\sigma_1 + K_2\sigma_2 + K_3\sigma_3 + K_4\sigma_4) \,. \tag{10.13.5}$$

Since $\sigma_1, \ldots, \sigma_4$ only have two values ($+1$ and -1), any function of them can be written as

$$L + L_1\sigma_1 + \ldots + L_{13}\sigma_1\sigma_3 + \ldots + L_{1234}\sigma_1\sigma_2\sigma_3\sigma_4 \,, \tag{10.13.6}$$

where all terms are linear in $\sigma_1, \ldots, \sigma_4$, there are 16 such terms, and $L, \ldots, L_{1234}$ are constant coefficients.

The function $W(\sigma_1, \sigma_2, \sigma_3, \sigma_4)$ is positive, so its logarithm is real and can be written in the form (10.13.6). Further, it is an even function of $\sigma_1, \ldots, \sigma_4$, so only the even terms in (10.13.6) occur. It must therefore be possible to find L, L_{12}, L_{13}, L_{14}, L_{23}, L_{24}, L_{34}, L_{1234} such that

$$W(\sigma_1, \sigma_2, \sigma_3, \sigma_4) = \exp\left[L + \sum_{1 \leqslant i < j \leqslant 4} L_{ij}\sigma_i\sigma_j + L_{1234}\sigma_1\sigma_2\sigma_3\sigma_4 \right].$$

(10.13.7)

(This is known as the 'star-square' transformation: it is a generalization of the star-triangle relation of Section 6.4.)

Substituting (10.13.7) back into (10.13.3), noting that each nearest-neighbour pair $\sigma_i\sigma_j$ occurs in two squares, we obtain

$$Z = \exp(NL) \sum_{\sigma}^{(\circ)} \exp\left[(L_{12} + L_{34}) \sum \sigma_i\sigma_j + (L_{23} + L_{14}) \sum \sigma_j\sigma_k \right.$$

$$\left. + L_{13} \sum \sigma_i\sigma_k + L_{24} \sum \sigma_j\sigma_l + L_{1234} \sum \sigma_i\sigma_j\sigma_k\sigma_l \right],$$

(10.13.8)

where the summations inside the exponential are over all vertical edges (i, j), horizontal edges (j, k), diagonal pairs (i, k) and (j, l), and squares (i, j, k, l), respectively.

Apart from the factor $\exp(NL)$, this is the partition function of an Ising-type model on the square lattice of open circles and dotted lines in Fig. 10.6, with nearest-neighbour, diagonal and four-spin interactions. This is exactly the formulation (10.3.1) of the eight-vertex model, the interaction energies J_h and J_v therein being given by

$$J_h/k_BT = L_{12} + L_{34}, \quad J_v/k_BT = L_{23} + L_{14}.$$

(10.13.9)

In general these are non-zero, so Jungling's model is equivalent to an eight-vertex model in electric fields (and vice-versa).

Yet more interesting is the fact that if J_1, J_2, J_3, J_4 satisfy the temperature-independent conditions

$$J_1 = J_3, \quad J_2 = -J_4,$$

(10.13.10)

then Jungling's model is equivalent to a *zero-field* eight vertex model.

To see this, note that $K_1 = K_3$ and $K_2 = -K_4$. The RHS of (10.13.5) is then unaltered by interchanging σ_1 with σ_3, or by interchanging and negating σ_2 and σ_4. It follows that (10.13.7) must take the form

$$W(\sigma_1, \sigma_2, \sigma_3, \sigma_4) = \exp[L + L_{13}\sigma_1\sigma_3 + L_{24}\sigma_2\sigma_4 + L_{12}(\sigma_1 + \sigma_3)(\sigma_2 - \sigma_4)$$

$$+ L_{1234}\sigma_1\sigma_2\sigma_3\sigma_4].$$

(10.13.11)

In particular, this implies that

$$L_{12} + L_{34} = L_{23} + L_{14} = 0 , \qquad (10.13.12)$$

so J_h and J_v are given by (10.13.9) to be zero.

This formulation of the eight-vertex model highlights a potential difficulty with universality. As explained in Section 1.3, the hypothesis admits that critical exponents may change discontinuously when a symmetry is broken. This is consistent with the argument of Kadanoff and Wegner (1971) mentioned in the previous section: in the presence of fields the eight-vertex model may have fixed Ising exponents, even though it does not have these in zero field.

In Jungling's formulation, this means that the model (10.13.1) has one set of exponents in general, but another set if $J_1 = J_3$ and $J_2 = -J_4$. Although we can now see that this special case has special symmetry, it is by no means obvious *a priori* that this is so. The symmetry is 'hidden'. Presumably such breaking of hidden symmetries occurs in other models: it could be hard to anticipate.

10.14   The *XYZ* Chain

Closely related to the zero-field eight vertex model is the problem of determining the eigenvalues of the operator

$$\mathscr{H} = -\tfrac{1}{2} \sum_{j=1}^{N} [J_x\sigma_j^x\sigma_{j+1}^x + J_y\sigma_j^y\sigma_{j+1}^y + J_z\sigma_j^z\sigma_{j+1}^z] , \qquad (10.14.1)$$

where J_x, J_y, J_z are constants and

$$\sigma_j^x = c_j, \quad \sigma_j^y = ic_js_j, \quad \sigma_j^z = s_j , \qquad (10.14.2)$$

and $s_1, \ldots, s_N, c_1, \ldots, c_N$ are the operators defined by (6.4.17). In direct product notation

$$\sigma_j^x = e \otimes \ldots . \otimes e \otimes c \otimes e \otimes \ldots . \otimes e$$

$$\sigma_j^y = e \otimes \ldots . \otimes e \otimes d \otimes e \otimes \ldots . \otimes e \qquad (10.14.3)$$

$$\sigma_j^z = e \otimes \ldots . \otimes e \otimes s \otimes e \otimes \ldots . \otimes e ,$$

where there are N terms in each product; c, d, s occur in position j; and e, c, d, s are the two-by-two Pauli spin matrices

$$e = \begin{pmatrix} 1 & 0 \\ 0 & 1 \end{pmatrix}, c = \begin{pmatrix} 0 & 1 \\ 1 & 0 \end{pmatrix}, d = \begin{pmatrix} 0 & -i \\ i & 0 \end{pmatrix}, s = \begin{pmatrix} 1 & 0 \\ 0 & -1 \end{pmatrix}. \quad (10.14.4)$$

The $\sigma_j^x, \sigma_j^y, \sigma_j^z, \mathcal{H}$ are all 2^N by 2^N matrices.

This operator $\mathcal{H}$ is the Hamiltonian of a one-dimensional quantum mechanical model of ferromagnetism: there are N spins, labelled $j = 1, \ldots, N$, on a line. With each spin j is associated the three-dimensional vector $\boldsymbol{\sigma}_j = (\sigma_j^x, \sigma_j^y, \sigma_j^z)$ of Pauli matrices; neighbouring spins j and $j + 1$ have interaction $-\frac{1}{2}\boldsymbol{\sigma}_j \cdot \boldsymbol{J} \cdot \boldsymbol{\sigma}_{j+1}$, where $\boldsymbol{J}$ is a three-by-three diagonal matrix with elements J_x, J_y, J_z. The partition function of the model is

$$Z_{\mathcal{H}} = \text{Trace} \exp(-\mathcal{H}/k_B T). \quad (10.14.5)$$

If $J_x = J_y = J_z$, this is the Heisenberg model (Heisenberg, 1928; Bloch, 1930, 1932). If $J_x = J_y = 0$, then $\mathcal{H}$ is diagonal and the model reduces to the nearest-neighbour Ising model (each spin is effectively either up or down, and lies in the z-direction). These models can be formulated on a lattice of any dimension, but it is the one-dimensional case that is represented by (10.14.1), and that is related to the two-dimensional eight-vertex model.

The case $J_x = J_y = 0$ is easily solved, being the one-dimensional Ising model of Chapter 2. The case $J_z = 0$ is known as the 'XY model', and is related to the Ising model. Explicit expressions for all the eigenvalues can be given (for finite N), and the partition function evaluated. This has been done by Lieb $et~al.$ (1961) and Katsura (1962).

The case $J_x = J_y$ is sometimes called the 'Heisenberg – Ising' model. Bethe (1931) gave the correct form of the eigenvectors of $\mathcal{H}$, Yang and Yang (1966) proved rigorously that Bethe's ansatz was correct, and derived the minimum eigenvalue in the limit of N large.

When Lieb (1967a, b, c) solved the ice-type models, he found that the eigenvectors of the transfer matrix were precisely those of the one-dimensional Heisenberg – Ising operator. He was therefore able to use many of Yang and Yang's results.

Sutherland (1970) showed directly that the transfer matrix of any zero-field eight-vertex model commutes with an XYZ operator $\mathcal{H}$. They therefore have the same eigenvectors. I was not aware of Sutherland's result when I solved the eight-vertex model (I did much of the work in the writing room of the P & O liner $Arcadia$, in the Atlantic and Indian Oceans. This was good for concentration, but not for communication). It should be obvious from Sections 10.4–10.6 that such commutation relations are closely linked with the solution of the problem.

In fact it can be shown, for any J_x, J_y, J_z (Baxter, 1971b, 1972c), that $\mathcal{H}$ is effectively a logarithmic derivative of an eight-vertex transfer matrix, and hence the minimum eigenvalue of $\mathcal{H}$ can be obtained. This will be done in this section.

The calculation of $Z_{\mathcal{H}}$ involves considering all the eigenvalues. This problem is very difficult, and in general the best that has been done is to reduce it to one of solving a set of non-linear integral equations (see the remarks and references regarding the correlation length in Section 10.10).

Relation Between $\mathcal{H}$ and V

The eight-vertex transfer matrix V is given by (9.6.1), where the weight function w is defined by (10.2.3), or equivalently (10.2.4). These definitions can in turn be written as

$$w(\mu, \alpha | \beta, \nu) = \tfrac{1}{2}\{[(a + c) + (a - c)\mu\alpha]\, \delta(\mu, \beta)\, \delta(\alpha, \nu)$$

$$+ [(b + d) - (b - d)\mu\alpha]\, \delta(\mu, -\beta)\, \delta(\alpha, -\nu)\}, \qquad (10.14.6)$$

where $\delta(\alpha, \beta) = 1$ if $\alpha = \beta$, $\delta(\alpha, \beta) = 0$ if $\alpha \neq \beta$.

First consider the case when

$$b = d = 0, \quad a = c = c_0 > 0. \qquad (10.14.7)$$

From (10.14.6) it is then true that

$$w(\mu, \alpha | \beta, \nu) = c_0\, \delta(\mu, \beta)\, \delta(\alpha, \nu). \qquad (10.14.8)$$

Substituting this into (9.6.1), the $\mu_1, \ldots, \mu_N$ summations can at once be performed. If V_0 is the matrix V for this case, then its elements $(\boldsymbol{\alpha}, \boldsymbol{\beta})$ are

$$(V_0)_{\alpha\beta} = c_0^N\, \delta(\alpha_1, \beta_2)\, \delta(\alpha_2, \beta_3) \ldots \delta(\alpha_N, \beta_1). \qquad (10.14.9)$$

Thus $c_0^{-N}V_0$ is the left-shift operator that takes an arrow configuration $\{\alpha_1, \ldots, \alpha_N\}$ to $\{\alpha_N, \alpha_1, \ldots, \alpha_{N-1}\}$.

Now perturb about this case and set

$$a = c_0 + \delta a, \quad b = \delta b, \quad c = c_0 + \delta c, \quad d = \delta d, \qquad (10.14.10)$$

where δa, δb, δc, δd are infinitesimal increments in a, b, c, d. Let δw, δV be the increments induced in the weight function w and the transfer matrix V. Then from (9.6.1)

$$\delta V_{\alpha\beta} = c_0^{N-1} \sum_{j=1}^{N} \ldots \delta(\alpha_{j-2}, \beta_{j-1})\, \delta w(\alpha_{j-1}, \alpha_j | \beta_j, \beta_{j+1})\, \delta(\alpha_{j+1}, \beta_{j+2})$$

$$\qquad (10.14.11)$$

(note the two sets of δs: increments and Kronecker symbols).

Pre-multiplying the matrix δV by V_0^{-1}, (10.14.9) and (10.14.11) give

$$(V_0^{-1}\,\delta V)_{\alpha\beta} = c_0^{-1} \sum_{j=1}^{N} \ldots \delta(\alpha_{j-1}, \beta_{j-1})\, \delta w(\alpha_j, \alpha_{j+1}|\beta_j, \beta_{j+1})$$

$$\times\; \delta(\alpha_{j+2}, \beta_{j+2}) \ldots \quad (10.14.12)$$

From (9.6.9), using the eight-vertex function w, it follows that

$$V_0^{-1}\,\delta V = c_0^{-1}(\delta U_1 + \delta U_2 + \ldots + \delta U_N)\,, \quad (10.14.13)$$

where δU_j is the increment in the vertex operator U_j.

Also, substituting the form (10.14.6) of w into (9.6.9), and using the definitions (6.4.17) of s_j and c_j, U_j can be written

$$U_j = \tfrac{1}{2}\{(a + c)\mathscr{I} + (a - c)s_j s_{j+1} + (b + d)c_j c_{j+1} - (b - d)s_j c_j s_{j+1} c_{j+1}\}$$

$$\quad (10.14.14)$$

or, using (10.14.2),

$$U_j = \tfrac{1}{2}\{(a + c)\mathscr{I} + (b + d)\,\sigma_j^x \sigma_{j+1}^x + (b - d)\,\sigma_j^y \sigma_{j+1}^y + (a - c)\,\sigma_j^z \sigma_{j+1}^z\}\,,$$

$$\quad (10.14.15)$$

where $\mathscr{I}$ is the identity operator.

The increment δU_j is given simply by replacing a, b, c, d in (10.14.15) by $\delta a, \delta b, \delta c, \delta d$. Doing this, (10.14.13) becomes

$$V_0^{-1}\,\delta V = (2c_0)^{-1} \sum_{j=1}^{N} \{(\delta a + \delta c)\mathscr{I} + (\delta b + \delta d)\,\sigma_j^x \sigma_{j+1}^x$$

$$+\; (\delta b - \delta d)\,\sigma_j^y \sigma_{j+1}^y + (\delta a - \delta c)\,\sigma_j^z \sigma_{j+1}^z\} \quad (10.14.16)$$

Apart from the additive term proportional to $\mathscr{I}$, this is an XYZ operator of the form (10.14.1).

To complete this identification, substitute the values of (10.14.10) of a, b, c, d into (10.4.6). To leading order in the increments, this gives

$$\Delta = \frac{\delta a - \delta c}{\delta b + \delta d}, \quad \Gamma = \frac{\delta b - \delta d}{\delta b + \delta d}, \quad (10.14.17)$$

so (10.14.16) can be written

$$V_0^{-1}\,\delta V = \frac{\delta b + \delta d}{2c_0} \sum_{j=1}^{N} \left\{ \frac{\delta a + \delta c}{\delta b + \delta d}\,\mathscr{I} + \sigma_j^x \sigma_{j+1}^x + \Gamma\,\sigma_j^y \sigma_{j+1}^y + \Delta\,\sigma_j^z \sigma_{j+1}^z \right\}.$$

$$\quad (10.14.18)$$

If J_x, J_y, J_z are related to Γ and Δ by

$$J_x : J_y : J_z = 1 : \Gamma : \Delta\,, \quad (10.14.19)$$

then it follows that

$$V_0^{-1} \, \delta V = c_0^{-1} \{ \tfrac{1}{2} N(\delta a + \delta c) \, \mathcal{I} - (\delta b + \delta d) \, \mathcal{H}/J_x \}, \quad (10.14.20)$$

where $\mathcal{H}$ is precisely the XYZ chain operator (10.14.1).

Suppose we keep Δ and Γ fixed while varying a, b, c, d. Then all transfer matrices V will commute with one another and, from (10.14.20), with $\mathcal{H}$. Thus if (10.4.6) and (10.14.19) are satisfied, with the same Γ and the same Δ, then the eight-vertex transfer matrix V commutes with the XYZ operator $\mathcal{H}$. They have the same eigenvectors. [This is Sutherland's result (1970).]

If c_0, δb, δd, J_x are positive, then from (10.14.20) it is apparent that the maximum eigenvalue $\Lambda_{\max}$ of V corresponds to the minimum eigenvalue of $\mathcal{H}$. Further, a, b, c, d are then positive, so $\Lambda_{\max}$ is precisely the eigenvalue needed in the calculation of the eight-vertex free energy.

Ground-State Energy E_0

Let NE_0 be the minimum eigenvalue of $\mathcal{H}$. Then E_0 is the ground-state energy per site. From (10.8.46) and (10.14.20), in the limit of N large, E_0 is given by

$$\delta(-f/k_B T) = c_0^{-1} \{ \tfrac{1}{2} (\delta a + \delta c) - (\delta b + \delta d) \, E_0/J_x \}, \quad (10.14.21)$$

where f is the eight-vertex free energy per site, and $\delta(-f/k_B T)$ is the increment induced in $-f/k_B T$.

Up to this point, no restriction has been made in this section on Δ, Γ, J_x, J_y, J_z, except that J_x be positive. Now let us consider the principal regime (10.7.5) of the eight-vertex model, wherein (from (10.4.6))

$$|\Gamma| < 1, \quad \Delta < -1. \quad (10.14.22)$$

From (10.14.19), this implies that

$$|J_y| < J_x < -J_z, \quad (10.14.23)$$

so I shall call this the 'principal regime' for the XYZ chain. From (10.7.1), the elliptic function parameters k, λ, v are real, and $0 < k < 1, 0 < \lambda < I'$. For given values of a, b, c, d, the free energy f is given by (10.4.21) and (10.4.23), (10.7.9) and (10.7.18), and (10.8.47) or (10.8.49).

Holding Γ and Δ fixed is equivalent to keeping k and λ unaltered. Without loss of generality we can require in this section that $c = 1$, so from (10.4.21) and (10.4.23)

$$a = \text{snh} \, \tfrac{1}{2} (\lambda - v) \, / \text{snh} \, \lambda, \quad b = \text{snh} \, \tfrac{1}{2} (\lambda + v) \, / \text{snh} \, \lambda, \quad (10.14.24)$$

$$c = 1, \quad d = k \, \text{snh} \, \tfrac{1}{2} (\lambda - v) \, \text{snh} \, \tfrac{1}{2} (\lambda + v).$$

This leaves v as a variable. The case (10.14.7) corresponds to $v = -\lambda$; incrementing v by δv gives

$$v = -\lambda + \delta v .\qquad (10.14.25)$$

From (15.5.1a), (15.5.5) and (10.4.20),

$$\frac{d}{dv} \operatorname{snh} v = \operatorname{cn} iv \ \operatorname{dn} iv ,\qquad (10.14.26)$$

for all complex numbers v. Using this formula to differentiate (10.14.24) with respect to v, then setting $v = -\lambda$, gives

$$\delta a/\delta v = -\tfrac{1}{2} \operatorname{cn} i\lambda \ \operatorname{dn} i\lambda /\operatorname{snh} \lambda, \quad \delta b/\delta v = \tfrac{1}{2}/\operatorname{snh} \lambda ,$$
$$\delta c/\delta v = 0, \quad \delta d/\delta v = \tfrac{1}{2} k \operatorname{snh} \lambda .\qquad (10.14.27)$$

[It is now easily verified that (10.4.17) and (10.14.17) are equivalent.]

Also evaluating $\delta(-f/k_B T)$ from (10.8.47) and (10.7.18), substituting the results into (10.14.21) gives

$$E_0 = \tfrac{1}{2} J_z - \frac{\pi \tau}{2I} \sum_{m=1}^{\infty} \frac{(x^m - x^{-m} q^m)^2 (1 - x^{2m})}{(1 - q^m)(1 + x^{2m})} ,\qquad (10.14.28)$$

where q, x are defined by (10.7.9), I is the complete elliptic integral of the first kind of modulus k, and

$$\tau = 2 J_x \operatorname{snh} \lambda/(1 + k \operatorname{snh}^2 \lambda) .\qquad (10.14.29)$$

An equivalent form can be obtained by using (10.8.49) instead of (10.8.47), namely

$$E_0 = \tfrac{1}{2}(-J_x + J_y + J_z)$$
$$- \frac{\pi \tau}{2I} \sum_{m=1}^{\infty} \frac{x^{-m}(x^{3m} - q^{m/2})(1 - q^{m/2} x^{-m})(1 - x^{2m})}{(1 - q^m)(1 + x^{2m})} .\qquad (10.14.30)$$

Define a parameter l by

$$l = 2 k^{\frac{1}{2}}/(1 + k) .\qquad (10.14.31a)$$

(This is the Landen-related elliptic modulus of (15.6.1).) Then eliminating y^2 (which is negative) between the two equations (10.4.12), using (10.4.9) and (10.14.19), gives

$$l = (J_x^2 - J_y^2)^{\frac{1}{2}}/(J_z^2 - J_y^2)^{\frac{1}{2}} .\qquad (10.14.31b)$$

Solving (10.4.17) for sn $i\lambda$, using (10.4.20) and (10.14.19), one obtains

$$k^{\frac{1}{2}} \operatorname{snh} \lambda = (J_x - J_y)^{\frac{1}{2}}/(J_x + J_y)^{\frac{1}{2}} .\qquad (10.14.31c)$$

Substituting this into (10.14.29) gives

$$\tau = k^{-\frac{1}{2}}(J_x^2 - J_y^2)^{\frac{1}{2}},\tag{10.14.31d}$$

which, using (10.14.31b), can be put into the form

$$\tau = (l/k)^{\frac{1}{4}}(J_x^2 - J_y^2)^{\frac{1}{4}}(J_z^2 - J_y^2)^{\frac{1}{4}}.\tag{10.14.32}$$

To summarize the results so far: if J_x, J_y, J_z lie in the principal regime (10.14.23), then the ground state energy per site of the XYZ operator (10.14.1) is given by (10.14.28) or (10.14.30), where k, λ, τ are defined by (10.14.31), q and x by (10.7.9).

Symmetries

The three Pauli matrices σ_j^x, σ_j^y, σ_j^z can be permuted by simple similarity transformations, so E_0 is a symmetric function of J_x, J_y and J_z. Suppose, as in (10.5.9), that N is even, and consider the similarity transformation

$$\mathcal{H} \rightarrow \sigma_1^z\sigma_3^z\sigma_5^z \ldots \mathcal{H}\, \sigma_1^z\sigma_3^z\sigma_5^z \ldots .\tag{10.14.33}$$

Since σ_j^z anti-commutes with σ_j^x and σ_j^y, but commutes with all the other Pauli operators, the effect of this on (10.14.1) is to negate J_x and J_y. Thus E_0 is unchanged by negating any two of J_x, J_y, J_z. These symmetries can be used to map any XYZ operator $\mathcal{H}$ into the principal regime (10.14.23). The ground state energy per site is then given by (10.14.28) or (10.14.30).

These symmetries are of course related to those of the eight-vertex model. From (10.4.6), (10.14.19) and (10.2.16)

$$J_x : J_y : J_z = ab + cd : ab - cd : \tfrac{1}{2}(a^2 + b^2 - c^2 - d^2)$$

$$= w_1^2 - w_2^2 + w_3^2 - w_4^2 : w_1^2 - w_2^2 - w_3^2 + w_4^2 : w_1^2 + w_2^2 - w_3^2 - w_4^2.\tag{10.14.34}$$

It is apparent that the eight-vertex symmetries (10.2.11)–(10.2.15), or equivalently and more obviously (10.2.17), merely re-arrange the terms on the RHS of (10.14.34), and possibly negate two of them.

Singularities of E_0

It follows from these results that E_0 is an analytic function of J_x, J_y and J_z except when the two numerically largest coefficients J_x, J_y, J_z have equal magnitude. The archetypal case is when $-J_z = J_x > |J_y|$. This is a boundary of the principal regime (10.14.23), and on it we see from (10.14.31) that $k = 1$. The behaviour near the boundary can be obtained by applying the

Poisson summation formula (15.8.1) to the series in (10.14.27). The working closely parallels that of (10.12.8)–(10.12.12) (indeed it can be obtained thereform by differentiating with respect to w). It gives, for $-J_z > J_x > |J_y|$,

$$E_0 = \tfrac{1}{2}(-J_x + J_y + J_z) - \frac{\pi\tau}{2I'} \Big\{ \hat{G}(0)$$

$$- 4 \sum_{n=1}^{\infty} (\cos 2n\mu - \cos n\pi \, \cos n\mu) \tan n\mu \, \frac{p^{2n}}{(1 - p^{2n})} \qquad (10.14.35)$$

$$- \frac{4\pi}{\mu} \sum_{n=1}^{\infty} \cot[(n - \tfrac{1}{2})\pi^2/\mu] \, p^{(2n-1)\pi/\mu} \Big/ \big(1 - p^{(2n-1)\pi/\mu}\big) \Big\},$$

where μ, p are defined by (10.12.5) and

$$\hat{G}(u) = \int_{-\infty}^{\infty} \frac{[\cosh(\pi - 2\mu)u - \cosh \mu u] \sinh \mu u}{\sinh \pi u \, \cosh \mu u} \exp(iku) \, du . \qquad (10.14.36)$$

The case $J_x > -J_z > |J_y|$, can be mapped into the principal regime by negating and interchanging J_x and J_z. As when interchanging w_1 and w_4 in the eight-vertex model, this leaves μ unchanged by negates p. In fact μ and p^2 are analytic across the boundary $-J_z = J_x$, while near it

$$p \simeq \frac{1}{16} |J_z^2 - J_x^2| \Big/ (J_x^2 - J_y^2) , \qquad (10.14.37)$$

so p vanishes linearly with $J_z + J_x$.

From (15.6.6), with k replaced by k', (10.14.31a) and (10.12.5):

$$I'(k/l)^{\frac{1}{2}} = \tfrac{1}{2} \pi \prod_{n=1}^{\infty} (1 - p^{2n})^2/(1 + p^{2n})^2 . \qquad (10.14.38)$$

From this and (10.14.32) is is apparent that negating and interchanging J_x and J_z leaves unaltered the factor $J_x/(\tau I')$ in (10.14.35), and this factor is analytic at $J_z = -J_x$. Thus all terms in (10.14.35) are analytic across this boundary, except for those in the last summation. The dominant singular term is

$$(E_0)_{\text{sing}} = (2\pi^2 \tau/\mu I') \cot(\pi^2/2\mu) \, p^{\pi/\mu} . \qquad (10.14.39)$$

When $J_z = -J_x$, then $k = 1$, $l = 1$, $I' = \tfrac{1}{2}\pi$, snh $\lambda = \tan \lambda$, so from (10.14.31c) and (10.12.5) μ is given by

$$J_y/J_x = \cos \mu, \quad 0 < \mu < \pi , \qquad (10.14.40)$$

while from (10.14.32)

$$\tau = (J_x^2 - J_y^2)^{\frac{1}{2}} = J_x \sin \mu , \qquad (10.14.41)$$

so (10.14.39) simplifies to

$$(E_0)_{\text{sing}} = 4\pi \mu^{-1} J_x \sin \mu \, \cot(\pi^2/2\mu) \, p^{\pi/\mu} , \qquad (10.14.42)$$

p being given by (10.14.37). Provided $\cot(\pi^2/2\mu)$ is finite and non-zero, it follows that E_0 has the power-law singularity

$$(E_0)_{\text{sing}} \sim |J_z + J_x|^{\pi/\mu} , \qquad (10.14.43)$$

at $J_z = -J_x$.

In fact, comparing (10.12.20) and (10.12.21) with (10.14.42) and (10.14.37), it is apparent (for all values of μ) that E_0 has the same singularity at $J_z + J_x = 0$ as the eight-vertex free energy has at $t = 0$.

Some General Comments on d-Dimensional Ising Models and $(d - 1)$-Dimensional Quantum Mechanical Models

Equation (10.14.20) relates the XYZ operator $\mathcal{H}$ with the eight-vertex model transfer matrix V. Here V is evaluated with a, b, c, d infinitesimally close to the values (10.14.7), for which the value V_0 of V is proportional to a simple shift operator.

As Suzuki (1976) has pointed out, such relations exist for many models. For instance, the layer-to-layer transfer matrix of the simple cubic Ising model is a matrix V with elements

$$V(\sigma, \sigma') = \exp\left\{ \sum_{i=1}^{M} \sum_{j=1}^{N} \left[\tfrac{1}{2}(K_1 \sigma_{ij}\sigma_{i+1,j} + K_1 \sigma'_{ij}\sigma'_{i+1,j} \right. \right.$$

$$\left. \left. + K_2 \sigma_{ij}\sigma_{i,j+1} + K_2 \sigma'_{ij}\sigma'_{i,j+1}) + K_3 \sigma_{ij}\sigma'_{ij} \right] \right\}. \qquad (10.14.44)$$

Here $\sigma = \{\sigma_{11}, \ldots, \sigma_{MN}\}$ denotes all spins in one layer of the lattice, $\sigma' = \{\sigma'_{11}, \ldots, \sigma'_{MN}\}$ all spins in the next layer.

Defining operators s_{ij}, c_{ij} analogously to (6.4.17), and using the identity (6.4.22), it follows that

$$(2 \sinh 2K_3)^{-\frac{1}{2}MN} V = \exp[\tfrac{1}{2}(K_1 A + K_2 B)] \exp[K_3^* C] \exp[\tfrac{1}{2}(K_1 A + K_2 B)] ,$$

$$(10.14.45)$$

where $\tanh K_3^* = \exp(-2K_3)$ and

$$A = \sum_{i=1}^{M} \sum_{j=1}^{N} s_{ij} s_{i+1,j} , \quad B = \sum_{i=1}^{M} \sum_{j=1}^{N} s_{ij} s_{i,j+1} ,$$

$$C = \sum_{i=1}^{M} \sum_{j=1}^{N} c_{ij} . \qquad (10.14.46)$$

When $K_1 = K_2 = K_3^* = 0$, the RHS of (10.14.45) is simply the identity operator. When K_1, K_2, K_3^* are all small, to first order

$$\ln[(2 \sinh 2K_3)^{-\frac{1}{2}MN} V] = \mathcal{H},$$
(10.14.47)

where now

$$\mathcal{H} = K_1 A + K_2 B + K_3^* C.$$
(10.14.48)

As in (10.14.2), s_{ij} and c_{ij} are the Pauli operators σ_{ij}^z and σ_{ij}^x, respectively. The RHS of (10.14.48) can therefore be regarded as a two-dimensional Heisenberg-type operator, in which the quantum-mechanical spins interact with one another only via their components in the z-direction, and an external field of strength K_3^* is applied in the x-direction.

If $K_3^* = 0$, the operator is diagonal and its eigenvalues are the energy levels of the two-dimensional Ising model. For this reason the operator $\mathcal{H}$ is known as the Hamiltonian of the two-dimensional Ising model in a *transverse* magnetic field (Stinchcombe, 1973; Oitmaa and Plischke, 1977; Pfeuty, 1977).

From (10.14.47), if we could evaluate the eigenvalues of the transfer matrix V of the three-dimensional Ising model, then we could also evaluate them for the two-dimensional Hamiltonian $\mathcal{H}$. Conversely, we might hope that solving the latter problem would lead to a solution of the former. Unfortunately neither has been solved exactly, though the approximate methods mentioned in Section 1.5 have been very successful.

The above arguments can easily be extended to arbitrary dimensions and to other lattices. However, it should be noted that the two-dimensional zero-field eight-vertex model has an extra property that does not so generalize: its transfer matrix V *always* commutes with some XYZ operator, even when V is far from its shift operator value V_0.

10.15   Summary of Definitions of Δ, Γ, k, λ, v, q, x, z, p, μ, w

The results of this chapter have inevitably been expressed in terms of elliptic function parameters such as q, x, p and μ. These have been defined as needed, for both the eight-vertex model and the XYZ chain, and some special cases have already been considered. For clarity, it seems helpful to summarize their definitions in general.

For the eight-vertex model, with Boltzmann weights a, b, c, d, define Δ and Γ by (10.4.6), i.e.

$$\Delta = \frac{a^2 + b^2 - c^2 - d^2}{2(ab + cd)}, \quad \Gamma = \frac{ab - cd}{ab + cd}.$$
(10.15.1a)

For the XYZ chain, with coefficients J_x, J_y, J_z, define them by (10.14.19), i.e.

$$J_x : J_y : J_z = 1 : \Gamma : \Delta . \tag{10.15.1b}$$

The eigenvectors of the eight-vertex transfer matrix, and of the XYZ Hamiltonian, are the same functions of Γ and Δ only.

Re-arrangement Procedure

Now map the models into their principal regimes. For the eight-vertex model this means using the procedure (i)–(iii) of Section 10.11. Let a, b, c, d be the original values of these parameters, and a_r, b_r, c_r, d_r their re-arranged values. Then a_r, b_r, c_r, d_r lie inside the principal regime (10.7.5), or on a boundary thereof (since the free energy, etc. are continuous functions of a, b, c, d, boundary cases can be handled by taking an appropriate limit), i.e.

$$c_r \geqslant a_r + b_r + d_r, \quad a_r \geqslant 0, \quad b_r \geqslant 0, \quad d_r \geqslant 0 . \tag{10.15.2}$$

Define Δ_r, Γ_r by (10.15.1a), with a, b, c, d replaced by a_r, b_r, c_r, d_r. Then it follows that

$$|\Gamma_r| \leqslant 1, \quad \Delta_r \leqslant -1 , \tag{10.15.3}$$

and equalities occur only on a boundary of the regime (10.15.2).

For the XYZ model the re-arrangement procedure is simpler. One merely permutes J_x, J_y, J_z, and possibly negates a pair of them, so as to bring them into the regime (10.14.23), or onto a boundary thereof. Let J_x^r, J_y^r, J_z^r be these re-arranged values. Then

$$|J_y^r| \leqslant J_x^r \leqslant -J_z^r . \tag{10.15.4}$$

Define Δ_r, Γ_r by (10.15.1b) with J_x, J_y, J_z replaced by J_x^r, J_y^r, J_z^r, i.e.

$$\Gamma_r = J_y^r/J_x^r, \quad \Delta_r = J_z^r/J_x^r . \tag{10.15.5}$$

then clearly these Δ_r, Γ_r also satisfy (10.15.3).

It was shown in (10.14.34) that the eight-vertex symmetries merely permute the quantities $ab + cd$, $ab - cd$, $\frac{1}{2}(a^2 + b^2 - c^2 - d^2)$, and possibly negate two of them. From (10.15.1) it follows that an equivalent definition of Γ_r, Δ_r for the eight-vertex model is:

Define Γ, Δ by (10.15.1a) and choose J_x, J_y, J_z to satisfy (10.15.1b), J_x being positive. Permute and pair negate these J_x, J_y, J_z to satisfy (10.15.4). Now define Γ_r, Δ_r by (10.15.5).

We shall find this alternative procedure helpful in the next section.

Elliptic Function Parameters

The elliptic function parameters k, λ are defined by (10.4.17), or equivalently by (10.4.9), (10.4.12) and (10.4.16), using the re-arranged values Γ_r, Δ_r of Γ, Δ. These equations can be written as

$$2k^{\frac{1}{2}}/(1 + k) = (1 - \Gamma_r^2)^{\frac{1}{2}}/(\Delta_r^2 - \Gamma_r^2)^{\frac{1}{2}}, \qquad (10.15.6a)$$

$$\mathrm{snh}\,\lambda = k^{-\frac{1}{2}}(1 - \Gamma_r)^{\frac{1}{2}}/(1 + \Gamma_r)^{\frac{1}{2}}, \qquad (10.15.6b)$$

where $\mathrm{snh}\,u = -i\,\mathrm{sn}\,iu$ and k, λ must satisfy

$$0 \le k \le 1, \quad 0 \le \lambda \le I'. \qquad (10.15.7)$$

I and I' being the complete elliptic integrals of the first kind of moduli k and $k' = (1 - k^2)^{\frac{1}{2}}$. Again, equalities occur in (10.15.7) only on regime boundaries.

The parameter v in the eight-vertex model can now readily be obtained from (10.4.21) and (10.4.23), using the re-arranged values of a, b, c, d, e.g.

$$\mathrm{snh}\,\tfrac{1}{2}(\lambda - v) = (a_r/c_r)\,\mathrm{snh}\,\lambda, \qquad (10.15.8)$$

and must satisfy

$$-\lambda \le v \le \lambda, \qquad (10.15.9)$$

The parameters q, x, z are then given by (10.7.9) and (10.7.18), i.e.

$$q = \exp(-\pi I'/I), \quad x = \exp(-\pi\lambda/2I), \quad z = \exp(-\pi v/2I), \qquad (10.15.10)$$

and satisfy

$$0 \le q \le x^2 \le 1, \quad x \le z \le z^{-1}. \qquad (10.15.11)$$

Finally, p, μ and w are given by (10.12.5), i.e.

$$p = \exp(-2\pi I/I'), \quad \mu = \pi\lambda/I', \quad w = \pi v/I', \qquad (10.15.12)$$

so

$$0 \le p \le 1, \quad 0 \le \mu \le \pi, \quad |w| \le \mu. \qquad (10.15.13)$$

The models are critical when, and only when, $k = 1$. When this is so, from Chapter 15 it follows that $I' = \frac{1}{2}\pi$, $I = \infty$, $q = 1$, $p = 0$, while from (10.15.6)–(10.15.13), λ, v, μ, w are finite, $\Delta_r = -1$ and $x = z = 1$.

10.16 Special Cases

There are three special cases of the eight-vertex model which were solved before the general zero-field model, namely the Ising model of Chapter

7, (Onsager, 1944), the 'free-fermion' model (Fan and Wu, 1970), and the 'ice-type' or 'six-vertex' model of Chapter 8 (Lieb, 1967a, b, c). These correspond respectively to the XZ, XY and Heisenberg – Ising cases of the XYZ chain. In each case the critical value of μ is either 0, $\frac{1}{2}\pi$ or π. Felderhof (1973) and Jones (1973, 1974) have considered these special cases in some detail.

Ising Model and XY Chain

As was shown in Section 10.3, the ordinary two-spin nearest-neighbour Ising model is a special case of the eight-vertex model, occurring when the condition (10.3.10) is satisfied, i.e. $cd = ab$. From (10.15.1a) this implies that

$$\Gamma = 0 . \qquad (10.16.1)$$

From (10.15.1b), this corresponds to J_y being zero, i.e. to the XZ case of the XYZ chain. Re-arranging J_x, J_y, J_z to satisfy (10.15.4) must leave J_y as the zero coefficient, so from (10.15.5)

$$\Gamma_r = 0 . \qquad (10.16.2)$$

From (10.15.6b), (15.4.32) and (10.15.12),

$$\lambda = \tfrac{1}{2}I', \quad \mu = \tfrac{1}{2}\pi . \qquad (10.16.3)$$

Thus μ lies exactly at the mid-point of its allowed range of values $(0, \pi)$.

The critical exponents α, β, ν, μ_s given by (10.12.24) are indeed the same as those in (7.12.12) and (7.12.14). Note that this is a case when π/μ is an even integer, so the free energy singularity contains a factor $\ln|t|$, as in (10.12.22b) and (7.12.10). Similarly, an extra factor $\ln|J_z + J_x|$ occurs in (10.14.43).

Free-Fermion Model

Fan and Wu (1970) used the Pfaffian method mentioned in Section 7.13 to solve the eight-vertex model for the case when

$$\omega_1\omega_2 + \omega_3\omega_4 = \omega_5\omega_6 + \omega_7\omega_8 \qquad (10.16.4)$$

Here $\omega_1, \ldots, \omega_8$ are the Boltzmann weights defined in (10.1.2). The method works even if the conditions (10.1.5) are not satisfied, i.e. it works for an eight-vertex model in a field.

For the zero-field case, from (10.2.1) the restriction (10.16.4) becomes

$$a^2 + b^2 = c^2 + d^2 , \tag{10.16.5}$$

so from (10.15.1), $\Delta = 0$ and $J_z = 0$. This model therefore corresponds to the XY chain. Re-arranging J_x, J_y, J_x to satisfy (10.15.4) must take J_y to zero, i.e. it must transform the XY chain to the XZ chain. Then (10.15.1b) gives

$$\Gamma_r = 0 , \tag{10.16.6}$$

which is the Ising case just discussed.

The zero-field free-fermion model can therefore be mapped to an Ising model, and has $\mu = \frac{1}{2}\pi$. However, the restriction (10.16.5) ensures that a, b, c, d always lie in the disordered regime III of Section 10.11, so there is no transition to an ordered state.

Six-Vertex Model and Heisenberg – Ising Chain

The six-vertex model is obtained from the eight-vertex by setting $d = 0$, so from (10.15.1) $\Gamma = 1$ and

$$J_x = J_y . \tag{10.16.7}$$

The model therefore corresponds to the Heisenberg – Ising chain, and the eigenvectors of the transfer matrix must be those of the Heisenberg – Ising Hamiltonian. Lieb (1967a, b, c) determined this correspondence directly and made considerable use of it.

Take $(J_x, J_y, J_z) = (1, 1, \Delta)$, thereby satisfying (10.15.1b). Permute and pair-negate them to satisfy (10.15.4). Define the elliptic function parameters, in particular k, λ and μ, by (10.15.5)–(10.15.13). There are three distinct cases, and it is readily seen that they correspond to the ferroelectrically ordered, disordered and anti-ferroelectrically ordered phases of Section 8.10:

(i) $\Delta > 1$ (ferroelectric order): $\Gamma_r = -1$, $\Delta_r = -\Delta$,
$k = 0$, $\lambda = I' = \infty$, $\mu = \pi$.
(ii) $-1 < \Delta < 1$ (disorder): $\Gamma_r = -\Delta$, $\Delta_r = -1$, $\qquad$ (10.16.8)
$k = 1$, $\lambda = \frac{1}{2}\mu$, $\Delta = -\cos\mu$.
(iii) $\Delta < -1$ (anti-ferroelectric order): $\Gamma_r = 1$, $\Delta_r = \Delta$,
$k = 0$, $\Delta = -\cosh\lambda$, $\mu = 0$.

The case (ii) is quite remarkable, since $k = 1$ throughout. This is the condition for the eight-vertex model to be critical. Thus in this phase of the six-vertex model the correlation length is infinite [for the free-fermion case of the six-vertex model this has been verified explicitly (Baxter, 1970a)]. The system is disordered only in the sense that there is no spontaneous polarization.

Conversely, a critical eight-vertex model can be mapped to a 'disordered' six-vertex model.

The six-vertex model has transitions at $\Delta = +1$ or -1. The former occur in ferroelectric models, such as the KDP model, and from (10.16.8) μ is then equal to π, which is its maximum possible value. As is shown in Section 8.11, this transition is first order, and the ordered state is frozen. This makes it difficult to interpret the general eight-vertex results, but note that (8.11.8) and (8.11.10) are correctly given by (10.12.24).

The $\Delta = -1$ transition occurs in anti-ferroelectric models, such as the F model. From (10.6.8), μ is then equal to its minimum value of zero. The derivation of the critical properties (10.12.22) and (10.12.23) is then invalid. A proper calculation of course gives the essential singularities of (8.11.14)–(8.11.25), and it is impossible to sensibly define critical exponents. Even so, it is worth noting that merely setting $\mu = 0$ in (10.12.24) does give the infinite exponent (8.11.18). Also, the 'exponent relations' (8.11.26) are in fact satisfied by the general eight-vertex model.

10.17 An Exactly Solvable Inhomogeneous Eight-Vertex Model

Until now in this chapter, the eight-vertex model has been taken to be homogeneous, i.e. the vertex energies $\varepsilon_1, \ldots, \varepsilon_8$ do not vary from site to site. For the solvable zero-field case, this means that the Boltzmann weights a, b, c, d, given by (10.1.2) and (10.2.1), are site-independent.

It is possible to weaken this requirement, in a very special way, and still be able to calculate the free energy, etc. by straightforward generalizations of the above methods. In fact, exactly the same equations apply, but the variables are interpreted slightly more generally.

Column Variation

To see this, first suppose that the Boltzmann weights a, b, c, d can vary from column to column, but not row to row. Let a_j, b_j, c_j, d_j be their values in column j. There are N columns in the lattice, so $j = 1, \ldots, N$.

Consider the star – triangle relation (10.4.1), which comes from (9.6.5)–(9.6.8). The matrices S and S' now depend on the lattice column j to which they correspond, i.e. to their position j in the matrix products in (9.6.5) and (9.6.6). Even so, it is still true that (9.6.7) implies the equality of (9.6.5) and (9.6.6), provided only that M is a single matrix, independent of j.

This corresponds to w'', and hence a'', b'', c'', d'' in Section 10.4, being independent of j. Since Δ'' and Γ'' are defined by (10.4.6) with a, b, c, d replaced by a'', b'', c'', d'', they are also independent of j. From (10.4.26), so therefore must be Δ, Γ, Δ', Γ'.

For each column j, define ρ, k, λ, v by (10.4.24). Then from (10.4.17), k and λ are independent of j, and are the same for the transfer matrix V' as they are for V.

Once (10.4.26) is satisfied, the star–triangle relation reduces to (10.4.30). Here u and u' may depend on j, but u'' may not. Let v_j (v_j') be the value of v for column j and transfer matrix $V(V')$. Then from (10.4.23) it follows that

$$v_j' - v_j = \text{independent of } j. \qquad (10.17.1)$$

To summarise: define the transfer matrix V by (9.6.1), where each weight function w depends on the lattice column j to which it corresponds, i.e. to its position in the product. For column j, define w by (10.2.3) and (10.4.24), where k and λ are independent of j and

$$v_j = v_j^0 + v. \qquad (10.17.2)$$

(The normalising factor ρ enters trivially into the equations and may be varied in any manner desired: here I shall regard it as constant for all lattice sites.)

Then two transfer matrices commute if they have the same values of $k, \lambda, v_1^0, \ldots, v_N^0$. Their eigenvectors are therefore independent of v: they depend only on k, λ and the differences of $v_1, \ldots, v_N$.

Regard V as a function $V(v)$ of v. Then again it satisfies (10.4.25).

The working of Section 10.5 now generalizes easily. The a, b, c, d in (10.5.1) should be replaced by a_j, b_j, c_j, d_j. However, Δ and γ in (10.5.2) are still independent of j, so p_j is still given by (10.5.14) and (10.5.8). The u and v in (10.5.11)–(10.5.21) should be replaced by u_j and v_j, still related by (10.4.23). The factors $\{\rho\, h[\frac{1}{2}(\lambda \pm v)]\}^N$ in (10.5.23) should be replaced in an obvious way by a product of $\rho_j h[\frac{1}{2}(\lambda \pm v_j)]$ over $j = 1, \ldots, N$; and incrementing v by $\pm 2\lambda'$ should be read as incrememting each v_j by $\pm 2\lambda'$. However, if v_j is put into the form (10.17.2), where each v_j^0 is regarded as a constant, then this is the same thing. We therefore again obtain (10.5.32), but now (10.5.24) is replaced by

$$\phi(v) = \prod_{j=1}^{N} \rho_j\, h[\tfrac{1}{2}(v_j^0 + v)]. \qquad (10.17.3)$$

Provided each v_j lies in the range $(-\lambda, \lambda)$ the working of Section 10.8 goes through virtually as written, the main change being that the definition (10.8.10) of $A(v)$ is replaced by

$$A(v) = \prod_{m=0}^{\infty} \prod_{j=1}^{N} \{1 - q^m \exp[-\pi(v_j^0 + v)/2I]\}. \tag{10.17.4}$$

Note also that the $v_1, \ldots, v_n$ in Section 10.8 are quite different from the $v_1, \ldots, v_N$ of this section.

Regard the RHS of (10.8.47) as a function of the Boltzmann weights a, b, c, d, the parameters q, x, z being defined as in Section 10.5. Write it as $-\psi(a, b, c, d)$, i.e.

$$\psi(a, b, c, d) = -\ln c$$
$$- \sum_{m=1}^{\infty} \frac{x^{-m}(x^{2m} - q^m)^2 (x^m + x^{-m} - z^m - z^{-m})}{m(1 - q^{2m})(1 + x^{2m})}. \tag{10.17.5}$$

Then for the inhomogeneous system being discussed, (10.8.44) becomes (in the limit of N large)

$$\ln \Lambda_{\max} = - \sum_{j=1}^{N} \psi(a_j, b_j, c_j, d_j), \tag{10.17.6}$$

a remarkably simple result!

The other properties discussed in Section 10.10 are all independent of v, so are valid as written for this inhomogeneous model.

Row Variation

Now consider a lattice in which the Boltzmann weights a, b, c, d can vary from row to row, as well as column to column. Let $a_{ij}, b_{ij}, c_{ij}, d_{ij}$ be their values for the site in row i and column j. For each site, define ρ, k, λ, v by (10.4.24) and require, for all i and j, that

$$k, \lambda = \text{independent of } i \text{ and } j, \tag{10.17.7}$$
$$v_{ij} = v_j^0 + v_i^1,$$

where v_{ij} is the value of v at site (i, j), $v_1^0, \ldots, v_N^0$ and $v_1^1, \ldots, v_M^1$ are some parameters which are at our disposal.

The transfer matrix V now depends on the row i to which it refers, so let us write it as V_i. Then (8.2.1) becomes

$$Z = \text{Trace } V_1 V_2 \ldots V_M. \tag{10.17.8}$$

Now note that k, λ and the differences of $v_{i1}, v_{i2}, \ldots, v_{iN}$ are independent of i. From the remarks after (10.17.2), it follows that the eigenvectors of V_i are independent of i: $V_1, \ldots, V_M$ all commute. Provided each v_{ij} lies in the range $(-\lambda, \lambda)$, each V_i has the same maximal eigenvector. Thus (8.2.4) generalizes to

$$Z \sim \Lambda_{\max}(1) \ldots \Lambda_{\max}(M), \qquad (10.17.9)$$

$\Lambda_{\max}(i)$ being the maximum eigenvalue of V_i. From (10.17.6) we therefore obtain

$$\ln Z = - \sum_{i=1}^{M} \sum_{j=1}^{N} \psi(a_{ij}, b_{ij}, c_{ij}, d_{ij}). \qquad (10.17.10)$$

Like (10.17.6), this is an amazingly simple formula: the total free energy $F = -k_B T \ln Z$ is the sum of the site free energies (but only in the limit of M, N large, and provided the conditions (10.17.7) are satisfied). Clearly this 'de-coupling' is connected with the commutation properties of the transfer matrices.

Again the v-independent properties (interfacial tension, correlation length, magnetization and polarization) must be the same as those of an homogeneous system with the same values of k and λ.

I regret that I know of no physically interesting inhomogeneous system satisfying (10.17.7). The staggered eight-vertex model (with different weights on the two sub-lattices) is extremely interesting as it contains as special cases the Ising model in a magnetic field (Wu and Lin, 1975), and the Potts and Ashkin – Teller models (see Chapter 12). Unfortunately it does not satisfy (10.17.7).

Even so, it can be mathematically useful to consider these inhomogeneous generalizations of the eight-vertex model. The derivation (Baxter, 1973c) of the spontaneous polarization of the anti-ferroelectric six-vertex model makes extensive use of the form of the dependence of the transfer matrix eigenvectors on $v_1^0, \ldots, v_N^0$. The remarks after (10.17.2) will also play a key role in Chapter 13 in establishing the multiplication properties of the corner transfer matrices.

11

KAGOMÉ LATTICE EIGHT-VERTEX MODEL

11.1 Definition of the Model

With very little extra work, the results of Chapter 10 can be generalized
to a particular class of eight-vertex models on the Kagomé lattice of Fig.
11.1. Not all Kagomé lattice eight-vertex models belong to this class: the
Boltzmann weights must satisfy the restrictions (11.1.7). Even so, the class
is interesting since it contains as special cases the triangular and honeycomb
Ising models, the triangular and honeycomb critical Potts models (see
Chapter 12), and the triangular three-spin model (Baxter and Wu, 1973,
1974).

The eight-vertex model can be defined for any graph or lattice with four
edges meeting at each site. (The word 'graph' is used here for any set of

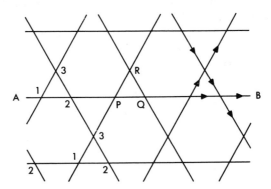

Fig. 11.1. The Kagomé lattice, showing the three types 1, 2, 3 of sites. Also shown
are three particular sites P, Q, R, and some typical right-pointing arrows.

276

sites and connecting edges; a 'lattice' is a regular graph.) Place arrows on the edges (one arrow per edge). Allow only configurations with an even number of arrows pointing into each site. At each site i there are eight possible arrangements of arrows: to arrangement j assign an energy ε_{ij} and a Boltzmann weight

$$\omega_{ij} = \exp(-\varepsilon_{ij}/k_B T), \qquad (11.1.1)$$

where k_B is Boltzmann's constant and T the temperature. Then the partition function is

$$Z = \sum_C \prod_i \omega_{i,j(i,C)}, \qquad (11.1.2)$$

where the sum is over all configurations C of arrows on the graph, the product is over all sites i, and $j(i, C)$ is the arrow arrangement at site i for configuration C.

For each site i, we can always label the eight arrow arrangements so that arrangement 2 is obtained from 1 by reversing all four arrows; and similarly for arrangements 4 and 3, 6 and 5, 8 and 7. Then the 'zero-field' condition is

$$\omega_{i1} = \omega_{i2}, \quad \omega_{i3} = \omega_{i4}, \quad \omega_{i5} = \omega_{i6}, \quad \omega_{i7} = \omega_{i8}. \qquad (11.1.3)$$

It is then convenient to write a_i, b_i, c_i, d_i for $\omega_{i1}, \omega_{i3}, \omega_{i5}, \omega_{i7}$, so that

$$\omega_{i1}, \ldots, \omega_{i8} = a_i, a_i, b_i, b_i, c_i, c_i, d_i, d_i. \qquad (11.1.4)$$

For the homogeneous square lattice, with the vertex arrow arrangements ordered as in Fig. 10.1, these a_i, b_i, c_i, d_i are independent of i and are the a,b,c,d of (10.2.1).

Now consider the Kagomé lattice. It is apparent from Fig. 11.1 that there are three types of sites. Let us call them simply 1, 2 and 3, and suppose that all sites of the same type have the same interaction energies and Boltzmann weights. We can then use b_1 to denote the value of b for all sites of type 1, and similarly for $a_1, c_1, d_1, a_2, b_2, c_2, d_2, a_3, b_3, c_3, d_3$.

For a site of type i, order the eight vertex arrow arrangements as in row i of Fig. 11.2. This ordering has the symmetry property that any row can be obtained by rotating the previous row anti-clockwise through 120°.

For each site of type i, it is useful to define a vertex weight function w_i analogously to (10.2.3). To connect with relevant equations in Chapters 9 and 10, it is convenient to do this in the following asymmetric way.

With each edge m associate an 'arrow spin' α_m, with value $+1$ (-1) if the corresponding arrow points generally to the right (left). (Some typical right-pointing arrows are shown in Fig. 11.1). Consider a particular site of the lattice, of type i, and let μ, α, β, ν be the arrow spins of the

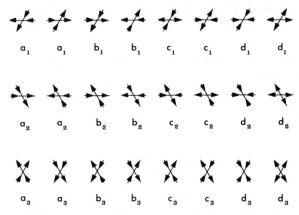

Fig. 11.2. The three types of vertex on the Kagomé lattice, with the eight arrow arrangements allowed on each. The corresponding Boltzmann weights are shown underneath.

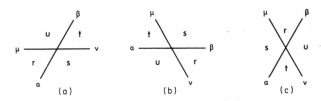

Fig. 11.3. The three types of vertex on the Kagomé lattice, showing the labelling μ, α, β, ν of the surrounding edge arrow-spins, and the labelling r, s, t, u of the surrounding faces.

Table 11.1. Values of $w_i(\mu, \alpha|\beta, \nu)$.

μ, $\alpha\|\beta$, ν	w_1	w_2	w_3
$+, +\|+, +$ and $-, -\|-, -$	a_1	a_2	b_3
$+, -\|-, +$ and $-, +\|+, -$	b_1	b_2	a_3
$+, -\|+, -$ and $-, +\|-, +$	c_1	c_2	c_3
$+, +\|-, -$ and $-, -\|+, +$	d_1	d_2	d_3
$\alpha\beta\mu\nu = -1$	0	0	0

surrounding edges, arranged as in Fig. 11.3. Let $w_i(\mu, \alpha|\beta, \nu)$ be the Boltzmann weight of the corresponding arrow configuration, as given by Fig. 11.2. Then the functions w_1, w_2, w_3 have the values given in Table 11.1.

Comparing this with (10.2.3), it is apparent that this is the same as defining each function w_i by (10.2.3), with a, b, c, d replaced by a_i, b_i, c_i, d_i, except that a_3 and b_3 are interchanged. The partition function can now be written in a form analogous to (10.2.6), namely

$$Z = \sum_\alpha \prod w_i(\alpha_l, \alpha_m|\alpha_p, \alpha_q), \qquad (11.1.5)$$

where the sum is over all choices $\alpha = \{\alpha_1, \alpha_2, \ldots\}$ of the arrow spins; the product is over all sites; and for each site the symbol i denotes its type, and l, m, p, q are the labels of the surrounding edges, arranged in the same way as μ, α, β, ν in Fig. 11.3.

Star – Triangle Restriction

It is obvious from Fig. 11.1 that there are two types of triangles in the Kagomé lattice: up-pointing and down pointing. Consider a triangle, of either type, and let $\alpha_1, \ldots, \alpha_6$ be the arrow spins on the external edges; $\beta_1, \beta_2, \beta_3$ the arrow spins on the internal edges, arranged as in Fig. 11.4. These two types of triangles contribute factors

$$w_1(\alpha_1, \alpha_2|\beta_2, \beta_3)\, w_3(\alpha_6, \beta_2|\alpha_5, \beta_1)\, w_2(\beta_1, \beta_3|\alpha_4, \alpha_3),$$

$$w_2(\alpha_6, \alpha_1|\beta_3, \beta_1)\, w_3(\beta_1, \alpha_2|\beta_2, \alpha_3)\, w_1(\beta_3, \beta_2|\alpha_5, \alpha_4),$$

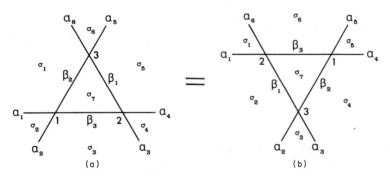

(a)   (b)

Fig. 11.4. The two types of triangles on the Kagomé lattice. The arrow spins $\alpha_1, \ldots, \alpha_6, \beta_1, \beta_2, \beta_3$ are associated with edges; the Ising spins $\sigma_1, \ldots, \sigma_7$ with faces. Equations (11.1.6) and (11.5.8) are the conditions that the total weights of the two triangles be equal.

respectively to the summand in (11.1.5). In each case all remaining factors in (11.1.5) are independent of β_1, β_2, β_3. The summations over β_1, β_2, β_3 may therefore be performed, giving an effective total weight for the triangle.

This weight is a function of $\alpha_1, \ldots, \alpha_6$. It will be shown in the next section that the Kagomé lattice eight-vertex model is solvable if the weight is the same for both types of triangle, i.e. if

$$\sum_{\beta_1,\beta_2,\beta_3} w_1(\alpha_1, \alpha_2|\beta_2, \beta_3)\, w_3(\alpha_6, \beta_2|\alpha_5, \beta_1)\, w_2(\beta_1, \beta_3|\alpha_4, \alpha_3)$$

$$= \sum_{\beta_1,\beta_2,\beta_3} w_2(\alpha_6, \alpha_1|\beta_3, \beta_1)\, w_3(\beta_1, \alpha_2|\beta_2, \alpha_3)\, w_1(\beta_3, \beta_2|\alpha_5, \alpha_4), \quad (11.1.6)$$

for all values of $\alpha_1, \ldots, \alpha_6$.

This is precisely the 'star – triangle' relation (9.6.8), with w, w'', w' replaced by w_1, w_2, w_3. It is therefore equivalent to the six equations (10.4.1), with $a, b, c, d, a'', b'', c'', d'', a', b', c', d'$ replaced by $a_1, b_1, c_1, d_1, a_2, b_2, c_2, d_2, b_3, a_3, c_3, d_3$. With this notation the equations (10.4.1) assume a more symmetric form, and can be written as

$$(a_i a_j - b_i b_j)c_k = (c_i c_j - d_i d_j)b_k,$$
$$\qquad\qquad\qquad\qquad\qquad\qquad\qquad\qquad (11.1.7)$$
$$(a_i b_j - b_i a_j)d_k = (c_i d_j - d_i c_j)a_k,$$

for all permutations (i, j, k) of $(1, 2, 3)$.

All the corollaries of (10.4.1) that were obtained in Section 10.4 can now be applied to (11.1.7). In particular, as in (10.4.6) define

$$\Delta_j = \tfrac{1}{2}(a_j^2 + b_j^2 - c_j^2 - d_j^2)/(a_j b_j + c_j d_j), \qquad (11.1.8)$$

$$\Gamma_j = (a_j b_j - c_j d_j)/(a_j b_j + c_j d_j),$$

for $j = 1, 2, 3$. Then from (10.4.26) it follows that $\Delta_1 = \Delta_2 = \Delta_3$ and $\Gamma_1 = \Gamma_2 = \Gamma_3$. Taking Δ and Γ to be the common values, we may therefore write

$$\Delta_j = \Delta, \quad \Gamma_j = \Gamma, \quad j = 1, 2, 3. \qquad (11.1.9)$$

Elliptic Function Parametrization

We can also apply the elliptic function parametrization of Section 10.4 to the equations (11.1.7). (We shall only need it in this chapter at the end of Section 11.7 and in Section 11.8.) Using (10.4.21), we can define k, λ, u_1, u_2, u_3 such that, for $j = 1, 2, 3$,

$a_j : b_j : c_j : d_j = \mathrm{snh}(\lambda - u_j) : \mathrm{snh}\, u_j : \mathrm{snh}\,\lambda : k\,\mathrm{snh}\,\lambda\,\mathrm{snh}\, u_j\,\mathrm{snh}(\lambda - u_j)$.

$$(11.1.10)$$

Here snh u is the hyperbolic sn function of argument u and modulus k, as defined by (10.4.20) and (15.1.4)–(15.1.6). From (10.4.17), it follows that

$$\Delta = -\mathrm{cn}\,i\lambda\;\mathrm{dn}\,i\lambda\,/(1 - k\,\mathrm{sn}^2 i\lambda) \qquad (11.1.11)$$

$$\Gamma = (1 + k\,\mathrm{sn}^2 i\lambda\,)/(1 - k\,\mathrm{sn}^2 i\lambda)\,.$$

The u_1, u_2, u_3 here correspond respectively to u, u'', $\lambda - u'$ in Section 10.4. (Interchanging a' and b' is equivalent to replacing u' by $\lambda - u'$.) Thus (10.4.30) becomes

$$u_1 + u_2 + u_3 = \lambda\,. \qquad (11.1.12)$$

This completes the parametrization. If the a_j, b_j, c_j, d_j satisfy (11.1.10) and (11.1.12), then the star–triangle restrictions (11.1.7) are satisfied. Conversely, (11.1.7) implies that there exist k, λ, u_1, u_2, u_3 satisfying (11.1.10) and (11.1.12).

11.2 Conversion to a Square-Lattice Model

In this section it will be shown that the effect of the star–triangle restrictions (11.1.7) is to ensure that certain properties of the Kagomé lattice model are the same as those of an associated square-lattice model. The results of Chapter 10 can then be used.

The argument here can be specialized to the Ising model (Baxter and Enting, 1978), or generalized to any graph made up of intersecting straight lines (Baxter, 1978a, b).

First consider any up-pointing triangle in the Kagomé lattice, e.g. the triangle PQR in Fig. 11.1. Label the surrounding arrow spins as in Fig. 11.4. Then the contribution of this triangle to the partition function (11.1.5), summed over arrow spins on internal edges, is the LHS of (11.1.6).

Replace this contribution by the RHS of (11.1.6). The partition function is now that of the graph shown in Fig. 11.5(a), in which the horizontal line AB has been shifted above the site R. The site P is still the intersection of AB with the SW–NE line, and still has weight function w_1. Similarly, Q and R lie on the same lines as before, and have the same weight functions (w_2 and w_3) as before.

This procedure not only leaves the partition function Z unchanged; it also leaves unchanged any correlation, e.g.

$$\langle \alpha_3 \alpha_4 \alpha_9 \ldots \alpha_{82} \rangle = Z^{-1} \sum_\alpha \alpha_3 \alpha_4 \alpha_9 \ldots \alpha_{82} \prod w_i(\alpha_l\,,\alpha_m | \alpha_p\,,\alpha_q)\,, \qquad (11.2.1)$$

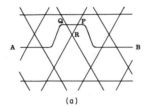 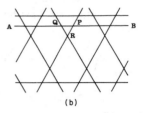

(a)   (b)

Fig. 11.5. The Kagomé lattice of Fig. 11.1 with: (a) the line AB shifted above R; (b) the line AB shifted up a complete row. The partition function, and all correlations in the lower half of the lattice, are unchanged.

provided none of the arrow spins α_3, α_4, α_9, ..., α_{82} correspond to edges inside the triangle PQR.

Suppose the lattice to be wound on a vertical cylinder and perform this procedure for each triangle that initially has its base on the line AB. The result is the graph in Fig. 11.5(b): the line AB has been shifted up one row.

Now repeat this procedure for the horizontal line above AB, then for AB itself, and so on until AB and all horizontal lines above it are at the top of the graph. Similarly, shift CD and all horizontal lines below it to the bottom of the graph. The end result is that the Kagomé lattice of Fig. 11.6(a) is changed to the graph of Fig. 11.6(b).

Consider any set of edges lying between (but not on) the initial lines AB and CD, i.e. adjacent to the middle row of sites. An example is the pair (j, k) in Fig. 11.6. At no stage of the transformation does a horizontal line pass over any of these edges. They are therefore external to all triangles involved in the many star–triangle transformations, so their correlations are unchanged.

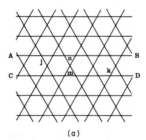 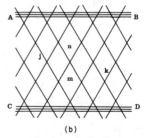

(a)   (b)

Fig. 11.6. (a) The Kagomé lattice, (b) the same lattice after all upper horizontal lines have been shifted to the top, and all lower ones to the bottom. The partition function, and correlations in the central row such as $\langle \alpha_j \alpha_k \rangle$ and $\langle \sigma_m \sigma_n \rangle$, are unchanged.

The eight-vertex models on the two graphs (a) and (b) in Fig. 11.6 therefore not only have the same partition function Z, but also the same correlations between spins in the middle row, e.g. $\langle \alpha_j \alpha_k \rangle$.

Let $2M$ be the number of horizontal lines. Then Fig. 11.6(b) consists of a central square lattice (drawn diagonally) region of $2M$ rows, 'framed' above and below by regions each containing M horizontal lines. All sites in the central region have weight function w_3.

In the limit of M large, edges j and k lie deep inside the square lattice region. We therefore expect the correlation $\langle \alpha_j \alpha_k \rangle$ to be that of the usual square lattice, drawn diagonally. In particular, this implies that

$$\langle \alpha_j \alpha_k \rangle = \text{function only of } a_3, b_3, c_3, d_3 \,, \qquad (11.2.2)$$

and similarly for any correlation between edge spins adjacent to the central row of sites.

Framing Boundaries

The result (11.2.2) is true even though the boundary 'framing' regions also become large. To see this, let V_3 be the row-to-row transfer matrix in the central region, and V_{12} the transfer matrix in the framing regions. Then

$$\langle \alpha_j \alpha_k \rangle = \frac{\psi^T V_{12}^M V_3^M s_k V_3 s_j V_3^M V_{12}^M \psi}{\psi^T V_{12}^M V_3^{2M+1} V_{12}^M \psi} \,, \qquad (11.2.3)$$

where s_j, s_k are diagonal matrices with diagonal entries α_j, α_k, and ψ is a vector whose entries are determined by the boundary conditions. Let Λ_{12} be the maximum eigenvalue of V_{12} and set

$$\phi = V_{12}^M \psi / \Lambda_{12}^M \,. \qquad (11.2.4)$$

Then (11.2.3) becomes

$$\langle \alpha_j \alpha_k \rangle = \phi^T V_3^M s_k V_3 s_j V_3^M \phi / \phi^T V_3^{2M+1} \phi \,. \qquad (11.2.5)$$

This is precisely the correlation inside a square lattice, with weight function w_3 and boundary conditions corresponding to the vector ϕ. When M is large, ϕ tends to a non-zero limit, namely the maximal eigenvector of V_{12}. From the Perron–Frobenius theorem (Frobenius, 1908), this vector has only non-negative entries, and so does the maximal eigenvector of V_3. They cannot therefore be orthogonal (unless the entries of one are zero for all non-zero entries of the other, which is not to be expected). This means that ϕ is not a pathological boundary condition on the square lattice, and the RHS of (11.2.5) can be evaluated for M large by the methods of

Section 2.2. These give

$$\langle \alpha_j \alpha_k \rangle = \zeta^T s_k (V_3/\Lambda_3) \, s_j \zeta / \zeta^T \zeta \, , \tag{11.2.6}$$

where Λ_3, ζ are the maximum eigenvalue and eigenvector of V_3. This result depends only on V_3, not on the boundary condition ϕ.

For simplicity I have assumed in this argument that V_{12} and V_3 (and the top and bottom boundary conditions) are symmetric: this is not a necessary restriction.

11.3   Correlation Length and Spontaneous Polarization

Return to (11.2.2). If edges j and k are far apart, but still lie in the central horizontal band in Fig. 11.6, then

$$\langle \alpha_j \alpha_k \rangle \sim \exp[-|j - k|/\xi_{KG}] \, , \tag{11.3.1}$$

where ξ_{KG} is the horizontal correlation length of the Kagomé lattice. From (11.2.2) it follows that

$$\xi_{KG} = \xi_{SQ}(a_3 \, , b_3 \, , c_3 \, , d_3) \, , \tag{11.3.2}$$

where $\xi_{SQ}(a \, , b \, , c \, , d)$ is the diagonal correlation length of the square lattice eight-vertex model with Boltzmann weights a, b, c, d as in row 3 of Fig. 11.2. This diagonal correlation length has not to my knowledge been evaluated (it would mean obtaining equations for the eigenvalues of the diagonal-to-diagonal transfer matrix, which should be possible). It presumably has the same critical behaviour as the row correlation length ξ given in Section 10.10, since near criticality correlations in all directions are expected to become similarly long-ranged.

Consider any site of type 3 in the central row of the Kagomé lattice in Fig. 11.6(a). Let μ, α, β, ν be the spins on the surrounding edges, arranged as in the last diagram in Fig. 11.3. These edges all lie between the lines AB and CD, so analogously to (11.2.2)

$$\langle \alpha \rangle, \langle \alpha\mu \rangle \, , \ldots , \langle \mu\alpha\beta\nu \rangle = \textit{functions only of } a_3, b_3, c_3, d_3,$$
$$\textit{being the same as those of the}$$
$$\textit{regular square lattice model}$$
$$\textit{with these weights, constructed}$$
$$\textit{on the interior lattice of}$$
$$\textit{Fig. 11.6(b).} \tag{11.3.3}$$

Thus all local correlations round a site of type 3 are the same as those of the corresponding square lattice. By symmetry, analogous equivalences apply for sites of type 1, and of type 2. Provided the restrictions (11.1.7) are satisfied, all local correlations of the Kagomé lattice model can therefore be expressed as square lattice correlations.

This means that for an edge j of the Kagomé lattice, adjacent to a site of type 3,

$$\langle \alpha_j \rangle = P_0(a_3, b_3, c_3, d_3), \qquad (11.3.4)$$

where $P_0(a, b, c, d)$ is the spontaneous polarization of the square lattice eight-vertex model with Boltzmann weights a, b, c, d. This is given by (10.10.24), q and x being defined in Section 10.15. It depends on a, b, c, d only via Δ and Γ, so from (11.1.8) the RHS of (11.3.4) is unchanged by replacing a_3, b_3, c_3, d_3 by a_1, b_1, c_1, d_1 or a_2, b_2, c_2, d_2. Using rotation symmetry, it also follows that (11.3.4) is true for all edges j. Thus $\langle \alpha_j \rangle$ has the same value for all edges of the Kagomé lattice.

11.4 Free Energy

While the framing regions in Fig. 11.6(b) do not contribute (for a large lattice) to central correlations, they certainly contribute to the partition function, and hence to the total Kagomé lattice free energy

$$F_{KG} = -k_B T \ln Z. \qquad (11.4.1)$$

In fact, in the limit of a large lattice we expect the bulk free energy F_{KG} to be simply the sum of the bulk free energies for the three regions in Fig. 11.6(b), the contributions from their boundaries being insignificant. Thus

$$F_{KG} = F_{SQ} + 2F_{FR}, \qquad (11.4.2)$$

where F_{SQ} is the total free energy of the central region, F_{FR} is the free energy of either the upper or lower framing regions.

Let N be the number of sites of type 1 in the lattice. Then (for N large) there are also N sites of type 2, and N of type 3. There are therefore N sites in the central region, all of type 3, so

$$F_{SQ} = N f(a_3, b_3, c_3, d_3), \qquad (11.4.3)$$

where $f(a, b, c, d)$ is the free energy per site of a regular square lattice eight-vertex model, as given by (10.8.47) and Section 10.15, and $k_B T$ is here to be regarded as some given constant.

An expanded picture of one of the framing regions is given in Fig. 11.7. Plainly this is also a square lattice, but with sites of type 1 and sites of type 2 on alternating columns. There are N sites altogether.

From (11.1.8), all sites have the same value of Δ and of Γ, so the eight-vertex model in either framing region is a column-inhomogeneous model of precisely the type discussed in Section 10.17. The total free energy is therefore (for N large)

$$F_{FR} = \tfrac{1}{2} N f(a_1, b_1, c_1, d_1) + \tfrac{1}{2} N f(a_2, b_2, c_2, d_2). \qquad (11.4.4)$$

Substituting the expressions (11.4.3) and (11.4.4) into (11.4.2) gives the properly symmetric and very simple result

$$F_{KG} = N[f(a_1, b_1, c_1, d_1) + f(a_2, b_2, c_2, d_2) + f(a_3, b_3, c_3, d_3)]. \qquad (11.4.5)$$

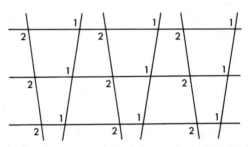

Fig. 11.7. Expanded section of one of the framing regions in Fig. 11.6(b), showing the two types of sites. This is a square lattice, with different weights on alternate columns.

11.5 Formulation as a Triangular-Honeycomb Ising Model with Two- and Four-Spin Interactions

Like the square-lattice model, the Kagomé lattice eight-vertex model can be formulated in terms of 'magnetic' spins on faces, instead of 'electric' arrows on edges.

The most symmetric way to do this is to regard the arrow configuration shown in Fig. 11.8(a) as a standard. (It is anti-ferroelectric: all vertices are of the fifth type in Fig. 11.2.) With each face r associate a spin σ_r, with values $+1$ and -1. Consider an edge j, with faces r and s on either side. Place an arrow on it according to the rule:

if $\sigma_r\sigma_s = +1$, point the arrow in the same direction as
the arrow on edge j in the standard configuration;
otherwise point it the opposite way.   (11.5.1)

Do this for all edges.

Imagine an observer walking round any particular site, going successively through the four faces. If he observes a change in spin from one face to the next, then the arrow on the intervening edge is non-standard, and conversely. When he returns to his starting point he must have seen an even number of changes, so there are an even number of non-standard arrows on the four edges at each site. Since the standard configuration is two in and two out, there must in any event be an even number of arrows into (and out of) each site. This is the eight-vertex condition.

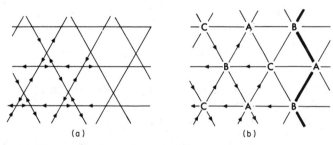

Fig. 11.8. (a) The standard anti-ferroelectric arrow configuration on the Kagomé lattice (not all arrows are shown); (b) the corresponding triangular lattice configuration, obtained by shrinking the up-pointing triangles of the Kagomé lattice down to points. The three sub-lattices A, B, C of the triangular lattice are also shown.

To each configuration of the face-spins, there therefore corresponds an arrow covering of the edges that satisfies the eight-vertex condition. Conversely, by the same reasoning as in Section 10.3, to each such arrow covering there correspond two face-spin configurations (differing from one another in the reversal of all spins). Using this one-to-two correspondence, (11.1.2) can be written

$$Z = \tfrac{1}{2} \sum_{\sigma} \prod_i \omega_{i, j(i, \sigma)} , \qquad (11.5.2)$$

where the sum is over all configurations $\sigma = \{\sigma_1 , \sigma_2 , \ldots\}$ of the face-spins, the product is over all sites i, and $j(i , \sigma)$ is the arrow arrangement at site i for spin-configuration σ.

Let r, s, t, u be the faces round a site of type i, arranged as in Fig. 11.3. Let $W_i(\sigma_r , \sigma_s , \sigma_t , \sigma_u)$ be the Boltzmann weight of the vertex arrow arrangement corresponding to the face spins having the values σ_r, σ_s, σ_t, σ_u. Then from rule (11.5.1) and Figs. 11.2, 11.3 and 11.8,

$$W_i(+,-,+,+) = W_i(+,+,+,-) = a_i,$$

$$W_i(-,+,+,+) = W_i(+,+,-,+) = b_i,$$

$$W_i(+,+,+,+) = W_i(-,+,-,+) = c_i,$$

$$W_i(-,-,+,+) = W_i(-,+,+,-) = d_i,$$

$$W_i(\sigma_r,\sigma_s,\sigma_t,\sigma_u) = W_i(-\sigma_r,-\sigma_s,-\sigma_t,-\sigma_u)$$

$$= W_i(\sigma_r,-\sigma_s,\sigma_t,-\sigma_u),$$

(11.5.3)

for $i = 1$, 2, 3 and $\sigma_r, \sigma_s, \sigma_t, \sigma_u = \pm 1$. This defines the function W_i, and (11.5.2) can be written as

$$Z = \tfrac{1}{2}\sum_\sigma \prod W_i(\sigma_r,\sigma_s,\sigma_t,\sigma_u),$$

(11.5.4)

the product being over all sites; i now denotes the type of the site, and r, s, t, u are the surrounding faces.

From (11.5.3), the function W_i can be written as

$$W_i(\sigma_r,\sigma_s,\sigma_t,\sigma_u) = M_i \exp[K_i\sigma_r\sigma_t + K_i'\sigma_s\sigma_u + K_i''\sigma_r\sigma_s\sigma_t\sigma_u],$$

(11.5.5)

where M, K_i, K_i', K_i'' are related to a_i, b_i, c_i, d_i by

$$a_i = M_i \exp(K_i - K_i' - K_i''), \quad b_i = M_i \exp(-K_i + K_i' - K_i''),$$

$$c_i = M_i \exp(K_i + K_i' + K_i''), \quad d_i = M_i \exp(-K_i - K_i' + K_i'').$$

(11.5.6)

Using this form (11.5.5) of W_i, it is apparent from (11.5.4) that Z (strictly $2Z$) is the partition function of an Ising model defined on the faces of the Kagomé lattice, with two-spin interactions between opposite faces at a site, and four-spin interactions between the four faces round a site.

Placing a dot at the centre of each face of the Kagomé lattice, and linking dots whose spins interact via a two-spin interaction, gives the lattice of Fig. 11.9. This consists of a honeycomb lattice interlacing a triangular one. The interaction coefficients K_i, K_i' associated with the various edges are shown. Let N be the number of sites of type 1 (or type 2, or type 3) in the original Kagomé lattice. Then the associated honeycomb lattice has $2N$ sites, the triangular one has N; each has $3N$ edges, and each edge of one is crossed by an edge of the other. Using (11.5.5), (11.5.4) can now be written as

$$Z = \tfrac{1}{2}(M_1M_2M_3)^N \sum_\sigma \exp\{\Sigma K_i\sigma_r\sigma_t + \Sigma K_i'\sigma_s\sigma_u + \Sigma K_i''\sigma_r\sigma_s\sigma_t\sigma_u\},$$

(11.5.7)

where the first sum inside the exponential is over all edges (r, t) of the honeycomb lattice; the second is over all edges (s, u) of the triangular lattice; and the third is over all honeycomb edges (r, t), (s, u) being the crossing triangular edge. In each case i (=1, 2 or 3) is the type of the edge, as in Fig. 11.9.

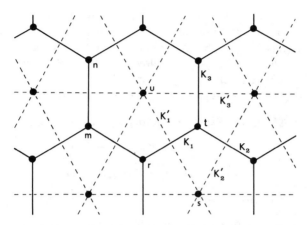

Fig. 11.9. The honeycomb-triangular lattice formed by placing a dot at the centre of each face of the Kagomé lattice, and linking dots whose Ising spins interact via a two-spin interaction. The corresponding interaction coefficients $K_1, \ldots, K_3'$ are indicated; r, s, t, u correspond to the four faces in Fig. 11.3(a); m and n correspond to the faces m and n in Fig. 11.6(a).

Thus Z is, to within a normalization factor, the joint partition function of a honeycomb and a triangular Ising model interacting via four-spin interactions on crossing edges.

Star – Triangle Restriction

For the solvable model, the interaction coefficients $K_1, \ldots, K_3''$ are not arbitrary: they must satisfy the restrictions (11.1.7), where a_i, b_i, c_i, d_i are given by (11.5.6).

These restrictions came from the star – triangle relation (11.1.6): it is interesting to go back to this and express it in terms of 'magnetic' spins on the faces of the Kagomé lattice, rather than 'electric' spins on the edges.

The relation (11.1.6) states that the two graphs in Fig. 11.4 have the same total weight, after summing over allowed configurations of arrows on the internal edges. To express this relation in 'magnetic' language, associate spins with the faces of the graphs, as in Fig. 11.4. Then the requirement that the two graphs have the same total weight is

$$\sum_{\sigma_7} W_1(\sigma_2, \sigma_3, \sigma_7, \sigma_1) \, W_2(\sigma_4, \sigma_5, \sigma_7, \sigma_3) \, W_3(\sigma_6, \sigma_1, \sigma_7, \sigma_5)$$

$$= \sum_{\sigma_7} W_1(\sigma_7, \sigma_4, \sigma_5, \sigma_6) \, W_2(\sigma_7, \sigma_6, \sigma_1, \sigma_2) \, W_3(\sigma_7, \sigma_2, \sigma_3, \sigma_4). \quad (11.5.8)$$

This equation must be true for all values of the external spins $\sigma_1, \ldots, \sigma_6$. The summations are over the values of the internal spin σ_7: this corresponds in (11.1.6) to summing over the allowed values of β_1, β_2, β_3 (only two sets of values are allowed by the eight-vertex rule).

Using (11.5.3), one can verify directly that the equations (11.5.8) are the same as (11.1.6). Using (11.5.5), they can be written explicitly as

$$\exp(K_1' \sigma_1 \sigma_3 + K_2' \sigma_3 \sigma_5 + K_3' \sigma_5 \sigma_1) \cosh(K_1 \sigma_2 + K_2 \sigma_4$$

$$+ K_3 \sigma_6 + K_1'' \sigma_1 \sigma_2 \sigma_3 + K_2'' \sigma_3 \sigma_4 \sigma_5 + K_3'' \sigma_5 \sigma_6 \sigma_1)$$

$$= \exp(K_1' \sigma_4 \sigma_6 + K_2' \sigma_6 \sigma_2 + K_3' \sigma_2 \sigma_4) \cosh(K_1 \sigma_5 + K_2 \sigma_1$$

$$+ K_3 \sigma_3 + K_1'' \sigma_4 \sigma_5 \sigma_6 + K_2'' \sigma_6 \sigma_1 \sigma_2 + K_3'' \sigma_2 \sigma_3 \sigma_4) \quad (11.5.9)$$

Since they are equivalent to (11.1.6), they imply (11.1.7) and (11.1.8), in particular $\Gamma_1 = \Gamma_2 = \Gamma_3$. From (11.1.9), this means that $a_i b_i / c_i d_i$ is therefore independent of i, and from (11.5.6) this implies

$$K_i'' = \text{independent of } i. \quad (11.5.10)$$

Thus all sites must have the same four-spin interaction coefficient, regardless of their type. Let us call this coefficient simply K''. Then by considering all 64 values of $\sigma_1, \ldots, \sigma_6$, we find that (11.5.9) is equivalent to the six equations:

$$\exp(2K_j' + 2K_k') = \frac{\cosh(K_i + K_j + K_k - K'')}{\cosh(-K_i + K_j + K_k + K'')},$$

$$\exp(2K_j' - 2K_k') = \frac{\cosh(K_i - K_j + K_k + K'')}{\cosh(K_i + K_j - K_k + K'')}, \quad (11.5.11)$$

for all permutations (i, j, k) of $(1, 2, 3)$.

These are the equations (11.1.7). They are not independent, since the second set can be easily deduced from the first. The first set, containing three equations, is plainly independent since it can be used to define K_1', K_2', K_3' for given values of K_1, K_2, K_3, K''. Alternatively, K_1', K_2', K_3', K'' can be regarded as given, and the equations solved for K_1, K_2, K_3.

Using (11.5.6), the corollaries (11.1.8)–(11.1.9) become, for $j = 1, 2, 3$,

$$\Delta = -\sinh 2K_j \sinh 2K'_j - \tanh 2K'' \cosh 2K_j \cosh 2K'_j , \quad (11.5.12a)$$

$$\Gamma = -\tanh 2K'' . \quad (11.5.12b)$$

Eliminating y^2 (taken to be negative) between the equations (10.4.12), and using (10.4.9), the elliptic modulus k is given by

$$2k^{\frac{1}{2}}/(1 + k) = l , \quad (11.5.13)$$

where

$$l^2 = (1 - \Gamma^2)/(\Delta^2 - \Gamma^2) . \quad (11.5.14)$$

Using (11.5.12), this last equation can be written

$$l^2 = \frac{(1 - v_j^2)^2 (1 - v'^2_j)^2 (1 - v''^2)^2}{16(1 + v_j v'_j v'')(v'' + v_j v'_j)(v_j + v'_j v'')(v'_j + v_j v'')} , \quad (11.5.15)$$

where

$$v_j = \tanh K_j, \quad v'_j = \tanh K'_j, \quad v'' = \tanh K'' . \quad (11.5.16)$$

Spontaneous Magnetization

The argument of Section 11.2 can be repeated in terms of face-spins, instead of edge arrow-spins. Instead of using (11.1.6), one uses (11.5.8). The resulting analogue of (11.2.2) and (11.3.3) is the following.

Let $\sigma_1 , \ldots , \sigma_m$ be any set of face spins of the Kagomé lattice, all lying between the lines AB and CD in Fig. 11.6(a). Suppose the restrictions (11.1.7), or equivalently (11.5.11), are satisfied. Then for a large lattice

$\langle \sigma_1 \ldots \sigma_m \rangle =$ *same as in the regular square-lattice eight-vertex*
model, with weights a_3, b_3, c_3, d_3, on the
interior lattice in Fig. 11.6(b).   (11.5.17)

Thus $\langle \sigma_1 \ldots \sigma_m \rangle$ is a function only of a_3, b_3, c_3, d_3, or equivalently of K_3, K'_3, K''.

In particular, this means that the expectation value of any face-spin σ_m is

$$\langle \sigma_m \rangle = M_0(a_3 , b_3 , c_3 , d_3) , \quad (11.5.18)$$

regardless of whether σ_m lies in a triangular or hexagonal face of the Kagomé lattice. Without ambiguity, we can therefore call $\langle \sigma_m \rangle$ the spontaneous magnetization.

Here $M_0(a_3, b_3, c_3, d_3)$ is the spontaneous magnetization of the square lattice eight-vertex model with Boltzmann weights a, b, c, d. It is given by (10.10,19). Like the spontaneous polarization, it depends on a, b, c, d only via Δ and Γ. These have the same value for all three types of lattice site, so (11.5.18) is unchanged by replacing a_3, b_3, c_3, d_3 by a_1, b_1, c_1, d_1, or by a_2, b_2, c_2, d_2. Indeed, it is obvious from the rotation symmetry of the lattice that this must be so.

Hyperbolic Trigonometry and the Elliptic Function Parametrization

Define two further parameters ω, Ω by

$$\coth 2K'' = \cosh \Omega, \quad -\Delta \coth 2K'' = \cosh \omega. \tag{11.5.19}$$

Then (11.5.12a) can be written as

$$\cosh \omega = \cosh 2K_j \cosh 2K_j' + \cosh \Omega \sinh 2K_j \sinh 2K_j'. \tag{11.5.20}$$

This is the same as the relation between the sides ω, $2K_j$, $2K_j'$ of an hyperbolic triangle, with angle $\pi + i\Omega$ between the sides $2K_j$, $2K_j'$ (Onsager, 1944, p. 135; Coxeter, 1947). Other corollaries of the star–triangle relations (11.5.11), e.g.

$$-\cosh 2K_2 \cosh 2K_3 + \coth 2K_1' \sinh 2K_2 \sinh 2K_3 \tag{11.5.21}$$
$$=\cosh 2K_1 \cosh 2K'' + \coth 2K_1' \sinh 2K_1 \sinh 2K'',$$

can be interpreted in terms of hyperbolic trigonometry. It seems likely that many of the properties of the star-triangle relation, notably the 'quadrilateral theorem' (Baxter, 1978a), could be conveniently interpreted in this way: as far as I know, this has not yet been done.

These ideas also provide an alternative approach to the elliptic function parametrization of Section 10.4. It is well known that the relation (11.5.20) can be simplified by introducing elliptic functions of modulus

$$l = \sinh \Omega / \sinh \omega, \tag{11.5.22}$$

(Greenhill, 1892, Paragraph 129). Onsager (1944, p. 144) refers to this as a uniformizing substitution.

This l is precisely that defined by (11.5.15), so it is related to the modulus k used in Chapter 10 by (11.5.13). From Section 15.6, l and k are therefore related by a Landen transformation, so both approaches lead effectively to the same elliptic function parametrization, as they should.

11.6 Phases

From (11.3.4) and (11.5.18), the spontaneous polarization and magnetization of the Kagomé lattice model are the same as those of a square-
lattice eight-vertex model with the same values of Δ and Γ. From Section
10.11, it follows that the Kagomé lattice model is ordered if $|\Delta| > 1$,
disordered if $|\Delta| < 1$.

The archetypal ordered regime is when

$$c_i > a_i + b_i + d_i, \quad a_i > 0, \quad b_i > 0, \quad d_i > 0, \qquad (11.6.1)$$

for $i = 1, 2, 3$. From (11.1.9), it follows that $\Delta < -1$. The ground state is
then the anti-ferroelectric arrow configuration shown in Fig. 11.8(a), or
the one obtained from it by reversing all arrows. Since we have used this
as a standard in relating face- and arrow-spins, our resulting face-spin
interpretation of this phase is ferromagnetic: in a ground state all spins on
triangular faces are the same, and all spins on hexagonal faces are the
same. There are four such ground states.

In addition to (11.6.1), there are seven other regimes in the available
parameter space in which the system is ordered (Baxter, 1978a, p. 337).
They can all be obtained from the archetypal case by reversing appropriate
sets of spins. For example, reversing all face-spins on up-pointing triangles
is equivalent to reversing all arrows on the sides of up-pointing triangles:
from Fig. 11.2 this interchanges a_i with d_i, and b_i with c_i, so maps (11.6.1)
to

$$b_i > a_i + c_i + d_i, \quad a_i > 0, \quad c_i > 0, \quad d_i > 0, \qquad (11.6.2)$$

with $\Delta > 1$.

Alternatively, reversing face-spins between alternate pairs of horizontal
lines is equivalent to reversing all horizontal arrows. This leaves $a_3, b_3, c_3,$
d_3 unaltered, but for $i = 1$ or 2 it interchanges a_i with b_i, c_i with d_i. Thus
it maps (11.6.1) to

$$d_1 > a_1 + b_1 + c_1, \quad d_2 > a_2 + b_2 + c_2, \quad c_3 > a_3 + b_3 + c_3, \qquad (11.6.3)$$

all weights being positive and $\Delta < -1$.

Two other mappings can obviously be obtained from this by using the
rotation symmetry. All eight ordered regimes can then be obtained by
combinations of these various mappings.

There is only one disordered regime, namely

$$0 < a_i, b_i, c_i, d_i < \tfrac{1}{2}(a_i + b_i + c_i + d_i), \qquad (11.6.4)$$

for $i = 1, 2, 3$, with $-1 < \Delta < 1$.

11.7 $K'' = 0$: The Triangular and Honeycomb Ising Models

In Section 10.3 we remarked that the square-lattice eight-vertex model factors into two independent Ising models if the four-spin interaction coefficient K'' is zero.

A similar factorization occurs for the Kagomé lattice model. From (11.5.10), we can set all four-spin coefficients K_i'' simultaneously to zero. The exponential in (11.5.7) then factors into two parts, one involving only spins on the honeycomb lattice of Fig. 11.9, the other only spins on the triangular lattice. It follows at once that

$$Z = \tfrac{1}{2}(M_1 M_2 M_3)^N Z_H(K_1, K_2, K_3) Z_T(K_1', K_2', K_3'), \qquad (11.7.1)$$

where $Z_H(K_1, K_2, K_3)$ is the partition function of the nearest-neighbour Ising model on the honeycomb lattice, with interaction coefficients K_1, K_2, K_3; and $Z_T(K_1', K_2', K_3')$ is the partition function of the nearest-neighbour Ising model on the triangular lattice, with interaction coefficients K_1', K_2', K_3'. The honeycomb lattice has $2N$ sites, the triangular has N.

The relations (11.5.11) can be written as

$$\frac{\cosh(K_1 + K_2 + K_3)}{\cosh(-K_1 + K_2 + K_3)} = \exp(2K_2' + 2K_3'), \qquad (11.7.2)$$

together with two other equations obtained by permuting the suffixes 1, 2, 3. These are precisely the Ising model star – triangle relations (6.4.8) (with R eliminated and K_j, L_j replaced by K_j', K_j).

This equivalence with the Ising model star – triangle relation is even clearer in the original equation (11.5.9). If the K_j'' therein vanish, it factors into two equations, one involving only the spins σ_1, σ_3, σ_5, the other involving σ_2, σ_4, σ_6. They are

$$2 \cosh(K_1\sigma_2 + K_2\sigma_4 + K_3\sigma_6)$$
$$= R \exp(K_1'\sigma_4\sigma_6 + K_2'\sigma_6\sigma_2 + K_3'\sigma_2\sigma_4), \qquad (11.7.3a)$$

$$2 \cosh(K_1\sigma_5 + K_2\sigma_1 + K_3\sigma_3)$$
$$= R \exp(K_1'\sigma_1\sigma_3 + K_2'\sigma_3\sigma_5 + K_3'\sigma_5\sigma_1), \qquad (11.7.3b)$$

where R is a common constant. Each of these equations is precisely the star – triangle relation (6.4.4) – (6.4.5).

From (11.5.12b), Γ is zero, so from (11.5.14) and (11.5.12a) we can choose

$$l^{-1} = -\Delta = \sinh 2K_j \sinh 2K_j', \quad j = 1, 2, 3. \qquad (11.7.4)$$

This is precisely the relation (6.4.13), with k, K_j, L_j therein replaced by l, K_j', K_j. From (6.4.16) and (6.3.5) it follows that

$$l^2 = \frac{(1 - v_1'^2)^2 (1 - v_2'^2)^2 (1 - v_3'^2)^2}{16(1 + v_1'v_2'v_3')(v_1' + v_2'v_3')(v_2' + v_3'v_1')(v_3' + v_1'v_2')}, \quad (11.7.5)$$

where v_1', v_2', v_3' are defined by (11.5.16) in terms of the triangular interaction coefficients K_1', K_2', K_3'.

Alternatively, using (6.4.12), l can be expressed in terms of the honeycomb lattice coefficients K_1, K_2, K_3 as

$$l^2 = \frac{16(1 + z_1z_2z_3)(z_1 + z_2z_3)(z_2 + z_3z_1)(z_3 + z_1z_2)}{(1 - z_1^2)^2 (1 - z_2^2)^2 (1 - z_3^2)^2}, \quad (11.7.6)$$

where

$$z_j = \exp(-2K_j), \quad j = 1, 2, 3. \quad (11.7.7)$$

[The similarity in form of (11.7.5) and the inverse of (11.7.6) is a reflection of the duality relation of Section 6.3. Why they should also resemble the eight-vertex single-site relation (11.5.15) I do not know.]

The R in (11.7.3) is the same as the R in Section 6.4, so from (6.4.14), changing k, L_j, K_j to l, K_j, K_j':

$$R^2 = 2l \sinh 2K_1 \sinh 2K_2 \sinh 2K_3$$
$$= 2/(l^2 \sinh 2K_1' \sinh 2K_2' \sinh 2K_3'). \quad (11.7.8)$$

Free Energy

In our present notation, the identity (6.4.7) becomes

$$Z_H(K_1, K_2, K_3) = R^N Z_T(K_1', K_2', K_3'). \quad (11.7.9)$$

Substituting this into (11.7.1), taking logarithms and using (11.4.1) and (11.4.5), gives

$$2 \ln Z_T(K_1', K_2', K_3') = \ln 2 - N \ln(M_1 M_2 M_3 R)$$
$$- (k_B T)^{-1} N (f_1 + f_2 + f_3), \quad (11.7.10)$$

writing f_j for $f(a_j, b_j, c_j, d_j)$, which is the free energy per site of the square lattice eight-vertex model with weights a_j, b_j, c_j, d_j.

These weights are given by (11.5.6). Since K_j'' is zero, the square lattice model is the product of two square lattice Ising models. Choosing each M_j

to be one, from (10.3.11) f_j is the free energy per site of the square lattice Ising model with interaction coefficients K_j, K_j'.

It is convenient to work with the dimensionless free energy per site

$$\psi = f/k_B T. \tag{11.7.11}$$

From (1.7.6), for a lattice of N sites this is related to the partition function Z by

$$\psi = - \lim_{N \to \infty} N^{-1} \ln Z. \tag{11.7.12}$$

For the square lattice Ising model with interaction coefficients K_j, K_j', ψ is a function only of these coefficients. Let us write it as $\psi_{SQ}(K_j, K_j')$. Similarly, for the triangular Ising model with interaction coefficients K_1', K_2', K_3', let us write it as $\psi_T(K_1', K_2', K_3')$. Then from (11.7.10) – (11.7.12),

$$\psi_T(K_1', K_2', K_3') = \tfrac{1}{2}[\ln R + \psi_{SQ}(K_1, K_1') + \psi_{SQ}(K_2, K_2')$$
$$+ \psi_{SQ}(K_3, K_3')]. \tag{11.7.13}$$

Also, using (11.7.9) and remembering that the honeycomb lattice herein has $2N$ sites, the dimensionless free energy per site of the honeycomb lattice Ising model, with interaction coefficients K_1, K_2, K_3, is

$$\psi_H(K_1, K_2, K_3) = \tfrac{1}{4}[-\ln R + \psi_{SQ}(K_1, K_1') + \psi_{SQ}(K_2, K_2')$$
$$+ \psi_{SQ}(K_3, K_3')]. \tag{11.7.14}$$

The function $\psi_{SQ}(K, K')$ has been obtained in Chapter 7, and (as a special case of the eight-vertex model) in Chapter 10. Replacing the K, L, k of Chapter 7 by the K', K, l herein, the equations (7.9.14), (7.9.16), (7.6.1) give, for all values of K and K',

$$\psi_{SQ}(K, K') = -(2\pi)^{-1} \int_0^\pi \ln\{2[\cosh 2K' \cosh 2K$$
$$+ (1 + l^{-2} - 2l^{-1}\cos 2\theta)^{\frac{1}{2}}]\} \, d\theta, \tag{11.7.15}$$

where, similarly to (11.7.4),

$$l^{-1} = \sinh 2K \sinh 2K'. \tag{11.7.16}$$

We are free to make any choice of K_1', K_2', K_3', or of K_1, K_2, K_3, so we can use (11.7.13) or (11.7.14) to obtain the free energy of any regular triangular or honeycomb Ising model. The other parameters are defined by the three equations (11.7.2), and by (11.7.8).

Nearest-Neighbour Correlations

When $K'' = 0$, (11.5.17) relates correlations of the triangular and honeycomb Ising models to those of the square. Consider the two faces m, n in Fig. 11.6. Let σ_m, σ_n be the corresponding Ising spins and apply (11.5.17) to $\langle \sigma_m \sigma_n \rangle$.

As in Section 11.5, place dots at the centre of each face and link dots whose spins interact via a two-spin interaction. Then Fig. 11.6(a) becomes Fig. 11.9, and m and n are vertical nearest neighbours on the honeycomb lattice, with interaction coefficients K_1, K_2, K_3. On the other hand, Fig. 11.6(b) becomes two interlaced square lattices: m and n are vertical nearest neighbours on the square lattice, with interaction coefficients K_3, K_3'. In both cases the edge (m, n) has interaction coefficient K_3. Since $\langle \sigma_m \sigma_n \rangle$ is the same for both, it follows that

$$g_H(K_3, K_1, K_2) = g_{SQ}(K_3, K_3') , \qquad (11.7.17)$$

where g_H, g_{SQ} are the nearest-neighbour correlations of the honeycomb and square lattices, respectively, and the first argument is the interaction coefficient of the edge under consideration.

Similarly, considering the correlation between the spins on the two faces in Fig. 11.6 that are next to both m and n, we obtain

$$g_T(K_3', K_1', K_2') = g_{SQ}(K_3', K_3) . \qquad (11.7.18)$$

The correlations g_{SQ}, g_H, g_T are derivatives of ψ_{SQ}, ψ_H, ψ_T. From (6.2.1) (with K, L replaced by K', K), (11.7.12) and (1.4.4),

$$g_{SQ}(K, K') = -\partial \psi_{SQ}(K, K')/\partial K . \qquad (11.7.19a)$$

Similarly,

$$g_H(K_1, K_2, K_3) = - \partial \psi_H(K_1, K_2, K_3)/\partial K_1 , \qquad (11.7.19b)$$

$$g_T(K_1', K_2', K_3') = - \partial \psi_T(K_1', K_2', K_3')/\partial K_1' , \qquad (11.7.19c)$$

where each differentiation is performed with the other arguments kept constant.

Alternative Derivation of the Ising Model Free Energy

The relations (11.7.17) and (11.7.18) are *not* mere consequences of (11.7.19). They contain more information.

To see this, note that (11.7.13) and (11.7.14) imply that

$$\psi_H(K_1, K_2, K_3) = \tfrac{1}{2}[- \ln R + \psi_T(K_1', K_2', K_3')] . \qquad (11.7.20)$$

(This is a simple corollary of (11.7.9).) Differentiate this equation with respect to K_3, keeping K_1 and K_2 constant.

The K_1', K_2', K_3' are defined by (11.7.2). From this, (11.7.4) and (11.7.8), one can verify that

$$\partial K_3'/\partial K_3 = -2w, \quad \partial \ln R/\partial K_3 = 2w_3, \tag{11.7.21}$$

$$\partial K_1'/\partial K_3 = 2w_2, \quad \partial K_2'/\partial K_3 = 2w_1,$$

where

$$w = l \sinh 2K_1' \sinh 2K_2' \sinh 2K_3', \tag{11.7.22}$$

$$w_r = w \coth 2K_r', \quad r = 1, 2, 3.$$

Differentiating (11.7.20), using (11.7.17)–(11.7.19), therefore gives

$$g_{SQ}(K_3, K_3') = w_3 + w_2 g_{SQ}(K_1', K_1)$$
$$+ w_1 g_{SQ}(K_2', K_2) - w g_{SQ}(K_3', K_3). \tag{11.7.23}$$

Since K_1, K_2, K_3 are independent, this is a three-variable equation for the two-variable function $g_{SQ}(K, K')$. Baxter and Enting (1978) have shown that it, together with (11.7.19a) and the symmetry property $\psi_{SQ}(K, K') = \psi_{SQ}(K', K)$, completely determines $g_{SQ}(K, K')$. The interested reader is referred to that paper for details: equation (8) therein (with K, L replaced by K', K) is the equation (11.7.23) above. Briefly, (11.7.23) implies that, for K and K' positive,

$$g_{SQ}(K, K') = \coth 2K \int_0^{2K} \frac{a(l) - b(l) \tanh^2 x}{(1 + l^2 \sinh^2 x)^{\frac{1}{2}}} \, dx, \tag{11.7.24}$$

where, consistently with (11.7.4),

$$l^{-1} = \sinh 2K \sinh 2K', \tag{11.7.25}$$

and $a(l)$, $b(l)$ are functions of l only. The RHS of (11.7.24) must tend to one as $K \to +\infty$, which implies a linear relation between $a(l)$ and $b(l)$.

The symmetry of $\psi_{SQ}(K, K')$, together with (11.7.19a), implies that $a(l)$, $b(l)$ satisfy certain differential equations. Solving them gives, for $0 < l < \infty$,

$$a(l) = [(1 + l) E(l_1) + (1 - l) I(l_1)]/\pi, \tag{11.7.26}$$

$$b(l) = 2(1 - l) I(l_1)/\pi,$$

where

$$l_1 = 2l^{\frac{1}{2}}/(1 + l), \tag{11.7.27}$$

and $I(l_1)$, $E(l_1)$ are the complete elliptic integrals of the first and second kinds, of modulus l_1, as defined in (15.5.9) and (15.5.13).

Now that $g_{SQ}(K, K')$ is determined, the free energy can be obtained from (11.7.19a).

The fascinating feature of this derivation is that it uses only the star–triangle relation, which is a local property of the Ising model. It uses it twice: once to establish (11.5.17) and the corollaries (11.7.17), (11.7.18); and again to establish (11.7.20). (Hilhorst $et\ al.$ (1978, 1979) and Knops and Hilhorst (1979) have shown that the star–triangle relation can also be used to obtain the critical properties of the Ising model via the renormalization group method.) This further underlines the significance of the star–triangle relation, at least for the Ising model. I am not sure whether the method can be generalized to the full eight-vertex model.

The result (11.7.24)–(11.7.27) of course agrees with the result of the transfer matrix calculation of Chapter 7, namely (7.9.16) and (11.7.15).

The critical singularities at $l = 1$ arises not from the integral in (11.7.24), but from the 'coefficients' $a(l)$ and $b(l)$. Near $l = 1$ these behave as

$$a(l) = \frac{2}{\pi} + \frac{1-l}{\pi}\ln\frac{4(1+l)}{1-l},$$

(11.7.28)

$$b(l) = \frac{2(1-l)}{\pi}\ln\frac{4(1+l)}{1-l}.$$

This makes it clear that all square, honeycomb and triangular Ising models have the same critical singularities in their internal energies, namely that of $b(l)$. The symmetric logarithmic divergence of the specific heat follows at once, so as in (7.12.12) the exponents α, α' are given by

$$\alpha = \alpha' = 0.$$

(11.7.29)

Magnetization

From (11.5.18), the triangular Ising model with coefficients K_1', K_2', K_3', the honeycomb model with coefficients K_1, K_2, K_3, and the square model with coefficients K_1, K_1' (or K_2, K_2', or K_3, K_3') all have the same spontaneous magnetization. We can therefore use the square-lattice result (7.10.50). With our present notation this is

$$M_0 = (1 - l^2)^{1/8} \quad \text{if} \quad |l| \leq 1,$$

(11.7.30)

$$= 0 \quad \text{if} \quad |l| > 1.$$

(Each Ising model is ordered if $|l| \leq 1$, disordered otherwise.)

Comparing this with (1.1.4), and noting that at the critical temperature $l - 1$ vanishes linearly with $T - T_c$, it follows that for all three models the exponent β is

$$\beta = 1/8. \qquad (11.7.31)$$

All other critical exponents are also expected to be the same for the triangular, honeycomb and square-lattice Ising models, and to have the values given in $(7.12.12) - (7.12.16)$. This is in agreement both with the scaling and universality hypotheses.

11.8   Explicit Expansions of the Ising Model Results

The results $(11.7.13)$ and $(11.7.15)$ are expressed in terms of elementary functions and integrals thereof. This form is easy to understand, but is not necessarily the most convenient to use. For some purposes, e.g. developing series expansions or even direct evaluations, it may be easier to use elliptic functions and their infinite product expansions.

Throughout this section it will be supposed that all the interaction coefficients $K_1, K_2, K_3, K'_1, K'_2, K'_3$ are non-negative. This means that for any lattice there are only two cases to consider: low temperature $(0 < l < 1)$, and high temperature $(l > 1)$.

Square Lattice: Low Temperature

First consider the square-lattice Ising model with interaction coefficients K_j, K'_j. This is equivalent to a square-lattice eight-vertex model with weights a_j, b_j, c_j, d_j given by $(11.5.6)$, M_j being one and K''_j being zero.

Replace the a, b, c, d of Chapter 10 by these a_j, b_j, c_j, d_j. Replace u, v, z by u_j, v_j, z_j. Then from $(10.4.21)$ and $(11.5.6)$, k, λ, u_j are given by

$$1 = (c_j d_j / a_j b_j)^{\frac{1}{2}} = k^{\frac{1}{2}} \operatorname{snh} \lambda , \qquad (11.8.1a)$$

$$\exp(-2K_j) = d_j / a_j = k^{\frac{1}{2}} \operatorname{snh} u_j , \qquad (11.8.1b)$$

$$\exp(-2K'_j) = d_j / b_j = k^{\frac{1}{2}} \operatorname{snh}(\lambda - u_j) . \qquad (11.8.1c)$$

As in Chapter 15, let I, I' be the complete elliptic integrals of the first kind of moduli $k, k' = (1 - k^2)^{\frac{1}{2}}$. Then from $(11.8.1a)$ and $(15.4.32)$, we obtain the result $(10.9.6)$, i.e.

$$\lambda = \tfrac{1}{2} I' . \qquad (11.8.2)$$

From $(10.4.23)$, $(10.7.9)$ and $(10.7.19)$,

$$u_j = \tfrac{1}{2}(\lambda + v_j), \qquad q = \exp(-\pi I'/I),$$

$$x = \exp(-\pi \lambda / 2I), \quad z_j = \exp(-\pi v_j / 2I) , \qquad (11.8.3)$$

so from (11.8.2)

$$q = x^4. \tag{11.8.4}$$

Provided $0 < l < 1$, which from the equation (11.7.4) means that $\sinh 2K_j \sinh 2K_j' > 1$, the free energy is given by (10.8.47). Using (11.5.6) and (11.8.4), the dimensionless free energy $\psi = f/k_B T$ is therefore

$$\psi_{SQ}(K_j, K_j') = - K_j - K_j'$$
$$- \sum_{m=1}^{\infty} \frac{x^{3m}(1 - x^{2m})^2(x^m + x^{-m} - z_j^m - z_j^{-m})}{m(1 - x^{8m})(1 + x^{2m})}. \tag{11.8.5}$$

We can write K_j and K_j' explicitly in terms of x and z_j by using the infinite product expansion (15.1.5) – (15.1.6) of the function $k^{\frac{1}{2}} \, \mathrm{sn} \, u$ in (11.8.1b) and (11.8.1c), together with the definitions (11.8.3) and (11.8.4). It is useful to define a parameter τ, and two functions $\phi(z)$ and $g(z)$ by

$$\tau = \sum_{m=1}^{\infty} \frac{x^{2m}(1 - x^{2m})^2}{m(1 - x^{8m})}, \tag{11.8.6a}$$

$$\phi(z) = \left(\frac{x}{z}\right)^{\frac{1}{2}} \prod_{n=1}^{\infty} \frac{(1 - x^{8n-7}z)(1 - x^{8n-1}z^{-1})}{(1 - x^{8n-5}z^{-1})(1 - x^{8n-3}z)}, \tag{11.8.6b}$$

$$g(z) = \sum_{m=1}^{\infty} \frac{x^{3m}(1 - x^{2m})(z^{-m} - x^{2m}z^m)}{m(1 - x^{8m})(1 + x^{2m})}, \tag{11.8.6c}$$

where x is regarded as a constant. Then, proceeding as above, we obtain

$$\exp(-2K_j) = \phi(z_j), \quad \exp(-2K_j') = \phi(z_j^{-1}), \tag{11.8.7a}$$

$$\psi_{SQ}(K_j, K_j') = - K_j - K_j' - \tau + g(z_j) + g(z_j^{-1}). \tag{11.8.7b}$$

This defines the function $\psi_{SQ}(K_j, K_j')$ for all non-negative numbers K_j, K_j' such that l, given by (11.7.4), is less than one. The parameters x and z_j are uniquely defined by (11.8.7a) and the restrictions (10.15.11), i.e.

$$0 < x < 1, \quad x < z_j < x^{-1}; \tag{11.8.8}$$

$\psi_{SQ}(K_j, K_j')$ is then given by (11.8.7b).

These equations can readily be used to evaluate the function ψ_{SQ}, or to expand $\psi_{SQ}(K_j, K_j') + K_j + K_j'$ in powers of $\exp(-2K_j)$ and $\exp(-2K_j')$. The parameter x is small for very low temperatures ($l \ll 1$), increasing to one at the critical temperature ($l = 1$).

(The suffix j is irrelevant in this and the next sub-section: it is included in anticipation of the sub-sections on the triangular and honeycomb Ising models.)

Square Lattice: High Temperature

The easiest way to obtain the high-temperature $(l > 1)$ result is to apply the duality transformation (6.2.14) to the result (11.8.7). This gives

$$\tanh K_j' = \phi(z_j), \quad \tanh K_j = \phi(z_j^{-1}), \tag{11.8.9a}$$

$$\psi_{SQ}(K_j, K_j') = -\ln[2\cosh K_j \cosh K_j'] - \tau + g(z_j) + g(z_j^{-1}). \tag{11.8.9b}$$

Again x and z_j satisfy (11.8.8). Now x is small for very high temperatures.

Triangular and Honeycomb Lattices: Low Temperature

The dimensionless free energies of the triangular and honeycomb lattices can now be obtained from (11.7.11) and (11.7.12). Define x and z_j by (11.8.7a), for $j = 1, 2, 3$. From (11.7.4) and (11.5.13), l and k, and hence q and x, are independent of j. Also, from (11.1.12) and (11.8.3), z_1, z_2, z_3 must satisfy the relation

$$z_1 z_2 z_3 = x^{-1}. \tag{11.8.10}$$

The main problem is to obtain a useful expression for $\ln R$ from (11.7.8). From (11.8.1b)

$$\sinh 2K_j = i(1 + k \operatorname{sn}^2 iu_j)/(2k^{\frac{1}{2}} \operatorname{sn} iu_j). \tag{11.8.11}$$

The moduli k and l are related by the Landen transformation (11.5.13). From (15.6.3), it follows that

$$\sinh 2K_j = i/[l \operatorname{sn}(i\hat{u}_j, l)], \tag{11.8.12}$$

where

$$\hat{u}_j = (1 + k)u_j. \tag{11.8.13}$$

Hence, from (11.7.4),

$$\sinh 2K_j' = -i \operatorname{sn}(i\hat{u}_j, l). \tag{11.8.14}$$

(This result can also be obtained directly from (11.8.1c). Comparing it and (11.8.12) with (7.8.5), we see that K_j, K_j', $\hat{u}_j$ correspond to the L, K, u of Chapter 7.)

Again we can use this infinite product expansion (15.1.5) – (15.1.6) of the sn function, only now the modulus is l rather than k. Using the relations (15.6.2), (11.8.3) and (11.8.4), it follows that

$$\sinh 2K_j' = \left(\frac{1}{lz_j}\right)^{\frac{1}{2}} \prod_{n=1}^{\infty} \frac{(1 - x^{4n-3}z_j)(1 - x^{4n-1}z_j^{-1})}{(1 - x^{4n-3}z_j^{-1})(1 - x^{4n-1}z_j)}. \tag{11.8.15}$$

Using this result in (11.7.8), together with (11.8.7) and (11.8.10), gives

$$R^2 \exp\left[2\sum_{j=1}^{3}(K_j' - K_j)\right]$$

$$= 2\left(\frac{x}{l}\right)^{\frac{1}{2}} \prod_{j=1}^{3} \prod_{n=1}^{\infty} \frac{(1 - x^{8n-5}z_j)^2 (1 - x^{8n-3}z_j^{-1})^2}{(1 - x^{8n-5}z_j^{-1})^2 (1 - x^{8n-3}z_j)^2}. \tag{11.8.16}$$

From (15.6.2) and (11.8.4), the nome of the modulus l is $q^{\frac{1}{2}} = x^2$. Using this modulus in (15.1.4a) therefore gives

$$2\left(\frac{x}{l}\right)^{\frac{1}{2}} = \prod_{n=1}^{\infty} \left(\frac{1 + x^{4n-2}}{1 + x^{4n}}\right)^2, \tag{11.8.17}$$

which can be re-arranged as

$$2\left(\frac{x}{l}\right)^{\frac{1}{2}} = \prod_{n=1}^{\infty} \frac{(1 - x^{8n-4})^4}{(1 - x^{8n-6})^2 (1 - x^{8n-2})^2}. \tag{11.8.18}$$

Taking logarithms of (11.8.18) and (11.8.16), Taylor expanding every term of the form $\ln(1 - x^{8n} \times \text{constant})$, then summing over n and comparing with (11.8.6), gives

$$\ln[2(x/l)^{\frac{1}{2}}] = 2\tau, \tag{11.8.19}$$

$$\ln R + \sum_{j=1}^{3}(K_j' - K_j) = \tau + \sum_{j=1}^{3}[g(z_j) - g(z_j^{-1})]. \tag{11.8.20}$$

Using this equation for $\ln R$, together with (11.8.7b) and (11.7.13), the dimensionless free energy of the triangular Ising model is

$$\psi_T(K_1', K_2', K_3')$$
$$= -K_1' - K_2' - K_3' - \tau + g(z_1) + g(z_2) + g(z_3), \tag{11.8.21}$$

where x, z_1, z_2, z_3 are defined in terms of K_1', K_2', K_3' by (11.8.10) and the second equation in (11.8.7a), i.e.

$$z_1 z_2 z_3 = x^{-1}, \quad \exp(-2K_j') = \phi(z_j^{-1}), \quad j = 1, 2, 3. \tag{11.8.22}$$

For the honeycomb lattice, using (11.7.14) and the first equation in (11.8.7a), we obtain

$$2\psi_H(K_1, K_2, K_3) = -K_1 - K_2 - K_3 - 2\tau$$
$$+ g(z_1^{-1}) + g(z_2^{-1}) + g(z_3^{-1}), \tag{11.8.23}$$

where

$$z_1 z_2 z_3 = x^{-1}, \quad \exp(-2K_j) = \phi(z_j), \quad j = 1, 2, 3. \tag{11.8.24}$$

The inequalities (11.8.8) must still be satisfied; x and l are small at very low temperatures and increase to one at the critical temperature.

For an isotropic system, $z_1 = z_2 = z_3 = x^{-\frac{1}{4}}$. The equation (11.8.22), or (11.8.24), then reduces to a single equation for x.

Triangular and Honeycomb Lattices: High Temperature

Above the critical temperature, the modulus l defined by (11.7.5) or (11.7.6) is greater than one. Similarly to the square lattice, the easiest way to handle this case is to apply the duality transformation (6.3.7) to the above low-temperature results. For the triangular lattice this gives

$$\psi_T(K_1', K_2', K_3') = -\ln[2 \cosh K_1' \cosh K_2' \cosh K_3']$$
$$- 2\tau + g(z_1^{-1}) + g(z_2^{-1}) + g(z_3^{-1}), \quad (11.8.25)$$

where z_1, z_2, z_3, x are defined by (11.8.8) and

$$z_1 z_2 z_3 = x^{-1}, \quad \tanh K_j' = \phi(z_j), \quad j = 1, 2, 3. \quad (11.8.26)$$

For the honeycomb lattice

$$2\psi_H(K_1, K_2, K_3) = -\ln[4 \cosh K_1 \cosh K_2 \cosh K_3]$$
$$- \tau + g(z_1) + g(z_2) + g(z_3), \quad (11.8.27)$$

where

$$z_1 z_2 z_3 = x^{-1}, \quad \tanh K_j = \phi(z_j^{-1}), \quad j = 1, 2, 3. \quad (11.8.28)$$

The parameter x is small at very high temperatures. Again $z_1 = z_2 = z_3 = x^{-\frac{1}{4}}$ for an isotropic system.

Magnetization

From (11.5.13), (15.6.2) and (11.8.4), the nome corresponding to the modulus l is $q^{\frac{1}{2}} = x^2$. Using (15.1.4b), the expression (11.7.30) for the spontaneous magnetization is therefore equivalent to

$$M_0 = \prod_{n=1}^{\infty} \frac{1 - x^{4n-2}}{1 + x^{4n-2}} \text{ if } |l| \leqslant 1,$$
$$= 0 \qquad \text{if } |l| > 1, \quad (11.8.29)$$

which agrees with (10.10.19).

This result applies to the square, triangular and honeycomb lattices. For $|l| \leq 1$, i.e. below the critical temperature, x is defined by (11.8.7), (11.8.22) or (11.8.24).

Combined Formulae for all Three Planar Lattices

The above results for the square, triangular and honeycomb lattices can all be combined into a simple form. It is intriguing that this should be so.

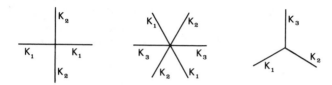

Fig. 11.10. The Ising model interaction coefficients for the square, triangular and honeycomb lattices, as used in (11.8.32)–(11.8.43).

Let $\bar{q}$ be the coordination number of the lattice, i.e. the number of neighbours of any site. For the square, triangular and honeycomb lattices, $\bar{q} = 4, 6$ and 3, respectively. Let $K_1, \ldots, K_{\bar{q}}$ be the Ising model interaction coefficients associated with the $\bar{q}$ edges at a site, as indicated in Fig. 11.10.

Note that this notation differs from the previous sub-sections: for the square lattice K_1 and K_2 replace the K_j and K_j' of (11.8.7) and (11.8.9), and $K_3 = K_1$, $K_4 = K_2$; for the triangular lattice, K_1, K_2, K_3 replace the K_1', K_2', K_3' of (11.8.22) and (11.8.36), and $K_4 = K_1$, $K_5 = K_2$, $K_6 = K_3$. For the honeycomb lattice there is no change.

With each K_r associate a parameter w_r, defined as follows:

Square lattice : $w_1, \ldots, w_4 = z_j, z_j^{-1}, z_j, z_j^{-1}$;

Triangular lattice : $w_1, \ldots, w_6 = z_1^{-1}, z_2^{-1}, z_3^{-1}, z_1^{-1}, z_2^{-1}, z_3^{-1}$;

Honeycomb lattice : $w_1, w_2, w_3 = z_1, z_2, z_3$.   (11.8.30)

Here z_j is the z_j in (11.8.7) and (11.8.9); z_1, z_2, z_3 are the z_1, z_2, z_3 of (11.8.22) – (11.8.28). The inequalities (11.8.8) therefore imply that

$$0 < x < 1, \quad x < w_r < x^{-1}, \quad r = 1, \ldots, \bar{q} .   (11.8.31)$$

The *low-temperature* $(l < 1)$ results (11.8.7), (11.8.21) – (11.8.24) can now all be written as

$$w_1 w_2 \ldots w_{\bar{q}} = x^{\bar{q}-4}, \quad \exp(-2K_r) = \phi(w_r), \quad r = 1, \ldots, \bar{q} ,   (11.8.32a)$$

$$\psi = -\tfrac{1}{2}\sum_r K_r - \tau + \tfrac{1}{2}\sum_r g(w_r^{-1})\,, \qquad (11.8.32b)$$

the sums being over $r = 1, \ldots, \bar{q}$ and τ, $\phi(z)$, $g(z)$ being defined by (11.8.6). Given $K_1, \ldots, K_{\bar{q}}$, the equations (11.8.31) and (11.8.32a) define $w_1, \ldots, w_{\bar{q}}$ and x. The dimensionless free energy per site of the lattice (square, triangular or honeycomb) is then given by (11.8.32b). The magnetization is again given by (11.8.29).

Similarly, the *high-temperature* $(l > 1)$ results (11.8.9), (11.8.25) – (11.8.28) can all be written as

$$w_1 w_2 \ldots w_{\bar{q}} = x^{\bar{q}-4}, \quad \tanh K_r = \phi(w_r^{-1})\,, \qquad (11.8.33a)$$

$$\psi = -\ln 2 - \tfrac{1}{2}\sum_r \cosh K_r + \tau + \tfrac{1}{2}\sum_r [g(w_r) - \tau]\,. \qquad (11.8.33b)$$

Similarity to the Bethe Lattice Formulae

These results for the anisotropic planar lattices are very similar in form to those for the anisotropic Bethe lattice of Section 4.9.

In fact, if in Section 4.9 we set $h = 0$ and replace t, x_r by x^2, $(x/w_r)^{\frac{1}{2}}$, respectively, then the Bethe lattice free-energy results (4.9.4), (4.9.6) become exactly (11.8.32), but with the definitions (11.8.6) replaced by

$$\tau = \ln(1 + x^2)\,,$$

$$\phi(z) = \left(\frac{x}{z}\right)^{\frac{1}{2}} \frac{1 - xz}{1 - x^2}\,, \qquad (11.8.34)$$

$$g(z) = \ln\frac{1 - x^4}{1 - x^3/z}\,.$$

In both (11.8.6) and (11.8.34) it is true that

$$\phi(x) = 1\,, \quad \phi(x^{-1}) = g(x^{-1}) = 0\,, \quad g(x) = \tau\,. \qquad (11.8.35)$$

The equation (4.9.5) for the magnetization becomes

$$M = (1 - x^2)/(1 + x^2)\,. \qquad (11.8.36)$$

This is not the same as (11.8.29), but again M is a function only of x^2.

For a ferromagnetic Bethe lattice model below its critical temperature, the inequalities (11.8.31) are also satisfied. [Above its critical temperature $x = w_1 = \ldots = w_{\bar{q}} = 1$, and the Bethe lattice model becomes rather trivial: ψ is exactly $-\ln 2 - \tfrac{1}{2}\Sigma \cosh K_r$, and M of course is zero.]

It is fascinating that there should be these correspondences between zero-field anisotropic Ising models on two-dimensional lattices, and on the infinite-dimensional Bethe lattice. I have used short series expansions to look for similar properties for three-dimensional models, and for planar models in a field, but with no success.

Critical Temperature

For the ferromagnetic Ising models on both the planar and Bethe lattices, the critical point occurs when $x, w_1, \ldots, w_{\bar{q}}$ differ infinitesimally from one. Set

$$x = \exp(-\delta), \quad w_r = \exp(-\alpha_r) \,. \tag{11.8.37}$$

Then from (11.8.6), (15.1.5) and (15.1.6), the planar function $\phi(w_r)$ is

$$\phi(w_r) = -i\,k^{\frac{1}{2}} \operatorname{sn}[I(\alpha_r + \delta)i/\pi] \,, \tag{11.8.38}$$

where k is the modulus of this section, with nome $q = x^4$. Thus

$$\pi I'/4I = \delta \,. \tag{11.8.39}$$

When $x = 1$, then $k = 1$ and $I' = \frac{1}{2}\pi$. Using (15.7.3a) and (11.8.39), (11.8.38) becomes

$$\phi(w_r) = \tan[\pi(\alpha_r + \delta)/8\delta] \,. \tag{11.8.40a}$$

This is the form of the planar function $\phi(w_r)$ in the limit when $\delta \to 0$, α_r/δ being kept constant.

On the other hand, for the Bethe lattice it is easily verified from (11.8.34) and (11.8.37) that in this limit

$$\phi(w_r) = (\alpha_r + \delta)/2\delta \,. \tag{11.8.40b}$$

In either case, (11.8.32a) must be satisfied, so

$$\alpha_1 + \ldots + \alpha_{\bar{q}} = (\bar{q} - 4)\delta \,, \tag{11.8.41a}$$

$$\exp(-2K_r) = \phi(w_r), \quad r = 1, \ldots, \bar{q} \,. \tag{11.8.41b}$$

Solving (11.8.41b) and (11.8.40) for α_r, then substituting into (11.8.41a), gives the following conditions for criticality for the Ising model on a lattice of coordination number $\bar{q}$, with interaction coefficients $K_1, \ldots, K_{\bar{q}}$:

$$\text{planar: } \operatorname{artan}(\zeta_1) + \ldots + \operatorname{artan}(\zeta_{\bar{q}}) = \pi(\bar{q} - 2)/4 \,, \tag{11.8.42}$$

$$\text{Bethe: } \zeta_1 + \ldots + \zeta_{\bar{q}} = \bar{q} - 2 \,,$$

where

$$\zeta_r = \exp(-2K_r), \quad r = 1, \ldots, \bar{q} \,. \tag{11.8.43}$$

For any particular planar lattice, (11.8.42) can be simplified. Let K_1, K_2 (K_1, K_2, K_3) be the interaction coefficients of the square (triangular or honeycomb) lattice. Then the criticality conditions are:

Square: $\zeta_1\zeta_2 + \zeta_1 + \zeta_2 = 1$,

Triangular: $\zeta_2\zeta_3 + \zeta_3\zeta_1 + \zeta_1\zeta_2 = 1$,   (11.8.44)

Honeycomb: $\zeta_1\zeta_2\zeta_3 - \zeta_2\zeta_3 - \zeta_3\zeta_1 - \zeta_1\zeta_2 - \zeta_1 - \zeta_2 - \zeta_3 + 1 = 0$,

For an *isotropic* lattice model, with all interaction coefficients equal to a common value K, the critical values of $\zeta = \exp(-2K)$ are, from (11.8.42),

$$\text{Planar: } \zeta = \tan[\pi(\bar{q} - 2)/4\bar{q}],$$ (11.8.45)

$$\text{Bethe: } \zeta = (\bar{q} - 2)/\bar{q}.$$

For the particular planar lattices these values are:

$$\text{Square: } \bar{q} = 4,\ \zeta = \sqrt{2} - 1 = 0.414214$$

$$\text{Triangular: } \bar{q} = 6,\ \zeta = 1/\sqrt{3} = 0.577350 \qquad (11.8.46)$$

$$\text{Honeycomb: } \bar{q} = 3,\ \zeta = 2 - \sqrt{3} = 0.267949.$$

The corresponding critical values of $K = J/k_BT$ are given in Table 11.2, together with numerical estimates for the three-dimensional lattices (Sykes *et al.* 1972; Gaunt and Sykes, 1973), and with the Bethe lattice values.

Table 11.2. Critical values of $K = J/k_BT$ for the Ising model on various lattices: $\bar{q}$ is the coordination number.

$\bar{q}$	Planar	Three-dimensional	Bethe
3	0.658479 (honeycomb)		0.549306
4	0.440687 (square)	0.36979 (diamond)	0.346574
6	0.274653 (triangular)	0.22169 (simple cubic)	0.202733
8		0.15741 (BCC)	0.143841
12		0.10209 (FCC)	0.091161

Since the Bethe lattice is infinite-dimensional (in the sense given in Section 4.2), for a given coordination number the three dimensional estimates should lie between the planar and the Bethe lattice results. They do.

The results (11.8.42) – (11.8.45) apply to a ferromagnetic Ising model, with all interaction coefficients non-negative. Any square, honeycomb or Bethe model can be mapped into this regime by reversing appropriate alternating layers of spins: in this way any interaction coefficient can be negated. For the triangular lattice only pairs of interaction coefficients can be negated: there are four critical surfaces in $(\zeta_1, \zeta_2, \zeta_3)$ space, namely that given in (11.8.44), and three others obtained from it by inverting any two of ζ_1, ζ_2, ζ_3.

11.9   Thirty-Two Vertex Model

An obvious generalization of the ice-type, or six-vertex, models on the square lattice is to place arrows on the edges of the triangular lattice so that at each site there are three arrows pointing in, and three pointing out. There are then 20 possible arrangements of arrows at a vertex, so this is known as the 'twenty-vertex' model. With each arrangement one associates a weight ω_j, where $j = 1, \ldots, 20$. The partition function is then

$$Z = \sum_C \prod_i \omega_{j(i,C)},$$ (11.9.1)

where the sum is over all allowed arrangements C of arrows on the edges of the triangular lattice, the product is over all sites i, and $j(i, C)$ is the arrow arrangement at site i for configuration C. This model has not been solved in general: only when certain conditions are satisfied by the weights (Baxter, 1969; Kelland, 1974a, b).

In the same way as one can generalize the six-vertex model to the eight-vertex by allowing the number of arrows into each vertex to be 0 or 4, as well as 2, so the twenty-vertex model can be generalized by allowing any odd number of arrows into each vertex. There are then 32 possible arrow arrangements at a vertex.

The arrows can be represented by bonds: leave an edge empty if the corresponding arrow points generally to the right (in the sense that the arrows in Fig. 11.1 point generally to the right), place a bond on the edge if the arrow points to the left. There are then an even number of bonds incident to each site. The 32 possible arrangements of bonds at a vertex are shown in Fig. 11.11. Also shown are their respective weights $\omega, \ldots, \bar{\omega}_{16}$.

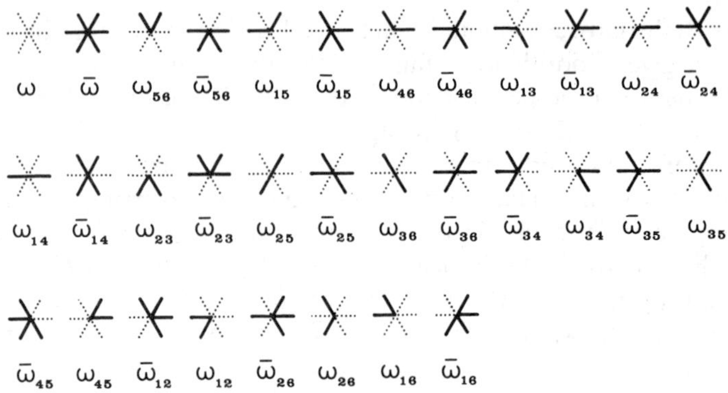

ω $\quad$ $\bar{\omega}$ $\quad$ ω_{56} $\quad$ $\bar{\omega}_{56}$ $\quad$ ω_{15} $\quad$ $\bar{\omega}_{15}$ $\quad$ ω_{46} $\quad$ $\bar{\omega}_{46}$ $\quad$ ω_{13} $\quad$ $\bar{\omega}_{13}$ $\quad$ ω_{24} $\quad$ $\bar{\omega}_{24}$

ω_{14} $\quad$ $\bar{\omega}_{14}$ $\quad$ ω_{23} $\quad$ $\bar{\omega}_{23}$ $\quad$ ω_{25} $\quad$ $\bar{\omega}_{25}$ $\quad$ ω_{36} $\quad$ $\bar{\omega}_{36}$ $\quad$ $\bar{\omega}_{34}$ $\quad$ ω_{34} $\quad$ $\bar{\omega}_{35}$ $\quad$ ω_{35}

$\bar{\omega}_{45}$ $\quad$ ω_{45} $\quad$ $\bar{\omega}_{12}$ $\quad$ ω_{12} $\quad$ $\bar{\omega}_{26}$ $\quad$ ω_{26} $\quad$ ω_{16} $\quad$ $\bar{\omega}_{16}$

Fig. 11.11. The 32 allowed bond arrangements at a vertex of the triangular lattice, and their associated weights.

Free-Fermion Case

Sacco and Wu (1975) considered this model and showed that it can be solved by the Pfaffian method of Section 7.13 provided that

$$\omega\bar{\omega} = \omega_{12}\bar{\omega}_{12} - \omega_{13}\bar{\omega}_{13} + \omega_{14}\bar{\omega}_{14} - \omega_{15}\bar{\omega}_{15} + \omega_{16}\bar{\omega}_{16}, \quad (11.9.2a)$$

and

$$\omega\bar{\omega}_{mn} = \omega_{ij}\omega_{kl} - \omega_{ik}\omega_{jl} + \omega_{il}\omega_{jk}, \quad (11.9.2b)$$

for all permutations i, j, k, l, m, n of $1, 2, \ldots, 6$ such that $m < n$ and $i < j < k < l$. There are 15 such permutations (corresponding to the 15 choices of m and n), and hence a total of 16 conditions.

Cases Reducible to the Solvable Kagomé Lattice Eight-Vertex Model

Another interesting class of solvable cases can be obtained from the Kagomé lattice eight-vertex model. In Fig. 11.1 the up-pointing triangles of the Kagomé lattice are drawn smaller than the down-pointing ones. Imagine this process continued until the up-pointing triangles become infinitesimal. The lattice then becomes the triangular lattice.

Each site of this triangular lattice is a 'city', consisting of three sites of the original Kagomé lattice. Such a city is shown in Fig. 11.4(a). The edges round it have spins $\alpha_1, \ldots, \alpha_6$ which are $+1$ if the edge is empty, -1 if it contains a bond. Summing over internal arrow or bond configurations within the triangle, the total Boltzmann weight of this city is the function of $\alpha_1, \ldots, \alpha_6$ occurring on the LHS of (11.1.6).

It follows that if $\omega, \ldots, \bar{\omega}_{16}$ are these total Boltzmann weights, for the appropriate values of $\alpha_1, \ldots, \alpha_6$, then the 32-vertex model is equivalent to the Kagomé lattice eight-vertex model. If the conditions (11.1.6), or equivalent (11.1.7), also hold, then the model can be solved as in Section 11.2–11.5.

Considering all 32 vertex arrangements, this will be so if

$$\omega_{25} = b_1 a_2 a_3 + c_1 d_2 d_3$$

$$\omega_{36} = a_1 b_2 a_3 + d_1 c_2 d_3$$

$$\omega = a_1 a_2 b_3 + d_1 d_2 c_3$$

$$\omega_{14} = c_1 c_2 c_3 + b_1 b_2 b_3$$

$$\omega_{15} = \omega_{24} = c_1 a_2 a_3 + b_1 d_2 d_3 = c_1 b_2 b_3 + b_1 c_2 c_3 \qquad (11.9.3a)$$

$$\omega_{46} = \omega_{13} = a_1 c_2 a_3 + d_1 b_2 d_3 = b_1 c_2 b_3 + c_1 b_2 c_3$$

$$\omega_{56} = \omega_{23} = a_1 a_2 c_3 + d_1 d_2 b_3 = b_1 b_2 c_3 + c_1 c_2 b_3$$

$$\omega_{12} = \omega_{45} = d_1 a_2 b_3 + a_1 d_2 c_3 = d_1 b_2 a_3 + a_1 c_2 d_3$$

$$\omega_{16} = \omega_{34} = b_1 d_2 a_3 + c_1 a_2 d_3 = a_1 d_2 b_3 + d_1 a_2 c_3$$

$$\omega_{35} = \omega_{26} = a_1 b_2 d_3 + d_1 c_2 a_3 = b_1 a_2 d_3 + c_1 d_2 a_3$$

and, for $1 \leqslant m < n \leqslant 6$,

$$\bar{\omega} = \omega, \quad \bar{\omega}_{mn} = \omega_{mn} . \qquad (11.9.3b)$$

If $a_1, \ldots, d_3$ can be found so as to satisfy (11.9.3), which includes the restrictions (11.1.7), then the spontaneous polarization of the 32-vertex model is given by (11.3.4), and from (11.4.5) its free energy per site, or vertex, is

$$f_{32V} = f(a_1, b_1, c_1, d_1) + f(a_2, b_2, c_2, d_2) + f(a_3, b_3, c_3, d_3) , \qquad (11.9.4)$$

where $f(a, b, c, d)$ is the free energy per site of the square lattice eight-vertex model with weights a, b, c, d, and $k_B T$ is here regarded as a constant, the same for each f.

From (11.1.8) – (11.1.12) it is apparent that there are only five degrees of freedom: k, u_1, u_2, u_3 and a single normalization factor for $\omega, \ldots, \bar{\omega}_{16}$. Thus (11.9.3) implies 27 restrictions on the weights of the 32-vertex model. Even so, this restricted model can still be interesting, as will be evident in the next section.

When $d_1 = d_2 = d_3 = 0$, the restricted model reduces to the solvable cases of the twenty-vertex model discussed by Kelland (1974b) and Baxter *et al*

(1978). In particular, these include the 'triangular KDP' model (Kelland, 1974a): this behaves very similarly to the square lattice ferroelectric model of Chapter 8, having a first order transition to a frozen ferroelectric state.

Formulation as an Honeycomb Lattice Ising Model with Multi-Spin Interactions

Like the eight-vertex model, the 32-vertex model can be regarded as an Ising model with multi-spin interactions. With each face m of the triangular lattice associate a spin σ_m, with values $+1$ or -1. As in Sections 10.3 and 11.5, establish a two-to-one correspondence between spin configurations $\{\sigma_1, \sigma_2, \dots\}$ and allowed arrow coverings of the lattice. This can be done by taking the arrow configuration in Fig. 11.8(b) to be the standard: if the spins on either side of an edge are equal (different), place an arrow on the edge pointing in the same (opposite) way as the standard. Do this for all edges. Then at each vertex, there must be an even number of non-standard arrows on the six incident edges, and hence an odd number of incoming (and outgoing) arrows.

Let $\sigma_1, \dots, \sigma_6$ be the six spins round a site, arranged as in Fig. 11.4(a), the triangle being shrunk to a point. Let $\overline{W}(\sigma_1, \dots, \sigma_6)$ be the function whose value for spin configuration $\sigma_1, \dots, \sigma_6$ is the weight of the corresponding arrow configuration at the site. Then (11.9.1) can be written as

$$Z = \tfrac{1}{2} \sum_\sigma \prod_i \overline{W}(\sigma_{1i}, \dots, \sigma_{6i}), \tag{11.9.5}$$

where the sum is over all values of the spins on the faces of the triangular lattice, the product is over all sites i, and $\sigma_{1i}, \dots, \sigma_{6i}$ are the face spins round site i.

Negating $\sigma_{1i}, \dots, \sigma_{6i}$ leaves unchanged the arrows into or out of site i. Thus $\overline{W}(\sigma_i, \dots, \sigma_6)$ is an even function, i.e.

$$\overline{W}(-\sigma_1, \dots, -\sigma_6) = \overline{W}(\sigma_1, \dots, \sigma_6). \tag{11.9.6}$$

Now use the dual lattice: the spins lie on the sites of the honeycomb lattice, the product in (11.9.5) is over all hexagonal faces i of this lattice, and $\sigma_{1i}, \dots, \sigma_{6i}$ are the six spins round face i. Thus the 32-vertex model is equivalent to an Ising-type model on the honeycomb lattice, with interactions between all six spins round each face. These interactions must be even, so that the face weight function $\overline{W}$ satisfies (11.9.6). This equivalence is quite general.

Now consider the solvable case of the 32-vertex model when the weights satisfy (11.9.3). One can of course obtain all the values of the function $\overline{W}$ by working back through the definitions of this subsection and using (11.9.3). For instance, $\overline{W}(+ , + , + , + , + , +)$ corresponds to the standard vertex configuration in Fig. 11.8(b). Replacing left-pointing arrows by bonds, from Fig. 11.11 this is the configuration with weight ω_{14}. Thus, using (11.9.3),

$$\overline{W}(+ , + , + , + , + , +) = \omega_{14} = c_1 c_2 c_3 + b_1 b_2 b_3. \quad (11.9.7a)$$

Other examples are

$$\overline{W}(- , + , + , + , + , +) = \omega_{46} = a_1 c_2 a_3 + d_1 b_2 d_3 \quad (11.9.7b)$$
$$\overline{W}(- , - , + , + , + , -) = \bar{\omega}_{36} = a_1 b_2 a_3 + d_1 c_2 d_3 .$$

More directly, we can remember that this case of the 32-vertex model is obtained by shrinking to points the up-pointing triangles of the Kagomé lattice. Doing this in the multi-spin Ising model formulation of Section 11.5, we obtain at once the Ising spin formulation of the 32-vertex model, the spins being those that remain in Section 11.5 after summing over those inside up-pointing triangles. The weight function $\overline{W}(\sigma_1 , \ldots , \sigma_6)$ is simply the total weight of the triangle in Fig. 11.4(a), summed over the centre spin σ_7. This is simply the LHS of (11.5.8). Writing the Kagomé lattice weight functions W_1, W_2, W_3 in the form (11.5.5), and using (11.5.10), it follows that

$$\overline{W}(\sigma_1 , \ldots , \sigma_6) = 2M_1 M_2 M_3 \exp(K_1' \sigma_1 \sigma_3 + K_2' \sigma_3 \sigma_5 + K_3' \sigma_5 \sigma_1) \times$$
$$\cosh(K_1 \sigma_2 + K_2 \sigma_4 + K_3 \sigma_6 + K'' \sigma_1 \sigma_2 \sigma_3 + K'' \sigma_3 \sigma_4 \sigma_5 + K'' \sigma_5 \sigma_6 \sigma_1) . \quad (11.9.8)$$

This is the LHS of (11.5.9), multiplied by the constant $2M_1 M_2 M_3$.

If $\overline{W}$ is given by (11.9.8), then the 32-vertex model is equivalent to the general Kagomé lattice eight-vertex model: they have the same partition function and, since neighbouring Ising spins σ_m, σ_n in the former are neighbouring Ising spins in the latter, the same spontaneous magnetization $\langle \sigma_m \rangle$ and polarization $\langle \sigma_m \sigma_n \rangle$.

If the restrictions (11.5.11) are also satisfied (which means that the LHS of (11.5.8) or (11.5.9) is equal to the RHS), then the Kagomé lattice model is the one solved in Sections 11.2 – 11.5. The total free energy, spontaneous magnetization and polarization of the 32-vertex model are therefore then given by (11.4.5), (11.5.18) and (11.3.4), respectively, where N is the number of vertices and $a_1 , \ldots , d_3$ are given by (11.5.6).

11.10   Triangular Three-Spin Model

Historical Introduction

The solution of the eight-vertex model (Baxter, 1971a, 1972b) excited interest in models with multi-spin interactions, notably the three-spin model on the triangular lattice.

In this model, at each site i of the triangular lattice there is a spin σ_i, with values $+1$ or -1. The energy of a given spin configuration is

$$E = -J \sum \sigma_i \sigma_j \sigma_k \,, \qquad (11.10.1)$$

where the sum is over all triangular faces (both up-pointing and down-pointing) of the triangular lattice. From (1.4.1), the partition function is

$$Z = \sum_{\sigma} \exp[K \sum \sigma_i \sigma_j \sigma_k] \,, \qquad (11.10.2)$$

where

$$K = J/k_B T. \qquad (11.10.3)$$

The dimensionless free energy per site is

$$\psi(K) = - \lim_{N \to \infty} N^{-1} \ln Z \,, \qquad (11.10.4)$$

where N is the number of sites of the lattice.

The triangular lattice can be divided into three sub-lattices A, B, C, as in Fig. 11.8(b), so that any triangular face (i, j, k) contains one site of type A, one of type B, and one of type C. From (11.10.1) it is obvious that negating all spins on just one sub-lattice is equivalent to negating J, and hence K. Without loss of generality we can therefore take K to be non-negative.

Wood and Griffiths (1972), and Merlini and Gruber (1972), considered this model and showed that it satisfies the duality relation

$$\psi(K^*) = 2K + \psi(K) - \ln(2 \cosh^2 K^*) \,, \qquad (11.10.5a)$$

where

$$\tanh K^* = \exp(-2K). \qquad (11.10.5b)$$

This is precisely the duality relation (6.2.14) of the square lattice isotropic Ising model (with $L = K$). The argument preceding (6.2.16) therefore applies: if there is just one critical point, then it must occur when $K = K_c$, where

$$\sinh 2K_c = 1, \quad K_c = 0.44068679 \dots. \qquad (11.10.6)$$

Griffiths and Wood (1973) used this argument and series expansions to estimate the critical exponents. They obtained $0.6 \leqslant \alpha' \leqslant 0.8$, $0.070 \leqslant \beta \leqslant 0.071$, $1.25 \leqslant \gamma' \leqslant 1.4$, and correctly guessed that $\alpha' = 2/3$. Their estimates of β and γ' were somewhat out: β is actually $1/12$ and (assuming scaling) γ' is $7/6$.

Suppose that a magnetic field term $-H \Sigma \sigma_j$ is added to the energy (11.10.1). Let $\langle \sigma_i \rangle_{N,H}$ be the average of a spin σ_i, evaluated as in (1.4.4) for a finite lattice of N sites in the presence of the field H, and let

$$\langle \sigma_i \rangle = \lim_{H \to 0^+} \lim_{N \to \infty} \langle \sigma_i \rangle_{N, H} , \qquad (11.10.7)$$

where the limit $N \to \infty$ means that the lattice becomes large in all directions.

This is the zero-field magnetization. Like the Ising and eight-vertex magnetization, it must be zero at sufficiently high temperatures.

This is not immediately obvious: the energy (11.10.1) is not unchanged by reversing all spins, so the usual Ising model argument of Section 1.7 (that M is an odd function of H, continuous for sufficiently high temperatures) does not apply. Instead, note from (11.10.1) that, when $H = 0$, E is unchanged by negating all spins on any two of the sub-lattices A, B, C.

Let σ_A, σ_B, σ_C denote all the spins on the A, B, C sub-lattices, respectively. Then to any total configuration $(\sigma_A, \sigma_B, \sigma_C)$ of spins there correspond three others that can be obtained by negating all spins on two sub-lattices. Thus the spin configurations can be grouped in equal-energy sets of four:

$$(\sigma_A , \sigma_B , \sigma_C) , (\sigma_A , -\sigma_B , -\sigma_C) , (-\sigma_A , \sigma_B , -\sigma_C) , (-\sigma_A , -\sigma_B , \sigma_C) .$$

$$(11.10.8)$$

For any single spin σ_i, the sum of its values over four such configurations is clearly zero. For a finite lattice, grouping configurations in such sets of four, it must therefore be true that

$$\langle \sigma_i \rangle_{N,0} = 0 . \qquad (11.10.9)$$

At sufficiently high temperatures the limits in (11.10.7) can be interchanged; for finite N, $\langle \sigma_i \rangle_{N,H}$ is a continuous function of H, so from (11.10.9)

$$\langle \sigma_i \rangle = 0 \text{ for sufficiently high temperatures} . \qquad (11.10.10a)$$

On the other hand, Merlini et al (1973), and Merlini (1973), used an argument due to Peierls (Peierls, 1936; Griffiths, 1972) to obtain lower bounds for $\langle \sigma_i \rangle_{N,H}$. They thereby showed that

$$\langle \sigma_i \rangle > 0 \text{ for sufficiently low temperatures} , \qquad (11.10.10b)$$

i.e. there is a non-zero spontaneous magnetization $\langle \sigma_i \rangle$. This proves that there must be a critical point: a temperature T_c at which $\langle \sigma_i \rangle$ just ceases to be zero as the temperature is decreased.

Very similar remarks apply to $\langle \sigma_i \sigma_j \rangle$, where i and j are nearest neighbours, evaluated by the double limiting procedure of (11.10.7). This must be zero for sufficiently high temperatures, and is expected to be non-zero for sufficiently low temperatures. By analogy with the multi-spin formulation of the eight-vertex model given in Section 10.3 and (10.10.22), it is convenient to call $\langle \sigma_i \sigma_j \rangle$ the 'polarization' of the three-spin model.

Baxter and Wu (1973, 1974) calculated the free energy of the three-spin model (with $H = 0$) directly, using the transfer matrix method and a generalized Bethe ansatz for the eigenvectors. Baxter *et al.* (1975) used series expansions to conjecture the exact expressions for the spontaneous magnetization $\langle \sigma_i \rangle$ and polarization $\langle \sigma_i \sigma_j \rangle$.

Baxter and Enting (1976) noted that these results were exactly the same as a particular eight-vertex model, and that the eight-vertex model also has a four-fold symmetry between spin configurations. Guided by this, they found a transformation of the triangular three-spin model into a square-lattice eight-vertex model. All the three-spin results could then be seen to be consequences of the eight-vertex ones.

Later, I was also able to show that the three-spin model is a special case of the solvable Kagomé lattice eight-vertex model (Baxter, 1978a). This equivalence is much simpler than the Baxter – Enting one, and is the one used in this section. From this point of view, the Baxter – Enting transformation provides a way of relating these particular square and Kagomé eight-vertex models: an alternative way to that of Section 11.2.

Equivalence to a Kagomé Lattice Eight-Vertex Model

In (11.10.2), perform the sum over all spins on one sub-lattice, say C. This can readily be done because each such spin interacts only with spins on sub-lattices A and B. This gives

$$Z = \sum_{\sigma_A, \sigma_B} \prod_i \overline{W}(\sigma_{1i}, \dots, \sigma_{6i}) , \qquad (11.10.11)$$

where

$$\overline{W}(\sigma_1, \dots, \sigma_6) = 2 \cosh K(\sigma_1\sigma_2 + \sigma_2\sigma_3 + \sigma_3\sigma_4 + \sigma_4\sigma_5 + \sigma_5\sigma_6 + \sigma_6\sigma_1) .$$

$$(11.10.12)$$

The sum in (11.10.11) is over all spins on the A and B sublattices. Taken together, these form an honeycomb lattice, as is evident in Fig. 11.8(b).

The product is over all faces i of this honeycomb lattice, $\sigma_{1i}, \ldots, \sigma_{6i}$ are the six spins round face i.

Apart from a factor of $\frac{1}{2}$, which is irrelevant for a large lattice, (11.10.11) is the same as the partition function (11.9.5) of the 32-vertex model. Further, a straightforward direct calculation, using the fact that $\sigma_1, \ldots, \sigma_6$ only have values $+1$ or -1, reveals that (11.10.12) can equivalently be written as

$$\overline{W}(\sigma_1, \ldots, \sigma_6) = 2 \cosh K(\sigma_2 + \sigma_4 + \sigma_6 + \sigma_1\sigma_2\sigma_3 + \sigma_3\sigma_4\sigma_5 + \sigma_5\sigma_6\sigma_1).$$

$$(11.10.13)$$

But this is precisely the function $\overline{W}$ given by (11.9.8), with

$$M_i = 1, \ K_i' = 0, \ K_i = K'' = K, \qquad (11.10.14)$$

for $i = 1, 2, 3$. Thus the three-spin model is equivalent to the Kagomé lattice eight-vertex model of Section 11.5, with these values of M_i, K_i, K_i', K''. It is interesting to note that these values are quite special: the triangular lattice edge interactions in (11.5.7) now vanish, and the remaining two- and four-spin interaction coefficients all have the same value.

The restrictions (11.5.11) are automatically satisfied, so from the remarks at the end of the previous section, the free energy per site of the three-spin model is

$$f_{3\mathrm{spin}} = f(a, b, c, d), \qquad (11.10.15)$$

where $f(a, b, c, d)$ is the free energy per site of the square-lattice eight-vertex model with weights a, b, c, d given by (11.5.6) and (11.10.14) for any particular value of i, i.e.

$$a, b, c, d = 1, \exp(-2K), \exp(2K), 1. \qquad (11.10.16)$$

Similarly, the spontaneous magnetization and polarization are

$$\langle \sigma_m \rangle = M_0(a, b, c, d),$$

$$\langle \sigma_m \sigma_n \rangle = P_0(a, b, c, d), \qquad (11.10.17)$$

where m and n are neighbouring sites.

The triangular three-spin model is therefore equivalent to the square-lattice eight-vertex model with weights a, b, c, d, in that $f, \langle \sigma_m \rangle, \langle \sigma_m \sigma_n \rangle$ are the same for both. Further, let $\sigma_1, \ldots, \sigma_m$ be any set of spins that all lie on the heavy vertical zig-zag line in Fig. 11.8(b). Then by using (11.5.17) and the above arguments, it is quite easy to show that $\langle \sigma_1 \ldots \sigma_m \rangle$ is the same for the three-spin model as for the eight-vertex model on the square lattice formed by removing all horizontal edges in Fig. 11.8(b). Thus the two models have the same correlation length ξ in the direction of the zig-zag line.

Ordered Phase

The properties f, M_0, P_0 of the square-lattice eight-vertex model can be obtained from Chapter 10. If $K > K_c$, where K_c is the critical value of K in (11.10.6), then from (11.10.16)

$$\sinh 2K > 1, \quad c > a + b + d. \tag{11.10.18}$$

Thus a, b, c, d lie in the principal regime (10.7.5), and the equations of Sections 10.4–10.10 can be used directly. No initial transformation of a, b, c, d is needed.

From (11.10.16), $ad = bc$ and $a = d$. From (10.4.21), this implies that

$$k\,\text{snh}^2(\lambda - u) = 1, \quad k\,\text{snh}\,\lambda\,\text{snh}\,u = 1. \tag{11.10.19}$$

From (10.7.1) and (10.4.23), k, λ and u are real, and $0 < u < \lambda < I'$. The elliptic function snh u defined by (10.4.20) is real, increases monotonically from 0 to ∞ as u increases from 0 to I', and from (15.2.5) and (15.2.6) it satisfies

$$\text{snh}(I' - u) = (k\,\text{snh}\,u)^{-1}. \tag{11.10.20}$$

From (11.10.19) it follows that

$$\lambda - u = \tfrac{1}{2}I', \quad \lambda + u = I', \tag{11.10.21}$$

so, using (10.4.23), (10.7.9) and (10.7.19),

$$\lambda = \tfrac{3}{4}I', \quad u = \tfrac{1}{4}I', \quad v = -\tfrac{1}{4}I', \tag{11.10.22}$$

$$x = q^{3/8}, \quad z = q^{-1/8}. \tag{11.10.23}$$

To relate these parameters to the interaction coefficient K, note from (10.4.21), (11.10.16) and (11.10.22) that

$$\exp(-2K) = \frac{b}{a} = \frac{\text{snh}\,u}{\text{snh}(\lambda - u)} = k^{\frac{1}{2}}\,\text{snh}(I'/4). \tag{11.10.24}$$

Define

$$p = q^{1/4}. \tag{11.10.25}$$

Then, using (10.4.20), (15.1.6) and (15.1.5) to expand $k^{\frac{1}{2}}\,\text{snh}(I'/4)$ as an infinite product, we obtain

$$\exp(-2K) = p^{\frac{1}{2}} \prod_{n=1}^{\infty} \frac{(1 - p^{8n-7})(1 - p^{8n-1})}{(1 - p^{8n-5})(1 - p^{8n-3})}, \tag{11.10.26}$$

which is an explicit relation betwen K and p.

The free energy is given by (10.8.47). Using (11.10.16), (11.10.23) and (11.10.25), this gives

$$-f/k_B T = 2K + \sum_{m=1}^{\infty} \frac{p^{3m}(1 - p^m)^3 (1 - p^{2m})}{m(1 - p^{8m})(1 + p^{3m})} . \qquad (11.10.27)$$

The summand can be written as the sum of two rational functions of p^m, having denominators $1 - p^{8m}$, $1 + p^{3m}$, respectively. Taylor expanding in powers of p^m, then summing over m and taking exponentials, gives

$$\exp(-f/k_B T) = \exp(2K) \prod_{n=1}^{\infty} \frac{(1 - p^{6n-3})(1 - p^{8n-4})^3 (1 - p^{8n})}{(1 - p^{6n})(1 - p^{8n-5})^2 (1 - p^{8n-3})^2} . \qquad (11.10.28)$$

For some purposes this form is more convenient than (11.10.27): it gives a power series in p in which all coefficients are integers.

From (10.10.19) and (10.10.24), using (11.10.23) and (11.10.25), the spontaneous magnetization and polarization are

$$M_0 = \prod_{n=1}^{\infty} \frac{1 - p^{6n-3}}{1 + p^{6n-3}} , \qquad (11.10.29)$$

$$P_0 = \prod_{n=1}^{\infty} \left(\frac{1 + p^{4n}}{1 - p^{4n}} \frac{1 - p^{3n}}{1 + p^{3n}} \right) . \qquad (11.10.30)$$

To summarize: if K is given, then p is defined by (11.10.26) subject to the inequality $0 < p < 1$, and f, M_0, P_0 are then given by (11.10.28) – (11.10.30) (Baxter *et al.* 1975).

It is possible to eliminate p and express $\exp(-f/k_B T)$, M_0, P_0 as algebraic functions of $\exp(-2K)$. The results are rather cumbersome and not particularly illuminating: they are given in Baxter and Wu (1973, 1974).

Disordered Phase

If $0 < K < K_c$, then a, b, c, d are not in the principal regime (10.7.5); rather they are in the disordered regime III of Section 10.11. In this regime there is no spontaneous magnetization or polarization, so

$$M_0 = P_0 = 0 . \qquad (11.10.31)$$

To obtain the free energy, we must use the rearrangement procedures (i)–(iii) of Section 10.11. These now imply the interchange of w_1 and w_4 in (10.2.16), which corresponds to replacing a, b, c, d by

$$a_r = \tfrac{1}{2}(a - b + c - d), \quad b_r = \tfrac{1}{2}(-a + b + c - d), \qquad (11.10.32)$$

$$c_r = \tfrac{1}{2}(a + b + c + d), \quad d_r = \tfrac{1}{2}(-a - b + c + d).$$

From (11.10.16) it follows that

$$a_r, b_r, c_r, d_r = \sinh 2K, 2\sinh^2 K, 2\cosh^2 K, \sinh 2K. \qquad (11.10.33)$$

Like a, b, c, d, these new weights satisfy $a_r d_r = b_r c_r$ and $a_r = d_r$. In fact (11.10.33) can be written as

$$(a_r, b_r, c_r, d_r) = \sinh 2K\,(1\,,\,\exp(-2K^*)\,,\,\exp(2K^*)\,,\,1)\,, \qquad (11.10.34)$$

where K^* is defined by (11.10.5b).

These are the same as the original Boltzmann weights (11.10.16), except that K therein has been replaced by K^*, and each is multiplied by $\sinh 2K$. The dimensionless free energy $\psi = f/k_B T$ is therefore

$$\psi(K) = -\ln \sinh 2K + \psi(K^*)\,, \qquad (11.10.35)$$

which is the duality relation (11.10.5a). Using this and the ordered-phase result (11.10.26), (11.10.28), it follows that if p is defined by

$$\tanh K = p^{\frac{1}{4}} \prod_{n=1}^{\infty} \frac{(1-p^{8n-7})(1-p^{8n-1})}{(1-p^{8n-5})(1-p^{8n-3})}\,,\ 0<p<1\,, \qquad (11.10.36)$$

then the free energy is given by

$$\exp(-f/k_B T) = 2\cosh^2 K \prod_{n=1}^{\infty} \frac{(1-p^{6n-3})(1-p^{8n-4})^3(1-p^{8n})}{(1-p^{6n})(1-p^{8n-5})^2(1-p^{8n-3})^2}. \qquad (11.10.37)$$

The parameters k, λ, u are now defined by (10.4.21), with a, b, c, d replaced by a_r, b_r, c_r, d_r. Since $a_r, \ldots, d_r$ differs from the $a, \ldots, d$ in (11.10.16) only by a normalization factor and the choice of K, the equations (11.10.19)–(11.10.23) remain valid. From (10.12.5) it follows that in both the ordered and disordered phases

$$\mu = \tfrac{3}{4}\pi, \quad w = -\tfrac{1}{4}\pi. \qquad (11.10.38)$$

Critical Behaviour

At $K = K_c$ the three-spin model has a critical point. Since f, M_0, P_0 are the same as for the square lattice eight-vertex model with $\mu = 3\pi/4$, from (10.12.24) the critical exponents α, α', β, β_e are

$$\alpha = \alpha' = \tfrac{2}{3}, \quad \beta = \beta_e = \tfrac{1}{12}. \qquad (11.10.39a)$$

The correlation length ξ is the same as the correlation length of the eight-vertex model in the diagonal direction. This is not the row or column correlation length of Chapter 10, but near the critical point it is expected that ξ diverges as in (1.7.25), with an exponent ν that is independent of the direction in which ξ is measured. Assuming this is so, then from (10.12.24) it must be true for the three-spin model that

$$\nu = \nu' = \tfrac{2}{3}. \tag{11.10.39b}$$

This agrees with the scaling relation (1.2.16). If one accepts the other scaling predictions, then the critical exponents γ, δ, η, and the interfacial tension exponent μ_s, are

$$\gamma = \tfrac{7}{6}, \quad \delta = 15, \quad \eta = \tfrac{1}{4}, \quad \mu_s = \tfrac{2}{3}. \tag{11.10.39c}$$

12

POTTS AND ASHKIN – TELLER MODELS

12.1 Introduction and Definition of the Potts Model

We have seen in Section 10.3 that the eight-vertex model is a generalization of the Ising model. There are of course an infinite number of other such generalizations. In this chapter I shall consider two of these: the Potts model and the Ashkin – Teller model, both in two dimensions. Neither has been solved exactly, but they can be expressed as staggered vertex models, and quite a lot is known about their critical behaviour.

R. B. Potts defined the former model in 1952, at the suggestion of C. Domb. He actually defined two models. The first is now known as the 'Z_N model', and supposes that at each site of a lattice there is a two-dimensional unit vector which can point in one of N equally spaced directions. Two adjacent vectors interact with interaction energy proportional to their scalar product.

The second model is the one that will be discussed here, and referred to simply as 'the Potts model'. This can be formulated on any graph $\mathcal{L}$, i.e. on any set of sites, and edges joining pairs of sites. For the sake of generality it is useful to do this: later on we shall specialize to the case when $\mathcal{L}$ is a two-dimensional lattice.

Let $\mathcal{L}$ have N sites, labelled $1, 2, \ldots, N$. With each site i associate a quantity σ_i which can take q values, say $1, 2, \ldots, q$. As in the Ising and eight-vertex model, let us call σ_i a 'spin'. Two adjacent spins σ_i and σ_j interact with interaction energy $-J\,\delta(\sigma_i, \sigma_j)$, where

$$\delta(\sigma, \sigma') = 1 \quad \text{if } \sigma = \sigma'$$
$$= 0 \quad \text{if } \sigma \neq \sigma' . \tag{12.1.1}$$

The total energy is therefore

$$E = -J \sum_{(i,j)} \delta(\sigma_i, \sigma_j), \tag{12.1.2}$$

where the summation is over all edges (i,j) of the graph. It follows from
(1.4.1) that the partition function is

$$Z_N = \sum_{\sigma} \exp\left\{ K \sum_{(i,j)} \delta(\sigma_i, \sigma_j)\right\}, \tag{12.1.3}$$

where

$$K = J/k_B T. \tag{12.1.4}$$

Here the σ-summation is over all values of all the spins $\sigma_1, \ldots, \sigma_N$. Thus
there are q^N terms in the summation.

For definiteness I have supposed that each σ_i takes the values
$1, \ldots, q$, but any q distinct numbers would be equally good. In particular,
for $q = 2$ we could let each σ_i take values $+1$ or -1. It is then true that
$\delta(\sigma, \sigma') = \frac{1}{2}(1 + \sigma\sigma')$; substituting this expression into (12.1.3) and comparing with (1.8.2), we see that the $q = 2$ Potts model (with K replaced
by $2K$) is equivalent to the zero-field Ising model.

In the next seven sections I shall show how the two-dimensional Potts
model can be solved at criticality (Temperley and Lieb, 1971; Baxter,
1973d; Baxter et al., 1976). There are a few other exactly solved cases:
$q = 1$ (which is trivial); $q = 2$ (the Ising model); the square-lattice model
with $q = 3$ and $K = -\infty$ (this is the three-colouring problem of Section
8.13); and the triangular-lattice model with $q = 4$ and $K = -\infty$, which is
a four-colouring problem (Baxter, 1970b).

12.2 Potts Model and the Dichromatic Polynomial

It has been shown (Kasteleyn and Fortuin, 1969; Fortuin and Kasteleyn,
1972; Baxter et al., 1976) that Z_N can be expressed as a dichromatic
polynomial (Tutte, 1967).

The argument is quite simple: set

$$v = \exp(K) - 1. \tag{12.2.1}$$

Then (12.1.3) can be written as

$$Z_N = \sum_{\sigma} \prod_{(i,j)} [1 + v\,\delta(\sigma_i, \sigma_j)]. \tag{12.2.2}$$

Let E be the number of edges of the graph $\mathscr{L}$. Then the summand in (12.2.2) is a product of E factors. Each factor is the sum of two terms (1 and $v\,\delta(\sigma_i, \sigma_j)$), so the product can be expanded as the sum of 2^E terms. Each of these 2^E terms can be associated with a bond-graph on $\mathscr{L}$. To do this, note that the term is a product of E factors, one for each edge. The factor for edge (i, j) is either 1 or $v\,\delta(\sigma_i, \sigma_j)$: if it is the former, leave the edge empty, if the latter, place a bond on the edge. Do this for all edges (i, j). We then have a one-to-one correspondence between bond-graphs on $\mathscr{L}$ and terms in the expansion of the product in (12.2.2).

Consider a typical graph G, containing l bonds and C connected components (regarding an isolated site as a component). Then the corresponding term in the expansion contains a factor v^l, and the effect of the delta functions is that all sites within a component must have the same spin σ. Summing over independent spins, it follows that this terms gives a contribution $q^C v^l$ to the partition function Z_N. Summing over all such terms, i.e. over all graphs G, we therefore have

$$Z_N = \sum_G q^C v^l. \tag{12.2.3}$$

The summation is over all graphs G that can be drawn on $\mathscr{L}$. The expression (12.2.3) is a dichromatic polynomial (Whitney, 1932; Tutte, 1967).

Note that q in (12.2.3) need not be an integer. We can allow it to be any positive real number, and this can be a useful generalization. For instance, regarding Z_N as a function of q and v (as well as N), we see that

$$\left(\frac{\partial}{\partial q}\ln Z\right)_{q=1} = \sum_G C\,v^l \bigg/ \sum_G v^l, \tag{12.2.4}$$

and this is just the mean number of components in the percolation problem, where each edge has probability $p = v/(1 + v)$ of being occupied. This is a famous unsolved problem (Essam, 1972).

If $K = -\infty$, then adjacent spins must be different, so from (12.1.3) it is apparent that Z_N is the number of ways of colouring the sites of $\mathscr{L}$ with q colours, no two adjacent sites having the same colour. This is a polynomial in q, known as the 'chromatic' polynomial. We see from the above that it is given by (12.2.3), with $v = -1$.

The edges of regular lattices can be grouped naturally into classes. For instance, the square lattice has edges which are either horizontal or vertical. It is then often convenient to generalize the Potts model so as to allow J (and hence K and v) to have different values, depending on the class to which the corresponding edge belongs. If J_r is the value of J for class r, then the required generalization of (12.2.3) is readily seen to be

$$Z_N = \sum_G q^C v_1^{l_1} v_2^{l_2} v_3^{l_3} \ldots, \tag{12.2.5}$$

where the summation is over all graphs G, C is the number of connected components in G, l_r is the number of bonds on edges of class r ($r = 1, 2, 3, \ldots$), and

$$K_r = J_r/k_B T, \quad v_r = \exp(K_r) - 1. \tag{12.2.6}$$

12.3 Planar Graphs: Equivalent Ice-Type Model

The Medial Graph $\mathscr{L}'$

The remarks of the previous two sections apply to any graph $\mathscr{L}$, whatever its structure or dimensionality. From now on let us specialize to $\mathscr{L}$ being a planar graph, i.e. one which can be drawn on a plane in such a way that no two edges cross one another, and no two sites coincide.

We can associate with $\mathscr{L}$ another graph $\mathscr{L}'$, as follows. Draw simple polygons surrounding each site of $\mathscr{L}$ so that:

(i) no polygons overlap, and no polygon surrounds another,
(ii) polygons of non-adjacent sites have no common corner,
(iii) polygons of adjacent sites i and j have one and only one common corner. This corner lies on the edge (i, j).

Let us take the corners of these polygons to be the sites of $\mathscr{L}'$, and the edges to be the edges of $\mathscr{L}'$. Hereinafter let us call these polygons the 'basic polygons' of $\mathscr{L}'$.

We see that there are two types of sites of $\mathscr{L}'$. Firstly, those common to two basic polygons. These lie on edges of $\mathscr{L}$ and have four neighbours in $\mathscr{L}'$. We term these 'internal' sites. Secondly, there can be sites lying on only one basic polygon. These have two neighbours and we term them 'external' sites. (The reason for this terminology will become apparent when we explicitly consider the regular lattices.)

The above rules do not determine $\mathscr{L}'$ uniquely, in that its shape can be altered, and external sites can be added on any edge. However, the topology of the linkages between internal sites is invariant, and the general argument of the following sections applies to any allowed choice of $\mathscr{L}'$. (For the regular lattices there is an obvious natural choice.) The graph $\mathscr{L}'$ is known as the 'medial' graph of $\mathscr{L}$ (Ore, 1967, pp. 47 and 124): a typical example is shown in Fig. 12.1.

It is helpful to shade the interior of each basic polygon, as in Fig. 12.1, and to regard such shaded areas as 'land', unshaded areas as 'water'. Then $\mathscr{L}'$ consists of a number of 'islands'. Each island contains a site of $\mathscr{L}$. Islands touch on edges of $\mathscr{L}$, at internal sites of $\mathscr{L}'$.

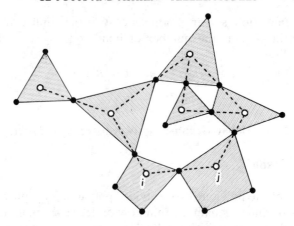

Fig. 12.1. A graph $\mathcal{L}$ (open circles and broken lines) and its medial graph $\mathcal{L}'$ (full circles and solid lines). The interior of each basic polygon is shaded, denoting 'land'.

Polygon Decompositions of $\mathcal{L}'$

We now make a one-to-one correspondence between graphs G on $\mathcal{L}$ and decompositions of $\mathcal{L}'$ as follows.

If G does not contain a bond on an edge (i, j), then at the corresponding internal site of $\mathcal{L}'$ separate two edges from the other two so as to separate the islands i and j, as in Fig. 12.2(a). If G contains a bond, separate the edges so as to join the islands, as in Fig. 12.2(b). Do this for all edges of $\mathcal{L}$.

The effect of this is to decompose $\mathcal{L}'$ into a set of disjoint polygons, an example being given in Fig. 12.3. (We now use 'polygon' to mean any simple closed polygonal path on $\mathcal{L}'$).

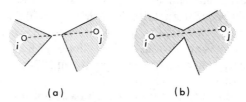

(a)   (b)

Fig. 12.2. The two possible separations of the edges at an internal site of $\mathcal{L}'$ (lying on the edge (i, j) of $\mathcal{L}$). The first represents no bond between i and j, the second a bond.

Clearly any connected component of G now corresponds to a large island in $\mathcal{L}'$, made up of basic islands joined together. Each such large island will have an outer perimeter, which is one of the polygons into which $\mathcal{L}'$ is decomposed. A large island may also contain lakes within; these correspond to faces of G and also have a polygon as outer perimeter.

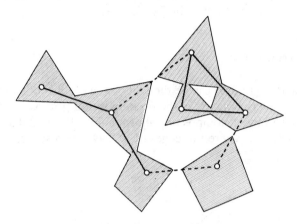

Fig. 12.3. A graph G on $\mathcal{L}$ (full lines between open circles represent bonds), and the corresponding polygon decomposition of $\mathcal{L}'$. To avoid confusion at internal sites, sites of $\mathcal{L}'$ are not explicitly indicated, but are to be taken to be in the same positions as in Fig. 12.1.

Each polygon is of one of these two types. Thus $\mathcal{L}'$ is broken into p polygons, where

$$p = C + S , \qquad (12.3.1)$$

C being the number of connected components in G, and S the number of internal faces.

The graph G has N sites and l bonds (where $l = l_1 + l_2 + l_3 + \ldots$) . The numbers C, S, N, l are not independent, but must satisfy Euler's relation

$$S = C - N + l . \qquad (12.3.2)$$

(Ore, 1967 p. 48: ν_f therein is the total number of faces, including the external infinite face, so $\nu_f = S + 1$.)

Eliminating C and S from the equations (12.2.5), (12.3.1) and (12.3.2), it follows that

$$Z_N = q^{N/2} \sum_{pd} q^{p/2} x_1^{l_1} x_2^{l_2} x_3^{l_3} \ldots , \qquad (12.3.3)$$

where

$$x_r = q^{-\frac{1}{2}}v_r \,, \qquad\qquad (12.3.4)$$

and the suffix 'pd' means that the sum is now taken to be over all polygon decompositions of $\mathscr{L}'$. Here p is the number of distinct polygons in the decomposition, and l_r is the number of internal sites of class r where the edges have been separated as in Fig. 12.2(b).

Arrow Coverings of $\mathscr{L}'$

The summand in (12.3.3) can be thought of as a product of various factors: a factor $q^{\frac{1}{2}}$ for every polygon in the decomposition (for example, there are four polygons in the decomposition shown in Fig. 12.3), and a factor x_r for every site of class r where the edges of $\mathscr{L}'$ have been separated as in Fig. 12.2(b).

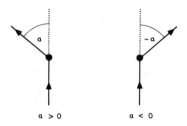

a > 0   a < 0

Fig. 12.4. Polygon corners of $\mathscr{L}'$ at which an observer moving in the direction of the arrows turns through an angle α to his left, or equivalently an angle $-\alpha$ to his right. Note that $-\pi < \alpha < \pi$, and the angle between the edges is $\pi - |\alpha|$.

The x_r factors are 'local', in that each depends only on what is happening at the appropriate site. The $q^{\frac{1}{2}}$ factors are not local in this sense, but we make them so by the following device.

Define quantities λ and z by the equations

$$q^{\frac{1}{2}} = 2 \cosh \lambda \,, \quad z = \exp(\lambda/2\pi) \,. \qquad (12.3.5)$$

Consider a polygon decomposition of $\mathscr{L}'$, such as that in Fig. 12.3. Each polygon is made up of edges of $\mathscr{L}'$, and has as many corners as it has edges. For instance, the polygon on the left side of Fig. 12.3 has 10 edges and 10 corners. Each polygon corner is the intersection of two polygon edges.

Place arrows on the edges of $\mathscr{L}'$ so that at each polygon corner there is one pointing in and one pointing out. Give each corner a weight z^α, where α is the angle to the left through which an observer moving in the direction of the arrows turns when passing through the corner. Since edges cannot overlap, α must lie in the interval $-\pi < \alpha < \pi$. Two typical examples are shown in Fig. 12.4.

Still considering a particular polygon decomposition of $\mathscr{L}'$, form the product over all corners of these weights z^α. Then sum this combined weight over all allowed arrangements of all the arrows.

The result has to be $q^{p/2}$. To see this, note that the arrows round a polygon must all point the same way: either all anti-clockwise or all clockwise. In the former case, the observer turns through a total angle 2π, so the product of the corner weights for this polygon is $z^{2\pi}$. In the latter case the angle is -2π and the combined weight is $z^{-2\pi}$. Both possibilities can occur independently for each polygon, so each polygon gives a total contribution $z^{2\pi} + z^{-2\pi}$. From (12.3.5) this is just $2 \cosh \lambda$, i.e. $q^{\frac{1}{2}}$. There are p polygons, so the total sum must indeed be $q^{p/2}$.

This means that we can write (12.3.3) as

$$Z_N = q^{N/2} \sum_{pd} x_1^{l_1} x_2^{l_2} \ldots \sum_{ac} \prod_m z^{\alpha_m}, \qquad (12.3.6)$$

where the suffixes 'pd' and 'ac' denote that the outer sum is over all polygon decompositions of $\mathscr{L}'$, and the inner sum is over all allowed arrow coverings of the edges of $\mathscr{L}'$. The product is over all polygon corners m, α_m being the corresponding angle α (as defined above). The weights z^{α_m} are local properties of the corners m; so is the rule that at each corner there must be one arrow in and one arrow out.

Ice-Type Model on $\mathscr{L}'$

Consider a particular internal site of $\mathscr{L}'$, lying on the edge (i, j) of $\mathscr{L}$. Let α, β, γ, δ be the angles between the four edges of $\mathscr{L}'$, as indicated in Fig. 12.5. There are two possible ways of separating the edges, as shown in Fig. 12.2. In each case there are four possible arrangements of arrows. The resulting eight possibilities are shown in Fig. 12.6, together with the product of the corresponding x_r and z^{α_m} factors. This product is the total contribution of this site configuration to the combined summand in (12.3.6).

Note that in each case there are two arrows into the site, and two arrows out of it. Thus the 'ice rule' of Section 8.1 is satisfied at all internal sites. If we ignore the way in which the edges are separated, then we obtain the usual six arrow arrangements allowed at a site (or vertex), as shown in Fig. 12.7.

The next trick is to interchange the two summations in (12.3.6). Start with the undecomposed graph $\mathscr{L}'$. Place arrows on its edges in all the ways that they can occur above. This means that the ice rule must be satisfied

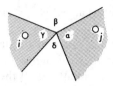

Fig. 12.5. A typical internal site of $\mathcal{L}'$, showing the angles between edges. Note that α and γ lie inside basic polygons ('islands;') of $\mathcal{L}'$, while β and δ lie outside.

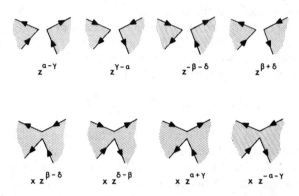

Fig. 12.6. The two possible separations of edges at an internal site of $\mathcal{L}'$, and the eight allowed arrangements of arrows thereon. The product of the corresponding x_r and z^α factors is shown underneath, using the notation of Fig. 12.5 and omitting the suffix of x_r.

at each internal site, and that there must be one arrow into (and one arrow out of) each external site.

Every such arrow covering can occur in the combined summation in (12.3.6), but some occur more than once. This is because arrow arrangements 5 and 6 in Fig. 12.7 can each arise in two ways. Arrangement 6 comes from either of the two right-hand possibilities in Fig. 12.6, arrangement 5 from the next two.

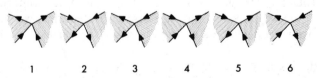

Fig. 12.7. The six possible arrangements of arrows at a site of $\mathcal{L}'$. Note that this figure is oriented so that the shaded areas ('land') are to the right and the left.

Suppose a particular allowed arrow covering of $\mathcal{L}'$ contains l vertices of types 5 and 6. Each vertex of type 1, 2, 3 or 4 corresponds to a unique separation of the edges at that vertex, but each vertex of type 5 or 6 corresponds to two possible choices of the edge separations. Thus the arrow covering corresponds to 2^l polygon decompositions, and occurs 2^l times in (12.3.6).

However, each choice can be made independently, so there is no problem in calculating the total contribution of this arrow covering to (12.3.6): one simply sums the appropriate two weights in Fig. 12.6. Thus (12.3.6) can be written as

$$Z_N = q^{N/2} \sum_{ac} \prod \text{(weights)} , \qquad (12.3.7)$$

where now the sum is over all allowed arrow coverings of $\mathcal{L}'$, and the product is over all sites of $\mathcal{L}'$. Each external site contributes a weight z^α to this product, where α is the angle in Fig. 12.4. Each internal site contributes a weight $\bar{\omega}_k$, where $k(=1,\ldots,6)$ is the arrow arrangement at the site, as listed in Fig. 12.7, and

$$\bar{\omega}_1, \ldots, \bar{\omega}_6 = z^{\alpha-\gamma}, z^{\gamma-\alpha}, x_r z^{\beta-\delta}, x_r z^{\delta-\beta},$$

$$z^{-\beta-\delta} + x_r z^{\alpha+\gamma}, z^{\beta+\delta} + x_r z^{-\alpha-\gamma}. \qquad (12.3.8)$$

Here r is the class of the internal site, and α, β, γ, δ are the angles shown in Fig. 12.5. It is important to note that the angles α and γ lie inside basic polygons ('islands') of $\mathcal{L}'$, while β and δ lie outside.

The sum in (12.3.7) is over all arrow coverings of $\mathcal{L}'$ such that each site has as many arrows pointing in as it has pointing out. For the internal sites this is the ice rule. Indeed, comparing (12.3.7) with (8.1.1) and (8.1.3), we see that $q^{-N/2} Z_N$ is the partition function of an 'ice-type' (or 'six-vertex') model, generalized to allow different weights on different sites, and to allow 'external' sites of coordination number two. Thus the Potts model on any planar graph $\mathcal{L}$ can be expressed as an ice-type model on the medial graph $\mathcal{L}'$.

Four-Colour Problem

As was remarked shortly after (12.2.4), if $v = -1$ then Z_N is the number of ways of colouring the planar graph $\mathcal{L}$ with q colours. It is fascinating to wonder whether the ice-type formulation (12.3.7) has any bearing on the famous four-colour problem (Ore, 1967; Saaty and Kainen, 1977), which was only recently solved (Appel and Haken, 1976, 1977; Appel et al., 1977) after tantalizing mathematicians for over a century.

Certainly $q = 4$ is a very special case: λ in (12.3.5) is real for $q \geq 4$; for $q < 4$ it is pure imaginary. In particular, for $q = 4$ and $v = -1$ we have that $z = 1$ and $x_r = -\frac{1}{2}$. The weights in (12.3.8) are therefore real, but $\bar{\omega}_3$ and $\bar{\omega}_4$ are negative. To obtain an alternative solution to the four-colour problem we would need to show that the negative contributions to the sum in (12.3.7) are numerically less than the positive ones.

Another intriguing point which suggests that our transformation may be relevant is the following. It is conjectured from numerical and other studies that the real zeros of the colouring polynomial of an arbitrary planar lattice cluster round limit points when the lattice becomes large. These limits are supposed to occur at the 'Beraha numbers' $q = [2 \cos(\pi/n)]^2$, $n = 2, 3, 4, \ldots$. (Beraha *et al.*, 1975, 1978; Beraha and Kahane, 1979; Tutte, 1970, 1973, 1974). From (12.3.5) we see that this corresponds simply to our parameter λ having the values $i\pi/2$, $i\pi/3$, $i\pi/4$, etc.

12.4 Square-Lattice Potts Model

Equivalent Ice-Type Model

For the interior of regular lattices there is an obvious natural choice of $\mathcal{L}'$, namely to take the sites of $\mathcal{L}'$ to be the mid-points of the edges of $\mathcal{L}$ and to take two sites of $\mathcal{L}'$ to be adjacent if any only if the corresponding edges of $\mathcal{L}$ meet at a common site and bound a common face. All sites of $\mathcal{L}'$ are then 'internal' except at the boundaries, which is the reason for our terminology. The square lattice ($\mathcal{L}$) is shown in Fig. 12.8, together with its resulting medial graph $\mathcal{L}'$. It has two classes of edges, horizontal and vertical, which we can call classes 1 and 2, respectively. Define a parameter

$$s = \exp(\lambda/4) . \tag{12.4.1}$$

Then from (12.3.7), (12.3.8) and (12.3.5), we see that $q^{-N/2}Z_N$ is the partition function of an ice-type model with weights

$$\bar{\omega}_1, \ldots, \bar{\omega}_6 = 1, 1, x_r, x_r, s^{-2} + x_r s^2, s^2 + x_r s^{-2}, \tag{12.4.2}$$

r being the class of the site of $\mathcal{L}'$, and the six arrow arrangements being labelled as in Fig. 12.7.

We can eliminate the fractional powers of e^λ by associating additional mutually inverse weights with the tips and tails of some of the arrows. With every arrow on a $SW - NE$ edge, associate a further weight s^{-1} with the site into which it points, and a weight s with the site it points out of.

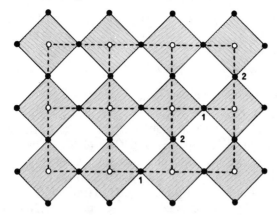

Fig. 12.8. The square lattice $\mathcal{L}$ (open circles and broken lines) and its medial lattice $\mathcal{L}'$ (full circles and lines). The two classes of edges of $\mathcal{L}'$, horizontal and vertical, are indicated by the numbers 1 and 2, respectively.

Obviously these weights cancel from Z_N, but the individual vertex weights ω_5, ω_6 are modified: those on sites of type 1 are multiplied by s^2, s^{-2}, respectively; those on sites of type 2 are multiplied by s^{-2}, s^2.

It must be noted that the weights in (12.4.2) are labelled as in Fig. 12.7, where the shaded areas are to the right and left. This is the way sites of type 1 are drawn in Fig. 12.8, but for sites of type 2 it is necessary to turn one of the figures through 90°.

It is more convenient to use the same direction throughout. Let $\omega_1, \ldots, \omega_6$ be the weights of the six arrow configurations shown in Fig. 12.9, always using the same orientation for this figure as for Fig. 12.8. For sites of type 1 this is the original labelling; for sites of type 2 we have that $\omega_1, \ldots, \omega_6 = \bar{\omega}_3, \bar{\omega}_4, \bar{\omega}_2, \bar{\omega}_1, \bar{\omega}_6, \bar{\omega}_5$. Allowing also for the multiplications by s^2 and s^{-2} mentioned above, we obtain:

$$\text{Type 1: } \omega_1, \ldots, \omega_6 = 1, 1, x_1, x_1, 1 + x_1 e^{\lambda}, 1 + x_1 e^{-\lambda}, \quad (12.4.3a)$$

$$\text{Type 2: } \omega_1, \ldots, \omega_6 = x_2, x_2, 1, 1, x_2 + e^{\lambda}, x_2 + e^{-\lambda}. \quad (12.4.3b)$$

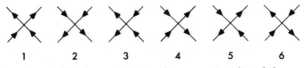

Fig. 12.9. The six possible arrangements of arrows at a site of the square lattice, using the same orientation for all sites.

The weights of the external sites at the right and left of Fig. 12.8 are now always one. Those at the bottom and top are $\exp(\lambda/2)$ if the arrows turn to the left, $\exp(-\lambda/2)$ if they turn to the right.

Alternative Derivation of the Equivalence

The equivalence of the square-lattice Potts model to this ice-type model was first obtained by Temperley and Lieb (1971). They used operators which form a rather elegant and interesting algebra. For this reason it seems worthwhile outlining this alternative approach.

Consider the Potts model on the square lattice $\mathscr{L}$. Let $\mathscr{L}$ have m rows and n columns. Then in the usual way (Sections 2.1, 7.2, 8.2) we can write the partition function as

$$Z_N = \xi^T VWVW \ldots V\xi, \tag{12.4.4}$$

there being m Vs and $m - 1$ Ws. Here ξ is the q^n-dimensional column vector whose entries are all unity, V is the transfer matrix that adds a row of horizontal edges to the lattice, and W adds a row of vertical edges. Both V and W are q^n by q^n matrices, with indices $\sigma = \{\sigma_1, \ldots, \sigma_n\}$ and $\sigma' = \{\sigma_1', \ldots, \sigma_n'\}$, and with elements

$$V_{\sigma,\sigma'} = \exp\left\{K_1 \sum_{j=1}^{n-1} \delta(\sigma_j, \sigma_{j+1})\right\} \prod_{j=1}^{n} \delta(\sigma_j, \sigma_j'), \tag{12.4.5a}$$

$$W_{\sigma,\sigma'} = \exp\left\{K_2 \sum_{j=1}^{n} \delta(\sigma_j, \sigma_j')\right\}. \tag{12.4.5b}$$

(Note that we do *not* impose cyclic boundary conditions on either the rows or columns of $\mathscr{L}$.)

Let us define matrices $U_1, \ldots, U_{2n-1}$ by

$$(U_{2i-1})_{\sigma,\sigma'} = q^{-\frac{1}{2}} \prod_{\substack{j=1 \\ \neq i}}^{n} \delta(\sigma_j, \sigma_j'), \tag{12.4.6a}$$

$$(U_{2i})_{\sigma,\sigma'} = q^{\frac{1}{2}} \delta(\sigma_i, \sigma_{i+1}) \prod_{j=1}^{n} \delta(\sigma_j, \sigma_j'). \tag{12.4.6b}$$

Thus U_{2i} is diagonal, with diagonal elements $q^{\frac{1}{2}}\delta(\sigma_i, \sigma_{i+1})$; while U_{2i-1} is of the form

$$U_{2i-1} = e \otimes e \otimes \ldots \otimes e \otimes g \otimes e \otimes \ldots \otimes e, \tag{12.4.7}$$

e being the unit q-by-q matrix, and g (occuring in position i in the product) being the q-by-q matrix with all entries equal to $q^{-\frac{1}{2}}$.

From (12.4.5) and (12.4.6) it is readily observed that

$$V = \exp\{q^{-\frac{1}{2}}K_1(U_2 + U_4 + \ldots + U_{2n-2})\}, \qquad (12.4.8a)$$

$$W = \prod_{j=1}^{n} (v_2 \mathscr{I} + q^{\frac{1}{2}}U_{2j-1}), \qquad (12.4.8b)$$

where $\mathscr{I}$ is the unit q^N-by-q^N matrix and $v_2 = \exp(K_2) - 1$, as in (12.2.1).
The expression (12.4.8a) can be put in a form similar to (12.4.8b) by using the fact that $U_i^2 = q^{\frac{1}{2}}U_i$. Alternatively, (12.4.8b) can be put in a form similar to (12.4.8a). We obtain

$$V = \prod_{j=1}^{n-1} \{\mathscr{I} + q^{-\frac{1}{2}}v_1 U_{2j}\}, \qquad (12.4.8c)$$

$$W = v_2^n \exp\{q^{-\frac{1}{2}}K_2^*(U_1 + U_3 + \ldots + U_{2n-1})\}, \qquad (12.4.8d)$$

where $v_1 = \exp(K_1) - 1$ and

$$\exp K_2^* = (v_2 + q)/v_2 = (e^{K_2} + q - 1)/(e^{K_2} - 1). \qquad (12.4.9)$$

The matrices $U_1, \ldots, U_{2n-1}$ satisfy the relations

$$\begin{aligned}
U_i^2 &= q^{\frac{1}{2}}U_i, & i &= 1, \ldots, 2n-1, \\
U_i U_{i+1} U_i &= U_i, & i &= 1, \ldots, 2n-2, \\
U_i U_{i-1} U_i &= U_i, & i &= 2, \ldots, 2n-1, \\
U_i U_j &= U_j U_i, & |i - j| &\geqslant 2.
\end{aligned} \qquad (12.4.10)$$

These relations define the algebra generated by $U_1, \ldots, U_{2n-1}$. In particular, they define all the eigenvalues of the complete transfer matrix VW (but not their degeneracies). They therefore define the maximum eigenvalue, and hence the large-m behaviour of the partition function.

Indeed, they define Z_N even for finite m. To see this, define the matrix

$$R = U_1 U_3 U_5 \ldots U_{2n-1}. \qquad (12.4.11)$$

Using the explicit representation (12.4.6a), or (12.4.7), we see that R is the q^n by q^n matrix, all of whose entries are equal to $q^{-n/2}$. Thus

$$R = q^{-n/2} \xi \xi^T, \qquad (12.4.12)$$

where ξ is the column vector in (12.4.4), all of whose entries are one. It is now obvious that for any q^n by q^n matrix X,

$$R X R = \tau(X) R, \qquad (12.4.13)$$

where $\tau(X)$ is a scalar.

More specifically, $\tau(X)$ is $q^{-n/2} \xi^T X \xi$. From (12.4.4) it follows that

$$Z_N = q^{n/2} \tau(VWVW \ldots V). \tag{12.4.14}$$

Now forget that $V, W, U_1, \ldots, U_{2n-1}$ were introduced as q^n by q^n matrices, and regard them simply as operators satisfying (12.4.8) and (12.4.10). Let X be any sum of products of $U_1, \ldots, U_{2n-1}$ and the identity operator $\mathcal{I}$. Then from (12.4.10) and (12.4.11) it can be established that (12.4.13) is true, and $\tau(X)$ can be evaluated. Since $VWVW \ldots V$ can be written as such a sum of products, it follows from (12.4.14) that Z_N can in principle be calculated in this way.

Of course I do not claim that this programme can readily be carried out for arbitrarily large m and n; only that it can in principle be done. This means that we do not have to use the representation (12.4.6): any set of operators $U_1, \ldots, U_{2n-1}$ that satisfy (12.4.10), and for which R is not identically zero, will be an equally good representation.

Temperley and Lieb (1971) showed that one such alternative representation is to take $U_1, \ldots, U_{2n-1}$ to be 4^n by 4^n matrices, with indices $\alpha = \{\alpha_1, \ldots, \alpha_{2n}\}$ and $\alpha' = \{\alpha_1', \ldots, \alpha_{2n}'\}$, where each α_j and α_j' takes the values $+1$ and -1, and the elements of U_i are

$$(U_i)_{\alpha,\alpha'} = \delta(\alpha_1, \alpha_1') \ldots \delta(\alpha_{i-1}, \alpha_{i-1}') h(\alpha_i, \alpha_{i+1})$$

$$\times h(\alpha_i', \alpha_{i+1}') \delta(\alpha_{i+2}, \alpha_{i+2}') \ldots \delta(\alpha_{2n}, \alpha_{2n}'), \tag{12.4.15}$$

for $i = 1, \ldots, 2n - 1$, where

$$h(+, +) = h(-, -) = 0,$$

$$h(+, -) = \exp(-\lambda/2), h(-, +) = \exp(\lambda/2), \tag{12.4.16}$$

and λ is given by (12.3.5).

Regard $\alpha_1, \ldots, \alpha_{2n}$ as representing a row of vertical, or rather near-vertical arrows: $\alpha_j = +$ if the arrow in column j points up, $\alpha_j = -$ if it points down. More specifically, let $\alpha_1, \ldots, \alpha_{2n}$ represent a typical row of arrows on the edges of $\mathcal{L}'$, such as the top row of edges in Fig. 12.8, labelled $1, \ldots, 2n$. Consider the operator $\mathcal{I} + x_1 U_{2j}$, using the α-representation (12.4.15). This acts on the arrows in positions $2j$ and $2j + 1$, and can be thought of as a 'vertex transfer matrix'. Its non-zero entries are precisely the weights $\omega_1, \ldots, \omega_6$ in (12.4.3a), corresponding to the six arrow arrangements in Fig. 12.9. Thus it is the vertex transfer matrix of a site of type 1 in the ice-type model. From (12.3.4) and (12.4.8c), we see that V is just the product of these matrices, for $j = 1, \ldots, n - 1$. Thus the V in (12.4.8) is the transfer matrix for a row of sites of type 1 in the ice-type model.

Similarly, from (12.4.8b), (12.3.4) and (12.4.3b) we find that $q^{-n/2}W$ is the transfer matrix for a row of sites of type 2 in the ice-type model. Noting that $N = mn$, it follows from (12.4.14) that

$$Z_N = q^{N/2} \times \text{partition function of the ice-type model}. \qquad (12.4.17)$$

We can verify that external sites have weights given by the rules following (12.4.3). The equivalence (12.3.7) therefore follows from the two representations (12.4.6) and (12.4.15) of the operators $U_1, \ldots, U_{2n-1}$.

Some General Comments on the Ice-Type Model

For general values of x_1 and x_2, the Potts model has not yet been solved. It is one of the most tantalizing unsolved models. For instance, as was remarked in Section 8.12, the homogeneous square-lattice ice-type model can be solved by the Bethe ansatz method of Chapter 8, even if the 'zero-field' restrictions (8.1.7) are violated. Then the definition (8.3.21) of Δ becomes

$$\Delta = (\omega_1\omega_2 + \omega_3\omega_4 - \omega_5\omega_6)/2(\omega_1\omega_2\omega_3\omega_4)^{\frac{1}{2}}, \qquad (12.4.18)$$

and the eigenvectors of the transfer matrix depend only on this Δ and the horizontal electric field E'.

The ice-type model we are considering here is not homogeneous: its weights are different on the two sub-lattices (types 1 and 2, respectively). Even so, from (12.4.3), (12.4.18) and (12.3.5) we see that

$$\Delta = -\cosh\lambda = -\tfrac{1}{2}q^{\frac{1}{2}} \qquad (12.4.19)$$

for both sites of type 1 and type 2. (Indeed, from (12.3.8) this relation is true for arbitrary planar graphs.) Thus Δ is uniform, but unfortunately the Bethe ansatz method of Chapter 8 still fails to work.

An intriguing point is that the ansatz fails even for $q = 1$ and $q = 2$, whereas the free energy can be calculated for these cases by other methods, the first case being trivial and the second being the Ising model. It is also worth noting that in this equivalence the Ising model corresponds to μ in (8.8.1), i.e. $\Delta = -\cos\mu$, having the value $\pi/4$. This contrasts with the fact that the Ising model is also equivalent to the free-fermion model, as was shown in Section 10.16. The free-fermion model is an eight-vertex generalization of an homogeneous six-vertex model with $\Delta = 0$ and $\mu = \pi/2$. There is therefore a relation between these $\mu = \pi/4$ and $\mu = \pi/2$ vertex models.

Duality

One advantage of this ice-type formulation of the Potts model is that it makes it very easy to show that the square lattice model satisfies a duality relation.

To see this, note from (12.4.3) that the weights on sites of type 2 are similar in form to those on sites of type 1: in fact we can interchange them by replacing x_1, x_2 by x_2^{-1}, x_1^{-1}, respectively, and then multiplying all type 1 weights by x_2, all type 2 weights by x_1.

On the other hand, for a large lattice $\mathcal{L}'$, we are free to choose which sub-lattice we designate as 1, and which as 2 (a simple way of saying this is to say that we can replace the black squares of a chequerboard by white ones, and vice versa). This affects the boundary conditions, but we do not expect this to affect the way Z_N grows exponentially with N. Keeping only such exponential factors, and regarding Z_N as a function of x_1 and x_2 (q being fixed), it follows that

$$Z_N(x_1, x_2) = (x_1 x_2)^N Z_N(x_2^{-1}, x_1^{-1}). \qquad (12.4.20)$$

More precisely, we expect the large-lattice limit

$$\psi = - \lim_{N \to \infty} N^{-1} \ln Z_N \qquad (12.4.21)$$

to exist. As in (6.2.10) and (1.7.6), this is the dimensionless free energy per site, being related to the usual free energy f by $\psi = f/k_B T$. Like Z_N, ψ can be regarded as a function of x_1 and x_2. Then (12.4.20) gives

$$\psi(x_1, x_2) = -\ln(x_1 x_2) + \psi(x_2^{-1}, x_1^{-1}). \qquad (12.4.22)$$

This is a duality relation, relating a high-temperature Potts model to a low-temperature one. It was first obtained by Potts (1952). We could also have obtained it by interchanging K_1 and K_2^* in (12.4.8a) and (12.4.8d), and replacing each U_i by U_{i+1}: apart from boundary conditions and scalar factors, this merely interchanges the two transfer matrices V and W.

We remarked in Section 12.1 that the $q = 2$ Potts model is equivalent to the Ising model: the relation (12.4.22) is then the square-lattice Ising model duality relation (6.2.14).

Location of the Critical Point

Now suppose that J_1 and J_2 are both positive. This means that the system is ferromagnetic: adjacent spins 'like' to be equal. From (12.2.6) and (12.3.4), K_1, K_2, v_1, v_2 and x_1, x_2 are all positive.

Obviously (12.4.22) relates the value of ψ at a point (x_1, x_2) to its value at (x_2^{-1}, x_1^{-1}). This mapping takes the domain $0 < x_1 x_2 < 1$ to the domain $x_1 x_2 > 1$. Every point on the line $x_1 x_2 = 1$ is self-dual.

We expect the ferromagnetic Potts model to be disordered at high temperatures (x_1 and x_2 small), all q possible spin-states being equally likely. At low temperatures (x_1 and x_2 large) we expect it to be ordered, one of the spin-states being preferred by all the spins. Somewhere in between we expect there to be a critical temperature T_c at which this spontaneous symmetry breaking just starts to occur. In the (x_1, x_2) plane this must be a line, separating the disordered and ordered regions. We expect $\psi(x_1, x_2)$ to be analytic, except possibly on this line.

We now argue as in Section 6.2. If $\psi(x_1, x_2)$ is non-analytic on a line inside the domain $0 < x_1 x_2 < 1$, then from the duality relation (12:4.22) it must also be non-analytic on a line inside $x_1 x_2 > 1$. The simplest possibility is that it is non-analytic only on the self-dual line $x_1 x_2 = 1$. (For an isotropic system, with $x_1 = x_2$, this corresponds to requiring that the critical temperature be unique.) Hintermann et al. (1978) have shown that this is in fact the case: the critical points of the square-lattice Potts model occur when

$$x_1 x_2 = 1. \tag{12.4.23}$$

12.5 Critical Square-Lattice Potts Model

Suppose that the condition (12.4.23) is satisfied, i.e. $x_2 = 1/x_1$. Then from (12.4.3) it is evident that the weights of type 2 are all equal to the corresponding weights of type 1, divided by x_1. Multiplying all the six weights at any given site by some factor is a trivial modification of the model: it merely multiplies the partition function by the same factor. The system is therefore effectively homogeneous: more precisely, using (12.3.7) and noting that there are N sites of type 2 in $\mathcal{L}'$,

$$Z_N = q^{N/2} x_1^{-N} Z'_{2N}, \tag{12.5.1}$$

where Z'_{2N} is the partition function of the ice-type model on the lattice $\mathcal{L}'$ with $2N$ sites, each with Boltzmann weights given by (12.4.3a).

Free Energy

Since it is homogeneous (i.e. all sites have the same weights), this ice-type model can be solved by the methods of Chapter 8. In fact, as was remarked at the end of Section 8.1, vertices of type 5 and 6 must occur in pairs, being respectively sinks and sources of horizontal arrows. Thus the weights ω_5 and ω_6 occur only in the combination $\omega_5\omega_6$. This means that the partition function is unchanged by replacing both ω_5 and ω_6 by their geometric mean. From (12.4.3a) and (8.3.3) (noting that the vertex ordering in Fig. 12.9 is the same as that in Fig. 8.2, after an appropriate rotation), it follows that Z'_{2N} is the partition function of a *zero-field* ice-type model on $\mathcal{L}'$, with weights

$$a = 1, \quad b = x_1, \quad c = (1 + 2x_1 \cosh \lambda + x_1^2)^{\frac{1}{2}}. \qquad (12.5.2)$$

The dimensionless free energy of the Potts model is defined by (12.4.21). From (12.5.1), (1.7.6) and (12.5.2), it is

$$\psi = -\ln(q^{\frac{1}{2}}/x_1) + 2f/k_B T, \qquad (12.5.3)$$

where $f/k_B T$ is the dimensionless free energy of the ice-type model. This has been calculated in Chapter 8. From (8.3.21) and (12.3.5),

$$\Delta = -\cosh \lambda = -\tfrac{1}{2}q^{\frac{1}{2}}, \qquad (12.5.4)$$

so there are two cases to consider: $0 > \Delta > -1$, and $\Delta < -1$. In the former case we see that $0 < q < 4$, and we use the results of Section 8.8; in the latter case($q > 4$), we use those of Section 8.9. Since ψ is continuous, the case $q = 4$ can be handled by taking the appropriate limits. Doing this, we find that

$$\psi = -\tfrac{1}{2}\ln q - \phi(x_1) - \phi(x_2), \qquad (12.5.5)$$

where x_1 and x_2 satisfy the criticality condition (12.4.23), q is regarded as a constant, and the function $\phi(x)$ is defined as follows:

$$0 < q < 4: \quad q^{\frac{1}{2}} = 2\cos\mu, \quad 0 < \mu < \pi/2,$$

$$x = \sin\gamma/\sin(\mu - \gamma), \quad 0 < \gamma < \mu, \qquad (12.5.6a)$$

$$\phi(x) = \tfrac{1}{2}\int_{-\infty}^{\infty} \frac{\sinh(\pi - \mu)t \sinh 2\gamma t}{t \sinh \pi t \cosh \mu t}\, dt\,;$$

$$q = 4: \quad x = \tau/(1 - \tau), \quad 0 < \tau < 1,$$

$$\phi(x) = \int_0^{\infty} y^{-1} \exp(-y)\, \text{sech}\, y \sinh 2\tau y\, dy\,: \qquad (12.5.6b)$$

$$q > 4: \quad q^{\frac{1}{2}} = 2\cosh\lambda, \quad \lambda > 0,$$

$$x = \sinh\beta/\sinh(\lambda - \beta), \quad 0 < \beta < \lambda, \qquad (12.5.6c)$$

$$\phi(x) = \beta + \sum_{n=1}^{\infty} n^{-1} \exp(-n\lambda)\, \text{sech}\, n\lambda \sinh 2n\beta.$$

For all values of q, this function $\phi(x)$ satisfies the identity

$$\phi(x) - \phi(x^{-1}) = \ln x . \qquad (12.5.6d)$$

Let γ_j, τ_j, β_j be the values of these parameters γ, τ, β when $x = x_j$. Then the criticality condition (12.4.23) implies that

$$\gamma_1 + \gamma_2 = \mu, \quad \tau_1 + \tau_2 = 1, \quad \beta_1 + \beta_2 = \lambda . \qquad (12.5.7)$$

Indeed, γ_1 and γ_2 are the expressions $\frac{1}{2}(\mu + w)$ and $\frac{1}{2}(\mu - w)$ in Section 8.8, while β_1 and β_2 are the $\frac{1}{2}(\lambda + v)$ and $\frac{1}{2}(\lambda - v)$ in Section 8.9. The λ of this chapter is the same as that of Chapter 8.

The integral in (12.5.6a) can be evaluated explicitly when μ is a rational fraction of π (e.g. $\pi/4$, $\pi/3$, $2\pi/5$). For the isotropic model, with $x_1 = x_2 = 1$, some of these cases have been tabulated by Temperley and Lieb (1971).

Internal Energy and Latent Heat

We can also calculate the internal energy of the Potts model at its critical point (Potts, 1952; Baxter, 1973d). To do this, we first return to considering the general Potts model, not necessarily satisfying the criticality condition (12.4.23). From (12.1.2), (12.1.3) and (1.4.4), the total average energy is

$$\langle E \rangle = k_B T^2 \frac{\partial}{\partial T} \ln Z_N , \qquad (12.5.8)$$

in agreement with (1.4.6). The square-lattice partition function depends on T via x_1 and x_2, and from (12.1.4), (12.2.6) and (12.3.4), x_r and T are related by

$$x_r = q^{-\frac{1}{2}}[\exp(J_r/k_B T) - 1] . \qquad (12.5.9)$$

Regarding Z_N as a function of x_1 and x_2 (q and N being kept constant), it follows that

$$\langle E \rangle = -q^{-\frac{1}{2}} \sum_{r=1}^{2} J_r \exp(J_r/k_B T) \frac{\partial}{\partial x_r} \ln Z_N . \qquad (12.5.10)$$

Now use the expression (12.3.7) for Z_N, where the vertex weights are given by (12.4.3). For a given arrow covering of the edges of $\mathscr{L}'$, let $n_k(n_k')$ be the total number of sites of type 1 (type 2) that are in the arrow configuration k shown in Fig. 12.9. Here $k = 1, \ldots, 6$. Then (12.3.7) can be written more explicitly as

$$Z_N = q^{N/2} \sum_{ac} x_1^{n_3 + n_4}(1 + x_1 e^{\lambda})^{n_5} (1 + x_1 e^{-\lambda})^{n_6}$$

$$\times x_2^{n_1' + n_2'}(x_2 + e^{\lambda})^{n_5'} (x_2 + e^{-\lambda})^{n_6'}, \qquad (12.5.11)$$

the sum being over all arrow coverings of $\mathcal{L}'$ that satisfy the ice rule at each site.

From (12.5.10), (12.5.11) and (1.4.4), it follows that

$$\langle E \rangle = -q^{-\frac{1}{2}} \sum_{r=1}^{2} J_r \exp(J_r/k_B T) I_r, \qquad (12.5.12)$$

where

$$I_1 = \frac{\langle n_3 \rangle + \langle n_4 \rangle}{x_1} + \frac{\exp(\lambda) \langle n_5 \rangle}{1 + x_1 \exp(\lambda)} + \frac{\exp(-\lambda) \langle n_6 \rangle}{1 + x_1 \exp(-\lambda)},$$

$$I_2 = \frac{\langle n_1' \rangle + \langle n_2' \rangle}{x_2} + \frac{\langle n_5' \rangle}{x_2 + \exp(\lambda)} + \frac{\langle n_6' \rangle}{x_2 + \exp(-\lambda)}, \qquad (12.5.13)$$

$\langle n_j \rangle$ and $\langle n_j' \rangle$ being as usual the average values of n_j and n_j'.

We can calculate I_1 and I_2 when the criticality condition (12.4.23) is satisfied, i.e. $x_1 x_2 = 1$. Then Z_N is given (for large N) by (12.4.21) and (12.5.5), i.e.

$$\ln Z_N = N[\tfrac{1}{2} \ln q + \phi(x_1) + \phi(x_2)], \qquad (12.5.14)$$

the function $\phi(x)$ being defined by (12.5.6), q being regarded as a constant. Using this expression for Z_N in (12.5.11), and differentiating logarithmically with respect to x_1 (remembering that $x_2 = 1/x_1$), we find that

$$N[\phi'(x_1) - x_2^2 \phi'(x_2)] = I_1 - x_2^2 I_2. \qquad (12.5.15)$$

$\phi'(x)$ being the derivative of $\phi(x)$.

This is one equation relating I_1 and I_2. We can obtain another by considering the symmetry relations between the 12 averages $\langle n_1 \rangle, \ldots, \langle n_6' \rangle$. As we noted at the beginning of this section, when $x_1 x_2 = 1$ we can renormalize the weights (12.4.3) so as to reduce them to the form (8.3.3), where a, b, c are given by (12.5.2). These renormalizations leave $\langle n_1 \rangle, \ldots, \langle n_6' \rangle$ unchanged, but they make it clear that the ice-type model (for $x_1 x_2 = 1$) has two symmetries: it is translation invariant (sites of type 1 have the same weights as sites of type 2), and it is unchanged by reversing all arrows (this is the 'zero-field' condition).

It is rigorously known (Brascamp et al., 1973) that if $c > a + b$, then each of these symmetries is spontaneously broken. As is explained in Section 8.10, the system is anti-ferroelectrically ordered and there is a spontaneous staggered polarization P_0. By considering a system in an infinitesimal staggered electric field (or with staggered fixed-arrow boundary conditions), we can define P_0 as $\langle \tau_i \rangle$. Here τ_i is the polarization of the electric dipole on edge i: it is defined as in Section 8.10, being $+1$ ('right') if the arrow points in the standard direction of Fig. 8.3, -1 ('wrong') if it points in the other.

This P_0 is the same for all the $4N$ edges of $\mathcal{L}'$, so

> $4 N P_0 = $ average number of 'right' arrows minus the
> average number of 'wrong' arrows .   (12.5.16)

Each arrow adjoins just one site of type 1, and each site of type 1 is the meeting-place of four arrows (a 'vertex'). Choose the standard configuration of Fig. 8.3 to correspond to all vertices of $\mathcal{L}'$ of type 1 being in the arrow arrangement 6 of Fig. 12.9. Then by comparing these figures it becomes apparent that arrow arrangements 1 to 4 each contain as many wrong arrows as right ones, arrangement 5 contains 4 wrong arrows, and 6 contains 4 right ones. Thus (12.5.16) can be written as

$$N P_0 = \langle n_6 \rangle - \langle n_5 \rangle .$$ (12.5.17)

Although the two symmetries of sub-lattice interchange and arrow-reversal are both spontaneously broken for $c > a + b$, the combined symmetry is not. For instance, the ground state shown in (8.3.3) is unchanged by reversing all arrows *and* then interchanging sites of type 1 with sites of type 2. It follows that

$$\langle n_1' \rangle = \langle n_2 \rangle, \quad \langle n_2' \rangle = \langle n_1 \rangle, \quad \langle n_3' \rangle = \langle n_4 \rangle,$$
$$\langle n_4' \rangle = \langle n_3 \rangle, \quad \langle n_5' \rangle = \langle n_6 \rangle, \quad \langle n_6' \rangle = \langle n_5 \rangle .$$ (12.5.18)

The equations (12.5.17) and (12.5.18) are still true when $|a - b| < c < a + b$. In this case the ice-type model is disordered in the sense that there is no spontaneous symmetry breaking (though the correlation length is infinite, as remarked in Section 8.10). This means that $\langle n_5 \rangle = \langle n_6 \rangle$, so P_0 in (12.5.17) is then zero.

There are N sites of type 1, each site must be in one of the six possible arrow arrangements, so $n_1 + \ldots + n_6 = N$. Forming the expression $x_1 I_1 + x_2 I_2$, where I_1 and I_2 are defined by (12.5.13), using the symmetry relations (12.5.18), together with (12.5.17) and (12.4.23), it follows that

$$x_1 I_1 + x_2 I_2 = N\{1 - 2 x_1 \zeta(x_1) P_0\},$$ (12.5.19)

where

$$\zeta(x) = \sinh \lambda / (1 + x^2 + 2x \cosh \lambda) .$$ (12.5.20)

This is the second equation for I_1 and I_2 that we needed. Remembering that $x_2 = 1/x_1$, from (12.5.20) and (12.5.6d) we can readily establish that

$$x_1 \, \zeta(x_1) = x_2 \, \zeta(x_2) \,, \tag{12.5.21}$$

$$x_1 \, \phi'(x_1) + x_2 \, \phi'(x_2) = 1 \,.$$

Solving (12.5.15) and (12.5.19) for I_1 and I_2, using the properties (12.5.21), it follows that

$$I_r = N[\phi'(x_r) - \zeta(x_r) \, P_0] \,, \tag{12.5.22}$$

for $r = 1, 2$. From (12.5.12), the internal energy per site $U = \langle E \rangle / N$ of the square lattice Potts model is therefore

$$U = q^{-\frac{1}{2}} \sum_{r=1}^{2} J_r \exp(J_r/k_B T) \, [-\phi'(x_r) \pm \zeta(x_r) \, P_0] \,. \tag{12.5.23}$$

I have introduced a $\pm$ sign to allow for the fact that the sign of the RHS of (12.5.17) is not yet determined. If the ice model symmetries are spontaneously broken in favour of arrow arrangement 6 (5) on sites of type 1 (2), then the RHS of (12.5.17) is positive, and P_0 therein is the usual spontaneous staggered polarization. On the other hand, if the symmetries are broken in favour of arrow arrangement 5 (6) on sites of type 1 (2), then P_0 is negated. From (12.4.3), the former situation occurs (for $\lambda > 0$) as $x_1 x_2$ approaches unity from below, the latter as $x_1 x_2$ approaches unity from above. The sign in (12.5.23) should therefore be chosen positive if T approaches the critical temperature T_c from above, negative if it approaches T_c from below.

The spontaneous polarization P_0 has been evaluated (Baxter, 1973c) and the relevant results are given in Section 8.10, notably in (8.10.9), (8.10.2) and (8.9.1). Using (12.5.4), we see that there are two cases to consider:

(i) $0 < q \le 4$: $0 > \Delta \ge -1$, λ pure imaginary.
   The ice-type model is disordered in the sense that there is no spontaneous symmetry breaking (though the correlation length is infinite). Thus P_0 is zero, and U is continuous across $T = T_c$. (We still expect U to be non-analytic at $T = T_c$: certainly this is so for $q = 2$, when the Potts model becomes the Ising model.)

(ii) $q > 4$: $\Delta < -1$, λ is real and is to be chosen positive. There is spontaneous symmetry breaking; P_0 is positive, being given by

$$P_0 = \prod_{m=1}^{\infty} [\tanh m\lambda]^2 \,. \tag{12.5.24}$$

From (12.5.23), the Potts model therefore has a first-order transition at $T = T_c$, with latent heat

$$L = 2q^{-\frac{1}{2}} \sum_{r=1}^{2} J_r \exp(J_r/k_B T) \, \zeta(x_r) \, P_0. \qquad (12.5.25)$$

For the isotropic model, with $J_1 = J_2 = J$ and $x_1 = x_2 = 1$, we can calculate $\phi'(x_1)$ from (12.5.21). Using (12.5.9) and (12.5.23), we obtain the simple formula

$$U_{av} = -(1 + q^{-\frac{1}{2}}) J. \qquad (12.5.26)$$

Here U_{av} is the average of the internal energy just below and just above T_c, i.e. $U_{av} = \frac{1}{2}(U_- + U_+)$. For $q \leq 4$, where there is no discontinuity in U, this is the internal energy at T_c. This result was obtained by Potts (1952) in his original paper.

12.6 Triangular-Lattice Potts Model

We can carry out a similar programme for the Potts model on the triangular and honeycomb lattices, i.e. we can locate the critical points, and at these points we can calculate the free energy and the internal energy (Baxter *et al.*, 1978).

Let Z_N^T and Z_N^H be the partition functions of the Potts model on the triangular and honeycomb lattices, respectively, where each lattice has N sites. These partition functions are given by (12.1.3), except that for each lattice there are three types of edges. Let us label them 1, 2, 3, and let K_1, K_2, K_3 be the corresponding values of K. Then each partition function depends on K_1, K_2, K_3, as well as on N and q.

The triangular lattice $\mathcal{L}$ is drawn in Fig. 12.10, together with its medial graph $\mathcal{L}'$ (which is a Kagomé lattice). From (12.3.7) and (12.3.8) we have that

$$Z_N^T(K_1, K_2, K_3) = q^{N/2} Z', \qquad (12.6.1)$$

where Z' is the partition function of an ice-type model on $\mathcal{L}'$, with weights

$$\bar\omega_1, \ldots, \bar\omega_6 = 1, 1, x_r, x_r, t^{-1} + x_r t^2, t + x_r t^{-2}. \qquad (12.6.2)$$

Here $r = 1$, 2 or 3 depending on the type of the site of $\mathcal{L}'$; the six allowed arrow arrangements are the first six shown in row r of Fig. 11.2. The lattice $\mathcal{L}'$ has $3N$ sites; t and x_r are defined by

$$t = \exp(\lambda/3), \quad x_r = q^{-\frac{1}{2}}[\exp(K_r) - 1]. \qquad (12.6.3)$$

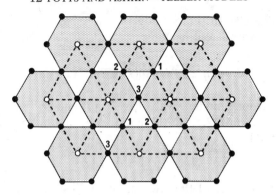

Fig. 12.10. The triangular lattice $\mathcal{L}$ (open circles and broken lines) and its medial Kagomé lattice $\mathcal{L}'$ (full circles and lines).

Triangular-Honeycomb Lattice Duality

Alternatively, consider the lattice $\mathcal{L}_D$ that is dual to $\mathcal{L}$. This is an honeycomb lattice of $2N$ sites, the sites lying in the unshaded triangular faces of $\mathcal{L}'$ in Fig. 12.10. The medial graph of $\mathcal{L}_D$ is also $\mathcal{L}'$ (apart from boundary effects), but now the shaded and unshaded faces in Fig. 12.10 are interchanged. By considering the Potts model on $\mathcal{L}_D$, and again using (12.3.7) and (12.3.8), we find that

$$Z_{2N}^H(L_1, L_2, L_3) = q^N Z'', \qquad (12.6.4)$$

where Z'' is the partition function of an ice-type model on $\mathcal{L}'$, with weights

$$\bar{\bar{\omega}}_1, \ldots, \bar{\bar{\omega}}_6 = y_r, y_r, 1, 1, y_r t^{-1} + t^2, y_r t + t^{-2}. \qquad (12.6.5)$$

The corresponding arrow arrangements are ordered as in (12.6.2), t is again given by (12.3.5) and (12.6.3), and

$$y_r = q^{-\frac{1}{2}}[\exp(L_r) - 1]. \qquad (12.6.6)$$

Suppose that $y_r = x_r^{-1}$. Then the weights (12.6.5) are the same as those in (12.6.2), except that they are all multiplied by y_r. Multiplying all weights of type r by any factor α merely multiplies the ice-model partition function by α^N. From (12.6.1) and (12.6.4), it follows that

$$Z_N^T(K_1, K_2, K_3) = (q^{-\frac{1}{2}} x_1 x_2 x_3)^N Z_{2N}^H(L_1, L_2, L_3), \qquad (12.6.7)$$

where

$$x_r = q^{-\frac{1}{2}}[\exp(K_r) - 1] = q^{\frac{1}{2}}/[\exp(L_r) - 1]. \qquad (12.6.8)$$

This is a duality relation, mapping a low-temperature (high-temperature) Potts model on the triangular lattice to a high-temperature (low-temperature) one on the honeycomb lattice. For the Ising model case, when $q = 2$, it is equivalent to the relation (6.3.7).

Location of the Critical Point

We can associate mutually inverse weights with the tips and tails of arrows, and incorporate these into the vertex weights. Obviously this leaves Z' unchanged, but alters $\bar{\omega}_1 , \ldots , \bar{\omega}_6$. If we use the same weights for all edges of the same type (e.g. for all horizontal edges of $\mathcal{L}'$), then $\bar{\omega}_1 , \ldots , \bar{\omega}_4$ are unchanged, $\bar{\omega}_5$ is multiplied by a factor α_r, and $\bar{\omega}_6$ is divided by α_r, where r is the type of the corresponding site, and $\alpha_1 \alpha_2 \alpha_3 = 1$.

We may ask whether we can use this freedom to make the ice-type model satisfy the 'zero-field' conditions $\bar{\omega}_1 = \bar{\omega}_2$, $\bar{\omega}_3 = \bar{\omega}_4$, $\bar{\omega}_5 = \bar{\omega}_6$, for $r = 1, 2, 3$. The first two conditions are automatically satisfied. The third implies that the product of the three weights $\bar{\omega}_5$ (for $r = 1 , 2 , 3$), is the same as that of the three weights $\bar{\omega}_6$. These products are unchanged by the additional arrow weights, so from (12.6.2) we must have

$$\prod_{r=1}^{3} (t^{-1} + x_r t^2) = \prod_{r=1}^{3} (t + x_r t^{-2}) . \tag{12.6.9}$$

From (12.3.5) and (12.6.3), t is related to q by

$$q^{\frac{1}{2}} = t^3 + t^{-3} . \tag{12.6.10}$$

Expanding both sides of (12.6.9), it follows that

$$q^{\frac{1}{2}} x_1 x_2 x_3 + x_2 x_3 + x_3 x_1 + x_1 x_2 = 1 . \tag{12.6.11}$$

This is a necessary condition for the ice-type model to be reducible to zero-field form. It is also sufficient.

What is intriguing is that (12.6.11) is also the condition for the triangular Potts model (or the dual honeycomb model) to be critical. This is not obvious: unlike the Ising model, the Potts model does not in general have a star-triangle relation, so we cannot establish a triangular – triangular duality relation like that of Section 6.5. However, Kim and Joseph (1974) showed that there is a star – triangle relation when (12.6.11) is satisfied (they actually considered the isotropic case, when $x_1 = x_2 = x_3$, but the argument readily generalizes). The resulting mapping takes the triangular Potts model back to itself (with the same interaction coefficients K_1, K_2, K_3). Thus the model is self-dual at the particular temperature specified by

(12.6.11). Kim and Joseph conjectured that this self-dual point is also the critical point. Baxter *et al.* (1978) generalized this argument to a triangular Potts model with additional three-site interactions on alternate triangular faces. They showed that this more general model always has a triangular – triangular duality relation, and that at the self-dual point the three-spin interaction vanishes and (12.6.11) is satisfied. Hintermann *et al.* (1978) have verified that the critical temperature is indeed this self-dual temperature.

Here I shall use the variables x_1, x_2, x_3, but is should be noted that it is quite natural to work instead with β_1, β_2, β_3, where β_r is the value of β in (12.5.6) when $x = x_r$. The condition (12.6.11) then takes the simple linear form

$$\beta_1 + \beta_2 + \beta_3 = \lambda . \tag{12.6.12}$$

(In fact β_r corresponds to u_r in Section 11.1, and (12.6.12) to (11.1.12).)

Critical Free Energy

Provided the criticality condition (12.6.11) is satisfied, we can calculate the free energy of the triangular lattice Potts model (and of the dual honeycomb-lattice model). One way to do this is to shrink the up-pointing triangles of the Kagomé lattice down to points. As is explained in Baxter *et al.* (1978), the ice-type model then becomes a '20-vertex' model on a triangular lattice. Kelland (1974b) has investigated the conditions under which such a 20-vertex model is solvable by the Bethe ansatz method of Chapter 8. It turns out that these conditions are precisely that the model correspond to the Kagomé lattice six-vertex model with weights given by (12.6.2) and (12.6.11). Thus Kelland's results enable us to calculate the free energy when, and only when, the 20-vertex model is equivalent to a critical triangular-lattice Potts model.

(There are also solvable 'free-fermion' cases of the 20-vertex model, but they are solved by other methods [Sacco and Wu, 1975].)

We can also obtain the free energy by using the results we obtained in Chapter 11. When the restriction (12.6.11) is satisfied, we can arrange the mutually inverse 'tips and tails' arrow weights so that (12.6.2) becomes

$$\bar{\omega}_1, \ldots , \bar{\omega}_6 = a_r, a_r, b_r, b_r, c_r, c_r, \tag{12.6.13}$$

where

$$a_r = 1, \quad b_r = x_r, \quad c_r = (1 + x_r t^3)^{\frac{1}{2}} (1 + x_r t^{-3})^{\frac{1}{2}}, \tag{12.6.14}$$

and again $r = 1$, 2 or 3 depending on the type of the site of $\mathcal{L}'$ that is being considered. The vertex arrow configurations corresponding to a_r, b_r, c_r are shown in Fig. 11.2.

The ice-type model is now a zero-field six-vertex Kagomé-lattice model. This is a special case of the eight-vertex model defined in Section 11.1, the weights d_1, d_2, d_3 being zero. Furthermore, when (12.6.11) is satisfied, we can verify that the six 'star-triangle' relations (11.1.7) are all satisfied.

The free energy is therefore given by (11.4.5). Remembering that the Kagomé lattice $\mathcal{L}'$ has $3N$ sites, this equation implies that (for N large)

$$Z' = Z_N^{sq}(x_1)\, Z_N^{sq}(x_2)\, Z_N^{sq}(x_3)\,, \tag{12.6.15}$$

where $Z_N^{sq}(x_r)$ is the partition function of a *square-lattice* ice-type model, with weights given by (12.6.14), all sites being of type r. Using the results of Chapter 8, or simply referring back to (12.5.3) – (12.5.6), we can verify that the dimensionless free energy per site of such a square-lattice model is

$$-N^{-1}\ln Z_N^{sq}(x) = -\phi(x)\,. \tag{12.6.16}$$

From (12.6.1) and (12.6.15), the dimensionless free energy per site of the critical triangular Potts model is therefore

$$\psi_T(K_1, K_2, K_3) = -N^{-1}\ln Z_N^T(K_1, K_2, K_3)$$

$$= -\tfrac{1}{2}\ln q - \phi(x_1) - \phi(x_2) - \phi(x_3)\,. \tag{12.6.17}$$

Critical Internal Energy

The equations (12.5.8) – (12.5.13) can readily be generalized to the triangular lattice, giving

$$\langle E \rangle = -q^{-\frac{1}{2}} \sum_{r=1}^{3} J_r \exp(J_r/k_B T)\, I_r\,, \tag{12.6.18}$$

where, using (12.6.1) – (12.6.3)

$$I_r = \frac{\langle n_3 \rangle + \langle n_4 \rangle}{x_r} + \frac{\exp(\lambda)\,\langle n_5 \rangle}{1 + x_r \exp(\lambda)} + \frac{\exp(-\lambda)\,\langle n_6 \rangle}{1 + x_r \exp(-\lambda)}\,. \tag{12.6.19}$$

Here $r = 1$, 2 or 3, and n_j in (12.6.19) is the number of sites of type r in $\mathcal{L}'$ that are in the arrow arrangement j of Fig. 11.2.

The ratio $\langle n_j \rangle/N$ is therefore the probability that a site of type r is in arrow arrangement j. This is a local correlation for this site, so from (11.3.3) it depends only on a_r, b_r, c_r, and hence only on x_r. (As usual, we regard q, and therefore λ and t, as constant.)

In fact, I_r must have the same value as for a square lattice of N sites, in which all sites have weights (12.6.2), r being the same for all sites. But this I_r has been calculated in (12.5.22). From (12.6.18), the internal energy per site $U = \langle E \rangle / N$ of the triangular lattice Potts model is therefore

$$U = q^{-\frac{1}{2}} \sum_{r=1}^{3} J_r \exp(J_r/k_B T) \left[-\phi(x_r) \pm \zeta(x_r) P_0 \right]. \qquad (12.6.20)$$

12.7 Combined Formulae for all Three Planar Lattice Potts Models

The critical free energy and internal energy of the honeycomb lattice Potts model can readily be obtained from those of the triangular lattice, by using the duality relation (12.6.7). The results are similar in form to those of the square and triangular lattices. In fact we can combine them into single formulae, just as we did in Section 11.8 for the Ising model. For all three lattices the criticality condition is

$$\prod_r \left[(1 + x_r \exp\{\lambda\}) / (1 + x_r \exp\{-\lambda\}) \right] = \exp(4\lambda) : \qquad (12.7.1)$$

the dimensionless free energy per site is then

$$\psi = f/k_B T = -\tfrac{1}{2} \ln q - \tfrac{1}{2} \sum_r \phi(x_r) , \qquad (12.7.2)$$

and the internal energy per site is

$$U = \tfrac{1}{2} q^{-\frac{1}{2}} \sum_r J_r \exp(J_r/k_B T) [-\phi'(x_r) \pm \zeta(x_r) P_0] , \qquad (12.7.3)$$

the upper (lower) signs being chosen if the critical temperature is approached from above (below).

The product in (12.7.1), and the sums in (12.7.2) and (12.7.3), are over all edges round a particular site: thus r has three values for the honeycomb lattice, four for the square lattice, and six for the triangular lattice. The J_r is the interaction coefficient for the rth edge in this sum, so with an obvious notation we have $J_3 = J_1$ and $J_4 = J_2$ for the square lattice; and $J_4, J_5, J_6 = J_1, J_2, J_3$ for the triangular lattice. As usual, x_r is defined by (12.3.4), i.e.

$$x_r = q^{-\frac{1}{2}} [\exp(J_r/k_B T) - 1] . \qquad (12.7.4)$$

The parameter λ is defined by (12.3.5) (it is pure imaginary if $q < 4$), and the functions $\phi(x)$, $\zeta(x)$ are defined by (12.5.6) and (12.5.20); P_0 is zero for $q \le 4$, while for $q > 4$ it is given by (12.5.24).

Let $\bar{q}$ be the coordination number of the lattice (3 for the honeycomb, 4 for the square, and 6 for the triangular lattice). Then r in (12.7.1)–

Then for an *isotropic* model, where $J_1 = J_2 = \ldots = J$, the criticality condition (12.7.1) becomes

$$\exp(-J/k_B T) = \sinh(\lambda - \theta)/\sinh(\lambda + \theta). \qquad (12.7.6a)$$

When $q = 2$, then $\lambda = i\pi/4$, and this equation reduces to the Ising model formula (11.8.45a).

We can complete the comparison with (11.8.45) by noting that the methods of Chapter 4 can readily be generalized to the q-component Potts model on the Bethe lattice of coordination number $\bar{q}$. The critical temperature of this model is given by

$$\exp(-J/k_B T) = (\bar{q} - 2)/(\bar{q} + q - 2). \qquad (12.7.6b)$$

12.8 Critical Exponents of the Two-Dimensional Potts Model

We have calculated the free energy f and internal energy U of the planar Potts models, but we have done so only for the zero-field model at the critical temperature T_c. (More accurately, we have calculated U in the limit when T approaches T_c from above, or from below.)

This tells us that the transition is first-order (with non-zero latent heat) for $q > 4$, and is continuous (no latent heat) for $q \leq 4$. It does not tell what the critical exponents are for $q \leq 4$. To obtain these directly we would have to solve the Potts model for general temperatures, which has not been done. (This is a very tantalising problem as I remark in Section 14.8.)

However, we do have some information. When $q = 2$, the Potts model becomes the Ising model, which was solved by Onsager (1944) and Yang (1952). The critical exponents α, β and δ are given in (7.12.12), (7.12.14) and (7.12.16). They are

$$\alpha = 0, \quad \beta = \tfrac{1}{8}, \quad \delta = 15 \quad \text{when } q = 2. \qquad (12.8.1a)$$

Alexander (1975) has argued that the three-state Potts model and the hard hexagon model (which each have three ordered states) should be in the same universality class and have the same critical exponents. From the hard hexagon results (14.7.12), (14.7.13), this implies that

$$\alpha = \tfrac{1}{3}, \quad \beta = \tfrac{1}{9}, \quad \delta = 14 \quad \text{when } q = 3. \qquad (12.8.1b)$$

Similarly, Domany and Riedel (1978) argue that the four-state Potts model and the three-spin model (each with four ordered states) should have the same exponents, so from (11.10.39),

$$\alpha = \tfrac{2}{3}, \quad \beta = \tfrac{1}{12}, \quad \delta = 15 \quad \text{when } q = 4 . \tag{12.8.1c}$$

We have seen in Sections 12.3–12.5 that the critical Potts model is equivalent to a zero-field six-vertex model, with weights given by (12.5.2). We can regard this as a special case of the eight-vertex model, in which $d = 0$. The Δ of (10.15.1) is given by (12.5.4): for $q < 4$ we see that $0 > \Delta > -1$. From (10.16.8) we see that this corresponds to a critical eight-vertex model, with μ given by $\Delta = -\cos \mu$. From (12.5.4) this implies

$$q^{\frac{1}{2}} = 2 \cos \mu, \quad 0 < \mu < \tfrac{1}{2}\pi . \tag{12.8.2}$$

This μ is the parameter that enters the formulae (10.12.24) for the critical exponents of the eight-vertex model. Den Nijs (1979) argued that the exponents of the Potts model should also depend in a simple way on μ, or more precisely on

$$y = 2\mu/\pi . \tag{12.8.3}$$

(This y lies between 0 and 1. Den Nijs and others refer to it as y_T^{8v}.) He conjectured that the critical exponent α of the Potts model (for $q \leq 4$) is

$$\alpha = (2 - 4y)/(3 - 3y) . \tag{12.8.4}$$

Similarly, Nienhuis et al. (1980), and Pearson (1980) have independently conjectured that

$$\beta = (1 + y)/12 . \tag{12.8.5}$$

The scaling relation (1.2.12) and (1.2.13) then predict that

$$\delta = (15 - 8y + y^2)/(1 - y^2) . \tag{12.8.6}$$

For $q = 1, 2, 3, 4$, the parameter y has the values $\tfrac{2}{3}, \tfrac{1}{2}, \tfrac{1}{3}, 0$, respectively. The conjectures (12.8.4)–(12.8.6) therefore agree with the values in (12.8.1), and also predict that

$$\alpha = -\tfrac{2}{3}, \quad \beta = \tfrac{5}{36}, \quad \delta = 18\tfrac{1}{5} \quad \text{when } q = 1 . \tag{12.8.7}$$

(As we remarked in (12.2.4), the $q = 1$ case is the percolation problem.) The conjectures are also consistent with numerical estimates of the exponents (Blumberg et al., 1980; Blöte et al., 1981), and with a renormalization-group perturbation expansion about $q = 4$ (Cardy et al., 1980). Very recently, Black and Emery (1981) have verified (12.8.4) by using renormalization-group methods: it seems likely that (12.8.5) and (12.8.6) are also exactly correct.

12.9   Square-Lattice Ashkin – Teller Model

Ashkin and Teller (1943) introduced their model as a generalization of the Ising model to a four component system. Each site of a lattice $\mathscr{L}$ is occupied by one of four kinds of atom: A, B, C or D. Two neighbouring atoms interact with an energy: ε_0 for AA, BB, CC, DD; ε_1 for AB, CD; ε_2 for AC, BD; and ε_3 for AD, BC.

Fan (1972b) showed that this model can be expressed in terms of Ising spins. With each site i associate *two* spins: s_i and σ_i. Let $(s_i , \sigma_i) = (+ , +)$ if there is an A atom at site i; $(+ , -)$ if a B atom; $(- , +)$ if C; and $(- , -)$ if D. Then the interaction energy for the edge (i , j) is

$$\varepsilon(i , j) = -J s_i s_j - J' \sigma_i \sigma_j - J_4 s_i \sigma_i s_j \sigma_j - J_0 , \qquad (12.9.1)$$

where

$$-J = (\varepsilon_0 + \varepsilon_1 - \varepsilon_2 - \varepsilon_3)/4$$

$$-J' = (\varepsilon_0 + \varepsilon_2 - \varepsilon_3 - \varepsilon_1)/4$$

$$-J_4 = (\varepsilon_0 + \varepsilon_3 - \varepsilon_1 - \varepsilon_2)/4 \qquad (12.9.2)$$

$$-J_0 = (\varepsilon_0 + \varepsilon_1 + \varepsilon_2 + \varepsilon_3)/4 .$$

As usual, we want to calculate the partition function. From (1.4.1), this is

$$Z_{AT} = \sum_s \sum_\sigma \exp\left[- \sum_{(i,j)} \varepsilon(i , j)/k_B T \right] , \qquad (12.9.3)$$

where k_B is Boltzmann's constant, T is the temperature. The summation inside the exponential is over all edges (i , j) of the lattice; the outer sums are over all values of all the spins $s_1, s_2, s_3, \ldots$ and $\sigma_1, \sigma_2, \sigma_3, \ldots$.

We shall find it convenient to use the dimensionless interaction coefficients

$$K = J/k_B T , \qquad K' = J'/k_B T , \qquad (12.9.4)$$

$$K_4 = J_4/k_B T , \qquad K_0 = J_0/k_B T ,$$

and the edge Boltzmann weights

$$\omega_i = \exp(-\varepsilon_i/k_B T) , \, i = 1, \ldots , 4 . \qquad (12.9.5)$$

From (12.9.2), we see that

$$\omega_0 = \exp(K + K' + K_4 + K_0) , \qquad \omega_1 = \exp(K - K' - K_4 + K_0) , \qquad (12.9.6)$$

$$\omega_2 = \exp(-K + K' - K_4 + K_0) , \qquad \omega_3 = \exp(-K - K' + K_4 + K_0) .$$

Equivalence to an Alternating Eight-Vertex Model

The above considerations apply to any lattice $\mathscr{L}$, planar or not. Now let us specialize to the case when $\mathscr{L}$ is the square lattice, with N sites. Then from (12.9.1) it is apparent that we can think of the Ashkin – Teller (AT) model as two square-lattice Ising models (the s-model and the σ-model), coupled via a four-spin interaction.

This is similar to the zero-field eight-vertex model, whose Hamiltonian is given by (10.3.1) with $J_v = J_h = 0$. However, the geometry is different: for the eight-vertex model the spins are arranged as in Fig. 10.4, the s-spins occupying the full circles, and the σ-spins the open circles. In the AT model the spins s_i and σ_i both lie on site i.

Even so, Wegner (1972) shows that the AT model could be expressed as an alternating eight-vertex model. The trick is to apply the duality transformation of Section 6.2 to the σ-spins only.

To do this, note that from (12.9.1), (12.9.3) can be written as

$$Z_{AT} = \sum_s \exp\left[\sum_{(i,j)} (Ks_i s_j + K_0)\right] Z_N\{L\} \qquad (12.9.7)$$

where

$$Z_N\{L\} = \sum_\sigma \exp\left[\sum_{(i,j)} L_{ij}\sigma_i\sigma_j\right], \qquad (12.9.8)$$

the edge-coefficient L_{ij} being given by

$$L_{ij} = K' + K_4 s_i s_j. \qquad (12.9.9)$$

Here N is the number of sites of the square lattice $\mathscr{L}$; L is the set of coefficients L_{ij}, one for each edge of $\mathscr{L}$.

Clearly each L_{ij} depends on the spins s_i and s_j, but for the moment let us regard $s_1, \ldots, s_N$ as fixed, and consider the expression on the RHS of (12.9.8). This is a standard square-lattice Ising model partition function, except that the system is inhomogeneous, having interaction coefficients L_{ij} that vary from edge to edge.

Now look at the duality relation (6.2.14). Using (6.2.10) (and ignoring boundary effects), this states that if $Z_N(K, L)$ is the partition function of a square-lattice Ising model with interaction coefficient K for vertical edges, L for horizontal edges, and with N sites, then

$$Z_N(K^*, L^*) = [2 \exp(-K - L) \cosh K^*$$

$$\times \cosh L^*]^N Z_N(K, L). \qquad (12.9.10)$$

The dual coefficients K^*, L^* are defined by (6.2.14a).

This relation applies to an homogeneous system. However, it is quite straightforward to generalize the working of Section 6.2 to an inhomogeneous system, with interaction coefficient L_{ij} on edge (i,j). The result is

$$Z_N\{L^*\} = \prod_{(i,j)} [2^{\frac{1}{2}} \exp(-L_{ij}) \cosh L_{ij}^*] \quad Z_N\{L\}, \qquad (12.9.11)$$

where

$$\tanh L_{ij}^* = \exp(-2L_{ij}). \qquad (12.9.12)$$

Here L denotes the set of all edge coefficients L_{ij}, and L^* the set of L_{ij}^*; $Z_N\{L\}$ is the original partition function for the lattice $\mathscr{L}$, as defined in (12.9.8); $Z_N\{L^*\}$ is the partition function for the lattice $\mathscr{L}_D$ that is dual to $\mathscr{L}$; L_{ij}^* is the interaction coefficient of the edge of $\mathscr{L}_D$ that is dual to the edge (i,j) of $\mathscr{L}$.

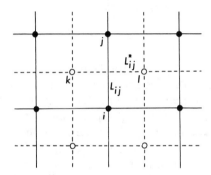

Fig. 12.11. The lattice $\mathscr{L}$ (solid circles and lines) and its dual $\mathscr{L}_D$ (open circles and broken lines); (k, l) is the edge of $\mathscr{L}_D$ that is dual to the edge (i, j) of $\mathscr{L}; L_{ij}^*$ and L_{ij} are the corresponding interaction coefficients of these two edges.

Let k, l be the sites of $\mathscr{L}_D$ such that (k,l) is the edge dual to (i,j), as indicated in Fig. 12.11. Then

$$Z_N\{L^*\} = \sum_t \exp\left[\sum_{(i,j)} L_{ij}^* t_k t_l\right], \qquad (12.9.13)$$

where $t_1, \ldots, t_N$ are spins on the sites of $\mathscr{L}_D$.

Using (12.9.11) to express $Z_N\{L\}$ in terms of $Z_N\{L^*\}$, substituting the result into (12.9.7) and using (12.9.13), we obtain

$$Z_{\mathrm{AT}} = \sum_s \sum_t \prod_{(i,j)} W_{ij}, \qquad (12.9.14)$$

the product being over all edges (i,j) of $\mathscr{L}$, with

$$W_{ij} = 2^{-\frac{1}{2}} \operatorname{sech} L_{ij}^* \exp[L_{ij} + K s_i s_j + K_0 + L_{ij}^* t_k t_l]. \qquad (12.9.15)$$

Remembering that L_{ij} is given by (12.9.9), where $s_i s_j = \pm 1$, and that L_{ij}^* is given by (12.9.12), we can write W_{ij} as

$$W_{ij} = \exp(As_i s_j + Bt_k t_l + Cs_i s_j t_k t_l + D), \qquad (12.9.16)$$

where A, B, C, D are defined by

$$a = \exp(A + B + C + D) = 2^{-\frac{1}{2}}(\omega_0 + \omega_1)$$

$$b = \exp(-A - B + C + D) = 2^{-\frac{1}{2}}(\omega_2 - \omega_3)$$

$$c = \exp(-A + B - C + D) = 2^{-\frac{1}{2}}(\omega_2 + \omega_3) \qquad (12.9.17)$$

$$d = \exp(A - B - C + D) = 2^{-\frac{1}{2}}(\omega_0 - \omega_1).$$

From (1.4.1), (12.9.14) and (12.9.16), Z_{AT} is therefore the partition function of a model with Hamiltonian

$$E(s, t) = -k_B T \sum_{(i,j)} (As_i s_j + Bt_k t_l + Cs_i s_j t_k t_l + D). \qquad (12.9.18)$$

Now take $\mathscr{L}$ to be the lattice of solid circles and lines in Fig. 10.4. Then $\mathscr{L}_D$ is the lattice of open circles and broken lines. The s-spins lie on $\mathscr{L}$, the t-spins on $\mathscr{L}_D$. We see that (12.9.18) is the Hamiltonian of a model with interactions between neighbouring s-spins, and between neighbouring t-spins, with a four-spin interaction between spins on crossing edges.

This Hamiltonian has the same general form as (10.3.1), with $J_v = J_h = 0$, with $J'' = k_B TC$, and with an extra constant energy $-k_B TD$ per edge. To avoid confusion, let us refer to the J and J' in (10.3.1) and Fig. 10.4 as J_{8V} and J'_{8V}: they are of course *not* the same as those in equations (12.9.1) – (12.9.4) above.

There is one significant difference between (12.9.18) and (10.3.1): in (12.9.18) A and B are associated with the solid and broken edges, respectively, of Fig. 10.4; in (10.3.1) J_{8V} and J'_{8V} are associated with the SW – NE and SE – NW edges of Fig. 10.4. To put (12.9.18) into the form (10.3.1) we must therefore allow J_{8V} and J'_{8V} to alternate from site to site. We can choose $(J_{8V}/k_B T, J'_{8V}/k_B T)$ to be (A, B) on one sub-lattice ($i + j$ even in (10.3.1)): it is then (B, A) on the other sub-lattice ($i + j$ odd). The Ashkin – Teller model is therefore equivalent to an alternating, or 'staggered' eight-vertex model. From (10.3.9) (including the extra constant energy $-D/k_B T$), the Boltzmann weights of this eight-vertex model are equal to the a, b, c, d of (12.9.17) on one sub-lattice. On the other sub-lattice they are equal to a, b, d, c.

It is of course intriguing that the Ashkin – Teller model should be equivalent to a staggered eight-vertex model. In many ways the equivalence is reminiscent of that found in Section 12.4 between the Potts model and a staggered six-vertex model. Indeed, both equivalences can be applied to the $q = 4$ Potts model (which is a special case of the Ashkin – Teller model, with $\varepsilon_1 = \varepsilon_2 = \varepsilon_3$): they lead to the same vertex model.

The form of the staggering in this eight-vertex model is particularly simple, consisting merely of the interchange of the weights c and d. From (10.15.1a) and (10.15.6), it follows that Δ, Γ, k and λ are not staggered, each having the same value for both sub-lattices. The staggering affects only the elliptic function parameter v, which is negated on going from one sub-lattice to another. Unfortunately it is still not possible to put k, λ, v into the form (10.17.7), which are the most general conditions under which the eight-vertex model has been solved. Thus the general Ashkin – Teller model remains unsolved.

Even so, the equivalence of the AT model to a staggered eight-vertex model does have some interesting consequences, as I hope to indicate in the remainder of this section.

Duality

Obviously, the partition function of the staggered eight-vertex model is unaltered by interchanging c and d on all sites (this merely interchanges the two sublattices). Write Z_{AT} as a function of the Boltzmann weights $\omega_0 , \dots , \omega_3$ in (12.9.6). Then from (12.9.17) it follows that

$$Z_{AT}(\omega_0' , \omega_1' , \omega_2' , \omega_3') = Z_{AT}(\omega_0 , \omega_1 , \omega_2 , \omega_3) , \qquad (12.9.19)$$

provided that

$$\omega_0' = \tfrac{1}{2}(\omega_0 + \omega_1 + \omega_2 + \omega_3) ,$$

$$\omega_1' = \tfrac{1}{2}(\omega_0 + \omega_1 - \omega_2 - \omega_3) , \qquad (12.9.20)$$

$$\omega_2' = \tfrac{1}{2}(\omega_0 + \omega_2 - \omega_3 - \omega_1) ,$$

$$\omega_3' = \tfrac{1}{2}(\omega_0 + \omega_3 - \omega_1 - \omega_2) .$$

This is a duality relation, relating a high-temperature AT model to a low-temperature one. It was obtained by Fan (1972a), who conjectured that the critical temperature might be given by the self-duality condition:

$$\omega_0 = \omega_1 + \omega_2 + \omega_3 . \qquad (12.9.21)$$

However, Wegner (1972) remarked that this is precisely the condition for the corresponding eight-vertex model to be homogeneous; since then $c = d$. The homogeneous eight-vertex model has been solved in Chapter 10. It is not in general critical (even when $c = d$), so nor is the AT model under the condition (12.9.21). (The spins $s_1, s_2, \ldots$ occur in both models, and it is reasonable to suppose that either model is critical if and only if the correlation $\langle s_i s_j \rangle$ decays as an inverse power of the distance r_{ij} between sites i and j, rather than decaying exponentially. This means that if one model is critical, then so is the other.)

Indeed, it is apparent that when $J_4 = 0$ in (12.9.1), then the AT model factors into two independent Ising models, one for the s-spins with coefficient J, the other for the σ-spins with coefficient J'. Provided $J \neq J'$, these will have different critical temperatures. The AT model therefore then has two critical temperatures. They lie on either side of the self-dual temperature defined by (12.9.21), and the mapping (12.9.20) takes one to the other.

Wegner (1972) argued that in general there should still be two such critical temperatures for $J_4 \neq 0$, and Wu and Lin (1974) have considered the possible location of the critical surfaces in (ω_1/ω_0, ω_2/ω_0, ω_3/ω_0) space. One currently unsolved problem is to locate these surfaces exactly. Presumably universality holds on these surfaces (except on certain special lines to be discussed below), in which case the critical exponents must be those of the $J_4 = 0$ case, i.e. of the Ising model.

Other Symmetry Properties

In addition to (12.9.19), Z_{AT} satisfies various other symmetry relations. It is unchanged by permuting J, J', J_4 in (12.9.1), since this corresponds merely to permuting s_i, σ_i, $s_i \sigma_i$ (any two of which can be regarded as independent spins) at each site. It is also unchanged by negating J' and J_4, since this corresponds to negating alternate σ-spins. From (12.9.4) and (12.9.6) it follows that

$$Z_{AT}(\omega_i, \omega_j, \omega_k, \omega_l) = Z_{AT}(\omega_0, \omega_1, \omega_2, \omega_3) \qquad (12.9.22)$$

for all permutations i, j, k, l of 0, 1, 2, 3. Thus Z_{AT} is unchanged by permuting the weights $\omega_0, \ldots, \omega_3$.

Another way of obtaining the duality and symmetry relations (12.9.19) and (12.9.22) is to define the eight-vertex 'w-weights' w_1, w_2, w_3, w_4 as in (10.2.16). The symmetry relations (10.2.17) are true even for an inhomogeneous eight-vertex model. Applying them, using (12.9.17), we again obtain (12.9.19) and (12.9.22).

Critical Isotropic AT Model

If $J = J'$, then the above argument that the AT model should have two critical temperatures breaks down. This is a particularly interesting case: let us refer to it as the 'isotropic' AT model. From (12.9.6) we see that it corresponds to imposing the condition

$$\omega_1 = \omega_2 \qquad (12.9.23)$$

on the AT model weights.

This still corresponds to a staggered eight-vertex model, in which the weights c, d are interchanged on alternate sites, so in general has not been solved. However, if the self-duality condition (12.9.21) is satisfied, then $c = d$ and the eight-vertex model becomes homogeneous. Using (10.2.16), its 'w-weights' are

$$w_1 = \tfrac{1}{2}(a + b) = 2^{\frac{1}{2}}\omega_1$$

$$w_2 = \tfrac{1}{2}(a - b) = 2^{-\frac{1}{2}}(\omega_1 + \omega_3) \qquad (12.9.24)$$

$$w_3 = \tfrac{1}{2}(c + d) = 2^{-\frac{1}{2}}(\omega_1 + \omega_3)$$

$$w_4 = \tfrac{1}{2}(c - d) = 0 .$$

From (10.2.17), Z_{AT} is unaltered by interchanging w_2 and w_4. Let a', b', c', d' be the resulting new eight-vertex weights. Then we see that

$$a' = b' = 2^{\frac{1}{2}}\omega_1, \quad c' = 2^{\frac{1}{2}}(\omega_1 + \omega_3), \quad d' = 0 . \qquad (12.9.25)$$

Since $d' = 0$, the model reduces to a six-vertex model (in fact to an F-model). Let f be the free energy per site of the AT model, defined as in (1.7.6) by

$$-f/k_B T = \lim_{N \to \infty} N^{-1} \ln Z_{AT} . \qquad (12.9.26)$$

Remembering that the vertex model has twice as many sites as the original AT model, from (8.3.3), (8.8.9), (8.8.17), (8.9.7) and (8.9.9) it follows that when the restrictions (12.9.21) and (12.9.23) are satisfied, then f is given by the following equations:

$$\omega_3 < \omega_1: \quad \cos(\mu/2) = \tfrac{1}{2}\Big(1 + \frac{\omega_3}{\omega_1}\Big), \quad 0 < \mu < \frac{2\pi}{3} ,$$

$$-f/k_B T = 2 \ln(2^{\frac{1}{2}}\omega_1) + \int_{-\infty}^{\infty} \frac{\tanh \mu x \sinh (\pi - \mu)x}{x \sinh \pi x} \, dx ; \qquad (12.9.27a)$$

$$\omega_3 = \omega_1: \quad -f/k_B T = 2 \ln(2^{\frac{1}{2}}\omega_1) + 4 \ln[\Gamma(\tfrac{1}{4})/2\Gamma(\tfrac{3}{4})] ; \qquad (12.9.27b)$$

$$\omega_3 > \omega_1: \quad \cosh(\lambda/2) = \tfrac{1}{2}\Big(1 + \frac{\omega_3}{\omega_1}\Big), \quad \lambda > 0 ,$$

$$-f/k_B T = 2 \ln(2^{\frac{1}{2}}\omega_1) + \lambda + 2 \sum_{m=1}^{\infty} m^{-1} \exp(-m\lambda) \tanh m\lambda . \qquad (12.9.27c)$$

Critical Exponents

The homogeneous eight-vertex model is critical if and only if the two middle numbers of the set $|w_1|$, $|w_2|$, $|w_3|$, $|w_4|$, arranged in decreasing order, are equal. In this case the symmetry relation (10.2.17) can be used to map it into a six-vertex model of the type discussed in Section 8.8, i.e. with $-1 < \Delta < 1$. The critical exponents are then given by (10.12.24). Let us give them a superfix '$8V$' to denote 'eight-vertex', and give the 'magnetic' exponents (corresponding to introducing a field $-H\Sigma\sigma_i$) a suffix 'm'. The 'electric' exponents (corresponding to a field $-E\Sigma\sigma_i\sigma_j$, where i, j are nearest-neighbours) already have a suffix 'e'. Then from (10.12.24)

$$\alpha^{8V} = 2 - 2y^{-1}, \quad \beta_m^{8V} = (8y)^{-1},$$

$$\beta_e^{8V} = (2 - y)/(4y),$$

(12.9.28)

where

$$y = 2\mu/\pi,$$

(12.9.29)

μ being the parameter defined in Section 8.8. [This y is the renormalization-group exponent that we used in (12.8.3).]

Applying these general considerations to the self-dual isotropic AT model, it follows from (12.9.24) that the model is critical provided that $\omega_3 < \omega_1$. Its free energy is then given by (12.9.27a), the μ therein being the same as that in (12.9.29).

As with the Potts model, we have only evaluated the free energy f at the critical point: we are not in a position to see how f varies with temperature or field, so we cannot directly determine any critical exponents. Certainly we cannot apply the homogeneous eight-vertex results (12.9.28) to the AT model, because the models are equivalent only at criticality.

Even so, Kadanoff (1977, 1979) and Kadanoff and Brown (1979) have used operator algebras and scaling theory (Kadanoff, 1976) to relate the critical exponents of the eight-vertex and isotropic Ashkin – Teller models. Knops (1980) obtained the same relations by using renormalization-group arguments; den Nijs (1981) has extended this approach, as well as justifying (1979) Enting's (1975) conjecture that $\delta_m = 15$ for the AT model. Zisook (1980) and Zittartz (1981) have checked some of these relations by developing perturbation expansions.

Altogether, they find that the Ashkin – Teller exponents are

$$\alpha^{AT} = (2 - 2y)/(3 - 2y), \quad \beta_m^{AT} = (2 - y)/(24 - 16y)$$

(12.9.30)

$$\beta_e^{AT} = (12 - 8y)^{-1}.$$

As with the eight-vertex model, there are two sets of critical exponents: 'magnetic' exponents corresponding to a field $-H\Sigma\sigma_i$, and 'electric' exponents corresponding to a field $-E\Sigma\sigma_i s_i$. I use the suffix m for the former, e for the latter. Thus β_m is the exponent of the order parameter $\langle\sigma_1\rangle$, β_e of $\langle\sigma_1 s_1\rangle$.

The same arguments that give (12.9.30) also imply the scaling relations (1.2.12) – (1.2.16), so we can use these to obtain γ_m, δ_m, η_m, γ_e, δ_e, η_e and ν. Both the eight-vertex and Ashkin – Teller models violate universality, having exponents that vary continuously with the parameter y. Both models satisfy the relations

$$\delta_m = (2 - \alpha - \beta_m)/\beta_m = 15 \,,$$

$$\beta_e = (1 - \alpha)/4 \,.$$

$$(12.9.31)$$

The criticality conditions (12.9.21) and (12.9.23) can be written in terms of the interaction coefficients K, K', K_4. From (12.9.6) they are

$$K' = K, \quad \exp(-2K_4) = \sinh 2K \,, \qquad (12.9.32a)$$

the restriction $\omega_3 < \omega_1$ is equivalent to

$$K_4 < K \,, \qquad (12.9.32b)$$

and the definition of μ in (12.9.27a) can be written as

$$\cos \mu = \tfrac{1}{2}[\exp(4K_4) - 1] \,, \qquad (12.9.33)$$

where $0 < \mu < 2\pi/3$.

Phase Diagram of the Isotropic AT Model

In fact (12.9.32) is not the only critical line of the isotropic AT model. The complete phase diagram has been obtained by Ditzian et al. (1980) and is surprisingly rich. It is shown in Fig. 12.12. There are give regions in (K_4, K) space: in I the system is ferromagnetically ordered, $\langle\sigma_1\rangle$, $\langle s_1\rangle$ and $\langle\sigma_1 s_1\rangle$ all being non-zero; in II these order parameters are all zero and the system is disordered; in III there is partial ordering, $\langle\sigma_1 s_1\rangle$ being non-zero, but $\langle\sigma_1\rangle$ and $\langle s_1\rangle$ vanishing; IV is similar to III, except that the order is anti-ferromagnetic, $\langle\sigma_1 s_1\rangle$ alternating from site to site; V is similar to I, except that $\langle\sigma_1\rangle$ and $\langle s_1\rangle$ are anti-ferromagnetically ordered.

The line EF, which is the boundary between regions I and II, is the critical line (12.9.32) discussed above. This is a line of continuously varying exponents, μ in (12.9.29) varying from 0 at F to $2\pi/3$ at E. F is the point (K_t, K_t), where $K_t = \tfrac{1}{4}\ln 3 = 0.2746\ldots$; E is where $K_4/K = -1$, $K \to \infty$.

The line $E'F'$ is also one of continuously varying exponents, and is obtained from EF simply by negating K. Indeed, negating all spins s_i and σ_i on one sub-lattice of $\mathcal{L}$ is simply equivalent to negating K, so the whole of Fig. 12.12 is symmetric about the K_4 axis.

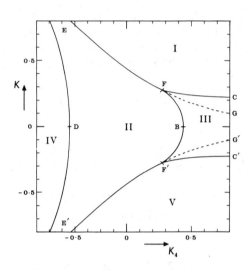

Fig. 12.12. Phase diagram of the isotropic Ashkin – Teller model, in (K_4, K) space.

The line EF continues onto the broken line FG. This is the self-dual line with $\omega_3 > \omega_1$: the system is not critical on this line segment; instead there are two critical lines FB and FC bifurcating from F. Their positions are not precisely known, but B must be the point $(K_c, 0)$, and C the point $(\infty, \frac{1}{2}K_c)$, where K_c is the Ising-model critical value of K, given by

$$\sinh 2K_c = 1, \quad K_c = 0.4406\ldots. \qquad (12.9.34)$$

The lines FB, FC map into one another under the duality relation (12.9.20). The critical exponents thereon are expected to be fixed, having Ising-model values.

Similarly, the position of the critical line EDE' is not precisely known, but D is the point $(-K_c, 0)$ and the exponents are also expected to be those of the Ising model.

13

CORNER TRANSFER MATRICES

13.1 Definitions

Notation

In Chapters 7–10, much use has been made of the row-to-row transfer matrix V. Multiplication by this matrix corresponds to adding a row to the lattice. Each element of V is the total Boltzmann weight of a row of the lattice, as in (7.2.2) and (8.2.2).

Another useful concept is the 'corner transfer matrix' (CTM), which corresponds to adding a quadrant to the lattice. In this section I shall define four such CTMs (one for each corner), and shall call them A, B, C, D. I shall also define four corresponding normalized matrices A_n, B_n, C_n, D_n; and four normalized and diagonalized matrices A_d, B_d, C_d, D_d. Here n and d are *not* indices, but merely denote 'normalized' and 'diagonalized', respectively.

The (σ, σ') element of one of these matrices will be denoted by a further double suffix $\sigma\sigma'$: e.g. $B_{\sigma\sigma'}$ and $(A_n)_{\sigma\sigma'}$ are the (σ, σ') elements of B and A_n, respectively.

The IRF Model

Corner transfer matrices can be defined for any planar lattice model with finite-range interactions, but for definiteness let us consider a square lattice model with interactions round faces. For brevity I shall call this the 'IRF' model. It is defined as follows.

To each site i of the square lattice associate a 'spin' σ_i. In this chapter we shall suppose that each σ_i has value $+1$ or -1; in the next chapter it will be more convenient to let then have values 0 or 1; in general they can take any desired set of values.

Let the total energy be

$$\mathscr{E} = \sum \varepsilon(\sigma_i , \sigma_j , \sigma_k , \sigma_l) , \qquad (13.1.1)$$

where the summation is over all faces of the lattice, and for each face the i, j, k, l are the surrounding sites, arranged as in Figure 13.1(a). From (1.4.1), the partition function is

$$Z = \sum \prod w(\sigma_i , \sigma_j , \sigma_k , \sigma_l) , \qquad (13.1.2)$$

where the product is over all faces of the lattice, the sum is over all values of all the spins, and

$$w(a , b , c , d) = \exp[-\varepsilon(a , b , c , d)/k_B T] . \qquad (13.1.3)$$

This $w(a , b , c , d)$ is the Boltzmann weight of the intra-face interactions between spins a, b, c, d.

Let N be the number of sites of the lattice and define

$$\kappa = \lim_{N \to \infty} Z^{1/N} . \qquad (13.1.4)$$

Then from (1.7.6), the free energy per site is

$$f = -k_B T \ln \kappa . \qquad (13.1.5)$$

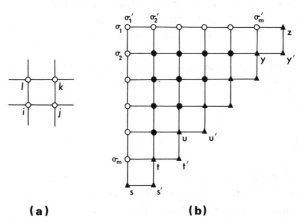

(a)   **(b)**

Fig. 13.1. (a) The ordering of the sites i, j, k, l round a face of the square lattice; (b) the quadrant lattice whose partition function is the $A_{\sigma \sigma'}$ in (13.1.8).

Also, from (1.4.4) the expectation value of a particular spin σ_1 is

$$\langle \sigma_1 \rangle = Z^{-1} \sum \sigma_1 \prod w(\sigma_i , \sigma_j , \sigma_k , \sigma_l) . \qquad (13.1.6)$$

The object of statistical mechanics is to calculate quantities such as κ and $\langle \sigma_1 \rangle$ in the limit of a large lattice. They are expected to be independent of the way in which the lattice becomes large, so long as it does so in all directions.

The function $w(a , b , c , d)$ is at this stage arbitrary, so this IRF model includes many models of particular interest in statistical mechanics. For instance, it includes the case of diagonal interactions together with a four-spin interaction in each face: this is the spin formulation (10.3.1) of the eight-vertex model. More generally, it includes the eight-vertex model in both 'magnetic' and 'electric' fields, and the Ising model in a field.

Ground State

The system will have one or more 'ground-states': these are configurations of all the spins on the lattice for which the energy $\mathscr{E}$, given by (13.1.1), is a minimum. More generally, they can be defined as the configurations which give the largest contribution to the sum-over-states in (13.1.2).

For a ferromagnetic Ising-type system there are at most two ground states: either all spins up, or all spins down. For a system which includes anti-ferromagnetic interactions, the ground state may consist of an alternating pattern of spins. This state is not translation-invariant, even though the energy function (13.1.1) is. There must be at least two ground states in this case, since applying a translational shift changes the spin configuration but not its energy.

In this chapter I refer often to the 'ground-state'. By this I mean a particular ground state, and it is important that the same ground state of the complete lattice be used throughout. For instance, formulae for $\langle \sigma_1 \rangle$ are given in (13.1.11), (13.1.14) and (13.5.15): if the ground state is not translation invariant, then $\langle \sigma_1 \rangle$ may depend (for sufficiently low temperature) on which ground state is used. In this case the translation invariance symmetry is 'spontaneously broken'.

The Matrices A, B, C, D

Consider the lattice in Fig. 13.1(b). Label the left-hand spins $\sigma_1 , \ldots , \sigma_m$, and the top ones $\sigma'_1 , \ldots , \sigma'_m$, as indicated. Clearly σ_1 and

σ_1' are both the upper-left corner spin, so

$$\sigma_1' = \sigma_1 . \qquad (13.1.7)$$

Fix the boundary spins, i.e. those on the sites shown as triangles in Fig. 13.1(b), to have their ground-state values. For instance, for the ferromagnetic Ising model they can all be chosen to be +1.

Let σ denote all the spins $\{\sigma_1, \ldots, \sigma_m\}$; and σ' all the spins $\{\sigma_1', \ldots, \sigma_m'\}$. Define

$$A_{\sigma,\sigma'} = \sum \prod w(\sigma_i, \sigma_j, \sigma_k, \sigma_l) \quad \text{if } \sigma_1 = \sigma_1', \qquad (13.1.8a)$$

$$= 0 \quad \text{if } \sigma_1 \neq \sigma_1', \qquad (13.1.8b)$$

where the product is now over the $\tfrac{1}{2}m(m + 1)$ faces in Fig. 13.1(b), and the sum is over all spins on sites denoted by solid circles. Note that the spins $\sigma_1, \ldots, \sigma_m'$ are *not* summed over, so the RHS of (13.1.8) is a function of σ and σ'.

Define $B_{\sigma,\sigma'}$ in the same way as $A_{\sigma,\sigma'}$, only with Fig. 13.1(b) rotated anti-clockwise through 90°, so that $\sigma_1, \ldots, \sigma_m$ lie on the bottom edge, and $\sigma_1', \ldots, \sigma_m'$ on the left. Similarly, define $C_{\sigma,\sigma'}$, $D_{\sigma,\sigma'}$ by rotating Fig. 13.1(b) twice more through 90°.

Now consider the lattice shown in Fig. 13.2. Divide it by two cuts into four quadrants of equal size, as indicated. Let σ_1 be the centre spin, and

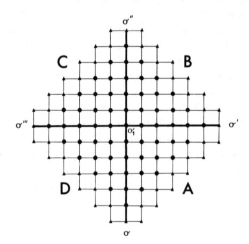

Fig. 13.2. The lattice with partition function (13.1.10). Boundary spins (on sites denoted by triangles) are fixed at their ground state values; σ_1 is the centre spin; σ is the set of all spins (including σ_1) on the lower half of the vertical heavy line; σ' is the set of spins (including σ_1) on the right half of the horizontal heavy line; and similarly for σ'', σ'''.

let σ, σ', σ'', σ''' be the sets of spins (including σ_1) on the corresponding half-cuts in Fig. 13.2. Then from the definition of A, B, C, D, the product

$$A_{\sigma,\sigma'}\, B_{\sigma',\sigma''}\, C_{\sigma'',\sigma'''}\, D_{\sigma''',\sigma}\,, \tag{13.1.9}$$

is the product of the Boltzmann weights of all the faces, summed over all spins other than those on the cuts. The partition function Z of the lattice is Fig. 13.2 is therefore

$$Z = \sum_{\sigma,\sigma',\sigma'',\sigma'''} A_{\sigma,\sigma'}\, B_{\sigma',\sigma''}\, C_{\sigma'',\sigma'''}\, D_{\sigma''',\sigma}\,. \tag{13.1.10}$$

The summation is over all spin-sets $\sigma, \ldots, \sigma'''$, subject only to the restriction that $\sigma_1 = \sigma_1' = \sigma_1'' = \sigma_1'''$, since each of these is the centre spin. However, this restriction can be ignored since from (13.1.8b) the summand vanishes unless it is satisfied. It follows that

$$Z = \text{Trace } ABCD\,. \tag{13.1.11}$$

From (13.1.6), the average value of σ_1 is the ratio of the RHS of (13.1.10), with an extra factor σ_1 in the summand, to its value without this factor. It follows that

$$\langle\sigma_1\rangle = \text{Trace } SABCD/\text{Trace } ABCD\,, \tag{13.1.12}$$

where S is the diagonal matrix whose element (σ, σ) is σ_1.

The matrices S, A, B, C, D are all block-diagonal, their elements (σ, σ') being zero unless $\sigma_1 = \sigma_1'$. The matrix S commutes with A, B, C and D. In particular, for the Ising-type models where each σ_j has value $+1$ or -1:

$$A, B, C, D = \begin{pmatrix} \blacktriangle & 0 \\ 0 & \blacktriangledown \end{pmatrix}, \quad S = \begin{pmatrix} I & 0 \\ 0 & -I \end{pmatrix}, \tag{13.1.13}$$

where I here is the identity matrix. In this case, $\langle\sigma_1\rangle$ is the magnetization M.

In (13.1.11) it is apparent that multiplication by A corresponds to introducing the lower-right quadrant, or 'corner', of the lattice. I therefore call A the 'lower-right corner transfer matrix'. Similarly B, C, D are respectively the upper-right, upper-left, lower-left CTMs.

The Normalized Matrices A_n, B_n, C_n, D_n

Let s, s', s'', s''' be the values of the spin-sets σ, σ', σ'', σ''' in Fig. 13.2 when all the spins are in the ground state configuration.

Set

$$\alpha = A_{ss'}, \quad \beta = B_{s's''},$$ (13.1.14)

$$\gamma = C_{s''s'''}, \quad \delta = D_{s'''s},$$

and define

$$A_n = \alpha^{-1}A, \quad B_n = \beta^{-1}B,$$ (13.1.15)

$$C_n = \gamma^{-1}C, \quad D_n = \delta^{-1}D.$$

These matrices A_n, B_n, C_n, D_n are the *normalized* corner transfer matrices. Their ground-state elements, e.g. $(A_n)_{ss'}$, are unity. We shall find them useful when considering the limit $m \to \infty$.

Many formulae involving CTMs are independent of the normalization of A, B, C, D: an obvious example is (13.1.12). Thus A, B, C, D therein can be replaced by A_n, B_n, C_n, D_n.

The Diagonal Matrices A_d, B_d, C_d, D_d

It is often useful to use the diagonal forms A_d, B_d, C_d, D_d of A, B, C, D (and of A_n, B_n, C_n, D_n), normalized so their maximum entries are unity. These are defined by

$$A_n = \alpha' P A_d Q^{-1}, \quad B_n = \beta' Q B_d R^{-1},$$ (13.1.16)

$$C_n = \gamma' R C_d T^{-1}, \quad D_n = \delta' T D_d P^{-1},$$

where α', β', γ', δ' are scalars; P, Q, R, T are non-singular matrices; and A_d, B_d, C_d, D_d are diagonal matrices whose maximum entries are unity.

The matrix P is the matrix of eigenvectors of $A_n B_n C_n D_n$, Q of $B_n C_n D_n A_n$, etc., and all matrices can be chosen to commute with S, i.e. to have the block-diagonal structure in (13.1.13). Then (13.1.12) can be written as

$$\langle \sigma_1 \rangle = \text{Trace } S A_d B_d C_d D_d / \text{Trace } A_d B_d C_d D_d.$$ (13.1.17)

For definiteness, I shall suppose that A_d, B_d, C_d, D_d are arranged so that their maximum entries are in the position $(1, 1)$. Then

$$(A_d)_{1,1} = (B_d)_{1,1} = (C_d)_{1,1} = (D_d)_{1,1} = 1.$$ (13.1.18)

The eigenvector matrices P, Q, R, T are not uniquely defined, since the normalization of the eigenvectors is arbitrary. This means that P, Q, R, T can be post-multiplied by diagonal matrices. This affects A_d, B_d, C_d, D_d, but not their product.

In some cases there is a natural unique choice of P, Q, R, T. For instance, for a ferromagnetic isotropic reflection-symmetric model (e.g. the isotropic nearest-neighbour Ising model), A, B, C, D are all equal and symmetric. It is then natural to take P, Q, R, T to be equal and orthornormal. A_d is then the matrix of eigenvalues of A, normalized to satisfy (13.1.18).

Once the diagonal matrices A_d, B_d, C_d, D_d are known, the magnetization is easily obtained from (13.1.17). The main thrust of this chapter is to show that A_d, B_d, C_d, D_d can be evaluated quite easily for certain models (notably the eight-vertex model), provided the lattice is infinitely large. In Section 13.8 it is also shown that self-consistent equations for A_d, B_d, C_d, D_d (and certain other matrices) can be written down. These equations are exact and infinite-dimensional, but they can be truncated to a finite set of approximate equations, and these can be used to obtain good numerical approximations to κ, or long series expansions of κ.

13.2 Expressions as Products of Operators

Consider the matrix U_i whose element (σ, σ') is

$$(U_i)_{\sigma,\sigma'} = \delta(\sigma_1, \sigma_1') \ldots \delta(\sigma_{i-1}, \sigma_{i-1}') \, w(\sigma_i, \sigma_{i+1}, \sigma_i', \sigma_{i-1})$$
$$\times \delta(\sigma_{i+1}, \sigma_{i+1}') \ldots \delta(\sigma_m, \sigma_m'). \tag{13.2.1}$$

As is indicated in Fig. 13.3, this corresponds to adding a single face to the lattice, going in the NE to SW direction. Indeed, U_i can be regarded as a 'face transfer matrix', or as a 'face operator'. It is analogous to the vertex

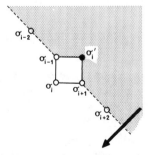

Fig. 13.3. A picture of the effect of pre-multiplying by the matrix U_i defined by (13.2.1). This corresponds to introducing the square shown, with its appropriate weight function w, and summing over the spin σ_i'. We use the convention that spins on open circles are fixed, while spins on solid circles are to be summed over.

operator of (9.6.9), the only difference being that here we are thinking in terms of 'spins on sites', rather than 'arrows on edges'.

Two operators U_i, U_j commute if i and j differ by two or more, i.e.

$$U_i U_j = U_j U_i \quad \text{if } |i - j| > 1 . \tag{13.2.2}$$

The corner transfer matrix A can be written as a product of face operators, one for each of the $\frac{1}{2}m(m + 1)$ faces in Fig. 13.1(b). To allow for the faces near the boundary, define U_m^s to be the operator U_m given by (13.2.1) with σ_{m+1} fixed to have the value s, i.e.

$$(U_m^s)_{\sigma, \sigma'} = \delta(\sigma_1, \sigma_1') \ldots \delta(\sigma_{m-1}, \sigma_{m-1}') \, w(\sigma_m, s, \sigma_m', \sigma_{m-1}) . \tag{13.2.3a}$$

Similarly, define U_{m+1}^{stz} to be U_{m+1} with σ_{m+1}, σ_{m+2}, σ_{m+1}' replaced by s, t, z, i.e.

$$(U_{m+1}^{stz}) = \delta(\sigma_1, \sigma_1') \ldots \delta(\sigma_m, \sigma_m') \, w(s, t, z, \sigma_m) . \tag{13.2.3b}$$

Thus U_{m+1}^{stz} is a diagonal matrix.

Let $s, t, \ldots, y, z$ and $s', t', \ldots y'$ be the boundary spins, arranged as in Fig. 13.1(b). Then it is easy to see that

$$A = U_3^{ss't} U_2^t U_3^{tt'z} \quad \text{for } m = 2 , \tag{13.2.4}$$

$$= U_4^{ss't} U_3^t U_2 U_4^{tt'y} U_3^x U_4^{yy'z} \quad \text{for } m = 3 ,$$

and in general that

$$A = \mathcal{F}_2^{ss't} \mathcal{F}_3^{tt'u} \ldots \mathcal{F}_{m+1}^{yy'z} , \tag{13.2.5}$$

where

$$\mathcal{F}_j^{ss't} = U_{m+1}^{ss't} U_m^t U_{m-1} U_{m-2} \ldots U_j . \tag{13.2.6}$$

13.3   Star – Triangle Relation

In Section 9.6 it was shown, using the 'electric' language of arrow spins on edges, that two six-vertex model transfer matrices commute provided the 'star – triangle' relation (9.6.8) is satisfied. This result was generalized to the eight-vertex model in Section 10.4, and in Section 11.5 it was expressed in the 'magnetic' language of Ising spins.

This last formulation can of course be derived directly. In fact it can be written down (but not necessarily solved) for any IRF model. Let the square lattice have N columns and be wound on a cylinder, so that column 1 follows column N. Then the row-to-row transfer matrix V has elements

$$V_{\sigma,\sigma'} = \prod_{j=1}^{N} w(\sigma_j, \sigma_{j+1}, \sigma'_{j+1}, \sigma'_j), \qquad (13.3.1)$$

where $\sigma_{N+1} = \sigma_1$, $\sigma'_{N+1} = \sigma'_1$, $\sigma = \{\sigma_1, \ldots, \sigma_N\}$, $\sigma' = \{\sigma'_1, \ldots, \sigma'_N\}$, and the weight function $w(a, b, c, d)$ is now arbitrary.

Let V' be similarly defined, with w replaced by w'. Then the elements of the matrix product $V V'$ are

$$(V V')_{\sigma\sigma'} = \sum_{\sigma''} V_{\sigma\sigma''} V'_{\sigma''\sigma'}$$

$$= \sum_{\sigma''_1, \ldots, \sigma''_N} \prod_{j=1}^{N} s(\sigma_j, \sigma''_j, \sigma'_j \,|\, \sigma_{j+1}, \sigma''_{j+1}, \sigma'_{j+1}), \qquad (13.3.2)$$

where

$$s(a, a'', a' \,|\, b, b'', b') = w(a, b, b'', a'')\, w'(a'', b'', b', a'). \qquad (13.3.3)$$

In fact $s(a, a'', a' \,|\, b, b'', b')$ is the weight of the two adjacent squares shown in Fig. 13.4.

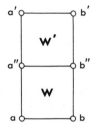

Fig. 13.4. The adjacent squares, the lower with weight function w, and the upper with weight function w'. Their combined weight is the $s(a, a'', a' \,|\, b, b'', b')$ of (13.3.3).

For given values of $\sigma_1, \ldots \sigma_N$ and $\sigma'_1, \ldots, \sigma'_N$, the RHS of (13.3.2) is a matrix product. Let $\mathbf{S}(a, a' \,|\, b, b')$ be the two-by-two matrix (for two-valued spins) with element $s(a, a'', a' \,|\, b, b'', b')$ in row a'' and column b''. Then (13.3.2) can be written

$$(V V')_{\sigma\sigma'} = \text{Trace } \mathbf{S}(\sigma_1, \sigma'_1 \,|\, \sigma_2, \sigma'_2)\, \mathbf{S}(\sigma_2, \sigma'_2 \,|\, \sigma_3, \sigma'_3)$$

$$\ldots \mathbf{S}(\sigma_N, \sigma'_N \,|\, \sigma_1, \sigma'_1). \qquad (13.3.4)$$

Define $\mathbf{S}'$ similarly, but with w and w' interchanged in (13.3.3). Then $V'V$ is given by (13.3.4), with $\mathbf{S}$ replaced by $\mathbf{S}'$. Clearly V and V' will commute if there exist two-by-two matrices $\mathbf{M}(a, a')$ such that

$$\mathbf{S}(a, a' \,|\, b, b') = \mathbf{M}(a, a')\, \mathbf{S}'(a, a' \,|\, b, b')\, [\mathbf{M}(b, b')]^{-1}, \qquad (13.3.5)$$

since the matrices $\mathbf{M}$ will cancel out of (13.3.4).

Post-multiply (13.3.5) by $\mathbf{M}(b, b')$. Write the element (c, d) of $\mathbf{M}(a, a')$ as $w''(c, a, d, a')$, and write the two-by-two matrix products explicitly. Then (13.3.5) becomes

$$\sum_c w(a, b, c, a'')\, w'(a'', c, b', a')\, w''(c, b, b'', b')$$

$$= \sum_c w''(a'', a, c, a')\, w'(a, b, b'', c)\, w(c, b'', b', a')\,, \qquad (13.3.6)$$

for all values of a, a', a'', b, b', b''.

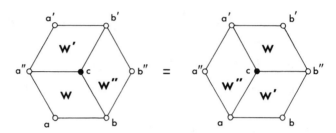

Fig. 13.5. Graphical representation of the generalized star–triangle relation (13.3.6): the partition functions of the two graphs are the two sides of the equation.

This equation can be represented as in Fig. 13.5. Define the operators U_i as in (13.2.1), and similarly define U_i' and U_i'' by replacing w therein by w' and w'', respectively. Then (13.3.6) is equivalent to the operator relation

$$U_{i+1}U_i'U_{i+1}'' = U_i''U_{i+1}'U_i\,, \qquad (13.3.7)$$

for $i = 2, \ldots, m - 1$.

Clearly this is a generalization to the IRF model of the star – triangle relations (6.4.27), (9.6.10) of the Ising and eight-vertex models.

For a given function w, (13.3.6) in general only admits trivial and uninteresting solutions for w' and w''. One obvious one is

$$w' = w, \quad w''(a, b, c, d) = \delta(a, c)\,, \qquad (13.3.8)$$

which corresponds merely to the fact that V commutes with itself.

Solvable Cases

We are interested in finding classes of commuting transfer matrices, and therefore in finding functions w such that (13.3.6) has infinitely many solutions for w' and w''. (One can of course always multiply w' and w'' by scalar factors, since these must cancel out of (13.3.6): this is not to be regarded as a new solution.)

We have already found one such family of solutions, namely the zero-field eight-vertex model. In fact (13.3.6) is the equation (11.5.8), with $\sigma_1, \ldots, \sigma_7$ replaced by b', a', a'', a, b, b'', c; W_2 by w; $W_3(a, b, c, d)$ by $w''(c, d, a, b)$, and $W_1(a, b, c, d)$ by $w'(b, c, d, a)$.

Let us define ρ, k, λ, u such that

$$w(a, b, a, b) = \rho \operatorname{snh} \lambda$$

$$w(a, b, -a, -b) = \rho k \operatorname{snh} \lambda \ \operatorname{snh} u \ \operatorname{snh}(\lambda - u) \qquad (13.3.9)$$

$$w(a, b, a, -b) = \rho \operatorname{snh}(\lambda - u)$$

$$w(a, b, -a, b) = \rho \operatorname{snh} u$$

for $a = \pm 1$ and $b = \pm 1$. Here $\operatorname{snh} u$ is the elliptic function of argument u and modulus k, defined by (10.4.20) and (15.1.6).

Now define w', w'' by (13.3.9), but with u replaced by u', u'', respectively. Then from (11.5.3) and (11.1.10), these u, u', u'' correspond to the u_2, $\lambda - u_1$, u_3 in Chapter 11. From (11.1.12) it follows that the star-triangle relation (13.3.6) is satisfied provided

$$u' = u + u''. \qquad (13.3.10)$$

Regard ρ, k, λ as fixed, and u as a complex variable. Then $w(a, b, c, d)$ is a function of u, as well as of the four spins a, b, c, d. Write it as $w[u | a, b, c, d]$, or simply $w[u]$. Then the solution of (13.3.6) is

$$w = w[u], \quad w' = w[u + u''], \quad w'' = w[u'']. \qquad (13.3.11)$$

Equivalently, writing the operator U_i as a function $U_i(u)$ of u, from (13.3.7) it follows that

$$U_{i+1}(u) \ U_i(u + u'') \ U_{i+1}(u'') = U_i(u'') \ U_{i+1}(u + u'') \ U_i(u), \qquad (13.3.12)$$

for all complex numbers u and u''. This is the relation (9.7.14), expressed in terms of spins-on-sites, rather than arrows-on-edges.

Rotating the lattice through $90°$ is equivalent to replacing a, b, c, d by b, c, d, a, and from (13.3.9) this corresponds to replacing u by $\lambda - u$. Thus

$$w[\lambda - u] = \textit{weight function } w \textit{ after rotating lattice} \\ \textit{through } 90°. \qquad (13.3.13)$$

Suppose, as in (10.7.1a), that

$$\rho > 0, \quad 0 < k < 1, \quad 0 < \lambda < I', \qquad (13.3.14)$$

where I' is the complete elliptic integral of the first kind of modulus $k' = (1 - k^2)^{\frac{1}{2}}$. Then from (13.3.9)

$$w[u \,|\, a, b, c, d] \geq 0 \quad \text{if } 0 \leq u \leq \lambda, \qquad (13.3.15)$$

for all values of the spins a, b, c, d. From (13.1.3), the Boltzmann weights w must be non-negative if the energies are real, so the 'physical' values of u are those lying on the interval $(0, \lambda)$ of the real axis.

In the next chapter, similar properties will be found for a restricted hard square model: the relations (13.3.8) are changed, but again it is possible to express the function w in terms of a complex variable u (and certain 'constants' k and λ) so that equations similar to (13.3.9)–(13.3.15) are satisfied. It is therefore interesting to consider the consequences of (13.3.10)–(13.3.15).

The star – triangle relation (13.3.6) implies that the two row-to-row transfer matrices V and V' commute. This result can be generalized to a column-inhomogeneous model of the type discussed in Section 10.17. Let the Boltzmann weight function w be different for different columns of the lattice, but in such a way that k and λ are the same for all columns. For the matrix V, let $u_1, \ldots, u_N$ be the values of u for columns $1, \ldots, N$. Similarly, let $u'_1, \ldots, u'_N$ be their values for V'. Then the derivation (13.3.1)–(13.3.6) of the commutativity of V and V' still applies, provided that u'', and hence w'' and M, is the same for all columns. From (13.3.10), it follows that

$$V V' = V'V \quad \text{if } u'_j - u_j \text{ is independent of } j. \qquad (13.3.16)$$

This in turn implies that the normalized eigenvectors of V depend only on the *differences* of $u_1, \ldots, u_N$.

Product Relation for CTMs

Now consider the lattice shown in Fig. 13.6, in which faces to the right of the centre line have weight function $w[u]$, while those to the left have weight function $w[v]$, and all faces have the same value of k, and of λ.

Let $\sigma = \{\sigma_1, \sigma_2, \ldots\}$ and $\sigma' = \{\sigma'_1, \sigma'_2, \ldots\}$, where $\sigma'_1 = \sigma_1$. Then σ and σ' together denote all the spins on the bottom row of the lattice. Define

$$\psi_{\sigma\sigma'} = \sum \prod w(\sigma_i, \sigma_j, \sigma_k, \sigma_l), \qquad (13.3.17)$$

where the product is over all the M faces of the lattice and the sum is over all spins on solid circles in Fig. 13.6.

We consider two possible boundary conditions. Firstly, apply the boundary conditions of Figs. 13.1 and 13.2. Then it is obvious from the definition (13.1.7) of A, and the corresponding definitions of B and C, that

$$\psi_{\sigma\sigma'} = [B(u)\, C(v)]_{\sigma\sigma'}, \qquad (13.3.18)$$

where the dependence of B on the parameter u, and of C on v, is explicitly exhibited.

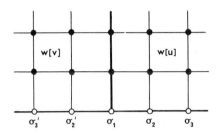

Fig. 13.6. Lattice with weight function $w[u]$ for faces to the right of the heavy line, $w[v]$ for faces to the left. Its partition function is the $\psi_{\sigma\sigma'}$ in (13.3.17).

Secondly, suppose instead that cylindrical boundary conditions are used, and fix the top row of spins to have values $\ldots, s_3', s_2', s_1, s_2, s_3, \ldots$. Let $s = \{s_1, s_2, \ldots\}$ and $s'' = \{s_1', s_2', \ldots\}$, with $s_1' = s_1$. Then

$$\psi_{\sigma\sigma'} = (V^r)_{\sigma\sigma'|ss'}, \qquad (13.3.19)$$

where r is the number of rows, V is the row-to-row transfer matrix of this section, σ and σ' together form the row-index of V in (13.3.19) while s and s' form the column-index.

In the limit of r large it follows that ψ is, apart from a normalization factor, the maximal eigenvector of V. From the remark following (13.3.16), this eigenvector depends on u and v only via their difference $u - v$. Thus

$$\psi_{\sigma\sigma'} = \tau'(u, v)\, [X'(u - v)]_{\sigma\sigma'}, \qquad (13.3.20)$$

where $\tau'(u, v)$ is a normalization factor, independent of σ and σ', and $[X'(u - v)]_{\sigma\sigma'}$ depends on u and v only via their difference $u - v$. For later comparison with (13.3.18), it is useful to regard $[X'(u - v)]_{\sigma\sigma'}$ as the element (σ, σ') of a matrix $X'(u - v)$. Since $\sigma_1 = \sigma_1'$, $X'(u - v)$ can be taken to have the block-diagonal structure (13.1.13): it is *not* the transpose of $X(u - v)$.

13.4   The Infinite Lattice Limit

To proceed further we must go to the limit when the lattice is infinitely large and the matrices are infinite dimensional. It is not easy to be mathematically rigorous in handling this limit, although tools are available (Ruelle, 1969). Here I shall rely heavily on physical intuition, much of it gained from considering low-temperature series expansions, where one perturbs about the ground state of the system.

The parameter α in (13.1.14) is itself a partition function, namely that of the lattice in Fig. 13.1(b), with all boundary spins fixed at their ground-state values. There are $\frac{1}{2}m(m + 1)$ faces, so from (13.1.4) we expect that

$$\kappa = \lim_{m \to \infty} \alpha^{2/m(m+1)} . \tag{13.4.1}$$

If a fixed number of the boundary spins σ, σ' are changed from their ground-state values, we expect this only to introduce an extra multiplicative factor which tends to a finite non-zero limit as $m \to \infty$. That is, we expect that the limit

$$\lim_{m \to \infty} (A_{\sigma\sigma'}/A_{ss'}) , \tag{13.4.2}$$

exists, provided there exists an integer r (independent of m) such that

$$\sigma_i = s_i \quad \text{and} \quad \sigma_i' = s_i' \quad \text{for} \quad i \geqslant r . \tag{13.4.3}$$

However, (13.4.2) is simply the element (σ, σ') of the matrix A_n defined by (13.1.14) and (13.1.15). We therefore expect that

$$\lim_{m \to \infty} (A_n)_{\sigma\sigma'} , \tag{13.4.4}$$

exists, and in this sense that the matrix A_n should tend to a limiting infinite-dimensional matrix as $m \to \infty$. Similarly, we expect B_n, C_n, D_n to tend to limits.

Arrange the columns of P, Q, R, T so that the diagonal entries of $A_d B_d C_d D_d$ in (13.1.16) are in numerically decreasing order. Then A_d, B_d, C_d, D_d can be chosen so that each has its entries in numerically decreasing order. For a wide class of choices of the function $w(a, b, c, d)$, and for sufficiently low temperatures, it appears that α', β', γ', δ', P, Q, R, T, A_d, B_d, C_d, D_d in (13.1.16) can all be chosen to tend to limits as $m \to \infty$. This is in the sense that matrix elements such as $P_{\sigma j}$ and $(A_d)_{jj}$ tend to limits, for fixed j and a fixed spin-set σ satisfying (13.4.3).

(High temperature regimes can be handled by using 'free-spin' boundary conditions, but let us concentrate on the low temperature case.)

For the ferromagnetically ordered eight-vertex model it will be found in Section 13.7 that $C_d = A_d$, $D_b = B_d$. and that the largest six diagonal elements of A_d (in the limit $m \to \infty$), are

$$1, s, s^2, s^3, s^3, s^4 , \qquad (13.4.5)$$

where $s = (xz)^{\frac{1}{2}}$, and x,z are the elliptic function parameters defined by (10.4.23), (10.7.9) and (10.7.18). (The elements of B_d are given by inverting z.) At low temperatures s is small: it increases to one as the temperature increases to its critical value.

More generally, for any IRF model at sufficiently low temperature, it seems that each of the corner transfer matrices has a discrete eigenvalue spectrum as $m \to \infty$. This is in the sense that for any $\varepsilon > 0$ there are only a finite number of eigenvalues numerically larger than ε.

This is quite different from the eigenvalue spectrum of the row-to-row transfer matrix V. If the lattice has N columns, then V usually has a unique maximum eigenvalue, then a band of N eigenvalues all close together, then another band of $\frac{1}{2}N(N-1)$ eigenvalues, etc. These bands become continuous in the limit $N \to \infty$. The normalized matrix of eigenvalues of V does *not* tend to a limit as $N \to \infty$.

13.5 Eigenvalues of the CTMs

Now let us return to Section 13.3 and suppose that w is such that the star – triangle relation (13.3.6) does admit a one-parameter class of solutions for w' and w''; that the members of this class are $w[u]$, where u is any complex number, and (13.3.9)–(13.3.12) are satisfied.

The relations (13.3.18) and (13.3.20) are then true. For a finite lattice, $\psi_{\sigma\sigma'}$ is different in the two equations, since different boundary conditions are imposed on the lattice. However, in the limit of an infinite lattice, we expect the boundary conditions to be irrelevant. Eliminating $\psi_{\sigma\sigma'}$, using (13.1.15) and absorbing the factors α and β into $\tau'(u, v)$, gives

$$B_n(u) \, C_n(v) = \tau'(u, v) \, X'(u - v) , \qquad (13.5.1)$$

where $B_n(u)$, $C_n(v)$ are the normalized corner transfer matrices of the infinite lattice.

Three equations analogous to (13.5.1) can be obtained by successively rotating the lattice through 90° intervals. In particular, rotating clockwise through 90° gives

$$A_n(u) \, B_n(v) = \tau(u, v) \, X(u - v) , \qquad (13.5.2)$$

where $\tau(u, v)$ is some scalar factor and $X(u - v)$ is a matrix that depends on u and v only via their difference $u - v$. I shall now regard (13.5.2) as the archetypal equation, and (13.5.1) as one of its rotated analogues.

There are problems with this equation, due to the infinite dimensionality of the matrices. The sum-over-elements involved in the matrix product on the LHS is probably not convergent, giving a divergent factor. However, this factor is common to all elements of $A_n(u) B_n(v)$, so can be absorbed into $\tau(u, v)$ and plays no role in the subsequent analysis. What I do expect to be true is that if s, s'' are the ground-state values of the spins σ, σ'' in Fig. 13.2, and if there exists an integer r (independent of m) such that

$$\sigma_j = s_j \quad \text{and} \quad \sigma_j'' = s_j'' \quad \text{for} \quad j \geq r, \tag{13.5.3a}$$

and if $A_n(u)$, $B_n(v)$ are again defined for finite m, then the limit

$$\lim_{m \to \infty} \{[A_n(u) B_n(v)]_{\sigma\sigma'}/[A_n(u) B_n(v)]_{ss''}\}, \tag{13.5.3b}$$

should exist and depend on u,v only via their difference $u - v$.

Also, I expect that there exist representations in which the appropriately normalized infinite dimensional matrices $A_n(u)$, $B_n(v)$ exist, together with the product $A_n(u) B_n(v)$.

From now on I shall therefore treat (13.5.2) as a normal matrix equation. Further, I shall assume that $A_n(u)$, $B_n(v)$ are not identically singular for all u,v, and shall as required assume completeness of eigenvector sets. Obviously all this is very non-rigorous. Even so, the assumptions appear to be justified, and the results to be exactly correct.

Symmetric Case

There is a wealth of information in (13.5.2). It is most easily explored when there is no spontaneous breaking of the translation invariance of the lattice, and when

$$w(a, b, c, d) = w(c, b, a, d) = w(a, d, c, b). \tag{13.5.4}$$

In this case (which includes the ferromagnetic eight-vertex model), the boundary spins $s, t, \ldots, z$ in Fig. 13.1 are all equal, A and B are symmetric matrices, and $C = A$, $D = B$. Also since B is obtained from A by rotating the lattice through $90°$, from (13.3.13) it follows that

$$B_n(v) = A_n(\lambda - v), \tag{13.5.5}$$

for all values of v.

Replacing v by $\lambda - v$ in (13.5.1) therefore gives

$$A_n(u) A_n(v) = \tau(u, \lambda - v) X(u + v - \lambda). \tag{13.5.6}$$

Interchanging u and v, and eliminating X, gives

$$\tau(v, \lambda - u) A_n(u) A_n(v) = \tau(u, \lambda - v) A_n(v) A_n(u). \quad (13.5.7)$$

By considering a representation in which $A_n(u)$ is diagonal, it can be seen that (13.5.7) implies

$$\tau(v, \lambda - u) = \tau(u, \lambda - v) \quad (13.5.8)$$

$$A_n(u) A_n(v) = A_n(v) A_n(u). \quad (13.5.9)$$

From (13.5.9) and (13.5.6), the matrices $A_n(u)$, $A_n(v)$, $X(u + v - \lambda)$ therefore commute and have common eigenvectors, independent of u and v. For some physical value of u, say $\frac{1}{2}\lambda$, let $p_1, p_2, p_3, \ldots$ be the eigenvectors of $A_n(u)$ arranged so that the corresponding eigenvalues $a_1(u), a_2(u), a_3(u), \ldots$ are in numerically decreasing order. Let $x_1(u), x_2(u), x_3(u), \ldots$ be the corresponding eigenvalues of $X(u)$. Define

$$A_d(u) = P^{-1} A_n(u) P/a_1(u),$$

$$X_d(u) = P^{-1} X(u) P/x_1(u). \quad (13.5.10)$$

Then $A_d(u), X_d(u)$ are diagonal matrices whose top-left entries are unity. Putting (13.5.6) into diagonal form by pre-multiplying by P^{-1} and post-multiplying by P, the (1,1) element gives

$$\tau(u, \lambda - v) = a_1(u) a_1(v)/x_1(u + v - \lambda), \quad (13.5.11)$$

and the equation then becomes

$$A_d(u) A_d(v) = X_d(u + v - \lambda). \quad (13.5.12)$$

Any given diagonal entry of (13.5.12) is a scalar equation of the same form, and must be true for all real numbers u, v in the interval $(0, \lambda)$. Differentiating logarithmically, it is readily verified that the general solution is, for $r = 1, 2, 3, \ldots$,

$$[A_d(u)]_{r,r} = m_r \exp(-\alpha_r u), \quad (13.5.13)$$

where m_r, α_r are constants, independent of u.

For a given model, the values of m_r, α_r can be determined from periodicity considerations and by considering special cases: this will be done in the next section for the ferromagnetically ordered eight-vertex model.

Using (13.5.5) and the fact that $C = A$, $D = B$, it is obvious that $A_n(u)$, $B_n(u)$, $C_n(u)$, $D_n(u)$ all commute with one another, so the P,Q,R,T in (13.1.13) can all be taken to be P. Then

$$C_d(u) = A_d(u), \quad D_d(u) = B_d(u) = A_d(\lambda - u). \quad (13.5.14)$$

Let S_r be the eigenvalue of S corresponding to the eigenvector p_r. Substituting the expressions (13.5.13) and (13.5.14) for the eigenvalues into (13.1.17), it follows that

$$\langle \sigma_1 \rangle = \sum_{r=1}^{\infty} S_r m_r^4 \exp(2\alpha_r\lambda) \Big/ \sum_{r=1}^{\infty} m_r^4 \exp(2\alpha_r\lambda) . \qquad (13.5.15)$$

Asymmetric Case

It is still possible to obtain explicit forms for A_d, B_d, C_d, D_d from (13.5.2) (and its rotated analogues), even if the symmetry conditions (13.5.4) and (13.5.5) are not satisfied. It is merely a little more tedious.

First replace u, v in (13.5.2) by $\lambda - v$, $\lambda - u$ and then eliminate $X(u - v)$. This gives

$$\tau(\lambda - v , \lambda - u) A_n(u) B_n(v) = \tau(u , v) A_n(\lambda - v) B_n(\lambda - u) . \qquad (13.5.16)$$

Set u equal to some fixed value u_0, and suppose that $A_n(u_0)$ is invertible, so that (13.5.16) can be solved for $B_n(v)$. Substitute the result back into (13.5.1) and post-multiply by $B_n^{-1}(\lambda - u_0) A_n^{-1}(u_0)$. The result is

$$\mathcal{A}(u) \mathcal{A}(\lambda - v) = \phi(u , v) Y(u - v) , \qquad (13.5.17)$$

where

$$\mathcal{A}(u) \quad = A_n(u) A_n^{-1}(u_0) ,$$

$$Y(u) \quad = X(u) B_n^{-1}(\lambda - u_0) A_n^{-1}(u_0) , \qquad (13.5.18)$$

$$\phi(u , v) = \tau(u , v) \tau(\lambda - v , \lambda - u_0)/\tau(u_0 , v) .$$

This equation (13.5.17) is precisely of the form (13.5.6), with v replaced by $\lambda - v$. It can be solved by exactly the same methods, giving

$$\mathcal{A}(u) = a_1(u) P A_d(u) P^{-1} , \qquad (13.5.19)$$

where P is independent of u and $A_d(u)$ is diagonal. The columns of P are arranged so that for physical values of u (or at any rate for some particular physical value), the diagonal entries of $A_d(u)$ are in numerically decreasing order. The top-left entry in $A_d(u)$ is unity. The (r , r) entry is

$$[A_d(u)]_{r,r} = m_r \exp(\alpha_r u) , \qquad (13.5.20)$$

where m_r, α_r are independent of u.

From (13.5.19) and the first of the three equations (13.5.18), it follows that

$$A_n(u) = a_1(u) P A_d(u) Q^{-1} , \qquad (13.5.21)$$

where Q, like P, is independent of u. [In fact $Q = A_n^{-1}(u_0)P$.] Substituting this form for $A_n(u)$ into (13.5.16), setting $u = u_0$, solving for $B_n(v)$, replacing v by u, and using the fact that $A_d(\lambda - u)$ and $A_d(u_0)$ commute, gives

$$B_n(u) = b_1(u)\, Q\, A_d(\lambda - u)\, R^{-1}, \qquad (13.5.22)$$

where $b_1(u)$ is a scalar factor, and R is independent of u.

Now consider the first rotated analogue of (13.5.2), namely (13.5.1). Substituting the form (13.5.22) of $B_n(u)$ into (13.5.1) is precisely analogous to substituting the form (13.5.21) of $A_n(u)$ into (13.5.2) and (13.5.16). The result analogous to (13.5.22) is

$$C_n(u) = c_1(u)\, R\, A_d(u)\, T^{-1}, \qquad (13.5.23)$$

where $c_1(u)$ is a scalar, and T is independent of u.

Similarly, the $C_n(u)\, D_n(v)$ analogue of (13.5.2) gives

$$D_n(u) = d_1(u)\, T\, A_d(\lambda - u)\, W^{-1}, \qquad (13.5.24)$$

where $d_1(u)$ is a scalar, and W is independent of u.

Substituting the forms (13.5.24), (13.5.21) of $D_n(u)$, $A_n(v)$ into the $D_n(u)\, A_n(v)$ analogue of (13.5.2) gives

$$A_d(\lambda - u)\, W^{-1} P\, A_d(v) = \tau''(u, v)\, X''(u - v), \qquad (13.5.25)$$

where $\tau''(u, v)$ is a scalar factor. Since $A_d(u)$ is diagonal, with element $(1,1)$ equal to unity, it follows from the $(1,1)$ element of (13.5.25) that $\tau''(u, v)$ can be taken to be unity. It is then readily apparent that the element (r, s) of $W^{-1}P$ satisfies

$$(W^{-1}P)_{rs} = 0 \quad \text{if } \alpha_r \neq \alpha_s. \qquad (13.5.26)$$

Thus $W^{-1}P$ is a block-diagonal matrix, independent if u.

Post-multiplying P, Q, R, T, W above by constant matrices with this block-diagonal structure, we can reduce $W^{-1}P$ to the unit matrix, and (13.5.21)–(13.5.24) to the form

$$A_n(u) = a_1(u)\, P\, A_d(u)\, Q^{-1}, \quad B_n(u) = b_1(u)\, Q\, B_d(\lambda - u)\, R^{-1},$$

$$C_n(u) = c_1(u)\, R\, C_d(u)\, T^{-1}, \quad D_n(u) = d_1(u)\, T\, D_d(\lambda - u)\, P^{-1}, \qquad (13.5.27)$$

where $A_d(u)$, $B_d(u)$, $C_d(u)$, $D_d(u)$ are all diagonal matrices, with elements given by (13.5.20). For each integer r, the value of α_r is the same for all four matrices, but m_r may be different. In every case $\alpha_1 = 0$, $m_1 = 1$.

As for the symmetric case, the values of α_r and the m_r can be obtained from periodicity conditions and special cases: this will be done in the next chapter for a restricted hard square model.

Substituting these forms (13.5.27), (13.5.20) into (13.1.17), we again obtain the formula (13.5.15) for the magnetization, except that now m_r^4 is to be replaced by the product of the values of m_r for A_n, B_n, C_n and D_n.

13.6 Inversion Properties: Relation for $\kappa(u)$

When $u = 0$ it is apparent from (13.3.9) that

$$w(a, b, c, d) = \rho \operatorname{snh} \lambda\, \delta(a, c). \tag{13.6.1}$$

Suppose the boundary conditions are such that $s = t = \ldots = z$ and $s' = t' = \ldots = y'$. Then from (13.2.1)

$$U_i = \rho \operatorname{snh} \lambda\, \mathcal{I}, \tag{13.6.2}$$

where $\mathcal{I}$ is the identity matrix and $i = 1, \ldots, m - 1$. Further from (13.2.3), (13.6.2) is also true for the boundary matrices U_m^s, U_{m+1}^{stz} that occur in (13.2.4)–(13.2.6). It follows that

$$A(0) = (\rho \operatorname{snh} \lambda)^{\frac{1}{2}m(m+1)} \mathcal{I}, \tag{13.6.3a}$$

and hence from (13.1.14) and (13.1.15) that

$$A_n(0) = \mathcal{I}. \tag{13.6.3b}$$

Setting $u = 0$ in the first of the equations (13.5.27), it follows that

$$Q = a_1(0)\, P A_d(0), \tag{13.6.4}$$

and hence that

$$A_n(u) = [a_1(u)/a_1(0)]\, P A_d(u)\, A_d^{-1}(0)\, P^{-1}. \tag{13.6.5}$$

From (13.5.13), the matrix $A_d(u)\, A_d^{-1}(0)$ is diagonal, with elements

$$[A_d(u)\, A_d^{-1}(0)]_{r,r} = \exp(-\alpha_r u), \tag{13.6.6}$$

α_r being independent of u.

Let $\mathcal{H}_d$ be the diagonal matrix with entries $\alpha_1, \alpha_2, \alpha_3, \ldots$, and set

$$\mathcal{H} = P \mathcal{H}_d P^{-1}. \tag{13.6.7}$$

Then from (13.6.5) and (13.1.15) it follows that

$$A(u) = \tau(u) \exp(-u\mathcal{H}), \tag{13.6.8}$$

where, noting that α can depend on u,

$$\tau(u) = \alpha\, a_1(u)/a_1(0). \tag{13.6.9}$$

As $m \to \infty$ we expect $a_1(u)$ and $\mathcal{H}$ to tend to limits. From (13.4.1), the partition function per site is therefore

$$\kappa(u) = \lim_{m \to \infty} [\tau(u)]^{2/m(m+1)}. \qquad (13.6.10)$$

These equations have been derived for the physical values of u, namely those on the interval $(0, \lambda)$ of the real axis. For these we expect the various large-m limits to exist. In particular, we expect that if the diagonal elements of $A_d(u) A_d^{-1}(0)$ are arranged in numerically decreasing order, the largest being normalized to unity, then any given element will tend to a limit as $m \to \infty$. This means that $\alpha_1, \alpha_2, \alpha_3, \ldots$ are non-negative and that $\mathcal{H}$ is non-negative definite.

However, it seems that these equations can be extended to small negative values of u, and that the resulting values of $\kappa(u)$ are those obtained by analytically continuing the function from positive values.

[For u negative, this $\kappa(u)$ is *not* that given by (13.1.4). This is because the order of the eigenvalues of $\exp(-u\mathcal{H})$ is reversed and it is now the smallest eigenvalue that is normalised to unity, instead of the largest.]

Suppose therefore that (13.6.8) is true for sufficiently small values of u, positive or negative. Then obviously

$$A(u) A(-u) = \tau(u) \tau(-u) \mathcal{I}. \qquad (13.6.11)$$

This result can be obtained directly. From (13.3.9) it is readily established that

$$\sum_{a'} w(u|a, b, a', d) w(-u|a', b, c, d) = g(u) \delta(a, a'), \qquad (13.6.12)$$

for all spins a, b, c, d and all complex numbers u, where

$$g(u) = \rho(u) \rho(-u) (\text{snh}^2 \lambda - \text{snh}^2 u), \qquad (13.6.13)$$

and for generality I now allow the normalization factor ρ in (13.3.9) to be some given function of u (but k and λ continue to be regarded as constants).

From (13.6.12) and (13.2.1) it follows that

$$U_i(u) U_i(-u) = g(u) \mathcal{I}. \qquad (13.6.14)$$

This is true for $i = 2, \ldots, m-1$ and also for U_m^s, provided s is fixed. It is not true for U_{m+1}^{stz} defined in (13.2.3b), but without doing violence to the boundary conditions this diagonal matrix can be replaced by one for which it is true. Since we are supposing that $s = t = \ldots = z$ and $s' = t' = \ldots = y'$, from (13.2.4)–(13.2.6) it follows that

$$A(u) A(-u) = [g(u)]^{\frac{1}{2}m(m+1)} \mathcal{I}, \qquad (13.6.15)$$

which is of the same form as the previous result (13.6.11). Comparing them, we obtain

$$\tau(u)\,\tau(-u) = [g(u)]^{\frac{1}{2}m(m+1)}.\tag{13.6.16}$$

Taking $\frac{1}{2}m(m+1)$ roots in this equation, using (13.6.10) and (13.6.13), gives

$$\kappa(u)\,\kappa(-u) = \rho(u)\,\rho(-u)\,[\operatorname{snh}^2\lambda - \operatorname{snh}^2 u]\,.\tag{13.6.17a}$$

A similar equation can be obtained by rotating the lattice through $90°$. From (13.3.13) this is equivalent to replacing the functions $\kappa(u)$, $\rho(u)$ by $\kappa(\lambda - u)$, $\rho(\lambda - u)$, respectively: (13.6.17a) then gives

$$\kappa(u)\,\kappa(2\lambda - u) = \rho(u)\,\rho(2\lambda - u)\,[\operatorname{snh}^2\lambda - \operatorname{snh}^2(\lambda - u)]\,.\tag{13.6.17b}$$

Let us set

$$\rho = \rho(u) = \rho'k^{\frac{1}{2}}\Theta(i\lambda)\,\Theta(iu)\,\Theta(i\lambda - iu)\,,\tag{13.6.18}$$

where ρ' is a constant. This ensures that the parametrization (13.3.9) of the Boltzmann weights is the same as (10.4.24) (with ρ replaced by ρ'), and that the weights are entire functions of u. Using the formula (15.4.19), (13.6.17) becomes

$$\kappa(u)\,\kappa(-u) = \rho'^2 h(\lambda - u)\,h(\lambda + u)\,,\tag{13.6.19a}$$

$$\kappa(u)\,\kappa(2\lambda - u) = \rho'^2 h(u)\,h(2\lambda - u)\,,\tag{13.6.19b}$$

where $h(u)$ is defined by (10.5.16), i.e.

$$h(u) = -i\,\Theta(0)\,H(iu)\,\Theta(iu)\,.\tag{13.6.20}$$

We have in fact already encountered the second equation (13.6.19). From (13.1.5), (10.8.46) and (10.4.23), the function $\Lambda(v)$ in Chapter 10 is related to $\kappa(u)$ by

$$\Lambda(v) = \{\kappa[\tfrac{1}{2}(\lambda + v)]\}^N\,,\tag{13.6.21}$$

where N is the number of columns in the lattice of Chapter 10. Replacing u in the second equation (13.6.19) by $\frac{1}{2}(\lambda + v)$, and taking Nth powers, the equation becomes

$$\Lambda(v)\,\Lambda(2\lambda - v) = \phi(\lambda + v)\,\phi(3\lambda - v)\,,\tag{13.6.22}$$

where the function $\phi(v)$ is defined by (10.5.24). However, this is precisely the relation (10.8.43).

The relations (13.6.19) are therefore certainly true for the eight-vertex model, which gives me greater confidence in the assumptions used to derive them. Analogous relations have been used by Stroganov (1979) for two special cases of the '81-vertex' model, and yet others will be used in the next chapter for a modified hard squares model.

More generally, for any IRF model we can always define an 'inverse' weight function $\overline{w}$ such that

$$\sum_f w(a,b,f,d)\,\overline{w}(f,b,c,d) = \delta(a,c)\,, \qquad (13.6.23)$$

for all spins a, b, c, d. Define $\overline{U}_i$ by (13.2.1), with w replaced by $\overline{w}$. Then (13.6.23) implies

$$U_i\overline{U}_i = \mathcal{I}\,, \qquad (13.6.24)$$

for $i = 2\,,\ldots,m$. Write κ as a function $\kappa[w]$ of w. Define $\kappa[\overline{w}]$ as the analytic continuation from w to $\overline{w}$. Then one can use arguments similar to those above to establish that

$$\kappa[w]\,\kappa[\overline{w}] = 1\,. \qquad (13.6.25)$$

[Probably the simplest way is to consider the diagonal-to-diagonal transfer matrix $(U_1U_3U_5\ldots U_{N-1})\,(U_2U_4U_6\ldots U_N)$. Then κ can be defined as the Nth root of the eigenvalue whose corresponding eigenvector has no negative entries. Inverting each U_i inverts all the eigenvalues, and hence inverts κ.]

This equation (13.6.25) is the generalization of (13.6.17a). The generalization of (13.6.17b) is obtained by rotating the lattice through 90°, i.e. by using the NW to SE inverse of w, instead of the NE to SW inverse $\overline{w}$.

In the next section it will be shown for the eight-vertex model that (13.6.17), together with some simple analyticity and periodicity properties, determines the function $\kappa(u)$, and hence the free energy. It is fascinating to speculate whether (13.6.25) does the same for any IRF model, e.g. the Ising model in a magnetic field. Unfortunately it seems that κ is then a much more complicated function and that (13.6.25), while true, is no longer sufficient to determine κ.

13.7 Eight-Vertex Model

Free Energy

Let us use the definition (13.6.18) of $\rho(u)$ and regard ρ' as a constant. Then the parametrization (13.3.9) of the Boltzmann weights is the same as that in (10.4.24), but with ρ therein replaced by ρ'. If (13.3.14) and (13.3.15) are satisfied, then the system is in an ordered ferromagnetic phase. We can therefore use the results of Section 10.8 to obtain the free energy, and hence $\kappa(u)$. From (13.6.15) and (10.8.44),

$$\ln \kappa(u) = \ln(\rho'\gamma/x) - \sum_{n=1}^{\infty} \frac{(x^{3n} + q^n x^{-n})(z^n + z^{-n})}{n(1 - q^n)(1 + x^{2n})}, \quad (13.7.1)$$

where q, x, z are given by (10.4.23), (10.7.9) and (10.7.18), i.e.

$$q = \exp(-\pi I'/I), \quad x = \exp(-\pi\lambda/2I), \quad z = x^{-1}\exp(-\pi u/I). \quad (13.7.2)$$

Since ρ', k, λ are regarded as constants, so are q and x. The parameter z varies with u. From the result (13.7.1) it can be seen that

$$\ln \kappa(u) = \text{analytic in a domain containing the vertical}$$
$$\text{strip } 0 \le \text{Re}(u) \le \lambda, \text{periodic of period } 2iI. \quad (13.7.3)$$

These analyticity and periodicity properties, together with the 'inversion' relations (13.6.17), actually define $\kappa(u)$.

To see this, note that (13.7.3) implies that $\ln \kappa(u)$, regarded as a function of z, is analytic in the annulus $x \le z \le x^{-1}$. It therefore has a Laurent expansion which converges in a domain containing this annulus, i.e. there exist coefficients c_n, independent of z and u, such that

$$\ln \kappa(u) = \sum_{n=-\infty}^{\infty} c_n z^n, \quad x \le z \le x^{-1}. \quad (13.7.4)$$

Also, from (10.8.6),

$$\ln h(u) = \ln \gamma + \pi u/2I - \sum_{n \neq 0} x^n z^n/n(1 - q^n), \quad (13.7.5)$$

where the sum is over all integer values of n, positive or negative but not zero, and u must lie in the strip $0 < \text{Re}(u) < I'$.

The equations (13.6.17) are equivalent to (13.6.19). Take logarithms of both sides of (13.6.19a). There exists a strip about the imaginary axis in the complex u plane inside which all functions can be expanded by using (13.7.4) or (13.7.5). This gives

$$\Sigma(c_n + x^{2n}c_{-n}) z^n$$
$$= 2\ln(\rho'\gamma/x) - \sum_{n \neq 0} (x^{3n} + q^n x^{-n}) z^n/n(1 - q^n). \quad (13.7.6a)$$

In both sums n runs from $-\infty$ to ∞, but $n = 0$ is excluded in the second. Similarly, (13.6.19b) gives

$$\Sigma(c_n + x^{-2n}c_n) z^n$$
$$= 2\ln(\rho'\gamma/x) - \sum_{n \neq 0} (x^n + q^n x^{-3n}) z^n/n(1 - q^n). \quad (13.7.6b)$$

for a vertical strip about $\mathrm{Re}(u) = \lambda$. Equating coefficients of z^n in (13.7.6a) and (13.7.6b), the resulting equations can be solved for c_n to give

$$c_0 = \ln(\rho' \gamma/x) \,, \tag{13.7.7}$$

$$c_n = c_{-n} = -(x^{3n} + q^n x^{-n})/[n(1 + x^{2n})(1 - q^n)] \quad \text{for } n \neq 0.$$

Substituting these expressions into (13.7.4) gives the result (13.7.1) of Chapter 10.

The analyticity and periodicity properties (13.7.3) could have been guessed from low-temperature series expansions, and $\kappa(u)$ then obtained by the above reasoning. Provided one is prepared to make these initial assumptions, this method is undoubtedly the simplest way yet known for evaluating $\kappa(u)$, and hence for obtaining the free energy of the eight-vertex model.

Magnetization

The ground state of the ordered ferromagnetic phase can be taken to be the configuration in which all spins have value ± 1. Since $w(a, b, c, d)$ is symmetric with respect to interchange of a and c, or b and d, it follows that the corner transfer matrices A, B, C, D are symmetric and that $C = A$ and $D = B$. This case has been discussed in (13.5.4)–(13.5.15). From (13.5.10) and (13.5.5).

$$A_n(u) = C_n(u) = P^{-1} A_d(u) P/a_1(u) \,, \tag{13.7.8}$$

$$B_n(u) = D_n(u) = P^{-1} A_d(\lambda - u) P/a_1(\lambda - u) \,.$$

The matrix $A_d(u)$ is diagonal, with elements of the form (13.5.13), where $m_1 = 1$ and $\alpha_1 = 0$. From (13.6.3b) it follows that $m_r = 1$ for all r, so the diagonal elements of $A_d(u)$ are

$$[A_d(u)]_{r,r} = \exp(-\alpha_r u) \,. \tag{13.7.9}$$

The α_1, α_2, . . . are constants, independent of u.

The Boltzmann weights (13.3.9) are periodic functions of u, of period $4iI$. There seems to be no problem in extending the reasoning of Section 13.5 to all complex numbers u in the vertical strip $0 < \mathrm{Re}(u) < \lambda$ (if we look at low-temperature series expansions, u only enters them via integer or half-integer powers of the z in (13.7.2)). This implies that $A_d(u)$ is also periodic of period $4iI$, so from (13.7.9) it follows that

$$[A_d(u)]_{r,r} = \exp(-\pi n_r u/2I) \,, \tag{13.7.10}$$

where n_1, n_2, . . . are integers.

These integers can be obtained by considering the case when $k \to 0$ while λ/I' and u remain fixed. The first and third weights in (13.3.9) then remain comparable with one another, while the other two become relatively negligible. From (15.1.6) and (13.7.2),

$$w(a, b, a, b) \to \tfrac{1}{2}\rho x^{-1},$$

$$w(a, b, -a, -b) \to 0,$$
$$(13.7.11)$$

$$w(a, b, a, -b) \to \tfrac{1}{2}\rho x^{-1} \exp(-\pi u/2I),$$

$$w(a, b, -a, b) \to 0.$$

From (13.2.1) and (13.2.3), the matrices U_i are therefore diagonal, with entries

$$(U_i)_{\sigma,\sigma} = \tfrac{1}{2}\rho x^{-1} \exp[-\pi u(1 - \sigma_{i-1}\sigma_{i+1})/4I], \qquad (13.7.12)$$

for $i = 2, \ldots, m + 1$, where σ_{m+1} and σ_{m+2} are to be given the ground state value of $+1$. From (13.2.4)–(13.2.6), the matrix A is also diagonal, so $A_d(u)$ is obtained by normalizing this matrix to have maximum element unity. This gives

$$[A_d(u)]_{\sigma,\sigma} = \exp\left[-\pi u \sum_{i=2}^{m+1} (i-1)(1 - \sigma_{i-1}\sigma_{i+1})/4I\right]. \quad (13.7.13)$$

Comparing this with the general formula (13.7.10), and replacing the single index r by the multiple index $\sigma = \{\sigma_1, \sigma_2, \ldots, \sigma_m\}$, we see that

$$n_\sigma = \tfrac{1}{2} \sum_{i=2}^{m+1} (i-1)(1 - \sigma_{i-1}\sigma_{i+1}). \qquad (13.7.14)$$

It follows that (13.7.13) is true throughout the ferromagnetic ordered phase, in the limit of m large. By this I mean that if we consider the rth largest diagonal element of $A_d(u)$ (for u positive), and let $m \to \infty$ while keeping r fixed, then this element tends to a limit, and this limit is given by (13.7.13). If σ is the spin set corresponding to this rth largest element, then there must be an integer j, independent of m, such that

$$\sigma_i = +1 \quad \text{for} \quad i > j. \qquad (13.7.15)$$

It is convenient to introduce a new set of spins $\mu_1, \ldots, \mu_m$ by

$$\mu_i = \sigma_i\sigma_{i+2}, \quad i = 1, \ldots, m, \qquad (13.5.16)$$

(taking $\sigma_{m+1} = \sigma_{m+2} = +1$ as before). Then $A_d(u)$ is a diagonal matrix whose rows and columns are labelled by $\mu = \{\mu_1, \ldots, \mu_m\}$, and whose diagonal entries are

$$[A_d(u)]_{\mu,\mu} = \exp\left[-\pi u \sum_{i=1}^{m} i(1 - \mu_i)/4I\right]. \qquad (13.7.17)$$

Also, since $\sigma_1 = \mu_1\mu_3\mu_5 \ldots$, the S in (13.1.13) is diagonal and has entries

$$S_{\mu,\mu} = \mu_1\mu_3\mu_5 \ldots . \qquad (13.7.18)$$

The matrices $A_d(u)$ and S are now direct products of two by two matrices. Set

$$s = (xz)^{\frac{1}{2}} = \exp(-\pi u/2I),$$

$$t = (x/z)^{\frac{1}{2}} = \exp(-\pi(\lambda - u)/2I]. \qquad (13.7.19)$$

Then, using (13.7.8) and (13.1.16),

$$A_d(u) = C_d(u) = \begin{pmatrix} 1 & 0 \\ 0 & s \end{pmatrix} \otimes \begin{pmatrix} 1 & 0 \\ 0 & s^2 \end{pmatrix} \otimes \begin{pmatrix} 1 & 0 \\ 0 & s^3 \end{pmatrix} \otimes \cdots$$

$$B_d(u) = D_d(u) = \begin{pmatrix} 1 & 0 \\ 0 & t \end{pmatrix} \otimes \begin{pmatrix} 1 & 0 \\ 0 & t^2 \end{pmatrix} \otimes \begin{pmatrix} 1 & 0 \\ 0 & t^3 \end{pmatrix} \otimes \cdots \qquad (13.7.20)$$

$$S = \begin{pmatrix} 1 & 0 \\ 0 & -1 \end{pmatrix} \otimes \begin{pmatrix} 1 & 0 \\ 0 & 1 \end{pmatrix} \otimes \begin{pmatrix} 1 & 0 \\ 0 & -1 \end{pmatrix} \otimes \cdots .$$

These equations are true only in the limit $m \to \infty$, when there are infinitely many terms in each of the direct products. I first conjectured them in 1976 (Baxter, 1976), but could not then prove them.
Substituting them into (13.1.17) gives

$$\langle \sigma_1 \rangle = \prod_{n=1}^{\infty} (1 - x^{4n-2})/(1 + x^{4n-2}), \qquad (13.7.21)$$

which is the formula (10.10.9) for the spontaneous magnetization of the eight-vertex model. We have therefore established this result, originally conjectured by Barber and Baxter (1973).

This reasoning does not easily generalize to the spontaneous polarization: the formula (10.10.24) is still a conjecture.

13.8 Equations for the CTMs

In this section I return to the general IRF model of Sections 13.1 and 13.2, and show that there are equations which relate κ and the CTMs, and in principle determine them. They have not so far proved particularly useful for exactly solvable models, but they have been used very successfully to obtain high-order series expansions (Baxter and Enting, 1979; Baxter *et al.*, 1980) and good numerical approximations (Baxter, 1968; Kelland, 1976; Tsang, 1979; Baxter and Tsang, 1980).

Define

$$A^* = \mathcal{F}_3^{tt'u} \mathcal{F}_4^{uu'v} \ldots \mathcal{F}_{m+1}^{yy'z} \tag{13.8.1}$$

$$A^{**} = \mathcal{F}_4^{uu'v} \ldots \mathcal{F}_{m+1}^{yy'z}.$$

Then from (13.2.5) and (13.2.6),

$$A = \mathcal{F}_2^{ss't} A^*, \quad A^* = \mathcal{F}_3^{tt'u} A^{**}, \tag{13.8.2}$$

$$\mathcal{F}_2^{ss't} = \mathcal{F}_3^{ss't} U_2.$$

For simplicity, suppose that the ground state is translation invariant, so that $s, t, u, \ldots$ and $s', t', \ldots$ are all equal. (This is not essential, but for a non-translation invariant system we must keep track of the boundary conditions appropriate to the various corner transfer matrices, and this complicates the notation.) Then eliminating $\mathcal{F}_2$ and $\mathcal{F}_3$ between the equations (13.8.2) gives

$$A = A^* (A^{**})^{-1} U_2 A^*. \tag{13.8.3}$$

These A^* and A^{**} are themselves corner transfer matrices, only with the spins shifted and m reduced. In fact

$$(A^*)_{\sigma_1 \ldots \sigma_m | \sigma_1' \ldots \sigma_m'} = \delta(\sigma_1, \sigma_1') A_{\sigma_2 \ldots \sigma_m | \sigma_2' \ldots \sigma_m'}, \tag{13.8.4a}$$

$$(A^{**})_{\sigma_1 \ldots \sigma_m | \sigma_1' \ldots \sigma_m'} = \delta(\sigma_1, \sigma_1') \delta(\sigma_2, \sigma_2') A_{\sigma_3 \ldots \sigma_m | \sigma_3' \ldots \sigma_m'} \tag{13.8.4b}$$

where the A on the RHS of (13.8.4a) is defined by (13.1.8) with m replaced by $m - 1$, and the A in (13.8.4b) has m replaced by $m - 2$.

Let us refer to the lattice quadrant in Fig. 13.1(b) as being of size m by m. Then (13.8.3) defines A for an m by m quadrant in terms of its values for $m - 1$ by $m - 1$ and $m - 2$ by $m - 2$ quadrants. It is a recursion relation.

There are of course analogous recursion relations for B, C and D, obtained from (13.8.3) by rotating the lattice successively through 90° intervals.

Recursion Relation for A_d, B_d, C_d, D_d

We are interested in calculating the diagonal forms A_d, B_d, C_d, D_d. There are at least two reasons for this: for those models which have been solved exactly, the diagonal forms have a very simple structure, e.g. (13.7.20); for other models only approximate calculations can be performed. In such approximations the corner transfer matrices must be truncated to manageable size. As will be discussed at the end of this section, it seems that a very good way to do this is to work with the diagonal forms of the matrices.

Rather than calculate the original CTMs A, B, C, D from (13.8.3) and its analogues, and then use (13.1.15) and (13.1.16) to obtain A_d, B_d, C_d, D_d, we can make the substitutions (13.1.15) and (13.1.16) directly into (13.8.3).

First let us establish some notation. For any 2^{m-1} by 2^{m-1} matrix X, with elements X_{ij}, define a 2^m by 2^m matrix X^*, with elements

$$X^*_{\sigma_1 i | \sigma'_1 j} = \delta(\sigma_1 , \sigma'_1) X_{ij}. \tag{13.8.5a}$$

Here the rows of X^* are labelled by the double index (σ_1 , i) the columns by (σ'_1 , j). In obvious notations we can write

$$X^* = e_1 \otimes X = \begin{pmatrix} X & 0 \\ 0 & X \end{pmatrix}, \tag{13.8.5b}$$

e_1 being the unit two-by-two matrix.

Similarly, for any 2^{m-2} by 2^{m-2} matrix X, define a 2^m by 2^m matrix X^{**} by

$$X^{**}_{\sigma_1 \sigma_2 i | \sigma'_1 \sigma'_2 j} = \delta(\sigma_1 , \sigma'_1) \, \delta(\sigma_2 , \sigma'_2) X_{ij}, \tag{13.8.6a}$$

i.e.

$$X^{**} = e_1 \otimes e_2 \otimes X = \begin{pmatrix} X & 0 & 0 & 0 \\ 0 & X & 0 & 0 \\ 0 & 0 & X & 0 \\ 0 & 0 & 0 & X \end{pmatrix}. \tag{13.8.6b}$$

These definitions are consistent with the equations (13.8.4) for A^* and A^{**}, being the obvious generalizations thereof.

Let P^*, Q^*, A_d^* be so defined, using the matrices P, Q, A_d appropriate to the $m - 1$ by $m - 1$ lattice quadrants. Similarly, define P^{**}, Q^{**}, A_d^{**} in terms of the $m - 2$ by $m - 2$ quadrants. Let $\tau_m = \alpha\alpha'$, where α, α' are the scalar factors in (13.1.15) and (13.1.16), evaluated for an m by m quadrant. Then substituting (13.1.15) and (13.1.16) into (13.8.3), using (13.8.4)–(13.8.6), we obtain

$$\kappa P_r A_d Q_r^{-1} = A_t, \tag{13.8.7}$$

where

$$A_t = A_d^*(Q_r^*)^{-1}(A_d^{**})^{-1} U_2 P_r^* A_d^*, \tag{13.8.8}$$

$$P_r = (P^*)^{-1}P, \qquad\qquad Q_r = (Q^*)^{-1}Q, $$

$$P_r^* = (P^{**})^{-1}P^*, \qquad\qquad Q_r^* = (Q^{**})^{-1}Q^*, \tag{13.8.9}$$

$$\kappa = \tau_m \tau_{m-2}/\tau_{m-1}^2. \tag{13.8.10}$$

All matrices in (13.8.7)–(13.8.10) are of dimension 2^m by 2^m. We have used the fact that P is of the block-diagonal form (13.1.13), so that P^{**} commutes with U_2. The suffix r can be regarded as standing for 'ratio', and t for 'total calculated corner transfer matrix'.

To understand this last remark, note that (13.8.7) is one of four relations which can be obtained from it by rotating the lattice through 90° intervals. This cyclicly permutes A, B, C, D and P, Q, R, T. The four equations are

$$\kappa P_r A_d Q_r^{-1} = A_t, \qquad \kappa Q_r B_d R_r^{-1} = B_t,$$

$$\kappa R_r C_d T_r^{-1} = C_t, \qquad \kappa T_r D_d P_r^{-1} = D_t. \tag{13.8.11}$$

These equations are precisely of the form (13.1.16). Since A_d, B_d, C_d, D_d are the diagonalized CTMs of the m by m lattice quadrants, it follows that A_t, B_t, C_t, D_t are also allowed representations of these CTMs. They are not in general diagonal, but appear to be 'more diagonal' than the original CTMs A, B, C, D: for instance, there seems to be no problem with the convergence of the relevant matrix products in the infinite-lattice limits. Since A_t is related to A by

$$A_t \propto (P^*)^{-1} A Q^*, \tag{13.8.12}$$

we can regard A_t as a 'partially diagonalized' form of A.

Suppose we have evaluated the 2^{m-1} by 2^{m-1} matrices A_d, B_d, C_d, D_d, P_r, Q_r, R_r, T_r appropriate to $m-1$ by $m-1$ quadrants, and the 2^{m-2} by 2^{m-2} matrices A_d, B_d, C_d, D_d appropriate to $m-2$ by $m-2$ quadrants. Then the definitions (13.8.5) and (13.8.6) give the 2^m by 2^m matrices A_d^*, P_r^*, Q_r^*, A_d^{**} in (13.8.8), so we can evaluate A_t. Similarly, using the rotation analogues of (13.8.8), we can evaluate B_t, C_t, D_t. The equations (13.8.11) can then be solved for the 2^m by 2^m matrices A_d, B_d, C_d, D_d, P_r, Q_r, R_r, T_r appropriate to m by m lattice quadrants.

(There is still some freedom, notably in the normalization of the column vectors of P_r, Q_r, R_r, T_r, but in any given example there is usually an obvious sensible choice of such factors. For instance, the simplest case is that of an isotropic reflection-symmetric model, when $A = B = C = D$ is symmetric. We can then choose $P_r = Q_r = R_r = T_r$ to be orthogonal, and

$A_d = B_d = C_d = D_d$ to have maximum eigenvalue unity. It is helpful to keep this simple case in mind.)

Also, for large m we expect the α' in (13.1.16) to tend to a limit, so from (13.4.1) it follows that $\tau_m = \kappa^{\frac{1}{2}m(m+1)}$, where κ is the partition function per site. The definitions (13.4.1) and (13.8.10) of κ are therefore equivalent.

We can therefore use the equations (13.8.8) and (13.8.11) to calculate κ and A_d, B_d, C_d, D_d, P_r, Q_r, R_r, T_r recursively for successively larger values of m.

Truncated Equations

Let us consider in a little more detail how we would solve (13.8.11), given A_t, B_t, C_t, D_t. Multiplying gives

$$\kappa^4 P_r A_d B_d C_d D_d P_r^{-1} = A_t B_t C_t D_t. \tag{13.8.13}$$

Thus one has to diagonalize $A_t B_t C_t D_t$: $\kappa^4 A_d B_d C_d D_d$ is the diagonal matrix of eigenvalues (κ^4 being the largest), and P_r is the matrix whose columns are the right-eigenvectors. Also, if P_s, Q_s, R_s, T_s are the inverses of P_r, Q_r, R_r, T_r, respectively, then P_s is the matrix whose rows are the left-eigenvectors of $A_t B_t C_t D_t$. Similar results for $Q_r, \ldots, T_s$ can be obtained by cyclically permuting A, B, C, D, and the definition (13.8.8) of A_t can be written as

$$A_t = A_d^* Q_s^* (A_d^{**})^{-1} U_2 P_r^* A_d^*. \tag{13.8.14}$$

If one keeps all such eigenvalues and eigenvectors, then the diagonal matrices A_d, B_d, C_d, D_d double in size at each recursion. However, we expect these to tend to infinite-dimensional limits. This is in the sense given in Section 13.4, namely that if their diagonal elements are arranged in numerically decreasing order, then any given such element (e.g. the 6th largest) should tend to a limit.

This suggests a self-consistent truncation of the equations, namely to keep only the larger half of the eigenvalues of $A_t B_t C_t D_t$, and the corresponding right- and left-eigenvectors. This means that we are solving the equations

$$\kappa P_r A_d = A_t Q_r, \tag{13.8.15a}$$

$$\kappa A_d Q_s = P_s A_t, \tag{13.8.15b}$$

$$Q_s Q_r = \mathcal{I}, \tag{13.8.16}$$

together with (13.8.14) and their rotated analogues.

If the number of eigenvalues thereby kept is n, then it follows from (13.8.5) and (13.8.14)–(13.8.16) that the dimensions of the various matrices are as follows:

$$A_d: n \times n; \quad A_d^*: 2n \times 2n; \quad A_d^{**}: 4n \times 4n;$$

$$A_t: 2n \times 2n; \quad P_r: 2n \times n; \quad Q_s: n \times 2n;$$

$$P_r^*: 4n \times 2n; \quad Q_s^*: 2n \times 4n; \quad U_2: 4n \times 4n; \quad (13.8.17)$$

and the dimensions are unchanged by cyclically permuting A, B, C, D and P, Q, R, T.

From (13.8.14), the matrices A_t, B_t, C_t, D_t are block-diagonal of the form (13.1.13). The diagonal elements of A_d (and of B_d, C_d, D_d) therefore fall into two sets: those from the block with $S = +1$, and those from the block with $S = -1$. Label the elements $i = 1, \ldots, n$, and let $\zeta_i = +1$ (-1) if the element comes from the $S = +1$ (-1) block. Then using (13.2.1), the elements of the various matrices can be written as:

$$(A_d)_{i|j} = a_i \delta(i, j) ; \quad (A_d^*)_{\sigma_1, i|\sigma_1', j} = a_i \delta(\sigma_1, \sigma_1') \delta(i, j) ;$$

$$(A_d^{**})_{\sigma_1, \sigma_2, i|\sigma_1', \sigma_2', j} = a_i \delta(\sigma_1, \sigma_1') \delta(\sigma_2, \sigma_2') \delta(i, j) ;$$

$$(A_t)_{\sigma_1, i|\sigma_1', j} = \delta(\sigma_1, \sigma_1') \bar{a}_{ij}(\sigma_1) ;$$

$$(P_r)_{\sigma_1, i|j} = \delta(\sigma_1, \zeta_j) a_i p_{ij}/a_j ;$$

$$\quad (13.8.18)$$

$$(Q_s)_{i|\sigma_1', j} = \delta(\zeta_i, \sigma_1') a_j \bar{q}_{ij}/a_i ;$$

$$(P_r^*)_{\sigma_1, \sigma_2, i|\sigma_1', j} = \delta(\sigma_1, \sigma_1') \delta(\sigma_2, \zeta_j) \, a_i p_{ij}/a_j ;$$

$$(Q_s^*)_{\sigma_1, i|\sigma_1', \sigma_2', j} = \delta(\sigma_1, \sigma_1') \delta(\zeta_i, \sigma_2') a_j \bar{q}_{ij}/a_i ;$$

$$(U_2)_{\sigma_1, \sigma_2, i|\sigma_1', \sigma_2', j} = \delta(\sigma_1, \sigma_1') \delta(i, j) w(\sigma_2, \zeta_i, \sigma_2', \sigma_1) .$$

Here the row- and column-indices are usually compound, and a vertical bar is used to separate them; e.g. P_r^* has row-index (σ_1, σ_2, i), and column-index (σ_1', j). The extra factors a_i/a_j and a_j/a_i are introduced for later convenience.

The matrix equations (13.8.14)–(13.8.16), together with their rotated analogues, are now finite-dimensional. Given a reasonable initial guess at the solution, they can be solved iteratively by calculating A_t, B_t, C_t, D_t from (13.8.14) and its analogues, then diagonalizing $A_t B_t C_t D_t$ and selecting the n largest eigenvalues (and the corresponding eigenvectors) to obtain the κ, A_d, B_d, C_d, D_d, P_r, Q_r, R_r, T_r, P_s, Q_s, R_s, T_s in (13.8.15) and its analogues. The equation (13.8.16) is then just a normalization condition on the various eigenvectors.

Once a solution is obtained for a given value of n, then selecting the $n + 1$ largest eigenvalues of $A_t B_t C_t D_t$ gives an initial guess for the next value of n. Thus one can in principle solve the equations systematically for $n = 1, 2, 3, \ldots$, at any rate for sufficiently low temperatures, when the iterative procedure converges and the initial guesses are quite good.

Accuracy of a Given Truncation

The equations (13.8.14)–(13.8.16) cannot usually be solved analytically, but for finite n they can be solved numerically on a computer, or they can be used to obtain a low-temperature series expansion of the solution. In the latter case, one expands κ and all the matrix elements in powers of some low-temperature variable x.

Of course, for any finite value of n the resulting κ, A_d, etc. do not have their true infinite lattice values, but we do expect them to converge thereto as $n \to \infty$.

It is therefore of interest to estimate the relative error in κ that is caused by using a finite-n truncation. Fortunately there seems to be a simple way of estimating this.

Note from (13.1.11) and (13.1.17) that a significant variable is

$$\rho_1 = \text{Trace } A_d B_d C_d D_d . \tag{13.8.19}$$

The n eigenvalues of $A_d B_d C_d D_d$ are contained in the $2n$ eigenvalues of $\kappa^{-4} A_t B_t C_t D_t$, and the largest of each is unity. Let

$$\lambda = \text{largest eigenvalue of } \kappa^{-4} A_t B_t C_t D_t$$
$$\text{omitted from } A_d B_d C_d D_d . \tag{13.8.20}$$

Then in some sense this λ is a measure of the relative error in ρ_1 caused by truncating the equations to finite n. Since $\langle \sigma_1 \rangle$ is a derivative of $\ln \kappa$ and from (13.1.17) is proportional to ρ_1, this suggests that

$$\text{relative error in } \kappa \simeq \lambda . \tag{13.8.21}$$

Of course this is a great over-simplification, since omitting one eigenvalue affects all the other eigenvalues, and indeed all matrix elements. However, in actual numerical calculations (13.8.21) seems in fact to be true (Tsang, 1979; Baxter and Tsang, 1980; Baxter, Enting and Tsang, 1980).

Further, in those series expansion calculations which have been performed (Baxter and Enting, 1979; Baxter et al., 1980), it has always been found that if $\lambda \sim x^p$, where p is some positive integer, then the relative error in κ is also of order x^p.

This means that quite long series expansions can be obtained from quite small values of n. For instance, from (13.7.20) and (13.7.19), for the zero-field eight-vertex model,

$$A_d B_d C_d D_d = \begin{pmatrix} 1 & 0 \\ 0 & x^2 \end{pmatrix} \otimes \begin{pmatrix} 1 & 0 \\ 0 & x^4 \end{pmatrix} \otimes \begin{pmatrix} 1 & 0 \\ 0 & x^6 \end{pmatrix} \otimes \dots, \quad (13.8.22)$$

where x is a parameter that is small at low temperatures, so can be used as the low-temperature expansion variable.

From (13.8.22), the first 14 eigenvalues of $A_d B_d C_d D_d$, in numerically decreasing order, are

$$1, x^2, x^4, x^6, x^6, x^8, x^8, x^{10}, x^{10}, x^{10}, x^{12}, x^{12}, x^{12}, x^{12}. \quad (13.8.23)$$

Keeping the first three of these, the largest eigenvalue omitted is of order x^6. Thus even the $n = 3$ truncation gives κ correctly to order x^5. Then $n = 10$ truncation gives κ to order x^{11}.

There are other examples for which the eigenvalues decrease still more rapidly: for the hard squares model, Baxter *et al.* (1980) were able to obtain the first 43 terms in the low-density expansion of κ, using only 13 by 13 matrices.

Variational Principle

The truncated equations (13.8.14)–(13.8.16) are in fact equivalent to a variational approximation for κ.

To see this, use the explicit forms (13.8.18) of the various matrices. Then (13.8.14) becomes

$$\bar{a}_{ij}(\sigma_1) = \sum_k w(\zeta_i, \zeta_k, \zeta_j, \sigma_1) \, \bar{q}_{ik} a_k p_{kj}. \quad (13.8.24)$$

Using this, (13.8.15) and (13.8.16) become

$$\kappa \, a_i p_{ij} b_j = \sum_{k,l} w(\zeta_i, \zeta_k, \zeta_l, \zeta_j) \, \bar{q}_{ik} a_k p_{kl} b_l q_{lj}, \quad (13.8.25a)$$

$$\kappa \, d_i \bar{q}_{ij} a_j = \sum_{k,l} w(\zeta_k, \zeta_l, \zeta_j, \zeta_i) \, \bar{p}_{ik} d_k \bar{q}_{kl} a_l p_{lj}, \quad (13.8.25b)$$

$$\sum_k \bar{q}_{ik} a_k b_k q_{kj} = a_i b_i \delta(i, j) \quad \text{if } \zeta_i = \zeta_j. \quad (13.8.26)$$

The three rotated analogues of these equations can be obtained by cyclically permuting a, b, c, d, and p, q, r, t, and the four arguments of the function w. The first such analogue of (13.8.25b) is

$$\kappa\, a_i \bar{r}_{ij} b_j = \sum_{k,l} w(\zeta_i,\, \zeta_k,\, \zeta_l,\, \zeta_j)\, \bar{q}_{ik} a_k \bar{r}_{kl} b_l q_{lj}\,. \tag{13.8.27}$$

Comparing this with (13.8.25a), and using the analogous comparisons, it is apparent that the equations permit solutions such that

$$\bar{r}_{ij} = p_{ij},\quad \bar{t}_{ij} = q_{ij},\quad \bar{p}_{ij} = r_{ij},\quad \bar{q}_{ij} = t_{ij}\,. \tag{13.8.28}$$

They then simplify to

$$\sum_k t_{ik} a_k b_k q_{kj} = a_i b_i \delta(i,j)\quad \text{if } \zeta_i = \zeta_j\,, \tag{13.8.29}$$

$$\sum_{k,l} w(\zeta_i,\, \zeta_k,\, \zeta_l,\, \zeta_j)\, t_{ik} a_k p_{kl} b_l q_{lj} = \kappa\, a_i p_{ij} b_j\,. \tag{13.8.30}$$

In both of these equations, i and j are integers from 1 to n. The first is true only if $\zeta_i = \zeta_j$, the second is true for all i and j.

It is useful to introduce an obvious new matrix notation. Let **a** be the n by n diagonal matrix with elements $a_i\, \delta(i,j)$, **p** the n by n matrix with elements p_{ij}, etc. Then (13.8.29) and (13.8.30) become

$$(\mathbf{tabq})_{ij} = (\mathbf{ab})_{ij}\quad \text{if } \zeta_i = \zeta_j\,, \tag{13.8.31a}$$

$$\sum_{k,l} w(\zeta_i,\, \zeta_k,\, \zeta_l,\, \zeta_j)\, t_{ik} (\mathbf{apb})_{kl} q_{lj} = \kappa\, (\mathbf{apb})_{ij}\,. \tag{13.8.31b}$$

These equations, and their rotated analogues, can be derived from a variational principal. Define

$$\rho_1 = \text{Trace } \mathbf{abcd}\,,$$

$$\rho_2 = \text{Trace } \mathbf{abqcdt}\,,$$

$$\rho_2' = \text{Trace } \mathbf{bcrdap}\,, \tag{13.8.32}$$

$$\rho_3 = \sum_{i,j,k,l} w(\zeta_i,\, \zeta_j,\, \zeta_k,\, \zeta_l)\, t_{ij}\, (\mathbf{apb})_{jk} q_{kl} (\mathbf{crd})_{li}\,,$$

and consider the quantity

$$\kappa_V = \rho_1 \rho_3 / (\rho_2 \rho_2')\,. \tag{13.8.33}$$

Differentiating κ_V logarithmically with respect to any element of $\mathbf{a}, \ldots, \mathbf{t}$, we find that the derivative vanishes if (13.8.31) are satisfied, together with their rotated analogues. Further, from (13.8.31) it is readily verified that $\rho_1 = \rho_2 = \rho_2' = \kappa^{-1}\rho_3$, so

$$\kappa_V = \kappa\,. \tag{13.8.34}$$

We therefore have a variational principle for the partition function per site κ: κ is the stationary value of κ_V. This is exact in the limit of n large, and a good approximation even for quite small n.

The equations (13.8.31)–(13.8.34) can be obtained (at least for sufficiently symmetric systems) from a variational approximation to the maximum eigenvalue of the row-to-row transfer matrix V. This is the way they were originally derived (Baxter, 1968; Kelland, 1976; Baxter, 1978c).

Matrices a, b, c, d not Necessarily Diagonal

The equations (13.8.31) are formally unchanged by the transformation

$$
\begin{aligned}
&\mathbf{a} \to \delta \mathbf{a} \alpha^{-1}, \quad \mathbf{b} \to \alpha \mathbf{b} \beta^{-1}, \\
&\mathbf{c} \to \beta \mathbf{c} \gamma^{-1}, \quad \mathbf{d} \to \gamma \mathbf{d} \delta^{-1}, \\
&\mathbf{p} \to \alpha \mathbf{p} \alpha^{-1}, \quad \mathbf{q} \to \beta \mathbf{q} \beta^{-1}, \\
&\mathbf{r} \to \gamma \mathbf{r} \gamma^{-1}, \quad \mathbf{t} \to \delta \mathbf{t} \delta^{-1},
\end{aligned}
\tag{13.8.35}
$$

where α, β, γ, δ are any non-singular matrices that are block-diagonal in the sense that their elements (i, j) are non-zero only if $\zeta_i = \zeta_j$. If the rows and columns of the various matrices are arranged so that $\zeta_i = +1$ for $i = 1, \ldots, n'$, and $\zeta_i = -1$ for $i = n' + 1, \ldots, n$, then it follows that α, β, γ, δ have the form

$$
\begin{pmatrix} \boxed{} & 0 \\ 0 & \boxed{} \end{pmatrix},
\tag{13.8.36}
$$

the top-left block being n' by n', the lower-right being $n - n'$ by $n - n'$.

The transformation (13.8.35) therefore destroys the strict diagonality of **a, b, c, d,** but they remain block-diagonal of the form (13.8.36). (This is actually equivalent to dropping the original requirement that A_d, B_d, C_d, D_d be diagonal, but still insisting that they be of the form (13.1.13).)

It can in fact be useful to work in a representation in which **a, b, c, d** are non-diagonal. In particular, in series calculations it is inconvenient to insist that **a, b, c, d** be completely diagonal, since this can introduce irrational coefficients. It is better to require merely that they be block-diagonal, all elements within a block having (to leading order) the same power-law dependence on the expansion variable x, and any element (i, j) being zero if $\zeta_i \neq \zeta_j$.

Graphical Interpretation

The equations (13.8.31) can be interpreted graphically. Consider the first lattice quadrant shown in Fig. 13.7. Regard i as denoting the spins on the left-hand edge, and ζ_i as being the top such spin. Similarly, let j denote the spins on the upper edge, and ζ_j the left-hand such spin. Let a_{ij} be the

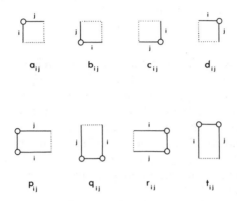

Fig. 13.7. Lattice segments and their corresponding partition functions, or total weights. The i and j denote all the spins on the corresponding edges.

element (i, j) of the matrix **a**. Since **a** is no longer necessarily diagonal, a_{ij} can be non-zero if $i \neq j$, but

$$a_{ij} \neq 0 \text{ only if } \zeta_i = \zeta_j . \tag{13.8.37}$$

Since ζ_i and ζ_j both denote the top-left spin, they must be equal. Regard a_{ij} as the 'total weight' of the lattice quadrant.

More generally, regard $a_{ij}, b_{ij}, c_{ij}, d_{ij}, p_{ij}, q_{ij}, r_{ij}, t_{ij}$ as weights of the corresponding lattice segments in Fig. 13.7, where in every case ζ_i is the spin at the circled end of the edge labelled i, and similarly for ζ_j.

Now consider the composite lattice segment shown on the left of Fig. 13.8. This consists of four pieces, with 'weights' $t_{ik}, a_{kl}, b_{lm}, q_{mj}$. Summing over all the spins internal to this composite segment is equivalent to summing over k, l, m, remembering that $\zeta_k = \zeta_l = \zeta_m$ is the spin on the site denoted by the solid circle. But this simply gives $(\mathbf{tabq})_{ij}$, so this is the total 'weight' of the left-hand figure in Fig. 13.8.

Similarly, the weight of the right-hand figure is $(\mathbf{ab})_{ij}$, so (13.8.31a) is represented graphically by Fig. 13.8(a).

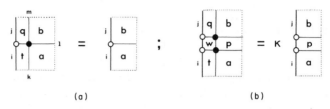

(a)   (b)

Fig. 13.8. Graphical representation of equations (13.8.31a) and (13.8.31b). Each figure represents its total weight, which is given by multiplying the segment weights (e.g. *tabq*) and summing over spins on internal lines.

Also, remembering that w is the Boltzmann weight of a square face of the lattice, (13.8.31b) is represented by Fig. 13.8(b). For infinitely large lattice segments these graphical equations have an obvious meaning: the partition function of a semi-infinite lattice is unchanged, apart from a normalization factor, by adding an extra column to the lattice; and this factor is independent of the choice of the spins on the left-hand edge.

The quantities ρ_1, ρ_3, ρ_2, ρ_2' are the weights of the composite lattice shown in Fig. 13.9, so the variational principle (13.8.33) can be interpreted graphically as indicated.

Since κ_V is stationary with respect to small perturbations of $a_{ij}, \ldots, t_{ij}$ from their exact values, (13.8.33) should give reasonably good approximations to κ even with quite simple choices of $a_{ij}, \ldots, t_{ij}$. An obvious choice is to take them to be the exact Boltzmann weights of finite lattice segments. If each long edge in Fig. 13.7 is taken to have r sites, then from (13.8.34) and Fig. 13.9 an approximation to κ is

$$\kappa = Z_{2r-1,2r-1}Z_{2r,2r}/(Z_{2r-1,2r}Z_{2r,2r-1}), \qquad (13.8.38)$$

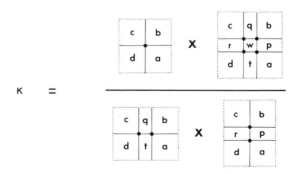

Fig. 13.9. Graphical representation of the variational principle (13.8.33)

where Z_{mn} is the partition function of a rectangular lattice of m rows and n columns. Appropriate boundary conditions should be applied: at low temperatures the boundary spins can be fixed at their ground-state values.

For a given size of the matrices $a, \ldots, t$, this approximation is nothing like as good as can be obtained by solving (13.8.31) and (13.8.32) exactly. Even so, it is moderately satisfactory, and has been discussed by Enting and Baxter (1977).

Throughout this section I have supposed that the ground state is translation invariant. If it is not, and the translation invariance is spontaneously broken, then one must define several corner transfer matrices A and a: one for every distinct position of the corner relative to the ground-state spin configuration of the infinite lattice. Similarly for the matrices B, C, $D, b, \ldots, t$.

The equations (13.8.31) can be extended to other planar lattices, notably the triangular lattice (Baxter and Tsang, 1980). (In some ways this is a simplification, since the equations are of lower degree.)

They can also be extended to three dimensions: one obvious way being to write down the generalization of Fig. 13.9, which will involve a cube sliced into 27 pieces by 6 cuts! Unfortunately the resulting equations are quite complicated and involve 'corner tensors' with three indices. There is no analogue of matrix diagonalization for these tensors, and as yet the equations have not been investigated.

14

HARD HEXAGON AND RELATED MODELS

14.1 Historical Background and Principal Results

From an historical point of view, an excellent example of the use of corner transfer matrices is provided by the hard hexagon model. This is a two-dimensional lattice model of a gas of hard (i.e. non-overlapping) molecules. In it, particles are placed on the sites of the triangular lattice so that no two particles are together or adjacent. A typical allowed arrangement of particles is shown in Fig. 14.1. If we regard each particle as the centre of a hexagon covering the six adjacent faces (such hexagons are shown shaded in Fig. 14.1), then the rule only allows hexagons that do not overlap: hence the name of the model.

For a lattice of N sites, the grand-partition function is

$$Z = \sum_{n=0}^{N/3} z^n g(n, N), \qquad (14.1.1)$$

where $g(n, N)$ is the allowed number of ways of placing n particles on the lattice, and the sum is over all possible values of n. (Since no more than 1/3 of the sites can be occupied, n takes values from 0 to $N/3$.) We want to calculate Z, or rather the partition-function per site of the infinite lattice

$$\kappa = \lim_{N \to \infty} Z^{1/N} \qquad (14.1.2)$$

as a function of the positive real variable z. This z is known as the 'activity'.

This problem can be put into 'spin'-type language by associating with each site i a variable σ_i. However, instead of letting σ_i take values $+1$ and -1, let it take the values 0 and 1: if the site is empty, then $\sigma_i = 0$; if it is full then $\sigma_i = 1$. Thus σ_i is the number of particles at site i: the 'occupation

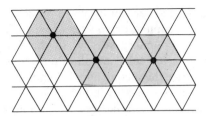

Fig. 14.1. A typical arrangement of particles (black circles) on the triangular lattice, such that no two particles are together or adjacent. The six faces round each particle are shaded: they form non-overlapping (i.e. "hard") hexagons.

number'. Then (14.1.1) can be written as

$$Z = \sum_{\sigma} z^{\sigma_1 + \ldots + \sigma_N} \prod_{(i,j)} (1 - \sigma_i \sigma_j) , \qquad (14.1.3)$$

where the product is over all edges (i, j) of the triangular lattice, and the sum is over all values (0 and 1) of all the occupation numbers $\sigma_1, \ldots, \sigma_N$.

This form of Z is very similar to the Ising model partition function (1.8.2). In fact it was shown in Section 1.9 that the general nearest-neighbour Ising model in a field is equivalent to the lattice gas with nearest-neighbour interactions. The hard hexagon model is a limiting special case of the latter.

We expect this model to undergo a phase transition from an homogeneous fluid state at low activity z to an inhomogeneous solid state at high activity z.

To see this, divide the lattice into three sub-lattices 1, 2, 3, so that no two sites of the same type are adjacent, as in Fig. 14.2. Then there are

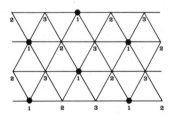

Fig. 14.2. The three sub-lattices of the triangular lattice: sub-lattice 1 consists of all sites of type 1, and similarly for sub-lattices 2 and 3. Adjacent sites lie on different sub-lattices. a close-packed arrangement of particles (black circles) is shown: all sites of one sub-lattice (in this case sub-lattice 1) are occupied, the rest are empty.

three possible close-packed configurations of particles on the lattice: either all sites of type 1 are occupied, or all sites of type 2, or all sites of type 3.

Suppose we fix the boundary sites as in the first possibility, i.e. all boundary sites of type 1 are full, and all other boundary sites are empty. Then for an infinite lattice the second and third possibilities give a negligible contribution to the sum-over-states in (14.1.3).

Clearly, sites on different sub-lattices are no longer equivalent. Let ρ_r be the local density at a site of type r, i.e., using (1.4.4),

$$\rho_r = \langle \sigma_l \rangle = Z^{-1} \sum_\sigma \sigma_l z^{\sigma_1 + \ldots + \sigma_N} \prod_{(i,j)} (1 - \sigma_i \sigma_j), \qquad (14.1.4)$$

where l is a site of type r.

When z is infinite, the system is close-packed with all sites of type 1 occupied, so $\rho_1 = 1$, $\rho_2 = \rho_3 = 0$. We can expand each ρ_r in inverse powers of z by considering successive perturbations of the close-packed state. For a site l deep inside a large lattice, this gives

$$\rho_1 = 1 - z^{-1} - 5z^{-2} - 34z^{-3} - 267z^{-4} - 2037z^{-5} - \ldots \qquad (14.1.5)$$
$$\rho_2 = \rho_3 = z^{-2} + 9z^{-3} + 80z^{-4} + 965z^{-5} + \ldots.$$

The system is therefore not homogeneous, since ρ_1, ρ_2, ρ_3, are not all equal. This contrasts with the low-activity situation: starting from the state with all sites empty and successively introducing particles, we obtain

$$\rho_1 = \rho_2 = \rho_3 = z - 7z^2 + 58z^3 - 519z^4 + 4856z^5 - \ldots \qquad (14.1.6)$$

To all orders in this expansion it is true that $\rho_1 = \rho_2 = \rho_3$.

The system is therefore inhomogeneous for sufficiently large z, and homogeneous for sufficiently small z. [Assuming the series converge: presumably a proof of this can be constructed using arguments similar to Peierls (1936).] There must be a critical value z_c of z above which the system ceases to be homogeneous. Since the homogeneous phase is typical of a fluid, and the ordered inhomogeneous phase is typical of a solid, the model can be said to undergo a fluid – solid transition at $z = z_c$.

Two related quantities of interest are the mean density

$$\rho = (\rho_1 + \rho_2 + \rho_3)/3 = z(\partial/\partial z) \ln \kappa, \qquad (14.1.7)$$

and the order parameter

$$R = \rho_1 - \rho_2 = \rho_1 - \rho_3. \qquad (14.1.8)$$

Note that R is by definition zero for $z \le z_c$. For $z > z_c$ we expect it to be positive.

Numerical Calculations

Several approximate numerical calculations were made of this model before it was solved exactly (Baxter, 1980). They are interesting in that they led to the exact solution.

Runnels and Combs (1966) calculated the maximum eigenvalue of the transfer matrix for lattices of finite widths. By extrapolating to an infinite width lattice they estimated $z_c = 11.12 \pm 0.1$.

Gaunt (1967) developed·the series expansions (14.1.5) and (14.1.6) to orders z^{-5}, z^8, respectively. From these he estimated

$$z_c = 11.05 \pm 0.15 . \tag{14.1.9a}$$

He also observed that $\kappa(z)$ appeared to have only two singularities in the complex z-plane, at $z = z_c$ and at $z = z_{NP}$, where NP stands for 'non-physical' and

$$z_{NP} = 0.0900 \pm 0.0003 \tag{14.1.9b}$$

He speculated that z_c and z_{NP} might be the two roots of some simple quadratic equation, so he formed their sum and product, giving

$$z_c + z_{NP} = 10.96 \pm 0.15 , \tag{14.1.10}$$
$$z_c z_{NP} = -0.995 \pm 0.014 .$$

Gaunt then conjectured that these numbers might be exactly 11 and -1, respectively, in which case z_c is given by

$$z_c^2 - 11z_c - 1 = 0, \quad z_c = \tfrac{1}{2}(11 + 5\sqrt{5}) = 11.09017 \ldots \tag{14.1.11}$$

Unfortunately he did not publish this conjecture. This was a pity, since we shall see that it is exactly correct.

Metcalf and Yang (1978) did some more finite width lattice calculations for the special case $z = 1$. They found that to four-figure accuracy

$$\ln \kappa = 0.3333 \tag{14.1.12}$$

and conjectured that $\ln \kappa$ was exactly 1/3.

Baxter and Tsang (1979) also looked at the case $z = 1$, but used the truncated corner transfer matrix equations (13.8.31), modified appropriately for the triangular lattice. We argued that since $z \ll z_c$, the CTM method should converge rapidly and give good numerical results. The results were indeed very encouraging for the CTM method: truncating the matrices to 2×2, 3×3, 5×5, 7×7 and 10×10 gave

$$\ln \kappa = 0.333\,050 \,,$$
$$0.333\,242\,657 \,,$$
$$0.333\,242\,721\,958 \,, \qquad\qquad (14.1.13)$$
$$0.333\,242\,721\,976\,1 \,,$$
$$0.333\,242\,721\,976\,1 \,,$$

respectively.

Table 14.1. The corner transfer matrix eigenvalues $a_1, \ldots a_{10}$, for the hard hexagon model with $z = 1$. The values are approximate, being calculated from finite truncations of the triangular lattice analogue of the matrix equation (13.8.31). The eigenvalues occur in groups of comparable magnitude, and it is sensible to include all members of a group. For this reason the truncations used are 2×2, 3×3, 5×5, 7×7 and 10×10. Each a_i is given for successively larger truncations, and clearly each is tending rapidly to a limit. This limit is its exact value for the infinite-dimensional corner transfer matrix.

i	ζ_i	a_i				
		2×2	3×3	5×5	7×7	10×10
1	0	1.0	1.0	1.0	1.0	1.0
2	1	0.7603903	0.7608436	0.7608440	0.7608440	0.7608440
3	0		−0.2548910	−0.2549635	−0.2549636	−0.2549636
4	0			0.06499191	0.6500641	0.06500641
5	1			0.04944546	0.4945972	0.04945974
6	0				−0.01657025	−0.01657427
7	1				−0.01260704	−0.01261043
8	0					0.004225773
9	0					0.004224830
10	1					0.003214313

Obviously Metcalf and Yang's conjecture was wrong, but some fascinating properties were emerging. In Table 14.1 are given the values of the eigenvalues a_i of the corner transfer matrix A, normalized so that the largest is unity, for each truncation.

These eigenvalues divide naturally into two classes, corresponding to the two diagonal blocks in (13.8.36). One of these blocks corresponds to the corner site being empty, the other to it being full. Let $\zeta_i = 0$ if a_i comes from the former block, and $\zeta_i = 1$ if a_i comes from the latter. The values of ζ_i are shown in Table 14.1.

In Table 14.2 are given the values of a_1a_4/a_3^2, $a_1^2a_5/(a_2a_3^2)$, $a_1^2a_6/a_3^3$ and $a_1^3a_7/(a_2a_3^3)$. It appears that these quantities are tending towards one as the matrices became larger. This is consistent with the assertion that in the limit of infinitely large matrices (which is when the equations are exact)

$$a_i = \alpha(\zeta_i) x^{n_i} \quad \text{for } i \geq 1, \tag{14.1.14}$$

where

$$x = a_3/a_1 \tag{14.1.15}$$

and n_i is a non-negative integer.

Table 14.2 Values of a_4/a_3^2, etc., for successively larger truncations.

	5×5	7×7	10×10
a_4/a_3^2	0.999 777 067	0.999 999 853	0.999 999 999
$a_5/(a_2a_3^2)$	0.999 711 560	0.999 999 539	0.999 999 999
a_6/a_3^3		0.999 757 797	0.999 999 849
$a_7/(a_2a_3^3)$		0.999 730 684	0.999 999 592

The reason this was fascinating is that the corresponding a_i of the eight-vertex model are all integer powers of some variable x, or s, as in (13.7.20). If the hard-hexagon model has a similar property, then perhaps it also can be solved exactly.

Dr Tsang therefore repeated the calculations for $z = 0.7$, and I used a series-expansion computer program to expand partially the first few a_i in powers of z (for small z) and of z^{-1} (for large z). Again our results were fully consistent with (14.1.14).

Exact Solution

Indeed, at this stage it was not difficult to guess the exact solution for the functions $\kappa(z)$ and $R(z)$. Defining x as in (14.1.15), I expanded z to order 30 in a power series in x (for $z < z_c$). Guided by the eight-vertex model results, I then put this expansion into the form

$$z = -x \prod_{n=1}^{\infty} (1 - x^n)^{c_n}, \tag{14.1.16}$$

and found that

$$c_1, \ldots, c_{29} = 5, -5, -5, 5, 5, 0, 5, -5, -5, 5, 0,$$
$$5, -5, -5, 5, 5, 0, 5, -5, -5, 5, 0,$$
$$5, -5, -5, 5, 0, 5, -5, -5, 5. \qquad (14.1.17)$$

It was then not hard to guess that

$$z = -x[H(x)/G(x)]^5, \qquad (14.1.18)$$

where

$$G(x) = \prod_{n=1}^{\infty} [(1 - x^{5n-4})(1 - x^{5n-1})]^{-1},$$
$$\qquad (14.1.19)$$
$$H(x) = \prod_{n=1}^{\infty} [(1 - x^{5n-3})(1 - x^{5n-2})]^{-1}.$$

The same computer run gave κ to order 29 in x. Writing it as a product like (14.1.16), it was not quite so obvious, but still very plausible, to guess that to all orders

$$\kappa = \frac{H^3(x)\, Q^2(x^5)}{G^2(x)} \prod_{n=1}^{\infty} \frac{(1 - x^{6n-4})(1 - x^{6n-3})^2(1 - x^{6n-2})}{(1 - x^{6n-5})(1 - x^{6n-1})(1 - x^{6n})^2} \qquad (14.1.20)$$

where

$$Q(x) = \prod_{n=1}^{\infty} (1 - x^n). \qquad (14.1.21)$$

These infinite products are of the type that occur in the elliptic theta functions (15.1.5). For large z, I computed z, κ, R only to relative order 9 in their x-expansions. Even so, this was enough to suggest that these functions could be written in terms of similar theta function products, namely

$$z = x^{-1}[G(x)/H(x)]^5, \qquad (14.1.22)$$

$$\kappa = \frac{x^{-1/3} G^3(x)\, Q^2(x^5)}{H^2(x)} \prod_{n=1}^{\infty} \frac{(1 - x^{3n-2})(1 - x^{3n-1})}{(1 - x^{3n})^2} \qquad (14.1.23)$$

$$R = Q(x)\, Q(x^5)/Q^2(x^3) \qquad (14.1.24)$$

As x decreases from 0 to -1, the z in (14.1.18) increased from 0 to z_c, where z_c is given by (14.1.11). Also, as x increases from 0 to $+1$, the z in (14.1.22) decreases from ∞ to z_c. This suggests that the guesses (14.1.18) and (14.1.20) apply throughout the fluid phase $0 < z < z_c$ (with $0 > x > -1$), while (14.1.22)–(14.1.24) apply throughout the solid phase $z > z_c$ (with $0 < x < 1$). The guesses also agree with Gaunt's conjecture (14.1.11) for the position of the critical point.

Having guessed the exact answer, the next step was to look for a way of deriving it. This calculation is given in Sections 14.2–14.7, and itself uses corner transfer matrices. It is not mathematically rigorous, in that certain analyticity properties of κ are assumed, and the results of Chapter 13 (which depend on assuming that various large-lattice limits can be interchanged) are used. However, I believe that these assumptions, and therefore (14.1.18)–(14.1.24), are in fact correct.

14.2   Hard Square Model with Diagonal Interactions

As is shown in Section 10.4, the first step in the solution of the eight-vertex model had been to set up a class of commuting row-to-row transfer matrices. Guided by this, I looked for lattice models whose transfer matrices commuted with that of the hard hexagon model. This led me to draw the triangular lattice as in Fig. 14.3(a). The hard hexagon model then becomes a square lattice model, in which nearest-neighbour sites, and next-nearest neighbour sites on NW – SE diagonals, cannot be simultaneously occupied.

I then generalized this model to one in which nearest-neighbour sites cannot be simultaneously occupied, and diagonally adjacent particles interact. This is a special case of the IRF model of Chapter 13: the partition function is given by (13.1.2), where each σ_i takes the values 0 or 1, and

$$w(a,b,c,d) = m\,z^{(a+b+c+d)/4}\,e^{Lac+Mbd}\,t^{-a+b-c+d}$$

$$\text{if } ab = bc = cd = da = 0\,, \quad (14.2.1)$$

$$= 0 \quad \text{otherwise}\,.$$

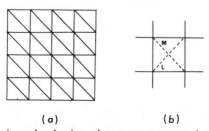

(a)   (b)

Fig. 14.3. (a) The triangular lattice, drawn as a square lattice with one set of diagonals. (b) The diagonals associated with the interaction coefficients L, M in (14.2.1).

Here a, b, c, d each take values 0 and 1; m is a trivial normalization factor; t cancels out of the partition function; L and M are diagonal interaction coefficients, as indicated in Fig. 14.3(b); z is the activity. The hard hexagon model is regained by taking $m = 1$, $L = 0$ and $M = -\infty$.

Star – Triangle Relation

Consider two such models, one with weight function w, the other with weight function w'. As is shown in Section 13.3, their row-to-row transfer matrices will commute if there exists a third function w'' such that the star-triangle relation (13.3.6) is satisfied, for all values 0, 1 of a, a', a'', b, b', b''. Take w' (w'') to be given by (14.2.1), with z, L, M, t replaced by z', L', M', t' (z'', L'', M'', t''). For convenience, interchange L' and M', invert t', and define

$$s = (z z' z'')^{\frac{1}{2}}/(t t' t'') .\qquad(14.2.2)$$

Then (13.3.6) reduces to just seven distinct equations, namely

$$(z' z'')^{\frac{1}{2}} = s + s^2 e^L ,\qquad(14.2.3a)$$

$$(z'' z)^{\frac{1}{2}} = s + s^2 e^{L'} ,\qquad(14.2.3b)$$

$$(z z')^{\frac{1}{2}} = s + s^2 e^{L''} ,\qquad(14.2.3c)$$

$$z(z' z'')^{\frac{1}{2}} e^M = s^2 + s^3 e^{L'+L''} ,\qquad(14.2.3d)$$

$$z'(z'' z)^{\frac{1}{2}} e^{M'} = s^2 + s^3 e^{L''+L} ,\qquad(14.2.3e)$$

$$z''(z z')^{\frac{1}{2}} e^{M''} = s^3 + s^3 e^{L+L'} ,\qquad(14.2.3f)$$

$$z z' z'' e^{M+M'+M''} = s^3 + s^4 e^{L+L'+L''} ,\qquad(14.2.3g)$$

which for the moment we shall refer to simply as (a)–(g). Forming $e^{L'}$(a)–e^L(b) gives the simple corollary

$$(z'^{\frac{1}{2}} e^{L'} - z^{\frac{1}{2}} e^L) z''^{\frac{1}{2}} = (e^{L'} - e^L) s .\qquad(14.2.3h)$$

Multiplying (c), (f), (g), by s, s^{-1}, s^{-1}, respectively, we see that the equations are homogeneous and linear in the five expressions

$$z''^{\frac{1}{2}}, \quad s, \quad s^2, \quad s^3 e^{L''}, \quad s^{-1} z'' e^{M''} ,\qquad(14.2.4)$$

with coefficients that are independent of s, z'', L'', M''.

For any five equations (or four equations not involving M''), the determinant of these coefficients must vanish: requiring this is equivalent to eliminating s, z'', L'', M'' between the equations. Doing this, we are left with the three equations

$$\Delta_i = \Delta_i', \quad i = 1, 2, 3, \tag{14.2.5}$$

where

$$\Delta_1 = z^{-\frac{1}{2}}(1 - z\, e^{L+M}),$$

$$\Delta_2 = z^{\frac{1}{2}}(e^L + e^M - e^{L+M}), \tag{14.2.6}$$

$$\Delta_3 = z^{-\frac{1}{2}}(e^{-L} + e^{-M} - e^{-L-M} - z\, e^{L+M}),$$

and Δ_1', Δ_2', Δ_3' are defined similarly, z, L, M being replaced by z', L', M'.

[The equations (a), (b), (d), (e) give $\Delta_1 = \Delta_1'$; (h), (c), (d), (e) give $\Delta_2 = \Delta_2'$; (h), (d), (e), (f), (g) give $\Delta_3 = \Delta_3'$.]

The three equations (14.2.5) are a sufficient condition for the star–triangle relation (14.2.3) to have a solution for s, z'', L'', M''. A corollary of (14.2.6) is

$$\Delta_1\Delta_2 - 1 = (\Delta_3 - \Delta_1 - \Delta_2)\, z^{\frac{1}{2}}\, e^{L+M}. \tag{14.2.7}$$

Suppose Δ_1, Δ_2, Δ_3, are given: normally (14.2.7) will then define $z^{\frac{1}{2}}\exp(L + M)$. The first equation (14.2.6) then gives $z^{\frac{1}{2}}$, and the second gives L and M. It follows that in general the only solutions of (14.2.5) are z', L', $M' = z$, M, L or z, L, M: these are not very interesting or useful, since they imply merely that the transfer matrix commutes with itself and its transpose.

However, suppose Δ_1, Δ_2, Δ_3 satisfy the constraints

$$\Delta_2 = \Delta_1^{-1}, \quad \Delta_3 = \Delta_1 + \Delta_1^{-1} \tag{14.2.8}$$

Then (14.2.7) no longer defines $z^{\frac{1}{2}}\exp(L + M)$, so (14.2.6) has infinitely many solutions for z, L, M. The transfer matrices corresponding to these solutions all commute.

From (14.2.6) the constraints (14.2.8) are both satisfied if

$$z = (1 - e^{-L})(1 - e^{-M})/(e^{L+M} - e^L - e^M). \tag{14.2.9}$$

Set $\Delta = \Delta_1$, i.e.

$$\Delta = z^{-\frac{1}{2}}(1 - z\, e^{L+M}). \tag{14.2.10}$$

If two models differ in their values of z, L, M, but have the same value of Δ and both satisfy (14.2.9), then their transfer matrices commute.

Note that (14.2.9) is satisfied for all z in the limit $L \to 0$ and $M \to -\infty$, which is the hard hexagon model. It is *not* so satisfied if L, $M \to 0$, which is the hard square model. Indeed, numerical solutions by Baxter *et al.* (1980) give no indication for hard squares of any simple property like (14.1.14) for the eigenvalues of the corner transfer matrix.

Elliptic Function Parametrization

Eliminating z between (14.2.9) and (14.2.10) gives

$$\Delta^{-2} e^{L+M} = (e^L - 1)(e^M - 1)(e^{L+M} - e^L - e^M). \qquad (14.2.11)$$

Given Δ, this is a symmetric biquadratic relation between e^L and e^M. As is shown in Section 15.10, such a relation can be parametrized in terms of elliptic functions, the general form being

$$e^L = \phi(v), \quad e^M = \phi(v - \lambda), \qquad (14.2.12)$$

where the function $\phi(v)$ is defined by

$$\phi(v) = \xi H(v + a) H(v - a)/[H(v + b) H(v - b)]. \qquad (14.2.13)$$

Here $H(v)$ is the elliptic theta function of argument v and modulus k, as defined by (15.1.5); k, λ, ξ, a, b are constants. (We have replaced the symbols u, l, η, λ, μ of Section 15.10 by v, k, λ, a, b, and have chosen the lower sign in (15.10.14).)

From (15.2.3), $\phi(v)$ is periodic of periods $2I$ and $2iI'$, i.e.

$$\phi(v + 2I) = \phi(v + 2iI') = \phi(-v) = \phi(v) \qquad (14.2.14)$$

where I and I' are the complete elliptic integrals of the first kind of moduli k and $k' = (1 - k^2)^{\frac{1}{2}}$, respectively.

[Given L, it follows that the first equation (14.2.12) has many solutions for v, obtainable from one another by incrementing v by integer multiples of $2I$ and $2iI$, and possible negating. However, all of these still give only two distinct solutions for e^M in (14.2.12): this is correct, since (14.2.11) is quadratic in this variable.]

Let v_0 be a value of v for which $e^L = 1$. Then from (14.2.12) and (14.2.13),

$$\xi = H(v_0 + b) H(v_0 - b)/[H(v_0 + a) H(v_0 - a)]. \qquad (14.2.15)$$

From this and (14.2.14), it follows that $\phi(v) - 1$ vanishes when $v = v_0 + 2mI + 2inI'$, so it contains a factor $H(v - v_0)$. Since it is even, it also contains a factor $H(v + v_0)$. Arguing as at the end of Section 15.3, or simply applying the identity (15.3.10), it follows that

$$e^L - 1 = \phi(v) - 1$$

$$= -\frac{H(a + b) H(a - b) H(v + v_0) H(v - v_0)}{H(a + v_0) H(a - v_0) H(v + b) H(v - b)}. \qquad (14.2.16)$$

To relate λ, a, b, v_0, consider some special values of L and M. If $e^L = 1$, then (14.2.11) gives $e^M = 0$ or ∞. From (14.2.16), v is either v_0

or $-v_0$. Associating these values of e^M and v, respectively, it follows from (14.2.12) and (14.2.13) that we can choose

$$a = v_0 - \lambda, \quad b = -v_0 - \lambda. \tag{14.2.17}$$

If $e^L = \infty$, then (14.2.11) gives both solutions for e^M to be 1, while (14.2.12) gives $v = b$ or $-b$. The RHS of (14.2.16) must therefore vanish for $v = -\lambda \pm b$, and this gives the extra condition

$$H(2\lambda + 2v_0) = 0. \tag{14.2.18}$$

The general solution of this is $v_0 = -\lambda + mI + inI'$, where m and n are integers. However, simultaneously incrementing v, v_0, a, $-b$ by $mI + inI'$ leaves $\phi(v)$ and $\phi(v - \lambda)$ unchanged, so without loss of generality we can choose

$$v_0 = -\lambda. \tag{14.2.19}$$

Substituting these forms of e^L and e^M back into (14.2.11) [using (14.2.12) on the LHS, (14.2.16) on the RHS], we obtain the relation

$$\Delta^{-2} H^6(\lambda) H^2(3\lambda) H(v + 2\lambda) H(v - 3\lambda)/H^4(2\lambda)$$
$$= H^4(2\lambda) H(v + \lambda) H(v - 2\lambda) - H^2(\lambda) H^2(3\lambda) H(v) H(v - \lambda). \tag{14.2.20}$$

This has to be an identity, true for all complex numbers v. Setting $v = 3\lambda$ and $v = 0$ gives

$$H(\lambda) H^3(3\lambda) = H^3(2\lambda) H(4\lambda), \tag{14.2.21}$$

$$\Delta^{-2} = H^8(2\lambda)/[H^5(\lambda) H^3(3\lambda)]. \tag{14.2.22}$$

These conditions ensure that the ratio of the RHS of (14.2.20) to the LHS is an entire doubly periodic function of v, equal to one when $v = 0$. From Liouville's theorem the ratio is therefore equal to one for all complex numbers v, and the identity is established.

Setting u, v, x, y = 0, λ, 2λ, 3λ in (15.3.10), we obtain the identity

$$H^3(2\lambda) H(4\lambda) - H(\lambda) H^3(3\lambda) = H^3(\lambda) H(5\lambda), \tag{14.2.23}$$

so the condition (14.2.21) is equivalent to

$$H^3(\lambda) H(5\lambda) = 0. \tag{14.2.24}$$

The solutions $\lambda = 2mI + 2inI'$ (where m, n are integers) of this equation are spurious, since from (14.2.12) and (14.2.14) they imply that $e^L = e^M$. It follows that

$$\lambda = (2mI + 2inI')/5, \tag{14.2.25}$$

where m and n are integers, not both divisible by 5.

We can choose λ to have any of these values. From now on let us take

$$\lambda = 2I/5 .\tag{14.2.26}$$

(Other choices merely lead to related parametrizations; for instance $\lambda = 2iI'/5$ is equivalent to using elliptic functions of conjugate modulus.)

Rather than work with the variable v and the elliptic theta function $H(v)$, it is convenient in this chapter to transform to the variable

$$u = \pi v/2I \tag{14.2.27}$$

and the function

$$\theta_1(u , q^2) = \sin u \prod_{n=1}^{\infty} (1 - 2 q^{2n} \cos 2u + q^{4n}) (1 - q^{2n})$$

$$= H(v)/(2q^{\frac{1}{4}}) . \tag{14.2.28}$$

(This notation is non-standard: the usual elliptic θ_1 function contains the factor $2q^{\frac{1}{4}}$. Since θ_1 enters our equations only via ratios of the form $\theta_1(u , q^2)/\theta_1(u' , q^2)$, this factor is irrelevant. It is convenient to remove it here, since we shall sometimes want q^2 to be negative: our present definition ensures that $\theta_1(u , q^2)$ then remains real.)

Using this definition to replace the functions H in (14.2.13), (14.2.15), (14.2.22) by θ_1, and writing $\theta_1(u , q^2)$ simply as $\theta_1(u)$, we finally obtain the parametrization

$$e^L = \theta_1\left(\frac{\pi}{5}\right) \theta_1\left(\frac{2\pi}{5} + u\right) \theta_1\left(\frac{2\pi}{5} - u\right) \Big/ \left[\theta_1\left(\frac{2\pi}{5}\right) \theta_1^2(u) \right]$$

$$e^M = \theta_1\left(\frac{\pi}{5}\right) \theta_1\left(\frac{3\pi}{5} - u\right) \theta_1\left(\frac{\pi}{5} + u\right) \Big/ \left[\theta_1\left(\frac{2\pi}{5}\right) \theta_1^2\left(\frac{\pi}{5} - u\right) \right] \tag{14.2.29}$$

$$\Delta^2 = \left[\theta_1\left(\frac{\pi}{5}\right) \Big/ \theta_1\left(\frac{2\pi}{5}\right) \right]^5 , \tag{14.2.30}$$

where we have used the identity $\theta_1(u) = \theta_1(\pi - u)$. From (14.2.9) it follows that

$$z = \theta_1^3\left(\frac{2\pi}{5}\right) \theta_1^2(u) \theta_1^2\left(\frac{\pi}{5} - u\right) \Big/ \left[\theta_1^3\left(\frac{\pi}{5}\right) \theta_1^4\left(\frac{2\pi}{5} + u\right) \right] . \tag{14.2.31}$$

We can use this parametrization to explicitly solve the full set of star – triangle relations (14.2.3). Let L', M', z' be given by (14.2.29) and (14.2.31), with u replaced by u'. Similarly, let L'', M'', z'' by obtained by replacing u by u''. Take q^2 to be the same throughout. Then all the equations (14.2.3) are satisfied, provided only that

$$u + u' + u'' = \pi/5 , \tag{14.2.32}$$

$$s = \theta_1^2\!\left(\frac{2\pi}{5}\right) \theta_1(u)\, \theta_1(u')\, \theta_1(u'') \Big/ \left[\, \theta_1^2\!\left(\frac{\pi}{5}\right) \theta_1\!\left(\frac{2\pi}{5} + u\right) \right.$$

$$\left. \times\, \theta_1\!\left(\frac{2\pi}{5} + u'\right) \theta_1\!\left(\frac{2\pi}{5} + u''\right) \right]. \tag{14.2.33}$$

Regions in the L, M Plane

We are interested in values of L and M such that z, as given by (14.2.9), is positive. These values lie in the unshaded regions in Fig. 14.4, the shaded regions corresponding to negative values of z.

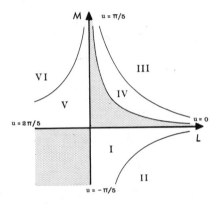

Fig. 14.4. The six regimes in the (L, M) plane, as listed in (14.2.37). Shaded areas are unphysical, since (14.2.9) gives z therein to be negative. Regimes I, III, V are disordered, II and VI have triangular ordering, IV has square ordering. The system is critical on the (I, II), (III, IV) and (V, VI) boundaries, where $|\Delta| = \Delta_c$. the values of u on the (L, M) axes are indicated.

We can regard (14.2.29) as a mapping from the variables L, M to the variables q^2, u. This is rather like a transformation from Cartesian to polar coordinates: q^2 increases from -1 to $+1$ as we go out radially from the origin through an unshaded region, while u increases as we move anti-clockwise round the origin. The three cases $-\pi/5 < u < 0$, $0 < u < \pi/5$, and $\pi/5 < u < 2\pi/5$ correspond respectively to the unshaded parts of the lower-right, upper right and upper-left quadrants.

We therefore take L, M to satisfy the restriction

$$z > 0 \tag{14.2.34}$$

where z is given by (14.2.9). We also take q^2 and u to satisfy

$$-1 < q^2 < 1, \quad -\pi/5 < u < 2\pi/5 . \tag{14.2.35}$$

The mapping from (L, M) to (q^2, u) is then one-to-one.

When $q = 0$ it is obvious from (14.2.28) and (14.2.30) that $\Delta = \pm\Delta_c$, where

$$\Delta_c = \left[\sin\left(\frac{\pi}{5}\right) \Big/ \sin\left(\frac{2\pi}{5}\right) \right]^{5/2}$$
$$= [\tfrac{1}{2}(1 + \sqrt{5})]^{-5/2} . \tag{14.2.36}$$

We shall need to distinguish the cases when $q^2 > 0$ from those when $q^2 < 0$. This leads us to divide the unshaded areas in Fig. 14.4 into six regions:

$$\begin{aligned}
&\text{I: } \Delta > \Delta_c, && q^2 < 0, \; -\pi/5 < u < 0 , \\
&\text{II: } 0 < \Delta < \Delta_c, && q^2 > 0, \; -\pi/5 < u < 0 , \\
&\text{III: } -\Delta_c < \Delta < 0, q^2 > 0, && 0 < u < \pi/5 , \\
&\text{IV: } \Delta < -\Delta_c, && q^2 < 0, \; 0 < u < \pi/5 , \\
&\text{V: } \Delta > \Delta_c, && q^2 < 0, \; \pi/5 < u < 2\pi/5 , \\
&\text{VI: } 0 < \Delta < \Delta_c, && q^2 > 0, \; \pi/5 < u < 2\pi/5 .
\end{aligned} \tag{14.2.37}$$

Regions V and VI differ from I and II only in the interchange of L and M. This is equivalent to merely rotating the lattice through 90°, so without loss of generality we hereinafter consider only regimes I, II, III and IV.

We can classify these regions as disordered or ordered by considering the following limits:

I: $L \to 0$, M finite: z, $\Delta^{-1} \to 0$.

II: L, $-M \to +\infty$: $z \sim \exp(-L - M)$; $\Delta \to 0$.

III: L, $M \to +\infty$: $z \sim \exp(-L - M)$; $-\Delta \to 0$.

IV: $z \to +\infty$, L and M finite: $-\Delta^{-1} \to 0$.

In these limits the dominant contribution to the partition function comes from the following states, respectively:

    I. the vacuum.

    II: a state such as that shown in Fig. 14.5(a), in which every third site is occupied. Forming a triangular lattice by adding diagonals as in

Fig. 14.3, this state becomes that of Fig. 14.2. There are three such states, corresponding to occupying any one of the three sub-lattices of the triangular lattice.

III: the vacuum.

IV: a close packed square-lattice state such as that shown in Fig. 14.5(b), in which every second site is occupied. There are two such states.

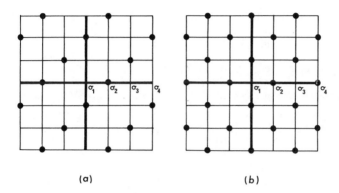

(a)   (b)

Fig. 14.5. (a) Typical ground-state in regime II, (b) in regime IV. Solid circles denote particles. The other ground states can be obtained by uniform translations, giving three ground-states of type (a), two of type (b). If we add diagonals to the lattice as in Fig. 14.3.(a), then the particles in (a) occupy one of the three sub-lattices of the resulting triangular lattice. We therefore refer to (a) as 'triangular ordering', (b) as 'square ordering'.

The heavy lines divide each lattice into four quadrants, corresponding as in Fig. 13.2 to the corner transfer matrices A, B, C, D (the shape of the outer boundary has been changed: this is irrelevant in the thermodynamic limit). The $\bar{\sigma}_1$, $\bar{\sigma}_2$, $\bar{\sigma}_3$, ... in (14.4.26) are the ground-state values of the σ_1, σ_2, σ_3, ... in the figures, so both figures correspond to taking $k = 2$ in (14.5.1).

The states in II and IV are ordered, in that the translation invariance of the lattice is spontaneously broken. We expect this to persist for finite values of L, M, z sufficiently close to the appropriate limits.

More strongly, we shall calculate the order parameter R given by (14.1.8). It is zero in regimes I and III, positive in regimes II and IV. Thus I and III are *disordered* regimes, while II and IV are *ordered*.

We shall also find that R vanishes on the boundary between regimes I and II, and between III and IV, and that the free energy is singular across this boundary. The system is therefore *critical* on this boundary, i.e., when $q^2 = 0$ and $\Delta = \pm \Delta_c$.

Boltzmann Weights

From (14.2.1), the Boltzmann weights of the allowed spin configurations around a face are:

$$\omega_1 = w(0,0,0,0) = m,$$

$$\omega_2 = w(1,0,0,0) = w(0,0,1,0) = m z^{\frac{1}{4}} t^{-1},$$

$$\omega_3 = w(0,1,0,0) = w(0,0,0,1) = m z^{\frac{1}{4}} t, \qquad (14.2.38)$$

$$\omega_4 = w(1,0,1,0) = m z^{\frac{1}{2}} t^{-2} e^L,$$

$$\omega_5 = w(0,1,0,1) = m z^{\frac{1}{2}} t^2 e^M.$$

Using the expressions (14.2.29), (14.2.31) for L, M and z, we can choose m and t so that

$$\omega_1 = m' \, \theta_1\left(\frac{2\pi}{5} + u\right) \Big/ \theta_1\left(\frac{2\pi}{5}\right),$$

$$\omega_2 = (m'/t') \, \theta_1(u) \Big/ \left[\theta_1\left(\frac{\pi}{5}\right) \theta_1\left(\frac{2\pi}{5}\right)\right]^{\frac{1}{2}},$$

$$\omega_3 = m' \, t' \, \theta_1\left(\frac{\pi}{5} - u\right) \Big/ \theta_1\left(\frac{\pi}{5}\right), \qquad (14.2.39)$$

$$\omega_4 = (m'/t'^2) \, \theta_1\left(\frac{2\pi}{5} - u\right) \Big/ \theta_1\left(\frac{2\pi}{5}\right),$$

$$\omega_5 = m' \, t'^2 \, \theta_1\left(\frac{\pi}{5} + u\right) \Big/ \theta_1\left(\frac{\pi}{5}\right).$$

The parameters m' and t' are related to the original m and t. They enter the working rather trivially, the partition function for a lattice of N faces being proportional to m'^N, and independent of t'.

Conjugate Modulus Parametrization

We shall assume and use certain analyticity and periodicity properties of the partition-function-per-site (κ), and of the eigenvalues of the corner transfer matrices. These properties are most easily expressed and understood by making a 'conjugate modulus' transformation from our variables q^2 and u to new variables x and w. [This variable w is not to be confused with the Boltzmann weight function $w(a,b,c,d)$.]

We use the relations (15.7.2) and (15.1.5) to write the infinite product in (14.2.28) in terms of the conjugate name q'. We shall find it convenient to express the results in terms of the function

$$f(w, q) = \prod_{n=1}^{\infty} (1 - q^{n-1}w)(1 - q^n w^{-1})(1 - q^n) \qquad (14.2.40)$$

Basically this is merely another way of writing the elliptic theta function. A useful symmetry property is

$$f(w, q) = f(qw^{-1}, q). \qquad (14.2.41)$$

We shall sometimes write $f(w, q)$ simply as $f(w)$, the particular nome q being understood.

If q^2 is positive we can convert (14.2.28) directly from (15.7.2a). If it is negative we first split the product in (14.2.28) into terms with n even and with n odd, and then use (15.7.2a) and (15.7.2d). This gives the identities

$$\theta_1(u, e^{-\varepsilon}) = \tfrac{1}{2}\left(\frac{2\pi}{\varepsilon}\right)^{\frac{1}{2}} \exp\left[\frac{\varepsilon}{8} - \frac{\pi^2}{2\varepsilon} + \frac{2u(\pi - u)}{\varepsilon}\right] f(e^{-4\pi u/\varepsilon}, e^{-4\pi^2/\varepsilon})$$

$$(14.2.42)$$

$$\theta_1(u, -e^{-\varepsilon}) = -\tfrac{1}{2}\left(\frac{\pi}{\varepsilon}\right)^{\frac{1}{2}} \exp\left[\frac{\varepsilon}{8} - \frac{\pi^2}{8\varepsilon} - \frac{u(2u + \pi)}{\varepsilon}\right] f(e^{2\pi u/\varepsilon}, -e^{-\pi^2/\varepsilon}).$$

Let us define parameters x, w, α as follows:

$$\text{I and IV} \quad (-1 < q^2 < 0): \qquad q^2 = -\exp(-\varepsilon)$$
$$x = -\exp(-\pi^2/5\varepsilon), \quad w = \exp(2\pi u/\varepsilon) \qquad (14.2.43)$$
$$\text{II and III} \quad (0 < q^2 < 1): \qquad q^2 = \exp(-\varepsilon)$$
$$x = \exp(-4\pi^2/5\varepsilon), \quad w = \exp(-4\pi u/\varepsilon).$$

Then by using the identities (14.2.42) in (14.2.30) and (14.2.39), we can define m' and t' so that

I and IV $(-1 < x < 0)$:

$$\Delta^{-2} = -x[f(x)/f(x^2)]^5, \qquad \omega_1 = f(xw)/f(x),$$
$$\omega_2 = \alpha r^{-1}(-x)^{\frac{1}{2}} f(w)/[f(x) f(x^2)]^{\frac{1}{2}}, \quad \omega_3 = r f(x^2 w)/f(x^2), \qquad (14.2.44a)$$
$$\omega_4 = r^{-2} w f(xw^{-1})/f(x), \qquad \omega_5 = r^2 f(x^2 w^{-1})/f(x^2),$$

II and III $(0 < x < 1)$:

$$\Delta^{-2} = x^{-1}[f(x^2)/f(x)]^5, \qquad \omega_1 = f(x^2w)/f(x^2) ,$$

$$\omega_2 = \alpha r^{-1}x^{\frac{1}{2}}f(w)/[f(x)f(x^2)]^{\frac{1}{2}}, \quad \omega_3 = rf(xw^{-1})/f(x) , \qquad (14.2.44b)$$

$$\omega_4 = r^{-2}wf(x^2w^{-1})/f(x^2), \qquad \omega_5 = r^2w^{-1}f(xw)/f(x) .$$

Here $f(w) = f(w, x^5)$ is the function defined by (14.2.40), with q therein replaced by x^5. The parameter r is proportional to t', and is at our disposal; $\alpha = \pm 1$ is chosen to ensure that ω_2 is positive.

From (14.2.33) and (14.2.39), it follows that

$$
\begin{array}{llll}
\text{I:} & 1 > w > x^2; & \text{II:} & 1 < w < x^{-1}; \\
\text{III:} & 1 > w > x; & \text{IV:} & 1 < w < x^{-2};
\end{array}
\qquad (14.2.45)
$$

$$\text{I and III:} \quad \alpha = +1; \qquad \text{II and IV:} \quad \alpha = -1 . \qquad (14.2.46)$$

One advantage of this parametrization is that x is small in the limit of extreme order or disorder. This means that the infinite product in the definition (14.2.40) of $f(w, x^5)$ is rapidly convergent, and it is easy to compare our results with high-density or low-density series expansions.

14.3 Free Energy

To recapitulate, the parametrization (14.2.44) comes from solving the star–triangle relation (13.3.6). If two models have the same value of x, but different values of u and r, then their row-to-row transfer matrices commute. The Boltzmann weights are entire functions of u.

These properties are very similar to those of the eight-vertex model (Section 10.4). Further, we put (14.2.3) into a symmetric form by interchanging L' and M', which from (14.2.29) is equivalent to replacing u' by $(\pi/5) - u'$. If we had not done this, then (14.2.32) would have been

$$u' = u + u'' . \qquad (14.3.1)$$

This is the same equation as (13.3.10). It follows that (13.3.16) and (13.5.16)–(13.5.27), with λ replaced by $\pi/5$, are valid for this model.

We now seek to calculate the free energy by the matrix-inversion trick given in Section 13.6. To do this we need analogues of the eight-vertex model equations (13.6.17a), (13.6.17b) and (13.7.3).

First Inversion Relation

Define the single-face transfer matrix U_i as in (13.2.1), where the Boltzmann weights $w(a, b, c, d)$ are given by (14.2.38). Restrict the spin-set $\sigma = \{\sigma_1, \ldots, \sigma_m\}$ to take only values in which no two adjacent spins are both unity. This restriction corresponds to ignoring forbidden spin configurations and can be written as

$$\sigma_i \sigma_{i+1} = 0 \quad \text{for } i = 1, 2, \ldots, m - 1. \tag{14.3.2}$$

Restrict $\sigma' = \{\sigma_1', \ldots, \sigma_m'\}$ similarly. Then for $m = 3$ there are five allowed values of σ and σ', namely $\{0, 0, 0\}$, $\{0, 0, 1\}$, $\{0, 1, 0\}$, $\{1, 0, 0\}$, and $\{1, 0, 1\}$. With this ordering, U_2 is the five-by-five matrix

$$U_2 = \begin{pmatrix} \omega_1 & 0 & \omega_2 & 0 & 0 \\ 0 & \omega_3 & 0 & 0 & 0 \\ \omega_2 & 0 & \omega_4 & 0 & 0 \\ 0 & 0 & 0 & \omega_3 & 0 \\ 0 & 0 & 0 & 0 & \omega_5 \end{pmatrix}. \tag{14.3.3}$$

(From now on we restrict all transfer matrices to act only between allowed states of a line of spins, i.e. states satisfying the restriction (14.3.2). This reduces the size of the matrices. This reduction is peculiar to the generalized hard-hexagon model, and is connected with its solvability, since it reduces the number of conditions implicit in the star – triangle relation (13.3.7).)

Clearly U_2 can be arranged as a block-diagonal matrix consisting of the blocks

$$\begin{pmatrix} \omega_1 & \omega_2 \\ \omega_2 & \omega_4 \end{pmatrix}, (\omega_3), (\omega_5). \tag{14.3.4}$$

More generally, so can any matrix U_i, for $m \geqslant 3$ and $2 \leqslant i \leqslant m - 1$. It follows that if we define

$$\bar{\omega}_1 = \omega_4/(\omega_1\omega_4 - \omega_2^2), \quad \bar{\omega}_2 = -\omega_2/(\omega_1\omega_4 - \omega_2^2),$$

$$\bar{\omega}_3 = \omega_3^{-1}, \qquad\qquad \bar{\omega}_4 = \omega_1/(\omega_1\omega_4 - \omega_2^2), \tag{14.3.5}$$

$$\bar{\omega}_5 = \omega_5^{-1},$$

and define $\bar{U}_i$ in the same way as U_i, but with $\omega_1, \ldots, \omega_5$ replaced by $\bar{\omega}_1, \ldots, \bar{\omega}_5$, then

$$U_i\bar{U}_i = \mathscr{I}, \tag{14.3.6}$$

where $\mathscr{I}$ is the identity matrix.

From the definition (14.2.40) of $f(x, q)$, with $q = x^5$, we can establish the identity

$$\frac{w f(x^2 w) f(x^2/w)}{f^2(x^2)} - \frac{x f^2(w)}{f(x) f(x^2)} = \frac{w f(xw) f(x/w)}{f^2(x)}. \qquad (14.3.7)$$

(The proof is similar to that of (15.3.10): take the ratio of the LHS to the RHS, and regard this as a function of w; show that it is analytic for $0 < |w| < \infty$, and is unchanged by replacing w by $x^5 w$; it is therefore bounded, and by a simple generalization of Liouville's theorem it is therefore a constant; setting $w = 1$ gives this constant to be unity.)

The Boltzmann weights ω_i (for $i = 1, \ldots, 5$) are defined by (14.2.44) as functions of r and w (regarding x as a given constant), so we can write them as $\omega_i(r, w)$. Substituting these definitions into (14.3.5) and using the identity (14.3.7), we find that

$$\bar{\omega}_i = \xi \, \omega_i(r^{-1}, w^{-1}), \quad i = 1, \ldots, 5, \qquad (14.3.8)$$

where the factor ξ is given by

I and IV: $\xi = f^2(x^2)/[f(x^2 w) f(x^2/w)]$ \qquad (14.3.9)

II and III: $\xi = f^2(x)/[f(xw) f(x/w)]$.

Thus replacing $\omega_1, \ldots, \omega_5$ by $\bar{\omega}_1, \ldots, \bar{\omega}_5$ is equivalent to inverting r and w in (14.2.44), and multiplying each weight by ξ.

Obviously we can now regard each matrix U_i as a function of r and w. Since U_i is linear in the weights $\omega_1, \ldots, \omega_5$, (14.3.8) implies that $\bar{U}_i = \xi U_i(r^{-1}, w^{-1})$. The relation (14.3.6) therefore gives

$$\xi \, U_i(r, w) \, U_i(r^{-1}, w^{-1}) = \mathcal{I}. \qquad (14.3.10)$$

This is a relation satisfied by the face transfer matrices $U_2, U_3, \ldots$, defined as functions of r and w by (13.2.1), (14.2.38) and (14.2.44). In fact it is the inversion relation (13.6.24). Define κ to be the Nth root of the eigenvalue of the diagonal-to-diagonal transfer matrix $(U_1 U_3 \ldots U_{N-1})$ $\times (U_2 U_4 \ldots U_N)$, choosing the eigenvalue corresponding to the eigenvector with all entries non-negative. Then it is easily seen that κ is independent of r (changing r is merely equivalent to a diagonal similarity transformation on the transfer matrix). It is still a function of w, and it follows from (14.3.10) that it satisfies

$$\xi \, \kappa(w) \, \kappa(w^{-1}) = 1. \qquad (14.3.11)$$

This is the 'inversion' relation (13.6.25). When w has the 'physical' values in (14.2.45), then the Boltzmann weights $\omega_1, \ldots \omega_5$ are all positive and κ is the partition-function per site, as in (13.1.4). For other values of w

(notably those obtained by passing through the point $w = 1$), it seems that this κ is the analytic continuation of the 'physical' $\kappa(w)$.

Second Inversion Relation

The inversion relation (14.3.11) and (14.3.12) is the analogue of the eight-vertex model relation (13.6.17a). We still need the analogue of (13.6.17b). This is a second inversion relation, obtained by working in the SE – NW direction, instead of the SW – NE.

Remember that $w(\sigma_i, \sigma_j, \sigma_k, \sigma_l)$ is the Boltzmann weight of the intra-face interactions between spins on sites i, j, k, l, where i, j, k, l are arranged as in Fig. 13.1(a). Clearly, rotating the lattice through 90° is equivalent to replacing $w(a, b, c, d)$ by $w(b, c, d, a)$. From (14.2.40) this is in turn equivalent to interchanging ω_2 with ω_3, and ω_4 with ω_5.

Let V_i be the SW – NE face transfer matrix. It is given by (13.2.1), with $w(a, b, c, d)$ replaced by $w(b, c, d, a)$. Its inverse $\bar{V}_i$ can be obtained similarly to the inverse $\bar{U}_i$ of U_i: all we have to do is interchange the suffixes 2 and 3, and the suffixes 4 and 5, in the equation (14.3.5). Doing this and repeating the working as far as (14.3.10), we obtain

$$\eta \, V_i(r, w) \, V_i(r_0^2/r, w_0^2/w) = \mathcal{I} \, , \qquad (14.3.12a)$$

where

$$r_0^2 = -x/\phi(x), x^{-1}\phi(x), x\phi(x), -x^{-1}/\phi(x) \, , \qquad (14.3.12b)$$

$$w_0 = -x^3, x^{-3/2}, x, x^{-2}$$

in regimes I, II, III, IV, respectively, and

$$\eta = f(x) f(x^2) \, r_0^2 / [\, |x| \, f(w) \, f(w_0^2/w)] \, , \qquad (14.3.13)$$

$$\phi(x) = f(x)/f(x^2) = H(x)/G(x) \, , \qquad (14.3.14)$$

$G(x)$ and $H(x)$ being defined as in (14.1.19).

Actually, there are an infinite number of relations of the form (14.3.12), corresponding to multiplying w_0 by an integer power of $x^{5/2}$. The particular ones given above ensure that w_0 is as close as possible to 1, while lying on the same side of unity as the physical values of w given in (14.2.45).

Just as (14.3.10) implies (14.3.11), so does (14.3.12) imply the relation

$$\eta \, \kappa(w) \, \kappa(w_0^2/w) = 1 \, . \qquad (14.3.15)$$

Analyticity of $\kappa(w)$

For 'physical' values of w, i.e. those satisfying (14.2.45), the function $\kappa(w)$ in (14.3.11) and (14.3.15) is the partition function per site (13.1.4). For other values it seems that it is the analytic continuation of the 'physical' $\kappa(w)$. More strongly, from series expansions it seems that

$$\ln[w^{-\mu} \kappa(w)] = analytic\ in\ an\ annulus$$

$$a < |w| < b\ containing\ the \qquad (14.3.16)$$

$$points\ w = 1\ and\ w = w_0.$$

Here $\mu = 0, \frac{1}{3}, 0, \frac{1}{2}$ in regimes I, II, III, IV, respectively.

This analyticity property is the analogue of the relation (13.7.3) for the eight-vertex model. I have not proved it, but it seems to be correct: arguments in its favour can be deduced from the corner transfer matrix equations (13.8.31); it also leads to results in regime I that agree with (14.1.18)–(14.1.20), and hence with the original order 29 hard hexagon series expansions mentioned in Section 14.1. Hereinafter I shall assume that (14.3.16) is correct.

Calculation of $\kappa(w)$

The equations (14.3.11), (14.3.12) and (14.3.16) are the analogues of (13.6.17a), (13.7.17b) and (13.7.3). Just as the latter set can be solved for the free energy of the eight-vertex model, so can the former be solved for the free energy of our generalized hard-hexagon model.

To do this, note that (14.3.16) implies

$$\ln[w^{-\mu} \kappa(w)] = \sum_{n=-\infty}^{\infty} c_n w^n, \qquad (14.3.17)$$

where the summation is convergent for $|w|$ in the neighbourhood of 1, and in the neighbourhood of w_0. In the neighbourhood of 1, (14.3.9) and (14.2.20) give

$$\ln \xi = d_0 + \sum_{n=1}^{\infty} d_n(w^n + w^{-n}), \qquad (14.3.18)$$

where, for $n > 0$,

I and IV: $d_0 = 2 \ln f(x^2), \quad d_n = (x^{2n} + x^{3n})/[n(1 - x^{5n})]$ (14.3.19)

II and III: $d_0 = 2 \ln f(x), \quad d_n = (x^n + x^{4n})/[n(1 - x^{5n})]$.

Taking logarithms of both sides of (14.3.11), substituting these Laurent expansions and equating coefficients of powers of w, we obtain

$$c_n + c_{-n} = -d_n, \quad n \geqslant 0. \tag{14.3.20}$$

Similarly, (14.3.15) gives

$$c_n + w_0^{-2n} c_{-n} = -d'_n - 2\mu \ln w_0 \, \delta_{n,0} \tag{14.3.21}$$

where the d'_n are the coefficients of the Laurent expansion of $\ln \eta$ in an annulus containing the point $w = w_0$. From (14.3.13), (14.2.45) and (14.2.46), these are given by

$$d'_0 = \ln\{r_0^2 w_0^{\alpha-1} f(x) f(x^2)/|x|\}, \tag{14.3.22}$$

$$d'_n = (1 + x^{5\alpha n} w_0^{-2n})/[\alpha n(1 - x^{5\alpha n})], \quad n \neq 0.$$

The equations (14.3.20) and (14.3.21) can now be solved for the coefficients c_n, giving

$$c_0 = -\tfrac{1}{2} d_0, \tag{14.3.23}$$

$$c_n = (w_0^{2n} d'_n - d_n)/(1 - w_0^{2n}) \quad \text{for } n \neq 0.$$

More explicitly, from (14.3.19) and (14.3.22), together with the above definitions of α and w_0, we obtain

$$c_0 = -\ln f(x^2) \quad \text{in regimes I and IV}, \tag{14.3.24a}$$

$$= -\ln f(x) \quad \text{in regimes II and III},$$

and, for $n \neq 0$,

$$\text{I:} \quad c_n = -x^{2n}(1 + x^n)/[n(1 - x^{5n})(1 + x^{3n})]$$

$$\text{II:} \quad c_n = -x^{3n}(1 - x^n + x^{2n} - x^{4n})/[n(1 - x^{5n})(1 - x^{3n})]$$

$$\text{III:} \quad c_n = -x^n(1 - x^n + x^{2n})/[n(1 - x^{5n})] \tag{14.3.24b}$$

$$\text{IV:} \quad c_n = -x^{4n}(1 + x^n)/[n(1 - x^{5n})(1 + x^{2n})].$$

The partition function per site, namely $\kappa(w)$, can now be obtained at once from (14.3.17), remembering that $\mu = 0, \tfrac{1}{3}. 0, \tfrac{1}{2}$ in regimes I, II, III, IV, respectively. The resulting series can be simplified by working with $\omega_1 \kappa/(\omega_4 \omega_5)$, rather than κ. Using (14.2.44), we find that

$$\text{I:} \quad \ln(\omega_1 \kappa/\omega_4 \omega_5) = -\ln w - \sum_{n=1}^{\infty} \frac{(x^n + x^{2n})(w^n - w^{-n})}{n(1 + x^{3n})}$$

$$\text{II:} \quad = \tfrac{1}{3} \ln w + \sum_{n=1}^{\infty} \frac{(x^n - x^{2n})(w^n - w^{-n})}{n(1 - x^{3n})}$$

III: $\ln (\omega_1 \kappa / \omega_4 \omega_5) = 0$

IV: $$= -\tfrac{1}{2} \ln w - \sum_{n=1}^{\infty} \frac{x^n (w^n - w^{-n})}{n(1 + x^{2n})} .$$ (14.3.25)

Taylor expanding the denominators in the summand, summing each term over n, and using the definition (14.2.40), these results can be written as

I: $\omega_1 \kappa / (\omega_4 \omega_5) = w^{-1} f(xw, x^6) f(x^2 w, x^6) / [f(xw^{-1}, x^6) f(x^2 w^{-1}, x^6)]$

II: $= w^{1/3} f(xw^{-1}, x^3) / f(xw, x^3)$

III: $= 1$

IV: $= w^{-\frac{1}{4}} f(xw, x^4) / f(xw^{-1}, x^4) .$ (14.3.26)

These product expressions (14.3.26) are valid for all values of w satisfying (14.2.45), even though the sums in (14.3.25) are not always convergent. (This lack of convergence is merely due to the fact that ω_4 and ω_1 have zeros at $w = x$ and $w = x^{-1}$ in regimes I and IV, respectively.)

14.4 Sub-Lattice Densities and the Order Parameter R

We can obtain the sub-lattice densities, defined by (14.1.4), by using the corner transfer matrix methods of Section 13.5. We can then calculate the order parameter R from (14.1.8).

Diagonal Form of the Corner Transfer Matrices

As we remarked after (14.3.1), the equation (13.3.10) is satisfied by our parametrization of the generalized hard hexagon model. Regarding A_n, B_n, C_n, D_n, as functions of u, we therefore again obtain the product relation (13.5.1) and its rotated analogues. Again this leads to the relations (13.5.27), where $A_d(u)$, $B_d(u)$, $C_d(u)$, $D_d(u)$ are all diagonal matrices of the form (13.5.20), i.e.

$$[A_d(u)]_{j,j} = m_j \exp(\alpha_j u) ,$$ (14.4.1)

and similarly for $B_d(u)$, $C_d(u)$, $D_d(u)$. The coefficient α_j is the same for all four matrices; m_j may be different.

We only need to calculate these coefficients α_j and m_j. This can be done by considering the case $u = 0$, by using the inversion property (14.3.12), and by considering the limit $x \rightarrow 0$; as will now be shown.

The Case $u = 0$ and the First Inversion Relation

When $u = 0$, it is clear from (14.2.43) that $w = 1$. From (14.2.44) and (14.2.38), it follows that the Boltzmann weight function (not to be confused with the variable w) is then

$$w(a, b, c, d) = r^{b+d-a-c} \, \delta(a, c) . \tag{14.4.2}$$

This in turn implies that when $u = 0$, then

$$A = C = A_n = C_n = L , \tag{14.4.3}$$

where L is the diagonal matrix with elements

$$L(\sigma_1, \sigma_2, \ldots \,|\, \sigma_1', \sigma_2', \ldots) = r^{\sigma_1} \, \delta(\sigma_1, \sigma_1') \, \delta(\sigma_2, \sigma_2') \ldots \tag{14.4.4}$$

Setting $u = 0$ in (13.5.27), it follows that

$$\begin{aligned} Q &= a_1(0) \, L^{-1} P \, A_d(0) , \\ T &= c_1(0) \, L^{-1} R \, C_d(0) . \end{aligned} \tag{14.4.5}$$

All the matrices are block-diagonal, their elements being zero unless $\sigma_1 = \sigma_1'$. It follows that L commutes with them all. Substituting the expressions (14.4.5) for Q and T back into (13.5.27), we can re-define the scalar factors $a_1(u)$, $b_1(u)$, $c_1(u)$, $d_1(u)$, and the diagonal matrices $A_d(u)$, $B_d(u)$, $C_d(u)$, $D_d(u)$, so that

$$A_n(u) = a_1(u) \, P A_d(u) \, P^{-1}, \quad B_n(u) = b_1(u) \, P B_d(\lambda - u) \, R^{-1}, \tag{14.4.6}$$
$$C_n(u) = c_1(u) \, R C_d(u) \, R^{-1}, \quad D_n(u) = d_1(u) \, R D_d(\lambda - u) \, P^{-1},$$

where $a_1(0) = c_1(0) = 1$ and

$$A_d(0) = C_d(0) = L . \tag{14.4.7}$$

These matrices may depend on the parameter r in (14.2.44), but the dependence is quite trivial: $A_n(u)$, $A_d(u)$, $C_n(u)$, $C_d(u)$ are all of the form

$$L \times \text{(matrix independent of } r); \tag{14.4.8}$$

$B_n(u)$, $B_d(u)$, $D_n(u)$, $D_d(u)$ are of this form, but with L replaced by L^{-1}; P and R are independent of r.

The elements of the diagonal matrices $A_d(u)$, $B_d(u)$, $C_d(u)$, $D_d(u)$ are all of the form (14.4.1), α_j being the same for all four matrices. Using (14.4.7) it follows that

$$[A_d(u)]_{j,j} = [C_d(u)]_{j,j} = r^{s_j} \exp(\alpha_j u) , \tag{14.4.9}$$

where α_j is independent of both u and r, and $s_j = 0$ or 1, being the value of σ_1 for the element (j, j). From (14.4.6), exhibiting explicitly the dependence on r of A_n and C_n, it follows that

$$A_n(r, u) A_n(r^{-1}, -u) = a_1(u) a_1(-u) \, \mathcal{I}, \qquad (14.4.10)$$

$$C_n(r, u) C_n(r^{-1}, -u) = c_1(u) c_1(-u) \, \mathcal{I}.$$

To within a scalar factor, the inverse of $A_n(r, u)$ is therefore $A_n(r^{-1}, -u)$; and similarly for C_n.

We could have predicted this inversion property directly from (14.3.10), using (14.2.43) and the fact that A and C are products of operators U_i, as in (13.2.4) and (13.2.5).

The Second Inversion Relation

The operator V_i is defined by (13.2.1), with the Boltzmann weight function $w(a, b, c, d)$ replaced by $w(b, c, d, a)$. This is in turn equivalent to interchanging ω_2 with ω_3, and ω_4 with ω_5. The corner transfer matrices B and D are then given by (13.2.4) and (13.2.5), with each U_i replaced by V_i, using the appropriate boundary spins $s, t, \ldots, y'$.

We want to use the second inversion property (14.3.12) to obtain equations for B and D analogous to (14.4.10), but we have to be careful. For a start, when $u = u_0$ then $w = w_0$, and it can be seen from (14.3.12b) and (14.2.44) that only in regimes III and IV is ω_3 then zero. Thus only in these regimes is V_i (and hence B and D) then diagonal. This means that we cannot in general construct equations analogous to (14.4.7).

More seriously, consider the particle configuration in Fig. 14.5a. This is one of the three possible ground-states of the system in regime II (the other two are obtained by first shifting all particles one site to the right, and then repeating). The upper-right corner transfer matrix is B, and it is obvious that in the ground-state limit ($x = 0$) it is mapping the spin-state

○ ○ ● ○ ○ ● ○ ○ ● . . .

(on the upper vertical heavy-line segment, starting at the centre) to the spin-state

○ ● ○ ○ ● ○ ○ ● ○ ○ . . .

(on the right horizontal heavy-line segment).

For an infinite system, this effect will persist to non-zero values of x, in that B will map a vector space $\mathcal{V}$ to a vector space $\mathcal{W}$. Here $\mathcal{V}$ is the space of functions $\psi(\sigma_1, \sigma_2, \ldots)$ subject to the condition that

$$\sigma_{3k} \to 1, \quad \sigma_{3k \pm 1} \to 0 \quad \text{as } k \to \infty; \qquad (14.4.11a)$$

while $\mathcal{W}$ is the corresponding space such that

$$\sigma_{3k+2} \to 1, \quad \sigma_{3k} \text{ and } \sigma_{3k+1} \to 0 \quad \text{as } k \to \infty. \qquad (14.4.11b)$$

Since $\mathcal{V}$ and $\mathcal{W}$ are distinct, it makes no sense to multiply B by itself. What we can do is note, using Fig. 14.5a, that D maps $\mathcal{W}$ to $\mathcal{V}$. Remembering that B_n differs from B only by a scalar factor, and similarly for D_n and D, from (14.3.12) and (14.2.43) it follows that

$$B_n(r, u) \, D_n(r_0^2/r, 2u_0 - u) \propto \mathcal{I}, \qquad (14.4.12)$$

where u_0 is the value of u when $w = w_0$. Substituting the expressions (14.4.6) for B_n and D_n into this equation, we obtain

$$B_d(\lambda - u) \, D_d(\lambda - 2u_0 + u) = \gamma(u) \, L_0^2 L^{-2}, \qquad (14.4.13)$$

where $\gamma(u)$ is a scalar factor and L_0 is the value of L when $r = r_0$.

From the remarks following (13.5.27) and (14.4.9), $B_d(\lambda - u)$ and $D_d(\lambda - u)$ are diagonal matrices whose (j, j) elements are of the form

$$[B_d(\lambda - u)]_{j,j} = m_j' \, r^{-s_j} \exp(-\alpha_j u),$$
$$[D_d(\lambda - u)]_{j,j} = m_j''' \, r^{-s_j} \exp(-\alpha_j u), \qquad (14.4.14)$$

where the coefficients m_j' and m_j''' are independent of r and u. Looking at the (j, j) element of (14.4.13), and substituting these expressions, we obtain

$$m_j' m_j''' = \gamma(u) \, r_0^{2s_j} \exp(2\alpha_j u_0) \qquad (14.4.15)$$

Clearly $\gamma(u)$ is independent of u, so it can be written simply as γ.

In (13.1.17) we expressed the local density $\langle \sigma_1 \rangle$ in terms of A_d, B_d, C_d and D_d. Since this equation was obtained from (13.5.27) and (13.1.12) (with A, B, C, D replaced by their normalized values A_n, B_n, C_n, D_n), it follows that the A_d, B_d, C_d, D_d in (13.1.17) are the $A_d(u)$, $B_d(\lambda - u)$, $C_d(u)$, $D_d(\lambda - u)$ of this section.

(The parameter λ plays no role in this section: effectively it is just part of the notation for B_d and D_d, considered as functions of u.)

We can therefore substitute the expressions (14.4.9) and (14.4.14) for A_d, B_d, C_d, D_d directly into (13.1.17). Doing this, we find that r and u cancel out of the resulting expression. (This is as it should be, since we can use arguments similar to those in (7.10.28)–(7.10.48) to write $\langle \sigma_1 \rangle$ as $\psi_L^T S \psi_R$, where S here is the diagonal operator with elements σ_1, and ψ_L and ψ_R are the left- and right-eigenvectors of the row-to-row transfer matrix V. This V is defined in (13.3.1): it is independent of r and its eigenvectors are independent of u. Hence $\langle \sigma_1 \rangle$, and all correlations within a single row, must be independent of r and u.)

We obtain

$$\langle \sigma_1 \rangle = \sum_j s_j m'_j m'''_j \Big/ \sum_j m'_j m'''_j \qquad (14.4.16a)$$

Using (14.4.14), this result can in turn be written as

$$\langle \sigma_1 \rangle = \sum_j s_j r_0^{2s_j} \exp(2\alpha_j u_0) \Big/ \sum_j r_0^{2s_j} \exp(2\alpha_j u_0) \qquad (14.4.16b)$$

From (14.4.9) it follows that

$$\langle \sigma_1 \rangle = \text{Trace } S\, A_d^2(r_0, u_0) \big/ \text{Trace } A_d^2(r_0, u_0). \qquad (14.4.17)$$

The Coefficients α_j

We still have to calculate the coefficients α_j. To do this, we first invoke a periodicity property, just as we did in obtaining (13.7.10) for the eight-vertex model.

From (14.2.43) it is apparent that incrementing u by $i\varepsilon$ (or $\frac{1}{2}i\varepsilon$ in regimes II and III) leaves w unchanged. From (14.2.44) this leaves the Boltzmann weights, and hence the diagonal matrices A_d, B_d, C_d, D_d, unchanged. Further, it seems that these matrices are analytic in a vertical strip containing the points $u = 0$ and $u = u_0$, so the expressions (14.4.9) and (14.4.14) must apply throughout this strip. These expressions must therefore be periodic of period $i\varepsilon$ (or $\frac{1}{2}i\varepsilon$), which implies that

$$\text{I and IV:} \qquad \alpha_j = 2\pi n_j/\varepsilon, \qquad (14.4.18)$$

$$\text{II and III:} \qquad \alpha_j = -4\pi n_j/\varepsilon,$$

where each n_j is an integer. From (14.4.9), the diagonal elements of $A_d(r, u)$ are therefore

$$[A_d(r, u)]_{j,j} = r^{s_j} w^{n_j}. \qquad (14.4.19)$$

Since the s_j and n_j are integers, they can be calculated by considering any suitable limiting or special case. In particular, consider the case when $x \to 0$ while w remains fixed. Then ω_2 in (14.2.44) tends to zero, so from (13.2.1) and (14.3.3) the matrices U_i are diagonal. So therefore is A. Further, from (14.2.43) the Boltzmann weight function is

$$w(a, b, c, d) = r^{b+d-2a} w^{a-\tau bd} \delta(a, c), \qquad (14.4.20)$$

where

$$\tau = 0 \quad \text{in regimes I and IV} \qquad (14.4.21)$$

$$= 1 \quad \text{in regimes II and III}.$$

Substituting this expression into (13.2.1), it follows that U_i is a diagonal matrix with elements

$$(U_i)_{\sigma,\sigma} = r^{\sigma_{i-1} + \sigma_{i+1} - 2\sigma_i} w^{\sigma_i - \tau\sigma_{i-1}\sigma_{i+1}} \qquad (14.4.22)$$

From (13.2.1)–(13.2.5), for a finite lattice A is given by

$$A = U_2 U_3^2 U_4^3 \ldots U_{m+1}^m, \qquad (14.4.23)$$

where σ_{m+1} and σ_{m+2} are to be given their ground-state values. Using (14.4.22), it follows that the diagonal elements of A are

$$[A]_{\sigma,\sigma} = r^{\sigma_1} w^{\sigma_2 + 2\sigma_3 + 3\sigma_4 + \ldots + m\sigma_{m+1}}$$

$$\times w^{-\tau(\sigma_1\sigma_3 + 2\sigma_2\sigma_4 + \ldots + m\sigma_m\sigma_{m+2})} \qquad (14.4.24)$$

From (13.1.15) and (14.4.6), the matrices A and A_d differ only by a scalar factor and a similarity transformation. Both are diagonal, so the similarity transformation can at worst only re-arrange the diagonal entries.

Let $\bar\sigma_1, \bar\sigma_2, \ldots, \bar\sigma_{m+2}$ be the ground-state values of $\sigma_1, \ldots \sigma_{m+2}$. Then it turns out that these values maximize (14.4.24) (with $r = 1$), so we can take A_d to be the diagonal matrix with entries

$$[A_d]_{\sigma,\sigma} = r^{s(\sigma)} w^{n(\sigma)}, \qquad (14.4.25)$$

where σ denotes the spin-set $\{\sigma_1, \ldots, \sigma_m\}$ and

$$s(\sigma) = \sigma_1, \qquad (14.4.26)$$

$$n(\sigma) = \sum_{i=1}^{m} i(\sigma_{i+1} - \tau\sigma_i\sigma_{i+2} - \bar\sigma_{i+1} + \tau\bar\sigma_i\bar\sigma_{i+2})$$

Plainly this result is of the expected form (14.4.19), the only difference being that the index j is replaced by σ. Clearly $s(\sigma)$ and $n(\sigma)$ are integers, so (14.4.25) is valid not just in the limit $x \to 0$, but for all x (provided we take the limit $m \to \infty$). From (14.4.17) it follows that

$$\langle \sigma_1 \rangle = \sum_{\sigma} \sigma_1 r_0^{2\sigma_1} w_0^{2n(\sigma)} \bigg/ \sum_{\sigma} r_0^{2\sigma_1} w_0^{2n(\sigma)}. \qquad (14.4.27)$$

This is a general formula for the density at a given site (site 1). In calculating the sum it should be remembered that $\sigma_1, \ldots, \sigma_m$ are not completely arbitrary: they must satisfy the requirement

$$\sigma_i\sigma_{i+1} = 0 \quad \text{for } i = 1, \ldots, m. \qquad (14.4.28)$$

As was remarked after (14.3.3), this condition is implicit in the above working, being built into the definition of the vector spaces on which the transfer matrices act. It corresponds simply to the fact that no two particles can be adjacent.

Symmetries

The corner transfer matrices A, B, C, D satisfy various symmetry relations. We have not used these in this section but it is helpful to be aware of them.

From (14.2.38), the Boltzmann weight function satisfies the relations

$$w(a,b,c,d) = w(c,b,a,d) = w(a,d,c,b) \qquad (14.4.29)$$

These are precisely the relations (13.5.4). They imply that the model is symmetric with respect to reflection through either diagonal. In regimes I, III and IV this symmetry is not spontaneously broken (regime IV is ordered, but from Fig. 14.5(b) it is apparent that each of the ground states has this symmetry). It follows that

$$\text{I, III, IV:} \quad A = A^T = C = C^T, \quad B = B^T = D = D^T. \qquad (14.4.30)$$

The matrices P and R in (14.4.7) are then equal and orthogonal.

The ground states in regime II are indicated in Fig. 14.5(a). It is apparent that these are symmetric on reflection through the SE – NW diagonal, but not through every SW – NE diagonal. This means that in general we only have the relations

$$\text{II:} \quad A = A^T, \quad C = C^T, \quad B = D^T. \qquad (14.4.31)$$

The matrices P and R are orthogonal, but not necessarily equal.

On the other hand, if the centre site lies on the preferred sublattice in Fig. 14.5(a), then the SW – NE reflection symmetry is in fact preserved and (14.4.30) still applies: this is the $k = 1$ case of regime II, as classified in the next section.

14.5 Explicit Formulae for the Various Cases: the Rogers – Ramanujan Identities

The sums in (14.4.27) can be evaluated, but there are seven different cases to consider. One reason for this is that $n(\sigma)$ has a different form in (14.4.26) depending whether τ is 0 or 1. Another reason is the boundary condition that σ_{m+1} and σ_{m+2} have their ground state values. In regimes I and III the ground state is unique:

$$\text{I and III:} \quad \bar{\sigma}_j = 0, \quad j = 1, 2, 3, \ldots \qquad (14.5.1a)$$

In regime IV the system has square ordering, as in Fig. 14.5(b). There are two ground states:

IV: $\bar{\sigma}_{2j+k} = 1$, $\bar{\sigma}_{2j+k+1} = 0$,   all integers j,   (14.5.1b)

where $k = 1$ for one ground state, and $k = 2$ for the other. In regime II the system has the triangular ordering of Fig. 14.5(a), the three ground states being:

II: $\bar{\sigma}_{3j+k} = 1$, $\bar{\sigma}_{3j+k\pm1} = 0$,   all integers j,   (14.5.1c)

where $k = 1, 2, 3$, for the three ground states, respectively.

To evaluate (14.4.27) we therefore first fix the regime and (if we are in II or IV) the value of k. We then perform the summations over $\sigma = \{\sigma_1, \sigma_2, \ldots \sigma_m\}$, subject to the condition (14.4.28). We find that the sums converge to limits as $m \to \infty$. The result is the density ρ, or (in regimes II and IV) the sub-lattice density ρ_k.

Performing the σ_1-summation explicity, (14.4.27) can be written as

$$\rho_k = \langle\sigma_1\rangle = r_0^2 F(1)/[F(0) + r_0^2 F(1)],   (14.5.2)$$

where

$$F(\sigma_1) = \sum_{\sigma_2,\sigma_3,\ldots,\sigma_m} w_0^{2n(\sigma)}   (14.5.3)$$

and the suffix k is redundant in regimes I and III. Our calculations therefore proceed in three stages: calculate $F(0)$ and $F(1)$ from (14.5.3); then calculate $\langle\sigma_1\rangle$ from (14.5.2); and (for the ordered regimes) calculate R from (14.1.8).

Regime I

From (14.4.21) and (14.3.12b):

$$\tau = 0,   w_0 = -x^3,   r_0^2 = -x\, G(x)/H(x),   (14.5.4)$$

where $-1 < x < 0$ and $G(x), H(x)$ are defined in (14.1.19). From (14.5.3), (14.4.26) and (14.5.1a), we have

$$F(\sigma_1) = \sum_{\sigma_2,\sigma_3,\ldots,\sigma_m} q^{\sigma_2 + 2\sigma_3 + 3\sigma_4 + \ldots},   (14.5.5)$$

where $q = w_0^2 = x^6$. Thus $0 < q < 1$.

First consider the ground state, with $\sigma_2, \sigma_3, \ldots$ all zero; then consider the states with one of them unity; then two of them unity; and so on.

Taking the limit $m \to \infty$ and remembering the restriction (14.4.28), we obtain

$$F(0) = 1 + \frac{q}{1-q} + \frac{q^4}{(1-q)(1-q^2)} + \cdots$$

$$+ \frac{q^{n^2}}{(1-q)(1-q^2)\cdots(1-q^n)} + \cdots \qquad (14.5.6)$$

$$F(1) = 1 + \frac{q^2}{1-q} + \frac{q^6}{(1-q)(1-q^2)} + \cdots$$

$$+ \frac{q^{n(n+1)}}{(1-q)(1-q^2)\cdots(1-q^n)} + \cdots$$

These series are well-known in the mathematical theory of partitions (Andrews, 1976, Chapter 7). From (14.5.5) it is fairly easy to see that $F(1)/F(0)$ (which is the ratio that enters (14.5.2)) is the simple continued fraction

$$1/(1 + q/(1 + q^2/(1 + q^3/ \cdots))) . \qquad (14.5.7)$$

What is by no means obvious, but was proved by Rogers (1894) and found by Ramanujan (1919), is that

$$F(0) = 1/[(1-q)(1-q^4)(1-q^6)(1-q^9)(1-q^{11})\cdots]$$

$$= G(q) , \qquad (14.5.8)$$

$$F(1) = 1/[(1-q^2)(1-q^3)(1-q^7)(1-q^8)(1-q^{12})\cdots]$$

$$= H(q) ,$$

where the functions $G(q)$ and $H(q)$ are those defined in (14.1.19). Thus these functions occur not only in the formula (14.5.4) for r_0^2, but also in our results for $F(0)$ and $F(1)$. Using the elliptic function identity (15.9.2), the expressions (14.5.8) can alternatively be written as

$$F(0) = [Q(q)]^{-1} \sum_{n=0}^{\infty} (-1)^n q^{5n(n+1)/2}(q^{-2n} - q^{2n+2}) ,$$

$$\qquad (14.5.9)$$

$$F(1) = [Q(q)]^{-1} \sum_{n=0}^{\infty} (-1)^n q^{5n(n+1)/2}(q^{-n} - q^{n+1}) ,$$

where

$$Q(q) = \prod_{n=1}^{\infty} (1 - q^n) . \qquad (14.5.10)$$

The identities implied by (14.5.6) and (14.5.8) are known as the Rogers–Ramanujan identities. There are many generalizations of these identities (Slater, 1951), and it is a remarkable fact that many of them occur naturally in the course of our present working. For instance, substituting (14.5.8) into (14.5.2), using (14.5.4) and $q = x^6$, we obtain

$$\rho = -x\, G(x)\, H(x^6)/[H(x)\, G(x^6) - x\, G(x)\, H(x^6)]\,. \qquad (14.5.11)$$

It turns out that the denominator in this expression can be written as a simple infinite product of the type that occur in the theta function expansions (15.1.5). Ramanujan stated (Birch, 1975, eq. 8), and Rogers (1921) proved that

$$H(x)\, G(x^6) - x\, G(x)\, H(x^6) = P(x)/P(x^3)\,, \qquad (14.5.12)$$

where

$$P(x) = \prod_{n=1}^{\infty} (1 - x^{2n-1})\,. \qquad (14.5.13)$$

Thus we finally find that the density is

$$\rho = -x\, G(x)\, H(x^6)\, P(x^3)/P(x^2)\,. \qquad (14.5.14)$$

It is fascinating that these Rogers–Ramanajuan type identities should occur in this problem, and it is of course very convenient to thereby simplify the results. This is particularly useful when we come to examine the critical behaviour when $|x| \to 1$: $G(x)$, $H(x)$ and $P(x)$ can all be related to elliptic functions, and their behaviour near $|x| = 1$ can be obtained from "conjugate modulus" identities such as (15.7.2). I know of no such straightforward techniques for handling the original expressions (14.5.6).

Regime II

Regime I is the simplest case to handle, but regime II is the most difficult. The function $n(\sigma)$ is more complicated and there are three ordered states to consider. These correspond to $k = 1, 2, 3$ in (14.1.5c), and each has its own $F(0)$, $F(1)$ and local sub-lattice density ρ_k.

From (14.4.21) and (14.3.12b),

$$\tau = 1, \quad w_0 = x^{-3/2}, \quad r_0^2 = x^{-1} H(x)/G(x)\,, \qquad (14.5.15)$$

where $0 < x < 1$. From (14.5.3), (14.4.26) and (14.5.1c) it follows that

$$F(\sigma_1) = \sum_{\sigma_2, \sigma_3, \ldots, \sigma_m} q^{\Sigma i(\sigma_i \sigma_{i+2} - \sigma_{i+1} + \bar{\sigma}_{i+1})}\,, \qquad (14.5.16)$$

where $q = x^3$, the inner summation is over integer values of i from 1 to m, $\bar{\sigma}_i$ is given by (14.5.1c), and $\sigma_1, \ldots, \sigma_m$ must satisfy (14.4.28).

We can develop recurrence relations to evaluate $F(0)$ and $F(1)$. Define

$$G_l(\sigma_l, \sigma_{l+1}) = \sum_{\sigma_{l+2}, \ldots, \sigma_m} q^{\Sigma i(\sigma_i \sigma_{i+2} - \sigma_{i+1} + \tilde{\sigma}_{i+1})} \qquad (14.5.17)$$

where now the inner summation is from $i = l$ to $i = m$. Then by considering explicitly the sum over σ_{l+2}, it is readily verified that

$$G_l(0,0) = \beta_l [G_{l+1}(0,0) + G_{l+1}(0,1)]$$

$$G_l(0,1) = \beta_l q^{-l} G_{l+1}(1,0) \qquad (14.5.18a)$$

$$G_l(1,0) = \beta_l [G_{l+1}(0,0) + q^l G_{l+1}(0,1)] ,$$

where

$$\beta_l = q^l \quad \text{if } (l - k + 1)/3 \text{ is an integer} , \qquad (14.5.18b)$$
$$\quad = 1 \quad \text{otherwise} ,$$

and that

$$F(0) = G_0(0,0) = G_0(1,0), \quad F(1) = G_0(0,1) . \qquad (14.5.19)$$

Each $G_l(\sigma, \sigma')$ tends to a limit as $m \to \infty$, and these limiting values satisfy

$$G_l(0,0) = (1 - q)^{-1} + \mathcal{O}(q^l) ,$$

$$G_l(0,1) = q(1 - q)^{-1}(1 - q^2)^{-1} + \mathcal{O}(q^l) , \qquad (14.5.20)$$

$$G_l(1,0) = 1 + \mathcal{O}(q^l)$$

provided that l is large and $(l - k)/3$ is an integer.

The recurrence relations (14.5.18), together with the large l boundary conditions (14.5.20), define the G_l. We can then obtain $F(0)$ and $F(1)$ from (14.5.19). Exhibiting the k-dependence by writing these as $F_k(0)$ and $F_k(1)$, we find that

$$F_1(0) = 1 + 2q + 2q^2 + 4q^3 + 5q^4 + 8q^5 + 11q^6 + \ldots$$

$$F_1(1) = 1 + q^2 + 2q^3 + 3q^4 + 4q^5 + 7q^6 + \ldots$$

$$F_2(0) = F_3(0) = 1 + q + 2q^2 + 3q^3 + 5q^4 + 7q^5 + 10q^6 + \ldots$$

$$F_2(1) = F_3(1) = q + q^2 + 2q^3 + 2q^4 + 4q^5 + 5q^6 + 8q^7 + \ldots$$

$$(14.5.21)$$

In regimes I, III and IV we can readily write $F(0)$ and $F(1)$ in explicit series forms like (14.5.6) (this can be done by regarding G_l as a function of l, and expanding it in powers of q^l, as in (14.5.35)). We can then use appropriate analogues of the Rogers – Ramanujan identities (14.5.6)–(14.5.8) so as to write $F(0)$ and $F(1)$ as simple products of theta functions.

In regime II this program is more complicated. However, Andrews (1981) has shown that each $F_k(0)$ and $F_k(1)$ can be written as a double series, and from this he has established that

$$F_1(0) = [Q(q)]^{-1} \sum_{n=0}^{\infty} (-1)^n q^{15n(n+1)/2} [q^{-4n} - q^{4n+4} + q(q^{-n} - q^{n+1})],$$

(14.5.22a)

$$F_1(1) = [Q(q)]^{-1} \sum_{n=0}^{\infty} (-1)^n q^{15n(n+1)/2} [q^{-7n} - q^{7n+7} - q(q^{-2n} - q^{2n+2})]$$

(14.5.22b)

$$F_2(0) = F_3(0) = [Q(q)]^{-1} \sum_{n=0}^{\infty} (-1)^n q^{15n(n+1)/2} (q^{-6n} - q^{6n+6}) \qquad (14.5.22c)$$

$$F_2(1) = F_3(1) = q [Q(q)]^{-1} \sum_{n=0}^{\infty} (-1)^n q^{15n(n+1)/2} (q^{-3n} - q^{3n+3}). \qquad (14.5.22d)$$

These expressions are similar to (14.5.9): the most obvious difference being that the first two involve the sum of two theta-function series, instead of just one.

First consider the cases $k = 2$ and 3. Using (15.9.2) and the definitions (14.5.8), (14.5.10) of the functions G, H, Q, we can write (14.5.22c) and (14.5.22d) as

$$F_2(0) = F_3(0) = Q(q^{15})/[Q(q) H(q^3)], \qquad (14.5.23)$$

$$F_2(1) = F_3(1) = q Q(q^{15})/[Q(q) G(q^3)].$$

From (14.5.15) *et seq*, we have that $r_0^2 = x^{-1} H(x)/G(x)$ and $q = x^3$. Substituting the expressions (14.5.23) into (14.5.2), it follows that

$$\rho_2 = \rho_3 = x^2 H(x) H(x^9)/[G(x) G(x^9) + x^2 H(x) H(x^9)]. \qquad (14.5.24)$$

Now we consult the list of Ramanujan's identities given by Birch (1975), and find from eq. (6) therein that

$$G(x) G(x^9) + x^2 H(x) H(x^9) = [Q(x^3)]^2/[Q(x) Q(x^9)] \qquad (14.5.25)$$

so (14.5.24) simplifies to

$$\rho_2 = \rho_3 = x^2 H(x) H(x^9) Q(x) Q(x^9)/[Q(x^3)]^2 \qquad (14.5.26)$$

The case $k = 1$ is more complicated, but from (14.5.22a) and (14.5.22b) we can establish that

$$F_1(0) = [G(x^9) Q(x^9) - H(x) Q(x)]/[x Q(x^3)] , \qquad (14.5.27)$$

$$F_1(1) = [G(x) Q(x) + x^2 H(x^9) Q(x^9)]/Q(x^3) .$$

(To do this we expand the numerators on the RHS as series, using the identities implied by (14.5.8) and (14.5.9). For $H(x)Q(x)$ and $G(x)Q(x)$ we break their series into three parts: terms with $n = 3r$, with $n = 3r + 1$, and with $n = 3r + 2$. After some cancellations, and remembering that $q = x^3$, we regain (14.5.22). In particular it follows that each RHS in (14.5.27) can be expanded in integer powers of x^3: something that is far from obvious.)

Substituting these expressions for $F_1(0)$ and $F_1(1)$ into (14.5.2), using (14.5.15), we obtain

$$\rho_1 = H(x) [G(x) Q(x) + x^2 H(x^9) Q(x^9)]/\{Q(x^9) [G(x) G(x^9)$$

$$+ x^2 H(x) H(x^9)]\} \qquad (14.5.28)$$

Again we can use Ramanujan's identity (14.5.25) to simplify the denominator, giving

$$\rho_1 = H(x) Q(x) [G(x) Q(x) + x^2 H(x^9) Q(x^9)]/[Q(x^3)]^2 \qquad (14.5.29)$$

Substituting these expressions for ρ_1 and ρ_2 into (14.1.8), the order parameter is

$$R = \rho_1 - \rho_2 = G(x) H(x) [Q(x)/Q(x^3)]^2 \qquad (14.5.30)$$

$$= Q(x) Q(x^5)/[Q(x^3)]^2$$

$$= \prod_{n=1}^{\infty} (1 - x^n) (1 - x^{5n})/(1 - x^{3n})^2 . \qquad (14.5.31)$$

This expression is rather similar to that for the order parameter of the eight-vertex model, namely (13.7.21).

Regime III

In this case the analogues of (14.5.15) and (14.5.16) are

$$\tau = 1, \quad w_0 = x, \quad r_0^2 = x H(x)/G(x) , \qquad (14.5.32)$$

$$F(\sigma_1) = \sum_{\sigma_2,\ldots,\sigma_m} q^{\Sigma i (\sigma_{i+1} - \sigma_i \sigma_{i+2})}, \tag{14.5.33}$$

where $0 < x < 1$, $q = x^2$ and $\sigma_{m+1} = \sigma_{m+2} = 0$. This is the same as (14.5.16), except that q is inverted and the $\bar{\sigma}_i$ are zero. We can therefore evaluate $F(0)$ and $F(1)$ by using the recursion relations (14.5.18), with q inverted and $\beta_l = 1$, together with (14.5.19). Again we take the limit of m large. The boundary conditions are then that for l large

$$G_l(0,0) = 1 + \mathcal{O}(q^l)$$

$$G_l(0,1) = \mathcal{O}(q^l) \tag{14.5.34}$$

$$G_l(1,0) = (1-q)^{-1} + \mathcal{O}(q^l).$$

We can expand the G_l in powers of q^l, and systematically evaluate the coefficients from (14.5.18) (with q replaced by q^{-1}). Doing this, we find that

$$G_l(0,0) = \sum_{n=0}^{\infty} q^{nl} a_n,$$

$$G_l(0,1) = \sum_{n=1}^{\infty} q^{nl + 2n - 2} a_{n-1} / (1 - q^{2n-1}), \tag{14.5.35}$$

$$G_l(1,0) = \sum_{n=0}^{\infty} q^{nl + n} a_n / (1 - q^{2n+1}),$$

where $a_0 = 1$ and

$$a_n = q^{3n-2} a_{n-1} / [(1 - q^n)(1 - q^{2n-1})] \tag{14.5.36}$$

for $n \geq 1$. Evaluating the a_n from this last relation, it follows from (14.5.19) that

$$F(0) = \sum_{n=0}^{\infty} q^{n(3n-1)/2} / [(1-q)(1-q^2)\ldots(1-q^n)$$

$$\times (1-q)(1-q^3)(1-q^5)\ldots(1-q^{2n-1})], \tag{14.5.37}$$

$$F(1) = \sum_{n=0}^{\infty} q^{3n(n+1)/2} / [(1-q)(1-q^2)\ldots(1-q^n)$$

$$\times (1-q)(1-q^3)(1-q^5)\ldots(1-q^{2n+1})].$$

Just as the regime I series (14.5.6) could be simplified by using the Rogers – Ramanujan identities, so can (14.5.37) be simplified by using the further identities (46) and (44) in the list compiled by Slater (1951). These give

$$F(0) = G(q^2)\, Q(q^2)/Q(q)\,, \tag{14.5.38}$$

$$F(1) = H(q^2)\, Q(q^2)/Q(q)\,.$$

From (14.5.2) and (14.5.32), it follows that

$$\rho = x\, H(x)\, H(x^4)/[G(x)\, G(x^4) + x\, H(x)\, H(x^4)]\,. \tag{14.5.39}$$

Ramanujan stated (Birch, 1975, eq. 2), and Rogers (1921) proved that

$$G(x)\, G(x^4) + x\, H(x)\, H(x^4) = [P(-x)]^2\,, \tag{14.5.40}$$

where $P(x)$ is defined by (14.5.13). Thus (14.5.39) simplifies to

$$\rho = x\, H(x)\, H(x^4)/[P(-x)]^2\,. \tag{14.5.41}$$

Regime IV

This regime is ordered and we have to distinguish the two cases $k = 1$ and $k = 2$ in (14.5.1b). From (14.4.21), (14.3.12b), (14.4.26) and (14.5.3),

$$\tau = 0,\quad w_0 = x^{-2},\quad r_0^2 = -x^{-1}\, G(x)/H(x)\,, \tag{14.5.42}$$

$$F(\sigma_1) = \sum_{\sigma_2,\dots,\sigma_m} q^{\Sigma i(\bar{\sigma}_{i+1} - \sigma_{i+1})} \tag{14.5.43}$$

where $-1 < x < 0$, $q = x^4$. The $\bar{\sigma}_i$ are defined by (14.5.1b) and the summation is as usual over all values (0 or 1) of $\sigma_2,\dots,\sigma_m$ that satisfy (14.4.28), where $\sigma_{m+1} = \bar{\sigma}_{m+1}$.

As in regimes II and III, we can set up recursion relations that define $F(0)$ and $F(1)$. Define

$$G_l(\sigma_l) = \sum_{\sigma_{l+1},\dots,\sigma_m} q^{\Sigma i(\bar{\sigma}_{i+1} - \sigma_{i+1})} \tag{14.5.44}$$

where now the inner sum is over $i = l,\dots,m$. Considering explicitly the contributions from $\sigma_{l+1} = 0$ and $\sigma_{l+1} = 1$, we find that

$$G_l(0) = \beta_l\,[G_{l+1}(0) + q^{-l}G_{l+1}(1)]\,, \tag{14.5.45a}$$

$$G_l(1) = \beta_l\, G_{l+1}(0)\,,$$

where

$$\beta_l = 1 \quad \text{if } l - k \text{ is even}\,, \tag{14.5.45b}$$

$$= q^l \quad \text{if } l - k \text{ is odd}\,.$$

Clearly

$$F(0) = G_1(0),\quad F(1) = G_1(1)\,. \tag{14.5.46}$$

Each $G_l(\sigma)$ tends to a limit as $m \to \infty$, and these limiting values satisfy the boundary condition

$$G_l(0) \to (1-q)^{-1}, \quad G_l(1) \to 1 \text{ as } l \to \infty, \qquad (14.5.47)$$

provided $l - k$ is even.

We can expand $G_l(0)$ and $G_l(1)$ in powers of q^l. Substituting the expansions into (14.5.45) and equating coefficients, we find that, for $l - k$ even,

$$G_l(0) = \sum_{n=0}^{\infty} q^{nl} a_n / (1 - q^{2n+1}),$$

$$G_l(1) = \sum_{n=0}^{\infty} q^{nl} a_n, \qquad (14.5.48)$$

where $a_0 = 1$ and

$$a_n = q^{2n-1} a_{n-1} / (1 - q^{2n-1})(1 - q^{2n}). \qquad (14.5.49)$$

This last equation can be solved sequentially for a_1, a_2, a_3, etc; $F(0)$ and $F(1)$ can be obtained from (14.5.46) and (14.5.48). (For $k = 2$ we need G_l for $l - k$ odd: this can readily be found from (14.5.45).) Exhibiting explicitly the dependence of $F(0)$ and $F(1)$ on k by writing them as $F_k(0)$ and $F_k(1)$, we find that

$$F_1(0) = \sum_{n=0}^{\infty} q^{n^2+n} / [(1-q)(1-q^2) \ldots (1-q^{2n+1})],$$

$$F_1(1) = \sum_{n=0}^{\infty} q^{n^2+n} / [(1-q)(1-q^2) \ldots (1-q^{2n})],$$

$$\qquad (14.5.50)$$

$$F_2(0) = \sum_{n=0}^{\infty} q^{n^2} / [(1-q)(1-q^2) \ldots (1-q^{2n})],$$

$$F_2(1) = \sum_{n=1}^{\infty} q^{n^2} / [(1-q)(1-q^2) \ldots (1-q^{2n-1})].$$

Again we look at the list of Rogers – Ramanujan-type identities compiled by Slater (1951). From her equations (94), (99), (98) and (96) we find that

$$F_1(0) = H(-q)/P(q),$$

$$F_1(1) = G(-q)/P(q),$$

$$\qquad (14.5.51)$$

$$F_2(0) = G(q^4)/P(q),$$

$$F_2(1) = q H(q^4)/P(q),$$

where again the functions G, H, P, Q are defined by (14.5.8), (14.5.13) and (14.5.10).

Substituting these results into (14.5.2), using (14.5.42) and $q = x^4$,

$$\rho_1 = G(x)\, G(-x^4)/[G(x)\, G(-x^4) - x\, H(x)\, H(-x^4)] \qquad (14.5.52)$$

$$\rho_2 = -x^3 G(x)\, H(x^{16})/[H(x)\, G(x^{16}) - x^3 G(x)\, H(x^{16})]$$

The first of these denominators does not appear to have been explicitly studied by Ramanujan, but he did state, and Watson (1933) proved, that

$$G(x)\, H(-x) + G(-x)\, H(x) = 2/[P(x^2)]^2 \qquad (14.5.53)$$

(this is eq. 23 of Birch, 1975). Further, Rogers (1894) showed that

$$G(-x^4) = Q(x^2)[H(x) + H(-x)]/[2Q(x^8)], \qquad (14.5.54)$$

$$H(-x^4) = Q(x^2)[G(x) - G(-x)]/[2xQ(x^8)].$$

From these three identities it follows that

$$G(x)\, G(-x^4) - x\, H(x)\, H(-x^4) = P(-x^2). \qquad (14.5.55)$$

Also, Ramanujan stated (eq. 5 of Birch, 1975), and Rogers (1921) proved that

$$H(x)\, G(x^{16}) - x^3 G(x)\, H(x^{16}) = P(-x^2). \qquad (14.5.56)$$

Using these last two identities, we can therefore simplify (14.5.52) to

$$\rho_1 = G(x)\, G(-x^4)/P(-x^2), \qquad (14.5.57)$$

$$\rho_2 = -x^3 G(x)\, H(x^{16})/P(-x^2).$$

Rogers (1894) proved that

$$H(x^{16}) = Q(x^2)[H(x) - H(-x)]/[2x^3 Q(x^8)]. \qquad (14.5.58)$$

Substituting this expression for $H(x^{16})$, and the expression (14.5.54) for $G(-x^4)$, into (14.5.57), we find that the mean total density is

$$\rho = \tfrac{1}{2}(\rho_1 + \rho_2) = \tfrac{1}{2} G(x)\, H(-x)\, [P(x^2)]^2, \qquad (14.5.59)$$

and the order parameter is

$$R = \rho_1 - \rho_2 = G(x)\, H(x)\, [P(x^2)]^2$$

$$= [Q(x^2)]^2\, Q(x^5)/\{Q(x)\, [Q(x^4)]^2\} \qquad (14.5.60)$$

$$= \prod_{n=1}^{\infty} (1 - x^{2n})^2 (1 - x^{5n})/[(1 - x^n)(1 - x^{4n})^2]$$

This has a similar form to the order parameter (14.5.30) in regime II, and the eight-vertex model order parameter (13.7.21), being a ratio of products of Q-functions.

This completes the derivation of the sub-lattice densities and order parameters of the generalized hard hexagon model. I have discussed the four regimes separately, but we can now see some common features: we can write down recursion relations defining $F_k(0)$ and $F_k(1)$. In regimes I, III, and IV these can be solved to give $F_k(0)$ and $F_k(1)$ as infinite series. We can then use the appropriate Rogers – Ramanujan-type identities, as listed by Slater (1951), to write $F_k(0)$ and $F_k(1)$ as infinite products of theta-function type. (In regime II this program is more difficult, but it still turns out that $F_k(0)$ and $F_k(1)$ can each be written as a sum of at most two theta-function products.) Further, when we substitute the results into (14.5.2), we find that the denominators can be simplified by using some of the Ramanujan identities listed by Birch (1975). Finally, in regimes II and IV the order parameter $R = \rho_1 - \rho_2$ is found to be a simple ratio of products of Q-functions.

It is fascinating that these Rogers – Ramanujan and Ramanujan-type identities occur so frequently in this working. With the benefit of hindsight, we can see signals of this in the star – triangle relations (14.2.3), in particular the elliptic function parametrization of (14.2.11). This led automatically to (14.2.29)–(14.2.30), and thence to (14.2.44). This last equation abounds in factors $f(x\,,x^5)$ and $f(x^2\,,x^5)$. From (14.2.40), (14.1.21) and (14.1.19), these are just the functions $H(x)\,Q(x)$ and $G(x)\,Q(x)$. The natural occurrence of these functions (particularly their ratio) should perhaps have warned us to expect the Rogers – Ramanujan identities to occur.

14.6 Alternative Expressions for the κ, ρ, R

Our results can be summarized as follows. Given a hard square model with activity z and interaction coefficients L, M satisfying (14.2.9), calculate Δ from (14.2.10). Determine from (14.2.37) the regime in which the model lies (if V or VI, interchange L and M). Calculate x and w (and m, t and r) from (14.2.38) and (14.2.44). Then the partition-function-per site κ is given by (14.3.26), and the density ρ (or, in regimes II and IV, the sub-lattice densities ρ_k and order parameter R) are given by the appropriate equations in Section 14.5.

All these results are expressed in terms of infinite products. These converge well when x is small, which is the condition for extreme disorder or extreme order. They can readily be compared with low-density or high-density series expansions. However, they converge poorly when $|x|$ is close to one, which is when $|\Delta|$ is close to Δ_c and the system is near-critical. It is then convenient to convert the products into forms which converge rapidly for $|x|$ close to one.

Part of this procedure has been performed already: we merely return from the 'conjugate modulus' equations (14.2.44) for w and x to the original equations (14.2.29)–(14.2.31) for q^2 and u.

Partition-Function-Per-Site κ

To convert the equations (14.3.26) for κ, we use the identity (14.2.42) in reverse, going from f-functions to θ_1-functions. Doing this we find we also need the elliptic theta function

$$\theta_4(u\,,q^2) = \prod_{n=1}^{\infty} (1 - 2q^{2n-1}\cos 2u + q^{4n-2})(1 - q^{2n}) \quad (14.6.1)$$

which satisfies the 'conjugate modulus' relation

$$\theta_4(u, e^{-\varepsilon}) = \left(\frac{2\pi}{\varepsilon}\right)^{\frac{1}{2}} \exp[-\frac{\pi^2}{2\varepsilon} + \frac{2u(\pi - u)}{\varepsilon}]f(-e^{-4\pi u/\varepsilon}, e^{-4\pi^2/\varepsilon}). \quad (14.6.2)$$

From (14.3.26), using (14.2.43) and (14.2.38), and then applying the identities (14.2.42) and (14.6.2), we obtain the following expressions for the partition-function-per-site in the four regimes:

I: $\kappa/(mze^{L+M}) = \dfrac{\theta_4\left(\dfrac{\pi}{6} - \dfrac{5u}{3}, p^{10/3}\right) \theta_1\left(\dfrac{\pi}{3} - \dfrac{5u}{3}, p^{10/3}\right)}{\theta_4\left(\dfrac{\pi}{6} + \dfrac{5u}{3}, p^{10/3}\right) \theta_1\left(\dfrac{\pi}{3} + \dfrac{5u}{3}, p^{10/3}\right)},$

II: $= \theta_1\left(\dfrac{\pi}{3} - \dfrac{5u}{3}, p^{5/3}\right)\Big/ \theta_1\left(\dfrac{\pi}{3} + \dfrac{5u}{3}, p^{5/3}\right),$

III: $= 1,$ (14.6.3)

IV: $= \theta_4\left(\dfrac{\pi}{4} - \dfrac{5u}{2}, p^5\right)\Big/ \theta_4\left(\dfrac{\pi}{4} + \dfrac{5u}{2}, p^5\right).$

Here $p = e^{-\varepsilon} = |q^2|$ (14.6.4)

where this is the q defined by (14.2.28) and (14.2.29). Thus we have now expressed our results for κ in terms of the parameters q^2 and u discussed in (14.2.29)–(14.2.39).

The parameter m is introduced in (14.2.1) as a simple factor that multiplies all Boltzmann weights. It therefore multiplies κ, so κ/m is independent of m. This means that (13.3.26) and (14.6.3) are correct for all values of m (notably $m = 1$), even though we made a particular choice of m in (14.2.44).

Sub-Lattice Densities ρ_k and the Order Parameter R

The results of Section 14.5 are all expressed in terms of the functions $G(x)$, $H(x)$, $Q(x)$, $P(x)$, where

$$G(x) = \prod_{n=1}^{\infty} [(1 - x^{5n-4})(1 - x^{5n-1})]^{-1}, \qquad (14.6.5a)$$

$$H(x) = \prod_{n=1}^{\infty} [(1 - x^{5n-3})(1 - x^{5n-2})]^{-1}, \qquad (14.6.5b)$$

$$Q(x) = \prod_{n=1}^{\infty} (1 - x^n), \qquad (14.6.5c)$$

$$P(x) = \prod_{n=1}^{\infty} (1 - x^{2n-1}). \qquad (14.6.5d)$$

We can use the identities (14.2.42) and (14.6.2) to convert these infinite products into forms which converge rapidly for x close to $+1$, or to -1. (To do this we start by considering (14.2.42) in the limit $u \to 0$, and taking cube roots of each side. This gives the conversion formulae for $Q(x)$. The remaining formulae can then be obtained directly.)

This procedure introduces two more functions, $G_1(x)$ and $H_1(x)$, defined by

$$G_1(x) = \left(2 \sin \frac{\pi}{5}\right)^{-1} \prod_{n=1}^{\infty} \left[1 - 2x^n \cos \frac{2\pi}{5} + x^{2n}\right]^{-1}, \qquad (14.6.5e)$$

$$H_1(x) = \left(2 \sin \frac{2\pi}{5}\right)^{-1} \prod_{n=1}^{\infty} \left[1 - 2x^n \cos \frac{4\pi}{5} + x^{2n}\right]^{-1}. \qquad (14.6.5f)$$

We obtain the identities

$$Q(e^{-\varepsilon}) = (2\pi/\varepsilon)^{\frac{1}{2}} \exp\left[\frac{\varepsilon}{24} - \frac{\pi^2}{6\varepsilon}\right] Q(\exp[-4\pi^2/\varepsilon]), \qquad (14.6.6a)$$

$$P(e^{-\varepsilon}) = 2^{\frac{1}{2}} \exp\left[-\frac{\varepsilon}{24} - \frac{\pi^2}{12\varepsilon}\right] \Big/ P(\exp[-2\pi^2/\varepsilon]), \qquad (14.6.6b)$$

$$G(e^{-\varepsilon}) = \exp\left[-\frac{\varepsilon}{60} + \frac{\pi^2}{15\varepsilon}\right] G_1(\exp[-4\pi^2/5\varepsilon]) , \qquad (14.6.6c)$$

$$H(e^{-\varepsilon}) = \exp\left[\frac{11\varepsilon}{60} + \frac{\pi^2}{15\varepsilon}\right] H_1(\exp[-4\pi^2/5\varepsilon]) , \qquad (14.6.6d)$$

$$Q(-e^{-\varepsilon}) = (\pi/\varepsilon)^{\frac{1}{2}} \exp\left[\frac{\varepsilon}{24} - \frac{\pi^2}{24\varepsilon}\right] Q(-\exp[-\pi^2/\varepsilon]) , \quad (14.6.6e)$$

$$P(-e^{-\varepsilon}) = \exp\left[-\frac{\varepsilon}{24} + \frac{\pi^2}{24\varepsilon}\right] P(-\exp[-\pi^2/\varepsilon]) , \qquad (14.6.6f)$$

$$G(-e^{-\varepsilon}) = \exp\left[-\frac{\varepsilon}{60} + \frac{\pi^2}{60\varepsilon}\right] H_1(-\exp[-\pi^2/5\varepsilon]) , \qquad (14.6.6g)$$

$$H(-e^{-\varepsilon}) = \exp\left[\frac{11\varepsilon}{60} + \frac{\pi^2}{60\varepsilon}\right] G_1(-\exp[-\pi^2/5\varepsilon]) . \qquad (14.6.6h)$$

Applying these identities to the formulae in Section 14.5 for ρ and R, we find that

I: $\quad \rho = H_1(-p) H_1(p^{2/3}) P(-p^{5/3})/P(-p^5) ,$

II: $\quad \rho_2 = \rho_3 = H_1(p) H_1(p^{1/9}) Q(p^5) Q(p^{5/9})/Q^2(p^{5/3}) ,$

$\quad\quad R = \rho_1 - \rho_2 = (3/\sqrt{5})p^{1/9}Q(p^5) Q(p)/Q^2(p^{5/3}) ,$

III: $\quad \rho = H_1(p) H_1(p^{1/4})/P^2(-p^{5/4}) ,$ $\qquad\qquad\qquad (14.6.7)$

IV: $\quad \rho_1 = H_1(-p) H_1(-p^{1/4})/P(-p^{5/2}) ,$

$\quad\quad \rho_2 = H_1(-p) H_1(p^{1/4})/P(-p^{5/2}) ,$

$\quad\quad \rho = \tfrac{1}{2}(\rho_1 + \rho_2) = H_1(-p) H_1(p^4)/P^2(p^5) ,$

$\quad\quad R = \rho_1 - \rho_2 = (2/\sqrt{5})p^{1/4}Q(-p) Q^2(p^{10})/[Q(-p^5) Q^2(p^5)] .$

In regime II there is no simple product formula for ρ_1, or for the mean density $\rho = (\rho_1 + \rho_2 + \rho_3)/3$, but one can establish that ρ is expandable in integer powers of $p^{1/3}$.

The parameter p in (14.6.3) and (14.6.7) is defined by (14.6.4), where ε is in turn defined by (14.2.43), and q^2 therein by (14.2.28)–(14.2.31). In particular, from (14.2.30) we can take p to be defined by

$$\Delta^{2/5} = H_1(\pm p)/G_1(\pm p) , \qquad (14.6.8)$$

choosing the positive sign in regimes II and III, the negative signs in regimes I and IV; p is non-negative.

Critical Singularities

Our results are now in a form where we can discuss the behaviour across the I–II and III–IV regime boundaries. From (14.2.37) these occur when $\Delta = \pm \Delta_c$ and $q^2 = p = 0$.

We have only solved the general model, i.e. the hard square model with diagonal interactions, when the constraint (14.2.9) is satisfied. This means that we cannot consider the full (L, M, z) parameter space: only the two-dimensional surface (14.2.9).

Consider a line in that surface, crossing the boundary line $\Delta = \pm \Delta_c$ non-tangentially at a point C. Consider κ, ρ and R as functions of position along the line. They are analytic except at C; R is identically zero on the disordered side of C (regimes I and III), positive on the ordered side (regimes II and IV). Thus C is a critical point.

The parameters u and q^2 are defined by (14.2.28)–(14.2.31). Both are analytic functions of position along the line, even at C. At C, u is non-zero, while q^2 vanishes linearly with position. We can therefore take q^2 (or $-q^2$) to be our 'deviation-from-criticality' variable, corresponding to t in Section 1.1.

More precisely, let us here define t to be given by the following equations:

$$\text{I: } t = -q^2 = p; \qquad \text{II: } t = -q^2 = -p; \qquad (14.6.9)$$

$$\text{III: } t = q^2 = p; \qquad \text{IV: } t = q^2 = -p.$$

Then t is positive for disordered regimes and negative for ordered ones, as in Section 1.1. It vanishes linearly at the critical point C. We want to obtain the leading behaviour of κ, ρ, R as functions of t, for t small.

This is readily done, using the results (14.6.3) and (14.6.7) together with the definitions (14.2.28), (14.6.1) and (14.6.5). We find that

I and II: $\dfrac{\kappa e^{-L-M}}{mz}$

$$= \frac{\sin[\pi - 5u)/3]}{\sin[(\pi + 5u)/3]} \{1 - 2\sqrt{3}\,|t|^{5/3} \sin(10u/3) + \mathcal{O}(t^{10/3})\},$$

III: $\quad = 1$, $\hspace{6cm} (14.6.10)$

IV: $\quad = 1 - 4(-t)^{5/2} \sin 5u + \mathcal{O}(t^5)$,

I and II: $\rho = \rho_c - \text{sgn}(t)\,|t|^{2/3}/\sqrt{5} + \mathcal{O}(t)$,

III: $\rho = \rho_c - t^{1/4}/\sqrt{5} + \mathcal{O}(t^{1/2})$, $\hspace{3cm} (14.6.11)$

IV: $\rho = \rho_c + \mathcal{O}(t)$

II: $R = (3/\sqrt{5})\,(-t)^{1/9}\{1 + t + 2(-t)^{5/3} + \mathcal{O}(t^2)\}$, $\hspace{2cm} (14.6.12)$

IV: $R = (2/\sqrt{5})\,(-t)^{1/4}\{1 - t + \mathcal{O}(t^2)\}$.

Here ρ_c, the critical density is

$$\rho_c = (5 - \sqrt{5})/10 = 0.27639\ldots. \qquad (14.6.13)$$

We expect R and the dominant singular parts of κ and ρ to behave for small t as

$$R \sim (-t)^\beta, \qquad \kappa_{\text{sing}} \sim t^{2-\alpha}, \qquad \rho_{\text{sing}} \sim t^{1-\bar{\alpha}}, \qquad (14.6.14)$$

where β, α, $\bar{\alpha}$ are the critical exponents of (1.1.14) and (1.7.9). From (14.6.10)–(14.6.12), this is the case for this model, and across the I–II boundary

$$\alpha = \bar{\alpha} = 1/3, \qquad \beta = 1/9 \qquad (14.6.15)$$

while across the III–IV boundary

$$\alpha = -1/2, \qquad \bar{\alpha} = 3/4, \qquad \beta = 1/4. \qquad (14.6.16)$$

For the general hard square model, with weight function (14.2.1), the mean density ρ is related to κ by (14.1.7), the differentiation being performed with L, M held fixed. This means that if we consider κ, ρ and R along a line parallel to the z-axis in (z, L, M) space, then α and $\bar{\alpha}$ in (14.6.14) must be the same. However, we are unable to do this, since we can only consider lines on the surface (14.2.9). Thus α and $\bar{\alpha}$ are not necessarily equal, and indeed we see that they differ along the III–IV boundary.

The ground state in IV is either that shown in Fig. 14.5(b), or the one obtained by shifting each particle one lattice space to the right. These are the ordered ground states of the usual hard-square model, which is (14.2.1) with $L = M = 0$. It seems likely that our critical III–IV boundary line lies on the same critical surface as the hard square model critical point ($z = 3.7962\ldots, L = M = 0$). Since the exponents (14.6.16) apply only to the surface (14.2.9), it is not surprising that they differ from those expected for hard squares, namely $\alpha = \bar{\alpha} = 0$, $\beta = 1/8$ (Baxter *et al.*, 1980). Even so, it is disappointing that they seem to have no connection at all.

14.7   The Hard Hexagon Model

We started this chapter by discussing the hard hexagon model, i.e., the triangular lattice gas with nearest-neighbour exclusion. In order to solve it, we generalized it to the hard-square model with diagonal interactions.

In Sections 14.2–14.6 we have considered this more general model; let us now return to the original hard hexagon model.

To do this, we let m, L, M in (14.2.1) tend to the values 1, 0, $-\infty$, respectively. From (14.2.29) it follows that

$$u \to -\pi/5, \qquad (14.7.1)$$

while from (14.2.10) and (14.2.30),

$$z = \Delta^{-2} = \left[\theta_1\!\left(\frac{2\pi}{5}, q^2\right) \Big/ \theta_1\!\left(\frac{\pi}{5}, q^2\right) \right]^5, \qquad (14.7.2)$$

the function θ_1 being defined by (14.2.28). If the activity z is given, then we can regard (14.7.1) and (14.7.2) as defining our two parameters u and q^2.

There are two cases to consider, corresponding to q^2 being negative and positive, respectively. From (14.7.2) these in turn correspond to $z < z_c$ and $z > z_c$, where

$$z_c = \left(\sin\frac{2\pi}{5} \Big/ \sin\frac{\pi}{5} \right)^5$$

$$= [\tfrac{1}{2}(1 + \sqrt{5})]^5$$

$$= \tfrac{1}{2}(11 + 5\sqrt{5}) = 11.09017 \ldots . \qquad (14.7.3)$$

We see that this is the value (14.1.11) conjectured by Gaunt (1967).

The case $z < z_c$ is the $u = -\pi/5$ limit of regime I, while $z > z_c$ is the corresponding limit of regime II. We define x as in (14.2.43). Then from (14.2.10) and (14.2.44), x is related to z by:

$$z < z_c: \quad z = -x[H(x)/G(x)]^5, \qquad (14.7.4)$$

$$z > z_c: \quad z = x^{-1}[G(x)/H(x)]^5,$$

where $G(x)$ and $H(x)$ are the Rogers–Ramanujan functions defined in (14.1.19) and (14.6.5). These are precisely the relations conjectured in (14.1.18) and (14.1.22), so the x of this section is the same as that of Section 14..

To obtain κ, we first re-normalize the weights $\omega_1, \ldots, \omega_5$ in (14.2.44) to ensure that ω_1 is unity. This means that m in (14.2.38) and (14.2.1) is unity. Using these re-normalized weights, and remembering that $f(w)$ in

(14.2.44) means $f(w, x^5)$, from (14.2.44) we have that

$$\text{I:} \quad \omega_4\omega_5/\omega_1 = \frac{f(xw^{-1}, x^5)f(x^2w^{-1}, x^5)f(x, x^5)}{f^2(xw, x^5)f(x^2, x^5)}$$

$$\text{II:} \quad \omega_4\omega_5/\omega_1 = \frac{f(x^2w^{-1}, x^5)f(xw, x^5)f(x^2, x^5)}{f^2(x^2w, x^5)f(x, x^5)}. \tag{14.7.5}$$

We now use (14.3.26) (which is true for all normalizations of $\omega_1, \ldots, \omega_5$) and take the limit $u \to -\pi/5$. From (14.2.43) this means that $w \to x^2$ in regime I, $w \to x^{-1}$ in regime II. In both cases the RHS of (14.3.26) has a simple pole at this value of w, but this is cancelled by a corresponding zero of (14.7.5). We find that κ is indeed given by (14.1.20) and (14.1.23).

For $z < z_c$, the density ρ is given at once by (14.5.14). For $z > z_c$, the sub-lattice densities ρ_1, ρ_2, ρ_3, and the order parameter R, are given by (14.5.29), (14.5.26) and (14.5.30). This last formula is the same as (14.1.24) so we see that all the conjectures of Section 14.1 have been verified.

Critical Behaviour

To study the behaviour near $z = z_c$, we use the alternative forms of the results, as given in the previous section, specializing them to the hard hexagon model.

The first step is simply to note that q^2 is given by (14.7.2). Using (14.2.28) and (14.7.3), this equation can be written explicitly as

$$z = z_c \prod_{n=1}^{\infty} \left[\left(1 - 2q^{2n}\cos\frac{4\pi}{5} + q^{4n}\right) \middle/ \left(1 - 2q^{2n}\cos\frac{2\pi}{5} + q^{4n}\right) \right]^5. \tag{14.7.6}$$

This defines q^2 for all real positive values of z; q^2 is negative for $z < z_c$, positive for $z > z_c$.

To obtain the small-q expansion of κ, we use (14.6.3) together with (14.2.29) and (14.2.31). We take $m = 1$ and let $u \to -\pi/5$. Using the functions $Q(x)$, $P(x)$, $G_1(x)$, $H_1(x)$ defined in (14.6.5), we find that:

$$z < z_c: \quad \kappa = \frac{3\sqrt{3}\, G_1^3(q^2)\, Q^2(q^2)\, Q(p^5)\, P^3(p^{10/3})}{5\, H_1^2(q^2)\, Q^3(p^{5/3})\, P(p^{10})},$$

$$z > z_c: \quad \kappa = \frac{3\sqrt{3}\, G_1^3(q^2)\, Q^2(q^2)\, Q(p^5)}{5\, H_1^2(q^2)\, Q^3(p^{5/3})} \tag{14.7.7}$$

where p is defined by (14.6.4), i.e. $p = |q^2|$. The critical value κ_c of κ is obtained by setting $q^2 = p = 0$ in either of these formulae, giving

$$\kappa_c = 3\sqrt{3}\,\sin^2(2\pi/5)/[10\sin^3(\pi/5)] \tag{14.7.8}$$
$$= [27(25 + 11\sqrt{5})/250]^{\frac{1}{2}} = 2.3144\ldots$$

For $z < z_c$, the density ρ is given by the first formula in (14.6.7). For $z > z_c$ the regime II formulae apply: in particular, the order parameter is

$$R = (3/\sqrt{5})p^{1/9}Q(p^5)\,Q(p)/Q^2(p^{5/3})\,. \tag{14.7.9}$$

From (14.7.6) it follows that q^2 is an analytic function of z at $z = z_c$, having a Taylor expansion of the form

$$q^2 = (z - z_c)/(5\sqrt{5}\,z_c) + \mathcal{O}[(z - z_c)^2]\,. \tag{14.7.10}$$

These results are exact. To obtain the critical behaviour we expand them in powers of q^2 and p, keeping only the first two or three terms. This gives

$$q^2 = (z - z_c)/(5\sqrt{5}\,z_c) + \mathcal{O}\{(z - z_c)^2\}\,,$$
$$\kappa = \kappa_c\{1 + \tfrac{5}{2}(\sqrt{5} - 1)q^2 + 3|q^2|^{5/3} + \mathcal{O}(q^4)\}\,, \tag{14.7.11}$$
$$\rho = \rho_c + \mathrm{sgn}(q^2)\,|q^2|^{2/3}/\sqrt{5} + \mathcal{O}(q^2)\,,$$
$$R = (3/\sqrt{5})\,(q^2)^{1/9}\{1 - q^2 + 2q^{10/3} + \mathcal{O}(q^4)\}\,.$$

Here $\rho_c = (5 - \sqrt{5})/10$ is the critical density of Section 14.6. The last equation (for R) applies only for $z > z_c$; the first three apply for both $z > z_c$ and $z < z_c$. Defining the critical exponents α, $\bar{\alpha}$, β in the usual way by (14.6.14), with t replaced by $z - z_c$, we again obtain (14.6.15), i.e.

$$\alpha = \bar{\alpha} = 1/3,\ \beta = 1/9\,. \tag{14.7.12}$$

Using these values, the scaling hypothesis relations (1.2.12)–(1.2.16) predict that the other critical exponents are:

$$\gamma = 13/9,\quad \delta = 14\,, \tag{14.7.13}$$
$$\mu = \nu = 5/6,\quad \eta = 4/15\,.$$

The results (14.1.18)–(14.1.24) involve elliptic functions of nomes x^5, x^6, x^3 and x. It should be possible to obtain algebraic relations between these functions (just as the Landen transformation of Section 15.6 relates elliptic functions of nomes q and q^2), and hence to eliminate x from the results. This program has been carried out by Joyce (1981) for the relation between z and the order parameter R, for $z > z_c$. He finds that

$$\psi^{18}(5 + 10s + s^2)^3 = (27s + \psi^6)(243s + \psi^6)^3, \qquad (14.7.14a)$$

where

$$\psi = \sqrt{5}R, \quad s = 125z/(z^2 - 11z - 1). \qquad (14.7.14b)$$

Thus R is an algebraic function of z.

14.8 Comments and Speculations

In this chapter we have used the "matrix inversion" trick to calculate the free energy, and have calculated the sub-lattice densities and order parameters by diagonalising the corner transfer matrices. Unlike the eight-vertex model calculation in Chapter 10, we have not obtained exact equations for all the eigenvalues of the row-to-row transfer matrix. As a result, we have not been able to calculate the interfacial tension and correlation length.

This has very recently been done (Baxter and Pearce, 1982). Regard m' and t' in (14.2.39) as fixed. Then the Boltzmann weights ω_i, and the row-to-row transfer matrix V, are functions of u. One can verify that

$$V(u)\, V(u + \lambda) = \phi(\lambda - u)\, \phi(\lambda + u)\, \mathcal{I} + \phi(u)\, V(u - 2\lambda), \qquad (14.8.1)$$

where $\mathcal{I}$ is the identity matrix and

$$\lambda = 2\pi/5, \quad \phi(u) = [m'\theta_1(u)/\theta_1(\lambda)]^N. \qquad (14.8.2)$$

As usual, the star-triangle relation implies that $V(u)$ and $V(v)$ commute, for all complex numbers u and v. We can therefore choose a representation in which $V(u)$ is diagonal, for all u. Then (14.8.1) is a functional relation for each eigenvalue. Together with the analyticity and quasi-periodic properties of $V(u)$, this relation in principle enables one to calculate every eigenvalue, for finite N. The free energy, interfacial tension and correlation length can thus be calculated. The results of course agree with (14.3.26). They also give

$$\mu = \nu = \nu' = 5/6 \qquad (14.8.3)$$

for the critical exponents of the original hard-hexagon model. Together with (14.7.12), it follows that the scaling relations (1.2.14a), (1.2.15) and (1.2.16) are satisfied.

We have seen in Sections 10.1 and 10.3 that the eight-vertex model contains as special cases the previously solved Ising and six-vertex models.

When F. Y. Wu and I solved the three-spin model in 1973 and 1974, it appeared then to be a quite distinct model. Howeve, as has been shown in Section 11.10, it can be expressed as a special case of the eight-vertex model.

Will history repeat itself for the hard hexagon model? More precisely, will a more general model be solved that contains both the eight-vertex and hard hexagon models as special cases? I doubt it. For one thing, the fact that the critical exponent δ is 15 for the former, and 14 for the latter, model, suggests that the two are quite distinct. On a more detailed level, the star – triangle relation (13.3.6), or equivalently (11.5.8), contains many more equations than unknowns. For the eight-vertex model the two spin-reversal symmetries reduce the number of equations by a factor of four, from 64 to 16. For the generalized hard hexagon model, the requirement that no two particles be adjacent eliminates 44 of the 64 equations: they just become $0 = 0$. The reasons for success in the two cases are therefore quite different, and its seems to me unlikely that one can trace a continuous path of solvable models from one to the other.

Can one extend these methods to Ising-type models in three dimensions? One can extend the star – triangle relation (13.3.6) or (11.5.8), getting a "tetrahedron relation", as has been shown by Zamolodchikov (1981). The trouble is that one then has 2^{14} equations to satisfy (instead of 2^6), and it is difficult to see where to begin. More seriously, one very useful property in two dimensions does not go over to three. If the model factors into two independent models, one on each sub-lattice of the square or simple cubic lattice, then the weight functions w factor. The planar star – triangle relation then factors into two identical relations (each being the original Ising model star – triangle relation of Section 6.4); but the three-dimensional tetrahedron relation factors into two non-identical relations, one of which is trivial and seems to preclude interesting solutions.

Even so, Zamolodchikov has found strong evidence that the tetrahedron relations do have some (non-factorizable) solutions. It will be fascinating to examine these and see if they correspond to interesting statistical mechanical models.

Of course one would still like to go much further in two dimensions. The Ising model in a magnetic field remains unsolved. Indeed the only models that have been solved in the presence of an appropriate symmetry-breaking field are the spherical model of Chapter 5, and the KDP ferroelectric model of Section 8.12. Only such solutions give a complete check on the scaling hypothesis (Section 1.2), and give the form of the scaling function $h_s(x)$.

It seems unlikely that the "commuting transfer matrix" trick can be used to solve the Ising and other models in the presence of a field, or even the non-critical Potts model. The only hope that occurs to me is that just as Onsager (1944) and Kaufman (1949) originally solved the zero-field Ising model by using the algebra of spinor operators, so there may be similar algebraic methods for solving the eight-vertex and Potts models. (Some credence to this hope is given by the fact that the diagonalized infinite-lattice corner transfer matrices of the eight vertex model have the simple direct product form (13.7.20).)

If so, it is conceivable that such methods might work for a staggered eight-vertex model, in which the weights are different on the two sublattices, but the combinations Δ and Γ in (10.4.6) are the same. In particular, one can still define a single parameter μ for this model, using (10.4.17) and (10.12.5). The case $\mu = \pi/2$ corresponds to $\Gamma = 0$: the model then factors into two independent staggered Ising models, and these can be solved algebraically. (Similarly, so can the "free fermion" case $\Delta = 0$: see Section 10.16.) In attempting to generalize such algebraic methods it would be natural to look first at other values of μ that are simple fractions of π, e.g. $\mu = \pi/3$, $3\pi/4$, etc.

If this could be done, it would indeed be a giant step forward. Many fascinating models can be expressed as special cases of such a staggered eight-vertex model, notably the non-critical Potts model (Section 12.4), the Ashkin–Teller model (Section 12.9), and even the Ising model in a magnetic field (Wu, 1979). Obviously it would be foolish to pin all one's hopes on such a possibility, which has evaded attainment for at least a decade. At the same time, I feel it is equally foolish to dismiss it out of hand.

14.9   Acknowledgements

I had enormous help in locating the various Rogers–Ramanujan-type identities used in Section 14.5, notably from George E. Andrews (who proved (14.5.22)), Richard Askey and David Bressoud (who showed that (14.5.55) is a consequence of (14.5.53) and (14.5.54), and carefully guided me through the relevant identities listed by Birch), and from Mike Hirschhorn and Geoff Joyce. I am extremely grateful for their encouragement and assistance.

15

ELLIPTIC FUNCTIONS

15.1 Definitions

The usual elliptic functions are functions of two variables, which we can take to be the *nome q* and the *argument u*. Usually q is regarded as a given real constant, with value between 0 and 1; while u is regarded as a variable, in general complex.

The *half-period magnitudes* I, I' (usually called K, K') are then given by

$$I = \tfrac{1}{2}\pi \prod_{n=1}^{\infty} \left(\frac{1 + q^{2n-1}}{1 - q^{2n-1}} \cdot \frac{1 - q^{2n}}{1 + q^{2n}} \right)^2 , \tag{15.1.1}$$

$$I' = \pi^{-1} I \ln(q^{-1}) , \tag{15.1.2}$$

so

$$q = \exp(-\pi I'/I) . \tag{15.1.3}$$

The *modulus k* and the *conjugate modulus k'* are given by

$$k = 4q^{\frac{1}{2}} \prod_{n=1}^{\infty} \left(\frac{1 + q^{2n}}{1 + q^{2n-1}} \right)^4 , \tag{15.1.4a}$$

$$k' = \prod_{n=1}^{\infty} \left(\frac{1 - q^{2n-1}}{1 + q^{2n-1}} \right)^4 . \tag{15.1.4b}$$

The *theta functions* are

$$H(u) = 2q^{\frac{1}{4}} \sin\frac{\pi u}{2I} \prod_{n=1}^{\infty} \left(1 - 2q^{2n}\cos\frac{\pi u}{I} + q^{4n}\right)(1 - q^{2n})$$

$$H_1(u) = 2q^{\frac{1}{4}} \cos\frac{\pi u}{2I} \prod_{n=1}^{\infty} \left(1 + 2q^{2n}\cos\frac{\pi u}{I} + q^{4n}\right)(1 - q^{2n})$$

$$\Theta(u) = \prod_{n=1}^{\infty} \left(1 - 2q^{2n-1}\cos\frac{\pi u}{I} + q^{4n-2}\right)(1 - q^{2n}) \tag{15.1.5}$$

$$\Theta_1(u) = \prod_{n=1}^{\infty} \left(1 + 2q^{2n-1}\cos\frac{\pi u}{I} + q^{4n-2}\right)(1 - q^{2n}).$$

The *Jacobian elliptic functions* are

$$\begin{aligned}
\operatorname{sn} u &= k^{-\frac{1}{2}} H(u)/\Theta(u), \\
\operatorname{cn} u &= (k'/k)^{\frac{1}{2}} H_1(u)/\Theta(u), \\
\operatorname{dn} u &= k'^{\frac{1}{2}} \Theta_1(u)/\Theta(u).
\end{aligned} \tag{15.1.6}$$

Multiplying (15.1.1) by (15.1.4b) gives the relation

$$\prod_{n=1}^{\infty} \left(\frac{1 - q^n}{1 + q^n}\right)^2 = \frac{2k'I}{\pi}. \tag{15.1.7}$$

Also, from (15.1.1), (15.1.2) and (15.1.4a),

$$kI' = 2q^{\frac{1}{2}} \ln(1/q) \prod_{n=1}^{\infty} \left(\frac{1 - q^{4n}}{1 - q^{4n-2}}\right)^2. \tag{15.1.8}$$

15.2　Analyticity and Periodicity

The theta functions H, H_1, Θ, Θ_1 are *entire* functions of u (i.e. they are analytic everywhere). Their zeros are all simple. In particular the zeros of $H(u)$, $\Theta(u)$ are given by

$$H(u) = 0 \quad \text{when } u = 2mI + 2inI', \tag{15.2.1}$$

$$\Theta(u) = 0 \quad \text{when } u = 2mI + i(2n - 1)I', \tag{15.2.2}$$

where m, n are any integers.

From (15.1.6); sn u, cn u and dn u are therefore *meromorphic* (i.e. their only singularities are poles). Their poles are all simple and occur when (15.2.2) is satisfied.

The function $H(u)$ satisfies the *quasi-periodic* relations

$$H(u + 2I) = -H(u) \,, \tag{15.2.3a}$$

$$H(u + 2iI') = -q^{-1} \exp(-\pi iu/I) \, H(u) \,. \tag{15.2.3b}$$

The other theta functions are related to $H(u)$ by

$$H_1(u) = H(u + I), \quad \Theta_1(u) = \Theta(u + I),$$

$$\Theta(u) = -iq^{\frac{1}{4}} \exp(\tfrac{1}{2}\pi iu/I) \, H(u + iI') \,, \tag{15.2.4}$$

$$\Theta_1(u) = q^{\frac{1}{4}} \exp(\tfrac{1}{2}\pi iu/I) \, H(u + I + iI') \,.$$

It follows that sn, cn, dn satisfy the relations:

$$\mathrm{sn}(-u) = -\,\mathrm{sn}\,u, \quad \mathrm{cn}(-u) = \mathrm{cn}\,u, \quad \mathrm{dn}(-u) = \mathrm{dn}\,u$$

$$\mathrm{sn}(u + 2I) = -\,\mathrm{sn}\,u$$

$$\mathrm{cn}(u + 2I) = -\,\mathrm{cn}\,u$$

$$\mathrm{dn}(u + 2I) = \mathrm{dn}\,u \tag{15.2.5}$$

$$\mathrm{sn}(u + 2iI') = \mathrm{sn}\,u$$

$$\mathrm{cn}(u + 2iI') = -\,\mathrm{cn}\,u$$

$$\mathrm{dn}(u + 2iI') = -\,\mathrm{dn}\,u \,,$$

and

$$\mathrm{sn}(u + iI') = (k\,\mathrm{sn}\,u)^{-1} \,,$$

$$\mathrm{cn}(u + iI') = -i\,\mathrm{dn}\,u/(k\,\mathrm{sn}\,u) \,, \tag{15.2.6}$$

$$\mathrm{dn}(u + iI') = -i\,\mathrm{cn}\,u/\mathrm{sn}\,u \,.$$

Any upright rectangle of width $2I$ and height $2iI'$ in the complex u-plane is known as a *period rectangle* (Fig. 15.1). Any function $f(u)$ satisfying the relations

$$f(u + 2I) = \pm f(u), \quad f(u + 2iI') = \pm f(u) \,, \tag{15.2.7}$$

is said to be *doubly periodic* (or perhaps anti-periodic). If such a function is known within and on a period rectangle, then its value at any point in the complex plane can be obtained by repeated use of (15.2.7). To within a sign, all values that it attains are attained within and on any period rectangle.

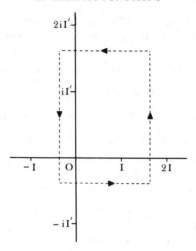

Fig. 15.1. A typical period rectangle (shown by broken lines) in the complex u-plane. It has width $2I$ and height $2I'$.

15.3   General Theorems

Theorem 15(a)

A well-known theorem in complex variable theory is *Liouville's theorem*, which states that if a function is entire and bounded, then it is a constant. A useful corollary for this chapter is:

> if a function is doubly periodic (or anti-periodic)
> and is analytic inside and on a period rectangle, then it
> is a constant. (15.3.1)

The proof is simple: since it is analytic in a closed region, it is certainly bounded. From the double periodicity it is therefore analytic and bounded everywhere. From Liouville's theorem it is therefore a constant.

Theorem 15(b)

> If a function $f(u)$ is doubly periodic (or anti-periodic)
> and meromorphic, and has n poles per period rectangle,
> then it also has just n zeros per period rectangle.
> (Multiple poles or zeros of order r being counted r
> times.) (15.3.2)

Proof: choose a period rectangle such that $f(u)$ has no poles or zeros on the boundary (since they are isolated, this must be possible). Consider the integral

$$\int_C [f'(u)/f(u)] \, du \, , \tag{15.3.3}$$

where C is the boundary of the rectangle, traversed anti-clockwise as in Fig. 15.1. The integrand is analytic on C and is strictly doubly periodic, so the contributions to (15.3.3) from the two sides (and the top and bottom) cancel. Thus (15.3.3) is zero.

On the other hand, if $f(u)$ has n poles and m zeros within the rectangle, then $f'(u)/f(u)$ has n poles with residue -1 and m poles with residue $+1$. These are its only singularities, so by Cauchy's integral formula

$$\int_C [f'(u)/f(u)] \, du = 2\pi i \, (m - n) \, . \tag{15.3.4}$$

Since the LHS is zero, it follows that $m = n$, which is the theorem.

Theorem 15(c)

If a function $f(u)$ is meromorphic and satisfies the (anti-) periodicity conditions

$$f(u + 2I) = (-1)^s f(u) \tag{15.3.5}$$

$$f(u + 2iI') = (-1)^r f(u) \, ,$$

where r, s are integers; and if $f(u)$ has just n poles per period rectangle, at $u_1, \ldots, u_n$ (counting a pole of order r as r coincident simple poles), then

$$f(u) = C \, e^{i\lambda u} \prod_{j=1}^{n} [H(u - v_j)/H(u - u_j)] \, , \tag{15.3.6}$$

where $C, \lambda, v_1, \ldots v_n$ are constants satisfying

$$v_1 + \ldots + v_n = u_1 + \ldots + u_n + (r + 2m)I - i(s + 2n)I' \, , \tag{15.3.7}$$

$$\lambda = \tfrac{1}{2}\pi(s + 2n)/I \, , \tag{15.3.8}$$

and m, n are integers.

Proof: from theorem 15(b), $f(u)$ has n zeros per rectangle. Let these be $v_1, \ldots v_n$ and let $\phi(u)$ be the product in (15.3.6), ignoring for the moment the restriction (15.3.7).

Now consider the function

$$[f'(u)/f(u)] - [\phi'(u)/\phi(u)] \, . \tag{15.3.9}$$

From (15.2.3) and (15.3.5), this is strictly doubly periodic. Since $f(u)$ and $\phi(u)$ have the same zeros and poles, the function (15.3.9) is analytic. From theorem 15(a), it is therefore constant.

Integrating, it follows that $f(u)$ must be of the form (15.3.6). The conditions (15.3.7) and (15.3.8) are necessary to ensure that (15.3.5) is satisfied.

This is a truly remarkable result: any double periodic meromorphic function must be expressible in the form (15.3.6). For this reason there is a bewildering array of identities between elliptic functions: many sums of products of such functions will satisfy the conditions of this theorem; if their zeros can be located, then they can be explicitly factored as in (15.3.6).

It is helpful to think of (15.3.6) as a 'generalization' of the fundamental theorem of algebra, which states that any polynomial of degree n can be factored into n linear forms.

These theorems are extremely powerful. They will be useful in the next section to obtain a number of algebraic identities satisfied by the elliptic functions, but perhaps the simplest example of the use of theorem 15(a) is the identity

$$H(u + x) H(u - x) H(v + y) H(v - y)$$
$$- H(u + y) H(u - y) H(v + x) H(v - x) \qquad (15.3.10)$$
$$= H(x - y) H(x + y) H(u + v) H(u - v).$$

To prove this, regard x, y, v as constants and u as a complex variable. Let $f(u)$ be the ratio of the LHS to the RHS. From (15.2.3), $f(u)$ is periodic of periods $2I$ and $2iI'$. It is meromorphic, with possible simple poles when $u = \pm v + 2mI + 2inI'$, for any integers m and n. Plainly the LHS of (15.3.10) vanishes when $u = v$ or $u = -v$, so these poles are removeable. By periodicity, so are all the others; $f(u)$ is therefore entire and doubly periodic. From theorem 15(a), it is therefore a constant. Setting $u = y$ gives this constant to the unity. This proves the identity (15.3.10).

Note that this derivation does not use the explicit formula (15.1.5) for $H(u)$. It needs only the following properties: (a) $H(u)$ is entire; (b) $H(u)$ satisfies the quasi-double-periodicity conditions (15.2.3); (c) $H(u)$ is odd, i.e. $H(-u) = -H(u)$.

15.4   Algebraic Identities

Relations Between sn, cn, dn

Consider the expression

$$[H_1^2(0)\, \Theta^2(u) - \Theta^2(0)\, H_1^2(u)]/H^2(u) . \qquad (15.4.1)$$

Clearly this is a meromorphic function of u, with possible double poles when $H(u) = 0$, i.e. when $u = 2mI + 2inI'$. However, the numerator clearly vanishes when $u = 0$, and since it is even, so does its derivative. Thus (15.4.1) is analytic at $u = 0$, and hence is analytic inside and on the period rectangle centred on the origin.

From (15.1.6) and (15.2.5) it is easily seen that (15.4.1) is (strictly) doubly periodic, so from theorem 15(a) it is a constant. Setting $u = I$ fixes this constant, giving

$$H_1^2(0)\, \Theta^2(u) - \Theta^2(0)H_1^2(u) = \Theta_1^2(0)H^2(u) \,. \tag{15.4.2}$$

From (15.1.5), the definitions (15.1.4) can be written

$$k^{\frac{1}{2}} = H_1(0)/\Theta_1(0) \,, \tag{15.4.3}$$
$$k'^{\frac{1}{2}} = \Theta(0)/\Theta_1(0) \,.$$

Dividing (15.4.2) by $H_1^2(0)\, \Theta^2(u)$, using (15.1.6) and re-arranging, it follows that

$$\mathrm{cn}^2 u + \mathrm{sn}^2 u = 1 \,. \tag{15.4.4}$$

Incrementing u by iI' and using (15.2.6), this also gives

$$\mathrm{dn}^2 u + k^2\, \mathrm{sn}^2 u = 1 \,. \tag{15.4.5}$$

These identities (15.4.4) and (15.4.5) make the elliptic functions particularly suitable for parametrizing expressions involving square roots of two quadratic forms. For instance, if one had the equation

$$y = x(1 - x^2)^{\frac{1}{2}} + (1 - k^2 x^2)^{\frac{1}{2}} \,, \tag{15.4.6}$$

an obvious parametrization would be to set

$$x = \mathrm{sn}\, u \,, \tag{15.4.7}$$

whereupon (15.4.4)–(15.4.6) would give

$$y = \mathrm{sn}\, u\ \mathrm{cn}\, u + \mathrm{dn}\, u \,. \tag{15.4.8}$$

This would ensure that x and y are both single-valued meromorphic functions of u, which can be a very convenient feature in carrying out some complicated calculation involving them. Unless $k^2 = 0$ or 1, such a parametrization cannot be performed in terms of elementary functions.

From (15.1.6), (15.4.3) and (15.2.4), it is readily verified that

$$\mathrm{sn}\, I = 1 \,, \quad \mathrm{dn}\, I = k' \,. \tag{15.4.9}$$

Setting $u = I$ in (15.4.5) therefore gives the relation

$$k^2 + k'^2 = 1 , \tag{15.4.10}$$

between the two moduli.

If $0 < q < 1$, then it is obvious from (15.1.4) that k and k' are positive. From (15.4.10) it follows that

$$0 < k < 1, \quad 0 < k' < 1 . \tag{15.4.11}$$

The Modified Amplitude Function

Another useful function for our purposes is

$$\text{Am}(u) = -i \ln[ik^{\frac{1}{2}} \text{sn}(u - \tfrac{1}{2}iI')] . \tag{15.4.12}$$

From (15.1.6) and (15.1.5),

$$\text{Am}(u) = -i \ln\left[z^{\frac{1}{2}} \prod_{n=1}^{\infty} \frac{(1 - q^{2n - 1\frac{1}{2}} z^{-1})(1 - q^{2n - \frac{1}{2}} z)}{(1 - q^{2n - 1\frac{1}{2}} z)(1 - q^{2n - \frac{1}{2}} z^{-1})} \right] , \tag{15.4.13}$$

where

$$z = \exp(i\pi u/I) . \tag{15.4.14}$$

Taking logarithms term by term and Taylor expanding each logarithm, the summation over n can be performed to give

$$\text{Am}(u) = \frac{\pi u}{2I} + 2 \sum_{m=1}^{\infty} \frac{q^{m/2}}{m(1 + q^m)} \sin(m\pi u/I) , \tag{15.4.15}$$

provided $|\text{Im}(u)| < \frac{1}{2}I'$.

For real u, this function is real, odd and monotonic increasing. It satisfies the quasi-periodic relation

$$\text{Am}(u + 2I) = \text{Am}(u) + \pi . \tag{15.4.16}$$

It is not meromorphic, since it has logarithmic branch cuts at $u = 2mI + i(n - \frac{1}{2})I'$.

The usual elliptic amplitude function is $\text{am}(u)$, which is given by (15.4.15) with q replaced by q^2. Such transformations from elliptic functions of nome q to nome q^2 are common: they are known as Landen transformations, and will be discussed in Section 15.6.

Addition Formulae for Theta Functions

The theta functions satisfy the following identities, for all complex numbers u and v. Each can be proved quite simply by regarding the ratio of the

LHS to the RHS as a function of u and verifying, using (15.2.4) and $H(0) = 0$, that this is entire and doubly periodic. From theorem 15(a) it is therefore constant. Setting $u = 0$, this constant is found to be unity.

$$H(u)\,\Theta(u)\,H_1(v)\,\Theta_1(v) - H(v)\,\Theta(v)\,H_1(u)\,\Theta_1(u)$$

$$= H(u - v)\,\Theta(u + v)\,H_1(0)\,\Theta_1(0)\,, \qquad (15.4.17)$$

$$H(u)\,\Theta_1(u)\,H_1(v)\,\Theta(v) - H_1(u)\,\Theta(u)\,H(v)\,\Theta_1(v)$$

$$= H(u - v)\,\Theta_1(u + v)\,H_1(0)\,\Theta(0)\,, \qquad (15.4.18)$$

$$H^2(u)\,\Theta^2(v) - \Theta^2(u)\,H^2(v) = H(u - v)\,H(u + v)\,\Theta^2(0)\,. \qquad (15.4.19)$$

Incrementing u by iI' in (15.4.19) and using (15.2.4) and (15.2.3) gives

$$\Theta^2(u)\,\Theta^2(v) - H^2(u)\,H^2(v) = \Theta(u - v)\,\Theta(u + v)\,\Theta^2(0)\,. \qquad (15.4.20)$$

Dividing each side of (15.4.17) by the corresponding side of (15.4.20) and using (15.1.6) and (15.4.3) gives the addition formula for sn:

$$\mathrm{sn}(u - v) = \frac{\mathrm{sn}\,u\ \mathrm{cn}\,v\ \mathrm{dn}\,v - \mathrm{cn}\,u\ \mathrm{dn}\,u\ \mathrm{sn}\,v}{1 - k^2\,\mathrm{sn}^2\,u\ \mathrm{sn}^2\,v}\,. \qquad (15.4.21)$$

Other addition formulae are:

$$-i\,\mathrm{sn}(iI' - u - v)\,\mathrm{cn}\,u\ \mathrm{cn}\,v - \mathrm{cn}(iI' - u - v)\,\mathrm{sn}\,u\ \mathrm{sn}\,v$$

$$= k^{-1}\,\mathrm{dn}(iI' - u - v)\,, \qquad (15.4.22)$$

$$\frac{\mathrm{sn}(a - u)\,\mathrm{sn}(a - v) - \mathrm{sn}\,u\ \mathrm{sn}\,v}{1 - k^2\,\mathrm{sn}\,u\ \mathrm{sn}\,v\ \mathrm{sn}(a - u)\,\mathrm{sn}(a - v)} = \mathrm{sn}\,a\ \mathrm{sn}(a - u - v)\,, \qquad (15.4.23)$$

$$\frac{\mathrm{sn}\,v\,\mathrm{sn}(a - v) - \mathrm{sn}\,u\ \mathrm{sn}(a - u)}{\mathrm{sn}(a - u)\,\mathrm{sn}\,v - \mathrm{sn}(a - v)\,\mathrm{sn}\,u} = \frac{\mathrm{sn}(a - u - v)}{\mathrm{sn}\,a}\,, \qquad (15.4.24)$$

$$\Theta(u)\,\Theta(v)\,\Theta(a - u)\,\Theta(a - v) - H(u)\,H(v)\,H(a - u)\,H(a - v)$$

$$= \Theta(0)\,\Theta(a)\,\Theta(u - v)\,\Theta(a - u - v)\,, \qquad (15.4.25)$$

$$H(v)\,H(a - v)\,\Theta(u)\,\Theta(a - u) - \Theta(v)\,\Theta(a - v)\,H(u)\,H(a - u)$$

$$= \Theta(0)\,\Theta(a)\,H(v - u)\,H(a - u - v)\,, \qquad (15.4.26)$$

$$\Theta(u)\,\Theta(v) - H(u)\,H(v) = -2q^{\frac{1}{4}}H[\tfrac{1}{2}(iI' + u - v)]\,H[\tfrac{1}{2}(iI' - u + v)]$$

$$\times\ H[\tfrac{1}{2}(iI' + u + v)]\,H[\tfrac{1}{2}(iI' - u - v)]/[H_1(0)\,\Theta_1(0)]\,, \qquad (15.4.27)$$

$$\Theta(u)\, H(v) + H(u)\, \Theta(v) = 2\, H[\tfrac{1}{2}(u + v)]\, \Theta[\tfrac{1}{2}(u + v)]$$

$$\times\, H_1[\tfrac{1}{2}(u - v)]\, \Theta_1[\tfrac{1}{2}(u - v)]/[H_1(0)\, \Theta_1(0)]\,, \tag{15.4.28}$$

$$H_1(u)\, H_1(v) - H(u)\, H(v) = 2\, H[\tfrac{1}{2}(I + u + v)]\, H_1[\tfrac{1}{2}(I + u + v)]$$

$$\times\, \Theta[\tfrac{1}{2}(I + u - v)]\, \Theta_1[\tfrac{1}{2}(I + u - v)]/[\Theta(0)\, \Theta_1(0)]\,, \tag{15.4.29a}$$

$$H_1(u)\, H(v) + H(u)\, H_1(v) = 2\, H[\tfrac{1}{2}(u + v)]\, H_1[\tfrac{1}{2}(u + v)]$$

$$\times\, \Theta[\tfrac{1}{2}(u - v)]\, \Theta_1[\tfrac{1}{2}(u - v)]/[\Theta(0)\, \Theta_1(0)]\,, \tag{15.4.29b}$$

Special Values of sn, cn, dn

From (15.1.5), (15.1.6), (15.2.6), (15.4.4) and (15.4.5), it is readily verified that

$$\operatorname{sn} 0 = 0,\quad \operatorname{cn} 0 = \operatorname{dn} 0 = 1\,, \tag{15.4.30}$$

$$\operatorname{sn} I = 1,\quad \operatorname{cn} I = 0,\quad \operatorname{dn} I = k'\,, \tag{15.4.31}$$

$$\operatorname{sn} \tfrac{1}{2}iI' = i\, k^{-\frac{1}{2}},\quad \operatorname{cn} \tfrac{1}{2}iI' = (1 + k^{-1})^{\frac{1}{2}},\quad \operatorname{dn} \tfrac{1}{2}iI' = (1 + k)^{\frac{1}{2}}. \tag{15.4.32}$$

15.5   Differential and Integral Identities

Consider the expression

$$X = \operatorname{sn}'(u)/(\operatorname{cn} u\ \operatorname{dn} u)\,, \tag{15.5.1a}$$

where the prime denotes differentiation with respect to u, q being kept constant. Differentiating (15.4.4) and (15.4.5) gives

$$X = -\operatorname{cn}'(u)/(\operatorname{sn} u\ \operatorname{dn} u)\,, \tag{15.5.1b}$$

$$= -\operatorname{dn}'(u)/(k^2 \operatorname{sn} u\ \operatorname{cn} u)\,. \tag{15.5.1c}$$

Substituting the expressions (15.1.6) for sn, cn, dn and remembering that the theta functions are entire, the above equations (15.5.1) give

$$X = \frac{\cdots}{H_1(u)\, \Theta_1(u)} = \frac{\cdots}{H(u)\, \Theta_1(u)} = \frac{\cdots}{H(u)\, H_1(u)}\,, \tag{15.5.2}$$

where in each case . . . stands for an entire function. From the first expression above for X, X has poles only when $H_1(u)$ or $\Theta_1(u)$ vanishes. From the second, it does *not* have poles when $H_1(u)$ vanishes. From the third, it does not have poles when $\Theta_1(u)$ vanishes.

It follows that X has no poles and is therefore entire. From (15.2.5) it is doubly periodic, so it is a constant.

Obtaining sn u from (15.1.6) and (15.1.5), then letting $u \to 0$, we obtain

$$\lim_{u \to 0} \frac{\operatorname{sn} u}{u} = \frac{\pi q^{\frac{1}{4}}}{k^{\frac{1}{2}} I \, \Theta(0)} \prod_{n=1}^{\infty} (1 - q^{2n})^3 . \tag{10.5.3}$$

Using (15.1.1), (15.1.4) and (15.1.5), this gives the simple result

$$\lim_{u \to 0} (\operatorname{sn} u)/u = 1 . \tag{15.5.4}$$

[In fact, the definition (15.1.1) of I can be regarded as chosen to ensure (15.5.4).]

Equation (15.5.4) ensures that $\operatorname{sn}'(0) = 1$, while from (15.1.6) and (15.4.3) it is obvious that $\operatorname{cn}(0) = \operatorname{dn}(0) = 1$. Evaluating the constant X by setting $u = 0$ in (15.5.1a) therefore gives

$$X = 1 . \tag{15.5.5}$$

Using (15.4.4) and (15.4.5) to express cn u and dn u in terms of sn u, (15.5.1a) becomes a first-order differential equation for sn u.

It can be integrated to give

$$u = \int_0^{\operatorname{sn} u} \frac{dt}{[(1 - t^2)(1 - k^2 t^2)]^{\frac{1}{2}}} . \tag{15.5.6}$$

Defining ϕ such that

$$\operatorname{sn} u = \sin \phi , \tag{15.5.7}$$

(10.5.6) can be written

$$u = \int_0^{\phi} \frac{d\alpha}{(1 - k^2 \sin^2 \alpha)^{\frac{1}{2}}} . \tag{15.5.8}$$

This is the usual integral form of the relation between u and sn u. Care has to be taken in choosing the path of integration and the sign of the integrand, but for u real and between 0 and 1 there is no problem: sn u, cn u, dn u are then all positive, and $0 < \operatorname{sn} u < 1$. Thus u is then given by the real integral (15.5.8), with positive integrand and $0 < \phi < \frac{1}{2}\pi$.

Now let $u \to I$. From (15.1.6), (15.2.4) and (15.4.3), sn $I = 1$. Hence $\phi = \frac{1}{2}\pi$ and (15.5.8) becomes

$$I = \int_0^{\pi/2} \frac{d\alpha}{(1 - k^2 \sin^2 \alpha)^{\frac{1}{2}}} , \tag{15.5.9}$$

which is the usual expression for I as the complete elliptic integral of the first kind, of modulus k.

If u is positive pure imaginary, then so is sn u; while cn u and dn u are real and of the same sign. Thus $-i \operatorname{sn} u$ increases monotonically with $\operatorname{Im}(u)$, and is finite for $0 \le \text{Im}(u) < I'$. In this case the appropriate path of integration in (15.5.8) is the positive imaginary axis. Again the integrand is positive.

From (15.1.5) and (15.1.6), sn u becomes infinite in the limit $u \to iI'$. Setting $\alpha = i\beta$, (15.5.8) gives

$$I' = \int_0^\infty \frac{\mathrm{d}\beta}{(1 + k^2 \sinh^2 \beta)^{\frac{1}{2}}} . \tag{15.5.10}$$

This is an integral expression for I'. It can be reduced to a more standard form by the substitution

$$\tan \gamma = \sinh \beta , \tag{15.5.11}$$

giving, using (15.4.10),

$$I' = \int_0^{\pi/2} \frac{\mathrm{d}\gamma}{(1 - k'^2 \sin^2 \gamma)^{\frac{1}{2}}} . \tag{15.5.12}$$

Comparing (15.5.9) and (15.5.12), it is obvious that the relation between I and k is the same as that between I' and k': I' is the complete elliptic integral of the first kind, of modulus k'.

The complete elliptic integral of the second kind, of modulus k, is

$$E = \int_0^{\pi/2} (1 - k^2 \sin^2 \gamma)^{\frac{1}{2}} \, \mathrm{d}\gamma . \tag{15.5.13}$$

Small-u Behaviour of sn u, $H(u)$

When $|u| \ll 1$ it is easily seen from (15.5.6) and (15.1.6) that

$$\text{sn } u \sim u, \quad H(u) \sim k^{\frac{1}{2}} \Theta(0) \, u . \tag{15.5.14}$$

15.6   Landen Transformation

Exhibit the dependence of q, I, I', sn u, cn u, dn u, etc. on the modulus k by writing them as q_k, I_k, I'_k, $\text{sn}(u, k)$, $\text{cn}(u, k)$, $\text{dn}(u, k)$, etc. If

$$l = 2k^{\frac{1}{2}}/(1 + k), \quad \hat{u} = (1 + k)u , \tag{15.6.1}$$

then by replacing u, k, t in (15.5.6) by $\hat{u}$, l, $(1 + k)t/(1 + kt^2)$, respectively (and noting that sn $I = 1$, sn $iI' = \infty$), it can be verified that

$$I_l = (1 + k) I_k, \quad I'_l = \tfrac{1}{2}(1 + k) I'_k, \quad q_l = q_k^{\frac{1}{2}}, \tag{15.6.2}$$

$$l \operatorname{sn}(\hat{u}, l) = 2k^{\frac{1}{2}} \operatorname{sn}(u, k)/[1 + k \operatorname{sn}^2(u, k)], \tag{15.6.3}$$

$$k^{\frac{1}{2}} \operatorname{sn}(u, k) = [1 - \operatorname{dn}(\hat{u}, l)]/[l \operatorname{sn}(\hat{u}, l)]. \tag{15.6.4}$$

Solving (15.6.1) for k, using (15.6.2) and replacing k, l by m, k, we obtain

$$m = (1 - k')/(1 + k'), \quad q_m = q_k^2. \tag{15.6.5}$$

From (15.1.1) and (15.1.4b), it follows that

$$I(k'/m')^{\frac{1}{2}} = \tfrac{1}{2}\pi \prod_{n=1}^{\infty} (1 - q^{4n})^2/(1 + q^{4n})^2. \tag{15.6.6}$$

15.7 Conjugate Modulus

Set

$$\chi(u) = (I/I')^{\frac{1}{2}} \exp[-\pi u^2/(4II')]. \tag{15.7.1}$$

Then

$$H(u, k) = -i\,\chi(u)\,H(iu, k'), \tag{15.7.2a}$$

$$\Theta(u, k) = \chi(u)\,H_1(iu, k'), \tag{15.7.2b}$$

$$H_1(u, k) = \chi(u)\,\Theta(iu, k'), \tag{15.7.2c}$$

$$\Theta_1(u, k) = \chi(u)\,\Theta_1(iu, k'), \tag{15.7.2d}$$

$$\operatorname{sn}(u, k) = -i\operatorname{sn}(iu, k')/\operatorname{cn}(iu, k'), \tag{15.7.3a}$$

$$\operatorname{cn}(u, k) = 1/\operatorname{cn}(iu, k'), \tag{15.7.3b}$$

$$\operatorname{dn}(u, k) = \operatorname{dn}(iu, k')/\operatorname{cn}(iu, k'). \tag{15.7.3c}$$

These identities can be proved by using theorem 15a. For instance, both sides of (15.7.2a) are entire functions of u, with simple zeros at $2mI + 2inI'$, for all integers m, n. Their ratio is therefore entire. Using (15.2.3) we can verify that it is doubly-periodic, so from theorem 15a it must be a constant.

As is often the case, it is much harder to obtain this constant than it is to obtain the rest of the equation. One way is to reason as follows.

The relations (15.7.2b)–(15.7.2d) can be obtained from (15.7.2a) by replacing u by $u + iI'$, $u + I$, $u + I + iI'$, respectively, and using (15.2.4). It follows that (15.7.2a)–(15.7.2d) are all valid, except possibly for the inclusion of some extra common factor on the RHS.

This factor must be independent of u, but may depend on k, or equivalently on

$$\varepsilon = I'/I.$$ (15.7.4)

Let us write the factor as $c(\varepsilon)$ and define a function

$$R(\varepsilon) = \varepsilon^{\frac{1}{4}} e^{-\pi\varepsilon/12} \prod_{n=1}^{\infty} (1 - e^{-2n\pi\varepsilon}).$$ (15.7.5)

From (15.1.5), (15.5.9) and (15.5.12),

$$q = \exp(-\pi\varepsilon), \quad q' = \exp(-\pi/\varepsilon)$$ (15.7.6)

where q is the nome corresponding to the modulus k, and q' is the nome corresponding to the conjugate modulus $k' = (1 - k^2)^{\frac{1}{2}}$. Taking the limit $u \to 0$ in (15.7.2a), including our still-to-be-determined factor $c(\varepsilon)$, using (15.1.5) and (15.7.5), it follows that

$$R^3(\varepsilon) = c(\varepsilon) R^3(\varepsilon^{-1}).$$ (15.7.7)

Similarly, multiplying (15.7.2a) and (15.7.2b) and taking the limit $u \to 0$, we obtain

$$R^2(\varepsilon/2) R^2(\varepsilon) = c^2(\varepsilon) R^2(2/\varepsilon) R^2(\varepsilon^{-1}).$$ (15.7.8)

We can obtain a third equation by replacing ε in (15.7.7) by $\varepsilon/2$. Eliminating $R(\varepsilon)$ and $R(\varepsilon/2)$ between the three equations gives

$$c(\varepsilon/2) = c^2(\varepsilon).$$ (15.7.9)

Further, from (15.7.7) it is obvious that $c(\varepsilon^{-1}) = 1/c(\varepsilon)$. Replacing ε in (15.7.9) by $2/\varepsilon$, and using this inversion property, it follows that

$$c(\varepsilon) = c^2(\varepsilon/2).$$ (15.7.10)

It follows at once that $c(\varepsilon) = c^4(\varepsilon)$. Since $c(\varepsilon)$ is real and non-zero, this implies that

$$c(\varepsilon) = 1.$$ (15.7.11)

Thus the factor multiplying the RHS of (15.7.2) is in fact unity: the equations are correct as written. The equations (15.7.3) follow from (15.7.2) and (15.1.6).

15.8 Poisson Summation Formula

$$\sum_{n=-\infty}^{\infty} f(n\delta) = \delta^{-1} \sum_{n=-\infty}^{\infty} g(2\pi n/\delta),$$ (15.8.1)

where

$$g(k) = \int_{-\infty}^{\infty} \exp(ikx) f(x) \, dx. \tag{15.8.2}$$

This identity is true for any function $f(x)$ that is analytic for real values of x, and for which the integral (15.8.2) is absolutely convergent (Courant and Hilbert, 1953, pp. 75–77). It can be used to express series such as (15.4.15) in a form which converges rapidly as $q \to 1$. This corresponds to going from elliptic functions of modulus k to ones of modulus k'.

✦

15.9 Series Expansions of the Theta Functions

$$H(u) = 2 \sum_{n=1}^{\infty} (-1)^{n-1} q^{(n-\frac{1}{2})^2} \sin[(2n-1)\pi u/2I], \tag{15.9.1a}$$

$$\Theta(u) = 1 + 2 \sum_{n=1}^{\infty} (-1)^n q^{n^2} \cos(n\pi u/I), \tag{15.9.1b}$$

$$H_1(u) = 2 \sum_{n=1}^{\infty} q^{(n-\frac{1}{2})^2} \cos[(2n-1)\pi u/2I], \tag{15.9.1c}$$

$$\Theta_1(u) = 1 + 2 \sum_{n=1}^{\infty} q^{n^2} \cos(n\pi u/I), \tag{15.9.1d}$$

$$\sum_{n=0}^{\infty} (-1)^n q^{n(n+1)/2}(z^{-n} - z^{n+1})$$

$$= \prod_{n=1}^{\infty} (1 - q^{n-1}z)(1 - q^n z^{-1})(1 - q^n). \tag{15.9.2}$$

To establish the identities (15.9.1), note from (15.1.5) that $H(u)$ is an entire function, odd and anti-periodic of period $2I$. It therefore has a Fourier expansion of the general form

$$H(u) = \sum_{n=1}^{\infty} h_n \sin[(2n-1)\pi u/2I]. \tag{15.9.3}$$

From (15.2.3b) it follows that $h_{n+1} = -q^{2n}h_n$, and hence that

$$h_n = 2c(-1)^{n-1} q^{(n-\frac{1}{2})^2}, \tag{15.9.4}$$

where c is some constant. Substituting this result into (15.9.3); replacing u by u, $u + iI'$, $u + I$, $u + I + iI'$, respectively; and using (15.2.4); we obtain the four identities (15.9.1a)–(15.9.1d), except that each has an extra factor c multiplying the RHS.

As in Section 15.7, it is the evaluation of c that causes the most problems. It is independent of u, but may depend on k, or equivalently q. Let us write it as $c(q)$ and define two functions

$$S(q) = \prod_{n=1}^{\infty} (1 - q^n)^2/(1 - q^{2n}) , \tag{15.9.5}$$

$$T(q) = 1 + 2 \sum_{n=1}^{\infty} (-1)^n q^{n^2} . \tag{15.9.6}$$

Setting $u = 0$ in (15.9.1b), including our still-to-be-determined factor $c(q)$, and using (15.1.5), we obtain

$$S(q) = c(q) \, T(q) . \tag{15.9.7}$$

Similarly, multiplying (15.9.1b) by (15.9.1d), setting $u = 0$ and using (15.1.5), we get

$$S^2(q^2) = c^2(q) \sum_{m=-\infty}^{\infty} \sum_{n=-\infty}^{\infty} (-1)^m q^{m^2+n^2} . \tag{15.9.8}$$

Set $m = r + s$, $n = r - s$. Then (15.9.8) becomes

$$S^2(q^2) = c^2(q) \sum_{r} \sum_{s} (-1)^{r+s} q^{2r^2+2s^2} . \tag{15.9.9}$$

Here r and s are either both integers, or both half-an-odd-integer. In the latter case the sum over s vanishes, the terms occurring in pairs of equal magnitude and opposite sign. Thus we can restrict the sum to all integer values (positive, zero or negative) of r and s. We then have

$$S^2(q^2) = c^2(q) \, T^2(q^2) . \tag{15.9.10}$$

Replacing q in (15.9.7) by q^2 and comparing with (15.9.10), it follows that

$$c(q^2) = c(q) . \tag{15.9.11}$$

However, it is obvious from (15.9.7) that $c(q)$ is Taylor expandable about $q = 0$ with leading term one. Substituting the Taylor series into (15.9.11) and equating coefficients, we find at once that

$$c(q) = 1 . \tag{15.9.12}$$

The identities (15.9.1) are therefore correct as written. The other identity (15.9.2) is a corollary of (15.9.1a), being obtained by using (15.1.5), setting $z = \exp(i\pi u/I)$, and replacing q by $q^{\frac{1}{4}}$.

15.10   Parametrization of Symmetric Biquadratic Relations

In the Ising, eight-vertex and hard hexagon models we encounter symmetric biquadratic relations, of the form

$$ax^2y^2 + b(x^2y + xy^2) + c(x^2 + y^2) + 2dxy + e(x + y) + f = 0. \quad (15.10.1)$$

Here x and y are variables (complex numbers), and a, b, c, d, e, f are given constants.

Any such relation can conveniently be parametrized in terms of elliptic functions. To see this, first apply the bilinear transformations

$$x \to (\alpha x + \beta)/(\gamma x + \delta), \quad y \to (\alpha y + \beta)/(\gamma y + \delta), \quad (15.10.2)$$

where α, β, γ, δ are numbers (in general complex) such that $\alpha\delta \neq \beta\gamma$. In general we can choose α, β, γ, δ so as to make b and e vanish in (15.10.1), and so that $a = f \neq 0$. (Exceptional cases can arise, but these can be handled by taking an appropriate limit.) Dividing (15.10.1) through by a, the biquadratic relation assumes the canonical form

$$x^2y^2 + 1 + c(x^2 + y^2) + 2dxy = 0. \quad (15.10.3)$$

This can be regarded as a quadratic equation for y. Its solution is

$$y = - \{dx \pm \sqrt{[-c + (d^2 - 1 - c^2)x^2 - cx^4]}\}/(c + x^2). \quad (15.10.4)$$

The argument of the square root is a quartic polynomial in x. It can be written as a perfect square by transforming from the variable x to the variable u, where

$$x = k^{\frac{1}{2}} \operatorname{sn} u, \quad (15.10.5)$$

$\operatorname{sn} u$ being the Jacobian elliptic sn function of argument u and modulus k, where

$$k + k^{-1} = (d^2 - 1 - c^2)/c. \quad (15.10.6)$$

Using (15.4.4) and (15.4.5), the argument of the square root is

$$- c[1 - (k + k^{-1})x^2 + x^4]$$
$$= - c(1 - \operatorname{sn}^2u)(1 - k^2 \operatorname{sn}^2u) = - c \operatorname{cn}^2u \operatorname{dn}^2u. \quad (15.10.7)$$

Define a parameter η by

$$c = - 1/(k \operatorname{sn}^2\eta). \quad (15.10.8)$$

Then from (15.10.6) it follows that we can choose the sign of η so that

$$d = \operatorname{cn} \eta \operatorname{dn} \eta /(k \operatorname{sn}^2\eta). \quad (15.10.9)$$

Substituting these expressions into (15.10.4), it follows that

$$y = k^{\frac{1}{2}} \frac{\operatorname{sn} u \operatorname{cn} \eta \operatorname{dn} \eta \pm \operatorname{sn} \eta \operatorname{cn} u \operatorname{dn} u}{1 - k^2 \operatorname{sn}^2 u \operatorname{sn}^2 \eta} \qquad (15.10.10)$$

Using the addition theorem (15.4.21), this result simplifies to

$$y = k^{\frac{1}{2}} \operatorname{sn}(u \pm \eta) . \qquad (15.10.11)$$

Thus y is given by an equation of the same form as the equation (15.10.5) for x, but with u replaced by $u \pm \eta$.

Put another way, if we transform from x, y to u, v according to the rule

$$x = k^{\frac{1}{2}} \operatorname{sn} u, \quad y = k^{\frac{1}{2}} \operatorname{sn} v , \qquad (15.10.12)$$

then the canonical biquadratic relation (15.10.3) simplifies to the pair of linear relations

$$v = u + \eta \quad \text{or} \quad v = u - \eta . \qquad (15.10.13)$$

We can now go back to the general biquadratic relation (15.10.1), using the transformation (15.10.2). This of course changes c and d, and it should be remembered that the c and d in (15.10.6)–(15.10.9) are those of (15.10.3). Even so, it is still true that there exist parameters k and η such that (15.10.1) reduces to

$$x = \phi(u), \quad y = \phi(u \pm \eta) , \qquad (15.10.14)$$

where the function $\phi(u)$ is defined by

$$\phi(u) = (\alpha k^{\frac{1}{2}} \operatorname{sn} u + \beta)/(\gamma k^{\frac{1}{2}} \operatorname{sn} u + \delta) . \qquad (15.10.15)$$

Define two further parameters λ, μ by

$$\operatorname{sn} \lambda = - k^{-\frac{1}{2}} \beta/\alpha, \quad \operatorname{sn} \mu = - k^{-\frac{1}{2}} \delta/\gamma . \qquad (15.10.16)$$

Then (15.10.15) can be written as

$$\phi(u) = (\alpha/\gamma) (\operatorname{sn} u - \operatorname{sn} \lambda)/(\operatorname{sn} u - \operatorname{sn} \mu) . \qquad (15.10.17)$$

Using (15.1.6) to express the sn function as a ratio of theta functions, then applying the identity (15.4.28) (with u negated), we obtain

$$\phi(u) = \text{constant} \times \frac{\tau(u - \lambda) \, \tau(u - 2I + \lambda)}{\tau(u - \mu) \, \tau(u - 2I + \mu)} , \qquad (15.10.18)$$

where the function $\tau(u)$ is defined by

$$\tau(u) = H(u/2) \, \Theta(u/2) . \qquad (15.10.19)$$

From (15.1.5), to within a constant factor, $\tau(u)$ is the elliptic H-function with q replaced by $q^{\frac{1}{2}}$ and u/I by $u/(2I)$. From (15.6.2), this means that they are related by a Landen transformation. In fact

$$\tau(u) = \text{constant} \times H(u', l) , \qquad (15.10.20)$$

where

$$l = 2k^{\frac{1}{2}}/(1 + k), \quad u' = \tfrac{1}{2}(1 + k)u . \qquad (15.10.21)$$

Modify each of u, λ, μ by subtracting l and then multiplying by $\tfrac{1}{2}(1 + k)$. Multiply η by $\tfrac{1}{2}(1 + k)$. Then (15.10.18) becomes

$$\phi(u) = \text{constant} \times \frac{H(u - \lambda , l) \, H(u + \lambda , l)}{H(u - \mu , l) \, H(u + \mu , l)} , \qquad (15.10.22)$$

and (15.10.14) is unchanged.

The general symmetric biquadratic relation (15.10.1) therefore can be reduced to the form (15.10.14), where $\phi(u)$ is given by (15.10.22), or equivalently by

$$\phi(u) = \text{constant} \times \frac{\text{sn}^2(u , l) - \text{sn}^2(\lambda , l)}{\text{sn}^2(u , l) - \text{sn}^2(\mu , l)} . \qquad (15.10.23)$$

The multiplicative constant herein, and the parameters l, λ, μ, η, are independent of x and y , being determined solely by the coefficients $a , \ldots , f$ in (15.10.1). In any specific case we can obtain these parameters by substituting the expressions (15.10.14) and (15.10.22) for x and y directly into (15.10.1), and then considering particular values of u .

There are many excellent books on elliptic functions. I mention Whittaker and Watson (1915, Chapters 20–22), Neville (1944) and Bowman (1953). I find particularly useful the identity list in Sections 8.110–8.197 of Gradshteyn and Ryzhik (1965): once one is familiar with the use of the theorems in Section 15.3, it is usually straightforward to verify any particular identity, as I hope I have managed to indicate.

REFERENCES

Abraham, D. B. (1973). *Phys. Lett.* **43A**, 163–4.

Abraham, D. B. (1979). *Phys. Rev.* **B19**, 3833–4.

Alexander, S. (1975). *Phys. Lett.* **A54**, 353–4.

Als-Nielsen, J., Birgeneau, R. J., Guggenheim, H. J. and Shirane, G. (1975). *Phys. Rev.* **B12** 4963–79.

Andrews, G. E. (1976). "The Theory of Partitions". Addison-Wesley, Reading, Massachusetts.

Andrews, G. E (1981). "The hard hexagon model and new Rogers – Ramanujan type identities". Penn. State University Math. Dept. Research Report 8167.

Appel, K. and Haken W. (1976). *Bull. Am. Math. Soc.* **82**, 711–2.

Appel, K. and Haken, W. (1977). *Illinois Jl. Math.* **21**, 429–90.

Appel, K., Haken, W. and Koch, J. (1977). *Illinois Jl. Math.* **21**, 491–567.

Ashkin, J. and Teller, E. (1943). *Phys. Rev.* **64**, 178–84.

Baker, G. A. (1977). *Phys. Rev.* **B15**, 1552–9.

Barber, M. N. (1979). *J. Phys. A: Math. Gen.* **12**, 679–88.

Barber, M. N. and Baxter, R. J. (1973). *J. Phys. C: Solid State Phys.* **6**, 2913–21.

Baxter, R. J. (1963). *Proc. Camb. Phil. Soc.* **59**, 779—87.

Baxter, R. J. (1964). *Phys. Fluids* **7**, 38–43.

Baxter, R. J. (1965). *Phys. Fluids* **8**, 687–92.

Baxter, R. J. (1968). *J. Math. Phys.* **9**, 650–4.

Baxter, R. J. (1969). *J. Math. Phys.* **10**, 1211–6.

Baxter, R. J. (1970a). *Phys. Rev.* **B1**, 2199–202.

Baxter, R. J. (1970b). *J. Math. Phys.* **11**, 784–9.

Baxter, R. J. (1970c). *J. Math. Phys.* **11**, 3116–24.

Baxter, R. J. (1971a). *Phys. Rev. Lett.* **26**, 832–3.

Baxter, R. J. (1971b). *Phys. Rev. Lett.* **26**, 834.

Baxter, R. J. (1972a). In "Phase Transitions and Critical Phenomena", (Domb, C. and Green, M. S., eds.), Vol. 1, pp 461–9. Academic Press, London

Baxter, R. J. (1972b). *Ann. Phys. (N.Y.)* **70**, 193–228.

Baxter, R. J. (1972c). *Ann. Phys. (N.Y.)* **70**, 323–37.

Baxter, R. J. (1973a). *Ann. Phys. (N.Y.)* **76**, 1–24, 25–47, 48–71.

Baxter, R. J. (1973b). *J. Stat. Phys.* **8**, 25–55.

Baxter, R. J. (1973c). *J. Phys. C: Solid State Phys.* **6**, L94–6; *J. Stat. Phys.* **9**, 145–182.

Baxter, R. J. (1973d). *J. Phys. C: Solid State Phys.* **6**, L445–8.

474

Baxter, R. J. (1976). *J. Stat. Phys.* **15**, 485–503.
Baxter, R. J. (1978a). *Phil. Trans. Roy. Soc. (London)* **289**, 315–46.
Baxter, R. J. (1978b). *Ann. Israel Phys. Soc.* **2**(1), 37–47.
Baxter, R. J. (1978c). *J. Stat. Phys.* **19**, 461–78.
Baxter, R. J. (1980). *J. Phys. A: Math. Gen.* **13**, L61–L70.
Baxter, R. J. and Enting, I. G. (1976). *J. Phys. A: Math. Gen.* **9**, L149–52.
Baxter, R. J. and Enting, I. G. (1978). *J. Phys. A: Math. Gen.* **11**, 2463–73.
Baxter, R. J. and Enting, I. G. (1979). *J. Stat. Phys.* **21**, 103–23.
Baxter, R. J. and Kelland, S. B. (1974). *J. Phys. C: Solid State Phys.* **7**, L403–6.
Baxter, R. J. and Pearce, P. A. (1982). *J. Phys. A: Math. Gen.* **15**.
Baxter, R. J. and Tsang, S. K. (1980). *J. Phys. A: Math. Gen.* **13**, 1023–30.
Baxter, R. J. and Wu, F. Y. (1973). *Phys. Rev. Lett.* **31**, 1294–7.
Baxter, R. J. and Wu, F. Y. (1974). *Aust. J. Phys.* **27**, 357–81.
Baxter, R. J., Sykes, M. F. and Watts, M. G. (1975). *J. Phys. A: Math. Gen.* **8**, 245–51.
Baxter, R. J., Kelland, S. B. and Wu, F. Y. (1976). *J. Phys. A: Math. Gen.* **9**, 397–406.
Baxter, R. J., Temperley, H. N. V. and Ashley, S. E. (1978). *Proc. Roy. Soc. (London)* **A358**, 535–59.
Baxter, R. J., Enting, I. G. and Tsang, S. K. (1980). *J. Stat. Phys.* **22**, 465–89.
Beraha, S. and Kahane, J. (1979). *J. Combinatorial Theory* **B27**, 1–12.
Beraha, S., Kahane, J. and Weiss, N. J. (1975). *Proc. Nat. Acad. Sci. (U.S.A.)* **72**, 4209–9.
Beraha, S., Kahane, J. and Weiss, N. J. (1978). *In* "Studies in Foundations and Combinatorics", (Rota, C. G., *ed.*), pp. 213–232. Academic Press, New York.
Berlin, T. H. and Kac, M. (1952). *Phys. Rev.* **86**, 821–35.
Bethe, H. A. (1931). *Z. Physik.* **71**, 205–26.
Bethe, H. A. (1935). *Proc. Roy. Soc. (London)* **A150**, 552–75.
Birch, B. J. (1975). *Math. Proc. Camb. Phil. Soc.* **78**, 73–9.
Birgeneau, R. J., Guggenheim, H. J. and Shirane, G. (1973). *Phys. Rev.* **B8**, 304–11.
Black, J. L. and Emery, V. J. (1981). *Phys. Rev.* **B23**, 429–32.
Bloch, F. (1930). *Z. Physik* **61**, 206–19.
Bloch, F. (1932). *Z. Physik.* **74**, 295–335.
Blöte, H. W. J., Nightingale, M. P. and Derrida, B. (1981). *J. Phys. A: Math. Gen.* **14**, L45–9.
Blumberg, R. L., Shlifer, G. and Stanley, H. E. (1980). *J. Phys. A: Math. Gen.* **13**, L147–52.
Bowman, F. (1953). "Introduction to Elliptic Functions, with applications". English Universities Press, London.
Bozorth, R. M. (1951). "Ferromagnetism". Van Nostrand, New York.
Bragg, W. L. and Williams, E. J. (1934). *Proc. Roy. Soc. (London)* **A145**, 699–730.
Brascamp, H. J., Kunz, H. and Wu, F. Y. (1973). *J. Math. Phys.* **14**, 1927–32.
Burley, D. M. (1972). *In* "Phase Transitions and Critical Phenomena", (Domb, C. and Green, M. S., eds.), Vol. 2, pp. 329–374. Academic Press, London.
Cardy, J. L., Nauenberg, M. and Scalapino, D. J. (1980). *Phys. Rev.* **B22**, 2560–8.
Chen, M. S., Onsager, L., Bonner, J. and Nagle, J. (1974). *J. Chem. Phys.* **60**, 405–19.

Courant, R. and Hilbert, D. (1953). "Methods of Mathematical Physics", Vol. 1. Interscience, New York.

Coxeter, H. S. M. (1947). "Non-Euclidean Geometry". Toronto University Press, Toronto.

den Nijs, M. P. M. (1979). *J. Phys. A: Math. Gen.* **12**, 1857–68.

den Nijs, M. P. M. (1981). *Phys. Rev.* **B23**, 6111–25.

Ditzian, R. V., Banavar, J. R., Grest, G. S. and Kadanoff, L. P. (1980). *Phys. Rev.* **B22**, 2542–53.

Domany, E. and Riedel, E. K. (1978). *J. Appl. Phys.* **49**(3), 1315–20.

Domb, C. (1960). *Adv. Phys.* **9**, 149–361.

Domb, C. (1974). *In* "Phase Transitions and Critical Phenomena" (Domb, C. and Green, M. S., eds.), Vol. 3, pp. 357–484. Academic Press, London.

Domb, C. and Hunter, D. L. (1965). *Proc. Phys. Soc. (London)* **86**, 1147–51.

Eggarter, T. P. (1974). *Phys. Rev.* **B9**, 2989–992.

Enting, I. G. (1975). *J. Phys. A: Math. Gen.* **8**, L35–8.

Enting, I. G. and Baxter, R. J. (1977). *J. Phys. A: Math. Gen.* **10**, L117–9.

Essam, J. N. (1972). *In* "Phase Transitions and Critical Phenomena" (Domb, C. and Green, M. S., eds.), Vol. 2, pp. 197–270. Academic Press, London.

Fan, C. (1972a). *Phys. Rev.* **B6**, 902–10.

Fan, C. (1972b). *Phys. Lett.* **39A**, 136–6.

Fan, C. and Wu, F. Y. (1970). *Phys. Rev.* **B2**, 723–33.

Felderhof, B. U. (1973). *Physica* **65**, 421–51; **66**, 279–97 and 509–26; *Phys. Lett.* **44A**, 437–8.

Fisher, M. E. (1961). *Phys. Rev.* **124**, 1664–72.

Fisher, M. E. (1966). *Phys. Rev. Lett.* **16**, 11–14.

Fisher, M. E. (1967). *Rep. Progr. Phys.* **30**, 615–730.

Fisher, M. E. (1969). *J. Phys. Soc. Japan (Suppl.)* **26**, 87–8.

Fisher, M. E. (1972). *Commun. Math. Phys.* **26**, 6–14.

Fisher, M. E. (1974). *Rev. Mod. Phys.* **46**, 597–616.

Fisher, M. E. and Felderhof, B. V. (1970). *Ann. Phys. (N.Y.)* **58**, 176–300.

Fortuin, C. M. and Kasteleyn, P. W. (1972). *Physica* **57**, 536–64.

Frobenius, F. G. (1908). Reprinted in Section 79 of "Ferdinand Georg Frobenius: Gesammelte Abhandlungen", Springer-Verlag, Berlin, 1968.

Gantmacher, F. R. (1959). "The Theory of Matrices". Chelsea, New York.

Gaudin, M. (1971). *Phys. Rev. Lett.* **26**, 1301–4.

Gaunt, D. S. (1967). *J. Chem. Phys.* **46**, 3237–59.

Gaunt, D. S. and Domb, C. (1970). *J. Phys. C: Solid State Phys.* **3**, 1442–61.

Gaunt, D. S. and Sykes, M. F. (1973). *J. Phys. A: Math., Nucl. Gen.* **6**, 1517–26.

Gibbs, J. W. (1902). "Elementary Principles in Statistical Mechanics". Reprinted by Dover, New York, 1960.

Glasser, M. L., Abraham, D. B. and Lieb, E. H. (1972). *J. Math. Phys.* **13**, 887–900.

Gradshteyn, I. S. and Ryzhik, I. M. (1965). "Table of Integrals, Series and Products". Academic Press, New York.

Greenhill, A. G. (1892). "Applications of Elliptic Functions". Reprinted by Dover, New York, 1959.

Griffiths, H. P. and Wood, D. W. (1973). *J. Phys. C: Solid State Phys.* **6**, 2533–54.

Griffiths, R. B. (1967). *Phys. Rev.* **158**, 176–87.

Griffiths, R. B. (1970). *Phys. Rev. Lett.* **24**, 1479–82.

Griffiths, R. B. (1972). *In* "Phase Transitions and Critical Phenomena" (Domb, C. and Green, M. S., eds.), Vol. 1, pp. 7–109. Academic Press, London.

Guggenheim, E. A. (1935). *Proc. Roy. Soc. (London)* **A148**, 304–12.

Hankey, A. and Stanley, H. E. (1972). *Phys. Rev.* **B6**, 3515–42.

Heisenberg, W. (1928). *Z. Physik* **49**, 619–36.

Hilhorst, H. J., Schick, M. and van Leeuwen, J. M. J. (1978). *Phys. Rev. Lett.* **40**, 1605–8.

Hilhorst, H. J., Schick, M. and van Leeuwen, J. M. J. (1979). *Phys. Rev.* **B19**, 2749–63.

Hintermann, A., Kunz, H. and Wu, F. Y. (1978). *J. Stat. Phys.* **19**, 623–32.

Hocken, R. and Moldover, M. R. (1976). *Phys. Rev. Lett.* **37**, 29–32.

Houtappel, R. M. F. (1950). *Physica* **16**, 425–55.

Hulthén, L. (1938). *Arkiv. Mat. Astron. Fysik* **26A**, No. 11.

Hurst, C. A. and Green, H. S. (1960). *J. Chem. Phys.* **33**, 1059–62.

Ising, E. (1925). *Z. Physik* **31**, 253–8.

Johnson, J. D. and Bonner, J. C. (1980). *Phys. Rev. Lett.* **44**, 616–9.

Johnson, J. D. and McCoy, B. M. (1972). *Phys. Rev.* **A6** 1613–26.

Johnson, J. D., Krinsky, S. and McCoy, B. M. (1972a). *Phys. Rev. Lett.* **29**, 492–4.

Johnson, J. D., McCoy, B. M. and Lai, C. K. (1972b). *Phys. Lett.* **38A**, 143–4.

Johnson, J. D., Krinsky, S. and McCoy, B. M. (1973). *Phys. Rev.* **A8**, 2526–47.

Jones, R. B. (1973, 1974). *J. Phys. A: Math. Nucl. Gen.* **6**, 928–50; **7**, 280–8 and 495–504.

Joyce, G. S. (1972). *In* "Phase Transitions and Critical Phenomena" (Domb, C. and Green, M. S., eds.), Vol. 2, pp. 375–442. Academic Press, London.

Joyce, G. S. (1981). Private communication.

Jüngling, K. (1975). *J. Phys. C: Solid State Phys.* **8**, L169–71.

Kac, M. and Thompson, C. J. (1971). *Physica Norvegica* **5**, 163–7.

Kac, M. and Ward, J. C. (1952). *Phys. Rev.* **88**, 1332–7.

Kac. M., Uhlenbeck, G. E. and Hemmer, P. L. (1963). *J. Math. Phys.* **4**, 216–247; (1964). *J. Math. Phys.* **5**, 60–84.

Kadanoff, L. P. (1966). *Physics* **2**, 263–72.

Kadanoff, L. P. (1976). *In* "Phase Transitions and Critical Phenomena" (Domb, C. and Green, M. S., eds.), Vol. 5A, pp. 1–34. Academic Press, London.

Kadanoff, L. P. (1977). *Phys. Rev. Lett.* **39**, 903–5.

Kadanoff, L. P. (1979). *Ann. Phys. (N.Y.)* **120**, 39–71.

Kadanoff, L. P. and Brown, A. C. (1979). *Ann. Phys. (N.Y)* **121**, 318–42.

Kadanoff, L. P. and Wegner, F. J. (1971). *Phys. Rev.* **B4**, 3989–93.

Kadanoff, L. P., Gotze, W., Hamblen, D., Hecht, R., Lewis, E. A. S., Palciauskas, V. V., Rayl, M., Swift, J., Aspnes, D. and Kane, J. (1967). *Rev. Mod. Phys.* **39**, 395–431.

Kadanoff, L. P., Houghton, A. and Yalabik, M. C. (1976). *J. Stat. Phys.* **14**, 171–203.

Kanamori, J. (1958). *Progr. Theor. Phys.* **20**, 890–908.

Kasteleyn, P. W. (1961). *Physica* **27**, 1209–25.

Kasteleyn, P. W. (1963). *J. Math. Phys.* **4**, 287–93.

Kasteleyn, P. W. and Fortuin, C. M. (1969). *J. Phys. Soc. Japan. Suppl.* **26**, 11–14.

Katsura, S. (1962). *Phys. Rev.* **127**, 1508–18.

Kaufman, B. (1949). *Phys. Rev.* **76**, 1232–43.

Kelland, S. B. (1974a). *J. Phys. A: Math., Nucl. Gen.* **7**, 1907–12.
Kelland, S. B. (1974b) *Austral. J. Phys.* **27**, 813–29.
Kelland, S. B. (1976). *Canad. J. Phys.* **54**, 1621–6.
Kikuchi, R. (1951). *Phys. Rev.* **81**, 988–1003.
Kim, D. and Joseph, R. I. (1974). *J. Phys. C: Solid State Phys.* **7**, L167–9.
Kirkwood, J. G. (1935). *J. Chem. Phys.* **3**, 300–13.
Knops, H. J. F. (1980). *Ann. Phys. (N.Y.)* **128**, 448–62.
Knops, H. J. F. and Hilhorst, H. J. (1979). *Phys. Rev.* **B19**, 3689–99.
Kramers, H. A. and Wannier, G. H. (1941). *Phys. Rev.* **60**, 252–62.
Kumar, K. (1974). *Austral. J. Phys.* **27**, 433–56.
Kunz, H. (1974). *Ann. Phys. (N.Y.)* **85**, 303–35.
Landau, L. D. and Lifshitz, E. M. (1968). "Statistical Physics". Pergamon, Oxford.
Lenard, A. (1961). *J. Math. Phys.* **2**, 682–93.
Lieb, E. H. (1967a). *Phys. Rev.* **162**, 162–72.
Lieb, E. H. (1967b). *Phys. Rev. Lett.* **18**, 1046–8.
Lieb, E. H. (1967c). *Phys. Rev. Lett.* **19**, 108–10.
Lieb, E. H. and Mattis, D. C. (1966). "Mathematical Physics in One Dimension", Academic Press, New York and London.
Lieb, E. H., Schultz, T. D. and Mattis, D. C. (1961). *Ann. Phys. (N.Y.)* **16**, 407–66.
Lieb, E. H. and Wu, F. Y. (1968). *Phys. Rev. Lett.* **20**, 1445–8.
Lieb, E. H. and Wu, F. Y. (1972). *In* "Phase Transitions and Critical Phenomena" (Domb, C. and Green, M. S., eds.), Vol. 1, pp. 321–490. Academic Press, London.
McCoy, B. M. and Wu, T. T. (1973). "The Two-Dimensional Ising Model". Harvard University Press, Cambridge, Mass.
Merlini, D. (1973). *Lett. Nuovo Cim.* **8**, 623–9.
Merlini, D. and Gruber, C. (1972). *J. Math. Phys.* **13**, 1814–23.
Merlini, D., Hinterman, A. and Gruber, C. (1973). *Lett. Nuovo Cim.* **7**, 815–8.
Metcalf, B. D. and Yang, C. P. (1978). *Phys. Rev.* **B18**, 2304–7.
Montroll, E. W. (1949). *Nuovo Cimento, Suppl.* **6**, 265–78.
Montroll, E. W., Potts, R. B. and Ward, J. C. (1963). *J. Math. Phys.* **4**, 308–22.
Muir, T. (1882). "A Treatise on the Theory of Determinants", p. 197. Macmillan, London.
Müller-Hartmann, E. and Zittartz, J. (1974). *Phys. Rev. Lett.* **33**, 893–7.
Muskhelishvili, N. I. (1953). "Singular Integral Equations". Noordhoff, Groningen.
Nagle, J. F. (1968). *J. Math. Phys.* **9**, 1007–19.
Nagle, J. F. (1969a). *J. Chem. Phys.* **50**, 2813–8.
Nagle, J. F. (1969b). *Commun. Math. Phys.* **13**, 62–7.
Nagle, J. F. and Temperley, H. N. V. (1968). *J. Math. Phys.* **9**, 1020–6.
Neville, E. H. (1944). "Jacobian Elliptic Functions". Oxford University Press, Oxford.
Nienhuis, B., Riedel, E. K. and Schick, M. (1980). *J. Phys. A: Math. Gen.* L189–92.
Nightingale, M. P. (1977). *Phys. Lett.* **59A**, 486–8.
Noble, B. (1958). "Methods Based on the Wiener – Hopf Technique". Pergamon, London.
Oitmaa, J. (1981). *J. Phys. A: Math. Gen.* **14**, 1159–68.
Oitmaa, J. and Plischke, M. (1977). *Physica B + C* **86–88**, 577–8.
Onsager, L. (1944). *Phys. Rev.* **65**, 117–49.

Onsager, L. (1949). *Nuovo Cimento (Suppl.)* **6**, 261.

Onsager, L. (1971). *In* "Critical Phenomena in Alloys, Magnets and Superconductors" (Mills, R. E., Ascher, E. and Jaffee, R. I., eds.). McGraw–Hill, New York.

Ore, O. (1967). "The Four-Colour Problem". Academic Press, New York.

Paley, R. E. A. and Wiener, N. (1934). "Fourier Transforms in the Complex Domain". American Mathematical Society, New York.

Pathria, R. K. (1972). "Statistical Mechanics". Pergamon, Oxford.

Pearce, P. A. and Thompson, C. J. (1977). *J. Stat. Phys.* **17**, 189–96.

Pearson, R. (1980). *Phys. Rev.* **B22**, 2579–80.

Peierls, R. (1936). *Proc. Camb. Phil. Soc.* **32**, 477–81.

Percus, J. K. (1962). *Phys. Rev. Lett.* **8**, 462–5.

Percus, J. K. and Yevick, G. J. (1958). *Phys. Rev.* **110**, 1–13.

Pfeuty, P. (1977). *Physica* B + C **86–88**, 579–80.

Potts, R. B. (1952). *Proc. Camb. Phil. Soc.* **48**, 106–9.

Potts, R. B. and Ward, J. C. (1955). *Progr. Theor. Phys. (Kyoto)* **13**, 38–46.

Ramanujan, S. (1919). *Proc. Camb. Phil. Soc.* **19**, 214–6.

Riedel, E. K. (1981). Physica 106*A*. 110–21.

Rogers, L. J. (1894). *Proc. Lond. Math. Soc.* **25**, 318–43.

Rogers, L. J. (1921). *Proc. Lond. Math. Soc.* **19**(2), 387–97.

Ruelle, D. (1969). "Statistical Mechanics: Rigorous Results". Benjamin, New York.

Runnels, L. K. (1967). *J. Math. Phys.* **8**, 2081–7.

Runnels, L. K. and Combs, L. L. (1966). *J. Chem. Phys.* **45**, 2482–92.

Rys, F. (1963). *Helv. Phys. Acta* **36**, 537–59.

Saaty, T. L. and Kainen, P. C. (1977). "The Four-Colour Problem". McGraw-Hill, New York.

Sacco, J. E. and Wu, F. Y. (1975). *J. Phys. A: Math. Gen.* **8**, 1780–7.

Schultz, T. D., Mattis, D. C. and Lieb, E. H. (1964). *Rev. Mod. Phys.* **36**, 856–71.

Slater, J. C. (1941). *J. Chem. Phys.* **9**, 16–33.

Slater, L. J. (1951). *Proc. Lond. Math. Soc.* **54**, 147–67.

Stanley, H. E. (1968). *Phys. Rev.* **176**, 718–22.

Stanley, H. E. (1971). "Introduction to Phase Transitions and Critical Phenomena". Clarendon, Oxford.

Stanley, H. E. (1974). *In* "Phase Transitions and Critical Phenomena" (Domb, C. and Green, M. S., eds.), Vol. 3, pp. 485–567. Academic Press, London.

Starling, S. G. and Woodall, A. J. (1953). "Electricity and Magnetism". Longman, London.

Stephen, M. J. and Mittag, L. (1972). *J. Math. Phys.* **13**, 1944–51.

Stephenson, J. (1970). *Phys. Rev.* **B1**, 4405–9.

Stinchcombe, R. B. (1973). *J. Phys. C: Solid State Phys.* **6**, 2459–83.

Stroganov, Y. G. (1979). *Phys. Lett.* **74A**, 116–8.

Sutherland, B. (1967). *Phys. Rev. Lett.* **19**, 103–4.

Sutherland, B. (1970). *J. Math. Phys.* **11**, 3183–6.

Suzuki, M. (1974). *Prog. Theor. Phys.* **51**, 1992–3.

Suzuki, M. (1976). *Prog. Theor. Phys.* **56**, 1454–69.

Sykes, M. F., Essam, J. W. and Gaunt, D. S. (1965). *J. Math. Phys.* **6**, 283–98.

Sykes, M. F., Gaunt, D. S., Roberts, P. D. and Wyles, J. A. (1972). *J. Phys. A: Gen. Phys.* **5**, 640–52.

Sykes, M. F., Gaunt, D. S., Essam, J. W. and Elliott, C. J. (1973a). *J. Phys. A: Math., Nucl. Gen.* **6**, 1507–16.

Sykes, M. F., Gaunt, D. S., Martin, J. L., Mattingley, S. R. and Essam, J. W. (1973b). *J. Math. Phys.* **14**, 1071–4.

Takahashi, M. (1973). *Prog. Theor. Phys.* **50**, 1519–36.

Takahashi, M. (1974). *Prog. Theor. Phys.* **51**, 1348–54.

Takahashi, M. and Suzuki, M. (1972). *Prog. Theor. Phys.* **48**, 2187–209.

Temperley, H. N. V. and Fisher, M. E. (1961). *Phil. Mag.* **6**, 1061–3.

Temperley, H. N. V. and Lieb, E. H. (1971). *Proc. Roy. Soc.* (*London*) **A322**, 251–80.

Thompson, C. J. (1965). *J. Math. Phys.* **6**, 1392–5.

Thompson, C. J. (1972). "Mathematical Statistical Mechanics". Princeton University Press, Princeton, N.J.

Tsang, S. K. (1979). *J. Stat. Phys.* **20**, 95–114.

Tutte, W. T. (1967). *J. Combinatorial Theory* **2**, 301–20.

Tutte, W. T. (1970). *J. Combinatorial Theory* **9**, 289–96.

Tutte, W. T. (1973). *Canad. J. Math.* **25**, 426–47.

Tutte, W. T. (1974). *Canad. J. Math.* **26**, 893–907.

van der Waals, J. D. (1873). "Over de Continuiteit van den Gas-en Vloeistoftoestand". Thesis, Leiden.

van Hove, L. (1950). *Physica* **16**, 137–43.

van Leeuwen, J. M. J. (1975). *Phys. Rev. Lett.* **34**, 1056–8.

van Leeuwen, J. M. J., Groeneveld, J. and de Boer, J. (1959). *Physica* **25**, 792–808.

Vdovichenko, N. V. (1965). *Soviet Physics JETP* **21**, 350–2.

Vicentini-Missoni, M. (1972). *In* "Phase Transitions and Critical Phenomena" (Domb, C. and Green, M. S., eds.), Vol. 2, pp. 39–78. Academic Press, London.

Walker, L. R. (1959). *Phys. Rev.* **116**, 1089–90.

Wannier, G. H. (1945). *Rev. Mod. Phys.* **17**, 50–60.

Watson, G. N. (1933). *J. Indian Math. Soc.* **20**, 57–69.

Wegner, F. J. (1972). *J. Phys. C: Solid State Phys.* **5**, L131–2.

Wegner, F. J. (1973). *Physica* **68**, 570–8.

Whitney, H. (1932). *Ann. Math.* (*N.Y.*) **33**, 688–718.

Whittaker, E. T. and Watson, G. N. (1915). "A Course of Modern Analysis". Cambridge University Press, Cambridge.

Widom, B. (1965). *J. Chem. Phys.* **43**, 3892–7 and 3898–905.

Wilson, K. G. (1971). *Phys. Rev.* **B4**, 3174–205.

Wilson, K. G. and Fisher, M. E. (1972). *Phys. Rev. Lett.* **28**, 240–3.

Wilson, K. G. and Kogut, J. (1974). *Phys. Reports* **12C**, 75–199.

Wood, D. W. and Griffiths, H. P. (1972). *J. Phys. C: Solid State Phys.* **5**, L253–5.

Wrege, D. E., Spooner, S. and Gersch, H. A. (1972). *American Institute of Physics Conference Proceedings*, **5**, 1334–8.

Wu, F. Y. (1971). *Phys. Rev.* **B4**, 2312–4.

Wu, F. Y. (1979). *J. Phys. C: Solid State Phys.* **12**, L637–40.

Wu, F. Y. and Lin, K. Y. (1974). *J. Phys. C: Solid State Phys.* **7**, L181–4.

Wu, F. Y. and Lin, K. Y. (1975). *Phys. Rev.* **B12**, 419–28.

Yang, C. N. (1952). *Phys. Rev.* **85**, 808–16.

Yang, C. N. and Yang, C. P. (1966). *Phys. Rev.* **150**, 321–27, 327–39; **151**, 258–64.

Yang, C. N. and Yang, C. P. (1969). *J. Math. Phys.* **10**, 1115–22.

Yang, C. P. (1967). *Phys. Rev. Lett.* **19**, 586–8.
Zamolodchikov, A. B. (1981). *Commun. Math. Phys.* **79**, 489–505.
Zisook, A. B. (1980). *J. Phys. A: Math. Gen.* **13**, 2451–5.
Zittartz, J. (1981). *Z. Physik* **B41**, 75–83.

INDEX